Lecture Notes in Computer Science 5681

Commenced Publication in 1973
Founding and Former Series Editors:
Gerhard Goos, Juris Hartmanis, and Jan van Leeuwen

Daniel Cremers   Yuri Boykov
Andrew Blake   Frank R. Schmidt (Eds.)

Energy Minimization Methods in Computer Vision and Pattern Recognition

7th International Conference, EMMCVPR 2009
Bonn, Germany, August 24-27, 2009
Proceedings

 Springer

Volume Editors

Daniel Cremers
Frank R. Schmidt
Universität Bonn, Institut für Informatik III
Römerstraße 164, 53117 Bonn, Germany
E-mail: {dcremers; schmidtf}@cs.uni-bonn.de

Yuri Boykov
Computer Science Department
University of Western Ontario
N6A 5A5 London, ON, Canada
E-mail: yuri@csd.uwo.ca

Andrew Blake
Microsoft Research, CB3 OFB Cambridge
J.J. Thomson Avenue, United Kingdom
E-mail: ablake@microsoft.com

Library of Congress Control Number: 2001012345

CR Subject Classification (1998): I.5, I.2.10, I.4, G.1.2, G.3, H.3.4, B.8

LNCS Sublibrary: SL 6 – Image Processing, Computer Vision,
Pattern Recognition, and Graphics

ISSN 0302-9743

ISBN 978-3-642-03640-8 Springer Berlin Heidelberg New York

springer.com

© Springer-Verlag Berlin Heidelberg 2009

Typesetting: Camera-ready by author, data conversion by Scientific Publishing Services, Chennai, India
Printed on acid-free paper   SPIN: 12733095   06/3180   5 4 3 2 1 0

Preface

Over the last decades, energy minimization methods have become an established paradigm to resolve a variety of challenges in the fields of computer vision and pattern recognition. While traditional approaches to computer vision were often based on a heuristic sequence of processing steps and merely allowed very limited theoretical understanding of the respective methods, most state-of-the-art methods are nowadays based on the concept of computing solutions to a given problem by minimizing respective energies.

This volume contains the papers presented at the 7th International Conference on Energy Minimization Methods in Computer Vision and Pattern Recognition (EMMCVPR 2009), held at the University of Bonn, Germany, August 24–28, 2009. These papers demonstrate that energy minimization methods have become a mature field of research spanning a broad range of areas from discrete graph theoretic approaches and Markov random fields to variational methods and partial differential equations. Application areas include image segmentation and tracking, shape optimization and registration, inpainting and image denoising, color and texture modeling, statistics and learning. Overall, we received 75 high-quality double-blind submissions. Based on the reviewer recommendations, 36 papers were selected for publication, 18 as oral and 18 as poster presentations. Both oral and poster papers were attributed the same number of pages in the conference proceedings.

Furthermore, we were delighted that three leading experts from the fields of computer vision and energy minimization, namely, Richard Hartley (Canberra, Australia), Joachim Weickert (Saarbrücken, Germany) and Guillermo Sapiro (Minneapolis, USA) agreed to further enrich the conference with inspiring keynote lectures.

We would like to express our gratitute to those who made this event possible and contributed to its success, in particular to a strong international Program Committee for providing excellent reviews and to the Hausdorff Center for Mathematics for providing financial support and guidance. We are particularly grateful to Heidi Georges-Hecking, Mohamed Souiai and the staff of the Hausdorff Center for administrative support.

It is our belief that this conference will help to advance the field of energy minimization methods and to further establish the mathematical foundations of computer vision.

August 2009

Daniel Cremers
Yuri Boykov
Andrew Blake
Frank R. Schmidt

Table of Contents

Segmentation and Tracking

Shape Optimization and Registration

Inpainting and Image Denoising

Color and Texture

Statistics and Learning

Multi-label Moves for MRFs with Truncated Convex Priors

Olga Veksler

Computer Science Department
University of Western Ontario
London, Canada
olga@csd.uwo.ca

Abstract. Optimization with graph cuts became very popular in recent years. As more applications rely on graph cuts, different energy functions are being employed. Recent evaluation of optimization algorithms showed that the widely used swap and expansion graph cut algorithms have an excellent performance for energies where the underlying MRF has Potts prior. Potts prior corresponds to assuming that the true labeling is piecewise constant. While surprisingly useful in practice, Potts prior is clearly not appropriate in many circumstances. However for more general priors, the swap and expansion algorithms do not perform as well. Both algorithms are based on moves that give each pixel a choice of only two labels. Therefore such moves can be referred to as binary moves. Recently, range moves that act on multiple labels simultaneously were introduced. As opposed to swap and expansion, each pixel has a choice of more than two labels in a range move. Therefore we call them multi-label moves. Range moves were shown to work better for problems with truncated convex priors, which imply a piecewise smooth labeling. Inspired by range moves, we develop several different variants of multi-label moves. We evaluate them on the problem of stereo correspondence and discuss their relative merits.

1 Introduction

Energy optimization with graph cuts [1,2,3] is increasingly used for different applications in computer vision and graphics. Some examples are image restoration [2], stereo and multi-view reconstruction [4,5,2,6,7], motion segmentation [8,9,10], texture synthesis [11], segmentation [12,13,14,15], digital photomontage [16]. Optimization with graph cuts either results in an exact minimum or an approximate minimum with non-trivial quality guarantees. This frequently translates into a result of high accuracy, given that the energy function is appropriate for the application.

A typical energy function to be minimized is as follows:

$$E(f) = \sum_{p \in \mathcal{P}} D_p(f_p) + \sum_{(p,q) \in \mathcal{N}} V_{pq}(f_p, f_q). \tag{1}$$

In Eq. (1), $\mathcal{L}$ is a finite set of labels, representing the property needed to be estimated at each pixel, such as intensity, color, etc. $\mathcal{P}$ is the set of sites that one needs to assign labels to. Frequently $\mathcal{P}$ is set of image pixels. The label assigned to pixel $p \in \mathcal{P}$ is denoted

D. Cremers et al. (Eds.): EMMCVPR 2009, LNCS 5681, pp. 1–13, 2009.

by f_p, and f is the collection of all pixel-label assignments. The first sum in Eq. (1) is the data term. In the data term, $D_p(f_p)$ is the penalty for pixel p to be assigned label f_p. The data term usually comes from the observed data. The second sum in Eq. (1) is the smoothness term, and it uses the prior knowledge about what the likely labelings f should be like. The sum is over ordered pixel pairs $(p, q) \in \mathcal{N}$. Often $\mathcal{N}$ is the 4 or 8 connected grid, however longer range interactions are also useful [5]. Without loss of generality, we assume that if $(p, q) \in \mathcal{N}$ then $p < q$.

Any choice for D_p is easy to handle. The choice of V_{pq} determines whether the energy function can be efficiently minimized. The V_{pq}'s often specify the smoothness assumptions on the labeling f. Different choices of V_{pq}'s correspond to different smoothness assumptions. A common choice is the Potts model, which is $V_{pq}(f_p, f_q) = w_{pq} \cdot \min\{1, |f_p - f_q|\}$. The coefficients w_{pq}'s can be different for each pair of neighboring pixels. Potts model penalizes any difference between f_p and f_q by the same amount. Intuitively, it corresponds to the prior knowledge that f should be piecewise constant, that is it consists of several pieces where pixels inside the same piece share the same label.

Another common choice is $V_{pq}(f_p, f_q) = w_{pq} \cdot \min\{T, |f_p - f_q|^a\}$. If $a = 1$ the model is called a truncated linear, and if $a = 2$, it is called a truncated quadratic. These V_{pq}'s correspond to the piecewise smooth assumption on f, that is the assumption that f consists of several pieces, where the labels between neighboring pixels inside each piece vary "smoothly"[1]. Parameter T is called a *truncation* constant. Without the truncation, that is if V_{pq} is the absolute linear or quadratic difference, the energy in Eq. (1) can be optimized exactly with a graph cut [17], but the corresponding energies are not discontinuity preserving. Energy in Eq. (1) is NP-hard to optimize if discontinuity preserving Potts, truncated linear or quadratic V_{pq}'s are used [2].

Recently, Szeliski et.al. [18] performed an experimental evaluation of several optimization methods popular for minimizing energies in Eq. (1) : the graph cut based expansion and swap algorithms [2], sequential tree-reweighted message passing (TRW-S) [19], and loopy belief propagation (LBP) [20]. They show that for Potts model, both expansion and swap algorithms have an excellent performance, they find an answer within a small percentage of the global minimum. TRW-S performs as well as graph cuts, but takes significantly longer to converge. An additional benefit of graph cuts over TRW-S is when longer range interactions are present. Szeliski et.al. [18] studied only the case when $\mathcal{N}$ is the 4-connected grid. Kolmogorov and Rother [21] performed a comparison between graph cuts and TRW-S when longer range interactions are present, and they concluded that graph cuts perform significantly better in terms of energy than TRW-S in this case. For the truncated linear V_{pq}'s the expansion and swap algorithms still perform relatively well, but for the truncated quadratic V_{pq} the energy value is noticeably worse than that of TRW-S.

Recently [22] developed a new type of moves, called the *range* moves for optimizing energies with truncated convex priors. A truncated quadratic and linear are examples of truncated convex prior. Informally, truncated convex priors correspond to assuming that f is piecewise smooth. The insight in [22] is that both expansion and swap algorithms give a pixel a choice of only two labels, but for problems where piecewise smoothness

---

[1] The term "smoothly" is used informally here.

assumptions are appropriate, to obtain a good approximation, a pixel should have a choice among more than two labels. The range moves that they develop act on a larger set of labels than the expansion and swap moves. Because of this property, we call the range moves *multi-label* moves. In [23] they use a similar idea to develop multi-label moves for an energy function useful for single-view scene reconstruction. The energy function in [23] is neither Potts nor truncated convex, and thus not directly related to our work.

We further explore the idea of multi-label moves for truncated convex priors. One can regard the range moves developed in [22] as a generalization of the swap move. In this paper, we develop a multi-label move that is can be regarded as a generalization of the expansion move. The optimal multi-label expansion move can be found only approximately. We explore an additional move that we call multi-label smooth swap. Note that simultaneously but independently, [24] developed a move similar to our multi-label expansion [25]. Their graph construction is very similar, with some minor differences in edge weights. They also do not find an optimal multi-label expansion move, but its approximation. The ideas that lay behind our multi-label moves (as, indeed, the ideas behind any move-making optimization algorithm) are related to the framework of majorization-minimization [26].

We evaluate our new multi-label moves on the energy functions arising in stereo correspondence, and discuss their relative merits as well as compare them with the range moves.

2   Prior Work

In this section, we briefly explain the prior work on optimization with graph cuts.

2.1   Assumptions on the Label Set

For the rest of the paper we assume that the labels can be represented as integers in the range $\{0, 1, ...k - 1\}$, which is necessary since the construction is based on that in [17].

2.2   Convex Priors

Ishikawa [17] develops a method to find the exact minimum of the energy function in Eq. (1) when V_{pq} are convex functions of the label differences. Specifically, $V_{pq}(l_1, l_2) = w_{pq} \cdot g(l_1 - l_2)$ is said to be convex if and only if for any integer x, $g(x + 1) - 2g(x) + g(x - 1) \geq 0$. It is assumed that $g(x)$ is symmetric[2]. In [27,28] they extend the results in [17] to handle a more general definition of convexity.

We follow [22] to describe the work in [17]. Ishikawa's method is based on computing a minimum cut in a particular graph. There are two special nodes in the graph, the source s and the sink t. For each $p \in \mathcal{P}$, we create a set of nodes $p_0, p_1, ..., p_k$, see Fig. 1 Node p_0 is connected with the source s with an edge of infinite capacity. Similarly, we connect p_k with the sink t with an edge of infinite capacity. This way p_0 is essentially identified with the source, and p_k with the sink. We connect node p_i to node p_{i+1} with a

---

[2] A function is symmetric if $g(x) = g(-x)$.

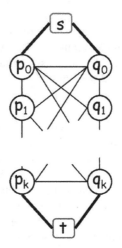

Fig. 1. Graph construction for convex V_{pq}. Thick links indicate edges of infinite capacity.

directed edge e_i^p for $i = 0, 1, ..., k - 1$. In addition, for $i = 1, ..., k$, node p_i is connected to node p_{i-1} with a directed edge of infinite weight. This ensures that for each p, only one of the edges of type e_i^p will be in the minimum cut. If an edge e_i^p is cut, then pixel p is assigned label i, thus any cut C induces a labeling f^C.

Furthermore, for any $(p, q) \in \mathcal{N}$, an edge e_{ij}^{pq} which connects node p_i to node q_j is created for $i = 0, ..., k$ and $j = 0, ..., k$. The weight of this edge is

$$w(e_{ij}^{pq}) = \frac{w_{pq}}{2}[g(i - j + 1) - 2g(i - j) + g(i - j - 1)]. \tag{2}$$

The edge weight defined by Eq. (2) is non-negative, since $g(x)$ is convex.

The weights e_i^p are defined as follows:

$$w(e_i^p) = D_p(i) - \sum_{q \in \mathcal{N}_p} w_{pq} \cdot h(i),$$

where $\mathcal{N}_p$ is the set of neighbors of pixel p, and $h(i) = -\frac{1}{2}[g(k + 1 - i) + g(i + 1)]$ Under this edge weights assignment, the cost of any finite cut C is exactly $E(f^C)$ plus a constant, see [22]. Therefore the minimum cut gives the optimal labeling.

Note that [29] develops an algorithm for minimizing energy with convex V_{pq} which is more memory and time efficient. However it can be used only when the D_p's are convex.

2.3 Expansion and Swap Algorithms

Boykov et.al. [2] develop the expansion and swap algorithms. These methods can be applied when V_{pq} is Potts, truncated linear or quadratic, but the answer is only approximate, since the energy is NP-hard to optimize in these cases [2].

Both the expansion and the swap algorithms find a local minimum of the energy function in the following sense. For each f, we define a set of "moves" $M(f)$. We say that f is a local minimum with respect to the set of moves, if $E(f') \geq E(f)$ for any $f' \in M(f)$.

Given a labeling f and a label pair (α, β), a move from f to f' is called an α-β swap if $f_p \neq f'_p \Rightarrow f_p, f'_p \in \{\alpha, \beta\}$. $M(f)$ is then defined as the collection of α-β swaps for all pairs of labels $\alpha, \beta \in \mathcal{L}$.

Given a labeling f and a label α, a move f' is called an α-expansion if $f_p \neq f'_p \Rightarrow f'_p = \alpha$. $M(f)$ is then defined as the collection of α-expansions for all labels $\alpha \in \mathcal{L}$.

The optimal α-expansion and the optimal α-β swap can be found as a minimum cut in a certain graph [2]. Thus the expansion and swap algorithms find a local minimum with respect to expansion or swap moves, correspondingly. Starting with an initial labeling f, optimal swap (or expansion) moves are found until convergence.

The energy with truncated linear V_{pq} can be optimized by both expansion and swap algorithms, whereas for truncated quadratic V_{pq}, only the swap algorithm applies directly. In practice, however, it is possible to apply the expansion algorithm with a "truncation trick" [30]. The resulting labeling is no longer guaranteed to be a local minimum with the respect to expansion moves, but the energy is guaranteed to go down.

2.4 Range Moves for Truncated Convex Priors

In this section we review the range moves of [22]. Based on the notion of convexity in [17], V_{pq} is truncated convex if there exists a symmetric function $g(x)$ such that $g(x+1) - 2g(x) + g(x-1) \geq 0$ and

$$V_{pq}(l_1, l_2) = w_{pq} \cdot \min\{g(l_1 - l_2), T\}. \tag{3}$$

Throughout the rest of the paper, we assume truncated convex V_{pq}'s.

Recall that $\mathcal{L} = \{0, 1, ..., k-1\}$. Let $\mathcal{L}_{\alpha\beta} = \{\alpha, \alpha+1, ..., \beta\}$, where $\alpha < \beta \in \mathcal{L}$. Given a labeling f, we say that f' is an α-β range move from f, if $f_p \neq f'_p \Rightarrow \{f_p, f'_p\} \subset \mathcal{L}_{\alpha\beta}$. The α-β range moves can be viewed as a generalization of α-β swap moves. An α-β swap move reassigns labels α, β among pixels that are currently labeled α and β. An α-β range move reassigns the labels in the range $\{\alpha, \alpha+1, .., \beta\}$ among the pixels that currently have labels in the range $\{\alpha, \alpha+1, .., \beta\}$.

In [22], they show how to find an optimal α-β range move if $|\alpha - \beta| \leq T$.[3] The basic idea is as follows. Let $\mathcal{T} = \{p | \alpha \leq f_p \leq \beta\}$. Notice that the truncated convex terms V_{pq} become convex when $p, q \in \mathcal{T}$, since for any $p, q \in \mathcal{T}$, $V_{pq}(f_p, f_q) = w_{pq}g(f_p - f_q) \leq w_{pq} \cdot T$. Non-convex term arise only on the boundary of $\mathcal{T}$, but they can be arranged in a graph construction by adding appropriate constants to edges e_i^p, see Section 2.2.

Just as with α-β swaps, the algorithm starts at some labeling f. Then it iterates over a set of label ranges $\{\alpha, .., \beta\}$ with $|\alpha - \beta| = T$, finding the best α-β range move f' and switching the current labeling to f'.

The α-β range move can be slightly generalized. As previously, let $|\alpha - \beta| = T$ and, as before, let $\mathcal{T} = \{p | \alpha \leq f_p \leq \beta\}$. Let

$$\mathcal{L}_{\alpha\beta t} = \{\alpha - t, \alpha - t + 1, ..., \beta + t - 1, \beta + t\} \cap \mathcal{L},$$

---

[3] If $|\alpha - \beta| > T$, α-β range move is NP-hard to find.

that is $\mathcal{L}_{\alpha\beta t}$ extends the range of $\mathcal{L}_{\alpha\beta}$ by t in each direction, making sure that the resulting range is still a valid range of labels in $\mathcal{L}$.

Let

$$M^{\alpha\beta t}(f) = \{f'|f'_p \neq f_p \Rightarrow f_p \in \mathcal{L}_{\alpha\beta}, f'_p \in \mathcal{L}_{\alpha\beta t}\}.$$

That is $M^{\alpha\beta t}(f)$ is a set of moves that change pixels labels from $\mathcal{L}_{\alpha\beta}$ to labels in $\mathcal{L}_{\alpha\beta t}$. Notice that $M^{\alpha\beta}(f) \subset M^{\alpha\beta t}(f)$. It is not possible to find the optimal move in $M^{\alpha\beta t}(f)$, but [22] shows how to find $\hat{f} \in M^{\alpha\beta t}(f)$ s.t. $E(\hat{f}) \leq E(f^*)$, where f^* is the optimal move in $M^{\alpha\beta}(f)$. Thus labeling $\hat{f}$ is not worse than the optimal move in $M^{\alpha\beta}(f)$, and if one is lucky, $E(\hat{f})$ could be significantly better than the optimal move in $M^{\alpha\beta}(f)$. In practice, t is set to a small constant. Let us call this generalized range move as α-β-t-range move.

3   Multi-label Moves

The key idea of the range moves in [22] is to allow a pixel to choose among several labels in a single move. This is in contrast to the swap and expansion moves, which allow each pixel a choice between only two labels. We are going to refer to moves that allow a choice of more than two labels as *multi-label*. Multi-label moves have already proven successful in [22,23,24]. There is a multitude of such moves possible. In this paper, we develop several different multi-label moves for truncated convex priors and compare their performance. To have a clear terminology, we are going to rename to the generalized α-β-t-range move with as multi-label α-β-t-swap. There is no need to rename the α-β-range move since it is a special case of α-β-t-range move with $t = 0$.

In [22], the idea was to find a subset of pixels $\mathcal{P}'$ and a subset of labels $\mathcal{L}'$ s.t. when the V_{pq} terms are restricted to $\mathcal{P}'$ and $\mathcal{L}'$, they are convex. The boundary terms are easy to implement, as shown in [22]. Throughout the remainder of this section, we are going to exploit different ways of selecting $\mathcal{P}'$ and $\mathcal{L}'$. The two new moves that we develop are called *multi-label expansion* and *multi-label smooth swap*.

In order to perform iterative energy optimization that reduces the energy of the current labeling f, it seems necessary to ensure that the labels of pixels in $\mathcal{P}'$ under labeling f are contained in $\mathcal{L}'$. This ensures that the current labeling f is also within the set of allowed moves, and the lowest energy move is not worse than the current labeling. For the multi-label smooth swap, we are able to enforce this condition. For multi-label expansion, we are not able to always enforce it. We will still guarantee though that the energy goes down at each iteration by simply rejecting any move whose energy is higher than that of the current labeling.

3.1   Multi-label Smooth Swap

Let f be a current labeling. Let $\mathcal{P}'$ be a subset of pixels of $\mathcal{P}$. We call $\mathcal{P}'$ a smooth subset under labeling f, if for any $(p, q) \in \mathcal{N}$, whenever $\{p, q\} \subset \mathcal{P}'$, then $|f_p - f_q| \leq T$, where T is the truncation constant in Eq. (3). In words, if a subset $\mathcal{P}'$ is smooth under f, then the label difference for any two pixels contained in $\mathcal{P}'$ is not larger than the truncation constant.

Let f be the current labeling and $\mathcal{P}'$ be a smooth subset under f. Let $\mathcal{L}(\mathcal{P}', f) = \{f_p | p \in \mathcal{P}'\}$, that is $\mathcal{L}(\mathcal{P}')$ is the collection of labels that pixels in $\mathcal{P}'$ have under labeling f.

Given a smooth subset $\mathcal{P}'$ under f, let $M_{smooth}(f, \mathcal{P}') = \{f' | f'_p \neq f_p \Rightarrow p \in \mathcal{P}'$ and $f'_p \in \mathcal{L}(\mathcal{P}', f)\}$. $M_{smooth}(f, \mathcal{P}')$ describes exactly the set of all multi-label smooth swap moves. In words, a smooth swap move takes a smooth set of pixels under f, collects their labels, and reassigns their labels among them.

Just as it was possible to generalize the multi-label swap move by extending the range of labels, it is possible to generalize the multi-label smooth swap. Let t be a constant for extending the range of labels $\mathcal{L}'(\mathcal{P}', f)$. Let us define the extended range of labels as

$$\mathcal{L}'(\mathcal{P}', f, t) = \{l \in \mathcal{L} | \exists l' \in \mathcal{L}'(\mathcal{P}', f) \text{ s.t. } |l - l'| \leq t\} \cap \mathcal{L}.$$

In words, to get $\mathcal{L}'(\mathcal{P}', f, t)$ we add to $\mathcal{L}'(\mathcal{P}', f)$ labels that are at distance no more than t from some label already in $\mathcal{L}'(\mathcal{P}', f)$. The intersection with $\mathcal{L}$ is performed to make sure that after the "padding", the augmented set is still contained in $\mathcal{L}$. Let the set of smooth swap moves augmented by t be denoted by $M_{smooth}(f, \mathcal{P}', t)$.

A multi-label smooth swap is naturally related to a multi-label swap. In a multi-label swap move, the pixels participating in a move have labels in a range limited by truncation, i.e. all the labels are between some α and β with $|\alpha - \beta| < T$. In a multi-label smooth swap, the domain of pixels participating in a move can be larger than that compared to the multi-label swap. That is the pixels participating in the move can have labels between some α and β with $|\alpha - \beta| > T$. The restriction is that in the pixels participating in a smooth swap must form a "smooth" component in the current labeling f, that is the labels of any two neighbors cannot differ by more than T.

There are two questions that remain to be answered: how to choose the smooth subsets $\mathcal{P}'$ and how to optimize with smooth swap moves. Let us first consider the question of optimization.

In general, it is not possible to find the optimal smooth swap move, given a smooth subset $\mathcal{P}'$ and the current labeling f. However, we are able to find a good swap move, the one that improves the current labeling f.

Let $\mathcal{S}$ be a subset of pixels in $\mathcal{P}$ and let us define:

$$E_{\mathcal{S}}(f) = \sum_{p \in \mathcal{S}} D_p(f_p) + \sum_{(p,q) \in \mathcal{N}, \{p,q\} \cap \mathcal{S} \neq \emptyset} V_{pq}(f_p, f_q).$$

In words, $E_{\mathcal{S}}(f)$ is the sum all the terms of the energy function which depend on pixels in $\mathcal{S}$. Let us further define:

$$E_{\mathcal{S}}^{open}(f) = \sum_{p \in \mathcal{S}} D_p(f_p) + \sum_{(p,q) \in \mathcal{N}, \{p,q\} \subset \mathcal{S}} V_{pq}(f_p, f_q).$$

In words, $E_{\mathcal{S}}^{open}(f)$ is the sum of all the terms of the energy function which depend *only* on pixels in $\mathcal{S}$. It is clear that for any $\mathcal{S} \in \mathcal{P}$, $E(f) = E_{\mathcal{S}}(f) + E_{\mathcal{P}-\mathcal{S}}^{open}(f)$.

Let f be the current labeling, and let $\mathcal{P}'$ be a smooth subset under f. Let f' be a smooth swap move from f, i.e. $f' \in M_{smooth}(f, \mathcal{P}', t)$.

We use basically the same construction as in Section 2.4. We construct a graph for pixels in $\mathcal{P}'$. However, the label range is $\mathcal{L}'(\mathcal{P}', f, t)$, and we identify it with label set

$\{0, 1, ..., |\mathcal{L}'(\mathcal{P}', f, t)| - 1\}$. Otherwise, the graph construction is identical to that in Section 2.4.

Let C be any finite cost cut in our graph. Notice that a cut of finite cost assigns labels (as described in Section 2.2) only to pixels in $\mathcal{P}'$. Let f^C be the labeling corresponding to the cut C, which we define as follows: $f_p^C = f_p$ for $p \notin \mathcal{P}'$, and for $p \in \mathcal{P}'$, f_p^C is equal to the label assigned to pixel p by the cut C. Let $w(C)$ be the cost of cut C. By graph construction, $w(C) = \tilde{E}_{\mathcal{P}'}(f^C) + K$, where K is a constant and $\tilde{E}(f)$ is the same energy as $E(f)$, except there is no truncation in V_{pq} terms for $p, q \in \mathcal{P}'$. That is for $p, q \in \mathcal{P}'$, $V_{pq}(f_p, f_q) = w_{pq} \cdot g(f_p - f_q)$ in the energy $\tilde{E}$.

For any f, $E_{\mathcal{P}'}(f) \leq \tilde{E}_{\mathcal{P}'}(f)$, since the only difference between E and $\tilde{E}$ is that the V_{pq} terms are not truncated in $\tilde{E}$ for $p, q \in \mathcal{P}'$. Recall that for any f, $E(f) = E_{\mathcal{P}'}(f) + E_{\mathcal{P}-\mathcal{P}'}^{open}(f)$. Also, $E_{\mathcal{P}-\mathcal{P}'}^{open}(f) = \tilde{E}_{\mathcal{P}-\mathcal{P}'}^{open}(f)$, since $\tilde{E}$ is not different from E outside of set $\mathcal{P}'$.

Let f be the current labeling. Notice that $E_{\mathcal{P}'}(f) = \tilde{E}_{\mathcal{P}'}(f)$, since V_{pq} terms in f do not need to be truncated on the set $\mathcal{P}'$. Let $\hat{C}$ be the minimum cost cut, and let $\hat{f}$ be its corresponding labeling, defined as above. Let f be the current labeling (notice that $f \in M_{smooth}(f, \mathcal{P}', t)$), and let C be the cut which corresponds to it in the graph. We have that $\tilde{E}_{\mathcal{P}'}(\hat{f}) + K = w(\hat{C}) \leq w(C) = \tilde{E}_{\mathcal{P}'}(f) + K$. Since $E_{\mathcal{P}'}(\hat{f}) \leq \tilde{E}_{\mathcal{P}'}(\hat{f})$ and $E_{\mathcal{P}'}(f) = \tilde{E}_{\mathcal{P}'}(f)$, we get that $E_{\mathcal{P}'}(\hat{f}) \leq E_{\mathcal{P}'}(f)$. Now, for any labeling f'', $E(f'') = E_{\mathcal{P}'}(f'') + E_{\mathcal{P}-\mathcal{P}'}^{open}(f'')$. We have that $E_{\mathcal{P}-\mathcal{P}'}^{open}(\hat{f}) = E_{\mathcal{P}-\mathcal{P}'}^{open}(f')$, and therefore we get that $E(\hat{f}) \leq E(f)$. This shows that the minimum cut gives a labeling $\hat{f}$ with energy not larger than the current labeling f. So if we cannot find the optimal smooth swap move, we can at least guarantee smooth swap move does not increase energy.

The question remains of how to find smooth subsets $\mathcal{P}'$. In general, given a current labeling f, we can partition it into a set of $\mathcal{P}_1, \mathcal{P}_2 ... \mathcal{P}_d$, s.t. $\bigcap_i \mathcal{P}_i = \mathcal{P}$ and each $\mathcal{P}_i$ is smooth. This partition can be performed by computing connected components. This partition is not unique, however. To remove bias due to visitation order, we compute connected components in random order. That is we pick a pixel p at random, compute a maximal smooth subset $\mathcal{P}_1$ containing p, then choose another pixel $q \notin \mathcal{P}_1$, compute a maximal smooth subset $\mathcal{P}_2$ containing q, and so on, until all pixels are partitioned into smooth subsets. Then we compute smooth swap moves for each $\mathcal{P}_i$. This is not the only way to proceed, but we found it to be effective. Computing all smooth swap moves for a partition $\mathcal{P}_1, \mathcal{P}_2 ... \mathcal{P}_d$ constitutes one iteration of the algorithm. We perform iterations until convergence.

The advantage of the multi-label smooth swap move over the multi-label swap is that it converges faster. If we start from a good solution (typically we start from the results of the binary expansion algorithm), the number of smooth subsets in a partition of $\mathcal{P}$ is small, so the number of moves is smaller compared to the multi-label swap. The disadvantage is that it gives energies that are slightly higher in practice.

3.2 Multi-label Expansion

We now develop a multi-label expansion move. Let α and β be two labels s.t. $\alpha < \beta$. The idea behind multi-label expansion move is similar to that of a binary expansion move. We wish to construct a move in which each pixel can either stay with its old

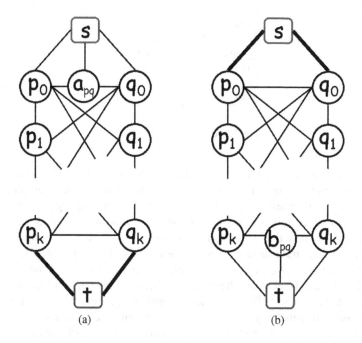

(a)   (b)

Fig. 2. Graph construction for multi-label expansion

label, or switch to a label in the set $\{\alpha, \alpha + 1, ..., \beta\}$. The name "expansion", as before, reflects the fact that labels in the set $\{\alpha, \alpha + 1, ..., \beta\}$ expand their territory.

Let $M^{\alpha\beta}(f) = \{f' | f'_p \neq f_p \Rightarrow f'_p \in \mathcal{L}_{\alpha\beta}\}$. That is $M^{\alpha\beta}(f)$ is exactly the set of all α-β multi-label expansion moves from labeling f. Unfortunately, the optimal expansion move cannot be computed exactly, so we are forced to approximate it.

Suppose that we are given a labeling f and we wish to approximate the optimal α-β expansion move, where $|\alpha - \beta| = T$. The construction is similar to that in Section 2.2. We identify label set $\{\alpha, \alpha + 1, ..., \beta\}$ with set $\{0, 1, ..., T\}$. One of the differences is that now all pixels participate in a move. First we build a graph exactly like in Sec. 2.4, except the links between the source and p_0 are not set to infinite, for all pixels p. We create an auxiliary pixel a_{pq} between each pair of neighboring pixels (p, q)[4]. We connect p_0 to a_{pq}, q_0 to a_{pq}, and s to a_{pq}, as illustrated in Fig. 2 (a). If the minimum cut severs edge between s and p_0, then p is assigned its old label in the move. Otherwise, the label assignment is exactly like in Sec. 2.4.

For the construction in 2.4 if we sever links e_i^p and e_j^q, then the cost of all the links e_{ij}^{pq} that have to be severed adds up to $C + V_{pq}(i, j)$. The costs of the new links that we create for the expansion algorithm are as in Fig. 3.

This construction insures that if a links between s and p_0 and between q_0 and q_1 are broken, then the cost of all edges severed corresponds exactly to $V_{pq}(f_p, \alpha)$ plus a constant, which is exactly what is needed. Similarly the correct thing happens if links between s and q_0 and between p_0 and p_1 are broken, and if the link between s and a_{pq}

---

[4] Note that auxiliary pixel is not necessary, see [3], but it clarifies the explanation.

link	weight
p_0 to a_{pq}	$V_{pq}(f_p, \alpha) + C/2$
q_0 to a_{pq}	$V_{pq}(\beta, f_q) + C/2$
s to a_{pq}	$V_{pq}(f_p, f_q)$+C
s to p_0	$D_p(f_p)$
s to q_0	$D_p(f_q)$

Fig. 3. Weights of the new links

is broken. Unfortunately in other cases, as long as the new links in Fig. 3 are involved, the V_{pq} value can be underestimated or overestimated. The minimum graph cut is not even guaranteed to reduce the energy from that of the current labeling f. Still in practice we found that many minimum cuts correspond to an assignment with a lower energy. Therefore, to make sure that the energy never goes up, if f' is the assignment returned by our approximate multi-label expansion, we first test if $E(f') < E(f)$, where f is the current labeling. If yes, we accept f' as the new current labeling. If no, we reject it.

As with the multi-label swap, the range of labels involved in multi-label expansion can be extended by some t. The construction changes appropriately, similar to what is done when extending the range of multi-label swap moves see Sec. 2.4.

In practice, we found the following version of the multi-label expansions to work better. Let $\mathcal{T} = \{p \in \mathcal{P}|f_p \leq \beta\}$ and let $\mathcal{B} = \{p \in \mathcal{P}|f_p \geq \alpha\}$. We perform the multi-label expansion on pixels in set $\mathcal{T}$ using the graph like in Fig. 2(a), and an expansion on pixel in set $\mathcal{B}$ using the graph like in Fig. 2(b), with symmetrically modified weights in Fig. 3 for the second case. The weights also have to be corrected because there are pixels not participating in the move, so the "border" conditions resulting from such pixels have to be incorporated into edge weights e_i^p, just like in Sec. 2.4. The improvement is probably due to the fact that more V_{pq}'s are correctly represented by this split graph construction. Another improvement is probably due to the fact that pixels on the border not participating in the move pull the energy in the right direction by having their V_{pq} terms correctly modeled through the edge weights e_i^p.

4   Experimental Results

In this section, we present our results on stereo correspondence for the Middlebury database stereo images[5]. We took four pairs of stereo images for evaluation, namely: *Venus, Sawtooth, Teddy, Cones*. This database was constructed by D. Scharstein and R. Szeliski, and these images are the top benchmark in evaluating the performance of stereo algorithms [31,32].

For stereo correspondence, $\mathcal{P}$ is the set of all pixels in the left image, $\mathcal{L}$ is the set of all possible stereo disparities. We take the disparity labels at sub-pixel precision, in quarter of a pixel steps. That is if $|f_p - f_q| = 1$, then the disparities of pixels p and q differ by 0.25 pixels. Let d_l stand for the actual disparity corresponding to the

---

[5] The images were obtained from www.middlebury.edu/stereo

	Venus	Sawtooth	Teddy	Cones
Swap	7,871,677	9,742,107	16,376,181	21,330,284
Expansion	8,131,203	9,418,529	15,829,221	21,020,174
α-β-2 swap	**7,188,393**	**9,371,745**	15,421,437	**20,490,753**
Multi-Label Smooth Range	7,193,823	9,373,126	15,616,999	20,515,493
α-β-2 Expansion	7,188,404	9,377,494	**15,408,234**	20,626,809

Fig. 4. Energies on Middlebury database. The minimum in each column is highlighted.

integer label l, for example label 2 corresponds to disparity 0.75. The data costs are $D_p(l) = \left| I_L(p) - [I_R(p - \overline{d_l}) \cdot (d_l - \underline{d_l}) + I_R(p - \underline{d_l})(\overline{d_l} - d_l)] \right|$, where $\underline{x}$ stands for rounding down, $\overline{x}$ stands for rounding up, and $p - x$ stands for the pixel that has the coordinates of pixel p shifted to the left by x.

We use the truncated quadratic $V_{pq}(f_p, f_q) = 100 \cdot \min\{(f_p - f_q)^2, 25\}$. Using spatially varying weights w_{pq} improves results of stereo correspondence, since it helps to align disparity discontinuities with the intensity discontinuities. We set all $w_{pq} = 10$, since the main purpose of our paper is to evaluate and compare the multi-label moves, and not to come up with the best stereo algorithm. Fig. 4 compares the energies obtained with the expansion algorithm, swap algorithm, multi-label swap moves (or range moves in terminology of [22]), multi-label expansion moves, and smooth swap moves.

From the Table 4, we can make the following conclusions. First let us consider the "binary" swap and expansion moves. The swap and expansion algorithms are clearly inferior when it comes to truncated convex priors. Even though the swap algorithm is guaranteed to find a best swap move and the expansion algorithm is not guaranteed to find the best move under the truncated quadratic model, expansion algorithm does performs better for all scenes except "Venus". This is probably explained by the fact that expansion moves are more powerful than the swap moves. Even if we do not find the optimal expansion, a good expansion may be better than the optimal swap.

Now let us discuss the multi-label moves. First of all, the running times for the multi-label swap move was on the order of minutes (from 5 to 10 minutes). The smooth range move achieved the energy very close to that of the multi-label swap, but its running time is about 2 or 3 times faster. The multi-label expansion move is almost always slightly worse that the multi-label swap, it is better only on the "Teddy sequence". One would expect a better performance from the expansion move, but since we cannot find the optimal one, only an approximate one, these results are not entirely surprising. The running time for the expansion move is much worse than for other multi-label moves, since the graphs are much bigger. Multi-label expansion takes about 9-10 times longer than multi-label swap.

We should mention that the running times of our algorithms can be significantly improved using the ideas in [33]. They employ techniques such as good initialization, reducing the number of unknown variables by computing partially optimal solutions, and recycling flow. All of these are directly transferable to the implementation of our multi-label moves. Their speed ups are around a factor of 10 or 15.

4.1 Discussion

In this paper develop and compare two new multi-label moves for energies with truncated convex prior, as well compare the new moves with the previously known multi-label moves called range moves. Clearly, there are more interesting multi-label moves that can be developed for multi-label energies. An interesting question is whether it is possible to discover automatically new multi-label moves with good properties for a given energy, rather than develop them by hand.

References

1. Ishikawa, H., Geiger, D.: Occlusions, discontinuities, and epipolar lines in stereo. In: Burkhardt, H.-J., Neumann, B. (eds.) ECCV 1998. LNCS, vol. 1406, p. 232. Springer, Heidelberg (1998)
2. Boykov, Y., Veksler, O., Zabih, R.: Fast approximate energy minimization via graph cuts. PAMI 23(11), 1222–1239 (2001)
3. Kolmogorov, V., Zabih, R.: What energy functions can be minimized via graph cuts? In: Heyden, A., Sparr, G., Nielsen, M., Johansen, P. (eds.) ECCV 2002. LNCS, vol. 2352, pp. 65–81. Springer, Heidelberg (2002)
4. Boykov, Y., Veksler, O., Zabih, R.: Markov random fields with efficient approximations. In: CVPR, pp. 648–655 (1998)
5. Kolmogorov, V., Zabih, R.: Computing visual correspondence with occlusions via graph cuts. In: ICCV, vol. II, pp. 508–515 (2001)
6. Kolmogorov, V., Zabih, R.: Multi-camera scene reconstruction via graph cuts. In: Heyden, A., Sparr, G., Nielsen, M., Johansen, P. (eds.) ECCV 2002. LNCS, vol. 2352, pp. 82–96. Springer, Heidelberg (2002)
7. Lempitsky, V., Boykov, Y., Ivanov, D.: Oriented visibility for multiview reconstruction. In: Leonardis, A., Bischof, H., Pinz, A. (eds.) ECCV 2006. LNCS, vol. 3953, pp. 226–238. Springer, Heidelberg (2006)
8. Wills, J., Agarwal, S., Belongie, S.: What went where. In: CVPR, vol. I, pp. 37–44 (2003)
9. Xiao, J., Shah, M.: Motion layer extraction in the presence of occlusion using graph cuts. PAMI 27(10), 1644–1659 (2005)
10. Schoenemann, T., Cremers, D.: High resolution motion layer decomposition using dual-space graph cuts. In: CVPR, pp. 1–7 (2008)
11. Kwatra, V., Schödl, A., Essa, I., Turk, G., Bobick, A.: Graphcut textures: Image and video synthesis using graph cuts. ACM Transactions on Graphics, SIGGRAPH 2003 22(3), 277–286 (2003)
12. Boykov, Y., Jolly, M.: Interactive graph cuts for optimal boundary and region segmentation of objects in n-d images. In: ICCV, vol. I, pp. 105–112 (2001)
13. Blake, A., Rother, C., Brown, M., Perez, P., Torr, P.: Interactive image segmentation using an adaptive GMMRF model. In: Pajdla, T., Matas, J(G.) (eds.) ECCV 2004. LNCS, vol. 3021, pp. 428–441. Springer, Heidelberg (2004)
14. Rother, C., Minka, T., Blake, A., Kolmogorov, V.: Cosegmentation of image pairs by histogram matching: Incorporating a global constraint into mrfs. In: CVPR, vol. I, pp. 993–1000 (2006)
15. Kolmogorov, V., Criminisi, A., Blake, A., Cross, G., Rother, C.: Probabilistic fusion of stereo with color and contrast for bilayer segmentation. PAMI 28(9), 1480–1492 (2006)
16. Agarwala, A., Dontcheva, M., Agrawala, M., Drucker, S., Colburn, A., Curless, B., Salesin, D., Cohen, M.: Iterative digital photomontage. In: ACM Transactions on Graphics, SIGGRAPH (2004)

17. Ishikawa, H.: Exact optimization for markov random fields with convex priors. PAMI 25(10), 1333–1336 (2003)
18. Szeliski, R., Zabih, R., Scharstein, D., Veksler, O., Kolmogorov, V., Agarwala, A., Tappen, M., Rother, C.: A comparative study of energy minimization methods for markov random fields with smoothness-based priors. IEEE Transacions on Pattern Analysis and Machine Intellegence 30(6), 1068–1080 (2008)
19. Kolmogorov, V.: Convergent tree-reweighted message passing for energy minimization. PAMI 28(10), 1568–1583 (2006)
20. Pearl, J.: Probabilistic reasoning in intelligent systems: networks of plausible inference. Morgan Kaufmann, San Francisco (1988)
21. Kolmogorov, V., Rother, C.: Comparison of energy minimization algorithms for highly connected graphs. In: Leonardis, A., Bischof, H., Pinz, A. (eds.) ECCV 2006. LNCS, vol. 3952, pp. 1–15. Springer, Heidelberg (2006)
22. Veksler, O.: Graph cut based optimization for mrfs with truncated convex priors. In: CVPR, pp. 1–8 (2007)
23. Liu, X., Veksler, O., Samarabandu, J.: Graph cut with ordering constraints on labels and its applications. In: CVPR, pp. 1–8 (2008)
24. Kumar, M.P., Torr, P.H.S.: Improved moves for truncated convex models. In: Koller, D., Schuurmans, D., Bengio, Y., Bottou, L. (eds.) Advances in Neural Information Processing Systems 21, pp. 889–896 (2009)
25. Torr, P.H.S.: In: Personal communication (2008)
26. Hunter, D.R., Lange, K.: A tutorial on MM algorithms. The American Statistician (58) (2004)
27. Schlesinger, D., Flach, B.: Transforming an arbitrary minsum problem into a binary one. Technical Report TUD-FI06-01, Dresden University of Technology (2006)
28. Darbon, J.: Global optimization for first order markov random fields with submodular priors. In: Brimkov, V.E., Barneva, R.P., Hauptman, H.A. (eds.) IWCIA 2008. LNCS, vol. 4958, pp. 229–237. Springer, Heidelberg (2008)
29. Kolmogorov, V.: Primal-dual algorithm for convex markov random fields. Technical Report MSR-TR-2005-117, Microsoft (2005)
30. Rother, C., Kumar, S., Kolmogorov, V., Blake, A.: Digital tapestry. In: CVPR, vol. I, pp. 589–596 (2005)
31. Scharstein, D., Szeliski, R.: A taxonomy and evaluation of dense two-frame stereo correspondence algorithms. IJCV 47(1-3), 7–42 (2002)
32. Scharstein, D., Szeliski, R.: High-accuracy stereo depth maps using structured light. In: CVPR, vol. I, pp. 195–202 (2003)
33. Alahari, K., Kohli, P., Torr, P.: Reduce, reuse, recycle: Efficiently solving multi-label mrfs. In: CVPR, pp. 1–8 (2008)

Detection and Segmentation of Independently Moving Objects from Dense Scene Flow

Andreas Wedel, Annemarie Meißner, Clemens Rabe,
Uwe Franke, and Daniel Cremers

Daimler Group Research, Sindelfingen, Germany
University of Applied Sciences, Stuttgart, Germany
Department of Computer Science, University of Bonn, Germany

Abstract. We present an approach for identifying and segmenting independently moving objects from dense scene flow information, using a moving stereo camera system. The detection and segmentation is challenging due to camera movement and non-rigid object motion. The disparity, change in disparity, and the optical flow are estimated in the image domain and the three-dimensional motion is inferred from the binocular triangulation of the translation vector. Using error propagation and scene flow reliability measures, we assign dense motion likelihoods to every pixel of a reference frame. These likelihoods are then used for the segmentation of independently moving objects in the reference image. In our results we systematically demonstrate the improvement using reliability measures for the scene flow variables. Furthermore, we compare the binocular segmentation of independently moving objects with a monocular version, using solely the optical flow component of the scene flow.

1 Introduction and Related Work

In this paper we present the segmentation of independently moving objects from stereo camera sequences, obtained from a moving platform. Classically, moving objects are separated from the stationary background by *change detection* (e.g. [1]). But if the camera is also moving in a dynamic scene, motion fields become rather complex. Thus, the classic change detection approach is not suitable as it can be seen in Fig. 1. Our goal is to derive a segmentation of moving objects for this general dynamic setting.

Fig. 1. From left to right: input image, difference image between two consecutive frames, motion likelihood, and segmentation result. With the motion likelihood derived from the scene flow, the segmentation of the moving object becomes possible although the camera itself is moving.

D. Cremers et al. (Eds.): EMMCVPR 2009, LNCS 5681, pp. 14–27, 2009.

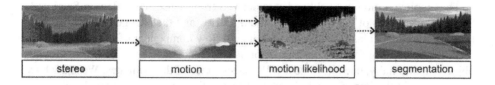

Fig. 2. The segmentation pipeline. Firstly, disparity and scene flow are computed; secondly, motion likelihoods are derived, and thirdly the image is segmented using graph cut.

We do not constrain the motion of the camera itself nor imply assumptions on the structure of the scene, such as rigid body motion. Rigid objects constrain the motion onto sub-spaces which yield efficient means to segment dynamic scenes using two views [2]. Another approach is used in [3], where the segmentation process is solved efficiently by incorporating a shape prior. If nothing about object appearance is known, the segmentation clearly becomes more challenging.

High-dynamic scenes with a variety of different conceivable motion patterns are especially challenging and reach the limits of many state-of-the-art motion segmentation approaches (e. g. [4,5]). This is a pity because the detection of moving objects implies certain scene dynamics. Although we do not constraint the camera motion, we assume that it is approximately known. In particular, we compute the fundamental matrix together with the scene flow, as proposed for the optical flow setting in [6,7]. From this, the motion of the camera is derived where the free scale parameter is fixed using the velocity sensor of the moving platform.

In [8] the authors use dense optical flow fields over multiple frames and estimate the camera motion and the segmentation of a moving object by bundle adjustment. The necessity of rather long input sequences however limits its practicability; furthermore, the moving object has to cover a large part of the image in order to detect its motion. The closest work related to our work is the work presented in [9]. It presents a monocular and a binocular approach to moving object detection and segmentation in high-dynamic situations using sparsely tracked features over multiple frames. In this paper we focus on moving object detection using only two consecutive stereo pairs, we use a dense scene flow fiels, and we show how per-pixel motion confidences are derived.

Fig. 2 illustrates the segmentation pipeline. The segmentation is performed in the image of a reference frame (left frame at time t) employing the graph cut segmentation algorithm [10]. The motion cues we use are derived from dense scene flow and calculated from the two stereo image pairs at time t-1 and t. Furthermore, we consider individual reliability measures for the variances of the flow vectors and the disparities at each image pixel. To our knowledge, the direct use of dense scene flow estimates for the detection and segmentation of moving objects is novel.

Paper Outline

In Section 2 we present the core graph cut segmentation algorithm. It minimizes an energy consisting of a motion likelihood for every pixel and a length term, favoring segmentation boundaries along intensity gradients.

The employed motion likelihoods are derived from dense scene flow in Section 3. Scene flow consists of the optical flow, the disparity, and the change of disparity over time. In the monocular setting, only the optical flow component of the scene flow is used. Compensating for the camera motion is a prerequisite step to detecting moving objects; additionally, one has to deal with inaccuracies in the estimates. We show how inaccuracies in the images can be modelled with reliability measures for the disparity and scene flow variables, and use error propagation to derive the motion likelihoods.

In Section 4 we compare the monocular method and the binocular method for the segmentation of independently moving objects in different scenarios. We systematically demonstrate that the consideration of inaccuracies, when computing the motion likelihoods for every pixel, yields increased robustness for the segmentation. Furthermore, we demonstrate the limits of the monocular and binocular segmentation methods and provide ideas for further research to overcome these limitations.

2   Segmentation Algorithm

The segmentation of the reference frame into moving and stationary parts can be expressed by a binary labelling of the pixels,

$$\mathcal{L}(\mathbf{x}) = \begin{cases} 1 & \text{if the pixel } \mathbf{x} \text{ is part of a moving object} \\ 0 & \text{otherwise.} \end{cases} \tag{1}$$

The goal is now to determine an optimal assignment of each pixel to *moving* or *non moving*. There are two competing constraints. Firstly, a point should be labelled *moving* if it has a high motion likelihood ξ_{motion} derived from the scene flow information and vice versa. Secondly, points should favour a labelling which matches that of their neighbors. Both constraints enter a joint energy of the form

$$E(\mathcal{L}) = E_{\text{data}}(\mathcal{L}) + \lambda E_{\text{reg}}(\mathcal{L}), \tag{2}$$

where λ weighs the influence of the regularization force. The data term is given by

$$E_{\text{data}} = -\sum_{\Omega} \left\{ \mathcal{L}(\mathbf{x})\, \xi_{\text{motion}}(\mathbf{x}) + \left(1 - \mathcal{L}(\mathbf{x})\right) \xi_{\text{static}}(\mathbf{x}) \right\} \tag{3}$$

on the image plane Ω, where ξ_{static} is a fixed prior likelihood of a point to be static. The regularity term favors labellings of neighboring pixels to be identical. This regularity is imposed more strongly for pixels with similar brightness:

$$E_{\text{reg}} = \sum_{\Omega} \left\{ \sum_{\hat{\mathbf{x}} \in \mathcal{N}_4(\mathbf{x})} g\left(I(\mathbf{x}) - I(\hat{\mathbf{x}})\right) |\mathcal{L}(\hat{\mathbf{x}}) - \mathcal{L}(\mathbf{x})| \right\}, \tag{4}$$

where $\mathcal{N}_4$ is the 4 neighborhood (upper, lower, left, right) of a pixel and $g(\cdot)$ is a positive, monotonically decreasing function of the brightness difference between neighboring pixels. Here, we set $g(z) = \frac{1}{z+\alpha}$ with a positive constant α.

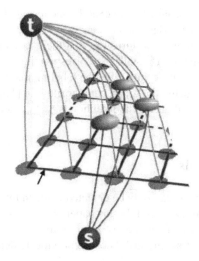

Fig. 3. Illustration of the graph mapping. Red connections illustrate graph edges from the source node s to the nodes, green connections illustrate graph edges from nodes to the target node t. Note, that the ξ_{motion} likelihood may be sparse due to occlusion. In the illustration only pixels with yellow spheres contribute to this motion likelihood. Black connections (indicated by the arrow) illustrate edges between neighboring pixels.

Graph Mapping

Summarizing the above equations, this yields for the energy (Equation 2)

$$\sum_{\Omega}\left\{-\mathcal{L}(\mathbf{x})\,\xi_{\text{motion}}(\mathbf{x})-\left(1-\mathcal{L}(\mathbf{x})\right)\,\xi_{\text{static}}(\mathbf{x})+\lambda\sum_{\hat{\mathbf{x}}\in\mathcal{N}_4(\mathbf{x})}\frac{|\mathcal{L}(\hat{\mathbf{x}})-\mathcal{L}(\mathbf{x})|}{|I(\mathbf{x})-I(\hat{\mathbf{x}})|+\alpha}\right\}. \quad (5)$$

Due to the combinatorial nature, finding the minimum of this energy is equivalent to finding the s-t-separating cut with minimum costs of a particular graph $\mathcal{G}(v, s, t, e)$, consisting of nodes $v(\mathbf{x})$ for every pixel $\mathbf{x}$ in the reference image and two distinct nodes: the source node s and the target node t [11]. The edges e in this graph connect each node with the source, target, and its $\mathcal{N}_4$ neighbors. The individual edge costs are defined as follows:

edge	edge cost
source link: $s \rightarrow v(\mathbf{x})$	$-\xi_{\text{motion}}(\mathbf{x})$
target link: $v(\mathbf{x}) \rightarrow t$	$-\xi_{\text{static}}(\mathbf{x})$
$\mathcal{N}_4$ neighborhood: $v(\hat{\mathbf{x}}) \leftrightarrow v(\mathbf{x})$	$\lambda\,\dfrac{1}{|I(\mathbf{x})-I(\hat{\mathbf{x}})|+\alpha}$

The cost of a cut in the graph is computed by summing up the costs of the cut (removed) edges. Removing the edges of an s-t-separating cut from the graph yields a graph where every node v is connected to exactly one terminal node, either the source s or the target t. If we define nodes that are connected to the source as static and those connected to the target as moving, it turns out that the cost of an s-t-separating cut is equal to the energy in Equation (5) with the corresponding labelling, and vice versa. Thus, the minimum s-t-separating cut yields the labeling that minimizes Equation (5). The minimum cut is found using the graph cut algorithm in [10].

In the next Section, we will derive the likelihoods $\xi_{\text{motion}}(\mathbf{x})$ from the disparity and scene flow estimates.

3   Motion Likelihoods

Independently moving objects can only be detected from an image sequences if at least two consecutive images are evaluated. In this paper we constraint ourselves to the minimum case of only two consecutive images. If more images are available, the detection task essentially becomes a tracking task because previously detected objects influence the current segmentation.

We analyze a monocular and a binocular camera setting, and derive likelihoods that pixels of a reference frame depict moving objects. In the monocular case, these constraints have been proposed in [12]. We will review the constraints and derive a Mahalanobis distance for every pixel in the image space which corresponds to the likelihood that the depicted object is moving. In the binocular case, the three-dimensional position for every pixel and its three-dimensional motion vector are reconstructed. Then the Mahalanobis distance of the three-dimensional translation vector yields a likelihood that the depicted object is moving.

3.1   Scene Flow Computation

The input for the motion likelihood is given by dense disparity and scene flow estimates $[d, u, v, p]$ for every pixel in the reference frame. The image position, $\mathbf{x} = [x, y]$, and the disparity, d, encode the three-dimensional position of a point. The optical flow (change of image position in between two frames), $[u, v]$, and the change in disparity, p, encode the scene flow motion information. Note, that for the monocular setting only the optical flow information, $[u, v]$, is used.

A variational approach to estimating this flow field was first proposed in [13]. The authors imposed regularity over all four variables and estimated all variables by minimizing a resulting single functional. Here we use the approach proposed in [14], where the authors split the position and motion estimation steps into two separate problems,

$$(A) \quad \Omega \to \mathbb{R}, \qquad [x, y] \mapsto d \tag{6}$$

$$\text{and } (B) \quad \Omega \times \mathbb{R} \to \mathbb{R}^3, \quad [x, y] \times d \mapsto [u, v, p]. \tag{7}$$

While (A) is the well-known disparity estimation step, (B) implies minimizing a scene flow energy, consisting of a data term and a smoothness term,

$$SF(u, v, p, d) = SF_{\text{data}}(u, v, p, d) + SF_{\text{smooth}}(u, v, p). \tag{8}$$

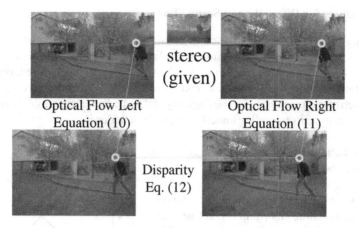

Fig. 4. Scene flow computation from two stereo image pairs. The stereo at the last time instance, t-1, is given by the semi-global matching algorithm. The data terms and smoothness term are described in the text in Equations (10 - 13).

The implicit dependancy of the variables u, v, p, and d on $[x, y]$ (e.g. $u(x, y)$) is left out in the notation to keep the notation uncluttered. Note, that the coupling between position and motion in such an approach is taken care of implicitly as the motion estimation step in (B) depends on the position estimation, which is the previously computed disparity map in (A).

The data term evaluates the gray value constancy of the scene flow,

$$SF_{\text{data}}(u, v, p, d) = \int_{\Omega} \left\{ E_{\text{sf-data-left}} + E_{\text{sf-data-right}} + E_{\text{sf-data-disp}} \right\} dx \, dy. \quad (9)$$

It evaluates the gray value constancy assumption for the optical flow field in the left image pair (I_L) and the right image pair (I_R):

$$E_{\text{sf-data-left}} = |I_L(x, y, t - 1) - I_L(x + u, y + v, t)| \quad (10)$$

$$E_{\text{sf-data-right}} = |I_R(x + d, y, t - 1) - I_R(x + d + u + p, y + v, t)|. \quad (11)$$

Additionally, the gray value constancy assumption for the stereo disparity field at time t is evaluated:

$$E_{\text{sf-data-disp}} = |I_L(x + u, y + v, t) - I_R(x + d + u + p, y + v, t)|. \quad (12)$$

The smoothness term minimizes the fluctuation in the scene flow field by penalyzing the flow field derivatives,

$$SF_{\text{smooth}}(u, v, p) = \int_{\Omega} E_{\text{sf-reg}} \, dx \, dy \quad \text{with} \quad E_{\text{sf-reg}} = |\nabla u| + |\nabla v| + |\nabla p|. \quad (13)$$

The resulting energy can be solved by calculus of variation. For the numerical solution scheme we refer to [14].

3.2   Variances for Disparity and Scene Flow

Computing the Mahalanobis distance implies that variances for the image position (monocular setting) or three-dimensional translation vector (binocular setting) need to be known. Although constant variances for the whole image may be used, our experiments show that individual variances yield more reliable segmentation results. Therefore, we derive such variances for the disparity and scene flow estimates for every pixel, depending on the corresponding underlying energy functional.

Disparity Reliability. The scene flow algorithm in [14] uses the semi-global matching algorithm [15] for the disparity estimation and a variational framework for the scene flow estimates. The core semi-global matching algorithm is pixel-accurate.

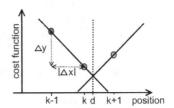

Fig. 5. The slope of the disparity cost function serves as a quality measure for the disparity estimate

Let k be the disparity estimate of the core SGM method for a certain pixel in the left image. The SGM method in [15] is formulated as an energy minimization problem. Hence, changing the disparity by ± 1 yields an increase in costs (yielding an increased energy). The minimum, however, may be located in between pixels, motivating a subsequent sub-pixel estimation step. Sub-pixel accuracy is achieved by a subsequent fit of a symmetric equiangular function (see [16]) in the cost volume. The basic idea of this step is illustrated in Figure 5. The costs for the three disparity assumptions k-1 px, k px, and k+1 px are taken and a symmetric first order function is fitted to the costs. This fit is unique and yields a specific sub-pixel minimum, located at the minimum of the function. Note, that this might not be the minimum of the underlying energy but is a close approximation, evaluating the energy only at pixel position.

The slope of this fitting function (the larger of the two relative cost differences between the current estimate and neighboring costs, Δy) serves as a quality measure for the goodness-of-fit. If the slope is low, the disparity estimate is not accurate in the sense that other disparity values could also be valid. If on the other hand the slope is large, the sub-pixel position of the disparity is expected to be quite accurate as deviation from this position increases the energy. Hence, the larger the slope, the better is the expected quality of the disparity estimate. Note that the costs mentioned here are accumulated costs that also incorporate smoothness terms.

Based on this observation an uncertainty measure is derived for the expected variance of the disparity estimate:

$$U_D(x, y, d) = \frac{1}{\Delta y} . \tag{14}$$

Scene Flow Reliability. For variational optic flow methods the idea of using the incline of the cost function or energy function as uncertainty measure becomes more complex than in the disparity setting. This is due to the higher dimensionality of the input and solution space. An alternative, energy-based confidence measure was proposed in [17]. The novel idea is that the reliability is inversely proportional to the local energy contribution in the energy functional, used to compute the optical flow. A large contribution

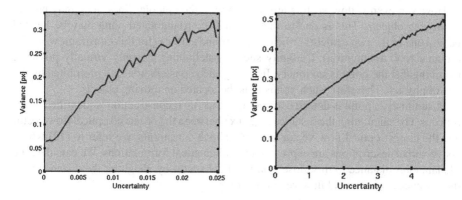

Fig. 6. Plots of the proposed reliability measures and corresponding variances for the disparity (VAR(d) vs. U_D, *left*) and for the scene flow u-component (VAR(u) vs. U_{SF}, *right*). The plots reveal that the proposed reliability measures are approx. proportional to the observed variances.

to the total energy implies low expected accuracy while the accuracy is expected to be good if the energy contribution is small. The authors show that this energy-based measure yields a better approximation of the *optimal* confidence for optic flow estimates than an image-gradient-based measure. The same idea is now applied to the scene flow case, yielding an expected variance of the scene flow estimate:

$$U_{SF}(x, y, d, u, v, p) = E_{\text{sf-data-left}} + E_{\text{sf-data-right}} + E_{\text{sf-data-disp}} + \lambda E_{\text{sf-reg}}. \quad (15)$$

Comparing Variances and Reliability Measures. To evaluate the reliability measures for the disparity and scene flow estimates, we plot the derived uncertainty measures against the observed error in Fig. 6 (for the disparity d and the u-component of the optical flow). To generate the plots a 400 frames long evaluation sequence, rendered with Povray and available in [18] together with the ground truth flow, is used.

The plots illustrate, that the proposed reliability measures are correlated to the true variances of the errors. Furthermore, the variance σ_z (for a scene flow component $z \in \{d, u, v, p\}$) can be approximated by a linear function of the reliability measure, denoted by γ_z, with fixed parameters a_z and b_z: $\sigma_z^2(\mathbf{x}) = a_z + b_z \gamma_z(\mathbf{x})$.

3.3 Monocular Motion Likelihood

For the monocular case we use the motion likelihood proposed for sparse data in [12]. There is a fundamental weakness of monocular three-dimensional reconstruction when compared to stereo methods – moving points cannot be correctly reconstructed by monocular vision. This is due to the camera movement between the two sequential images. Thus, optical flow vectors are triangulated, assuming that every point belongs to a static object. Such triangulation is only possible, if the displacement vector itself does not violate the fundamental matrix constraint. Needless to say that every track violating the fundamental matrix constraint belongs to a moving object and the distance to the fundamental rays directly serves as a motion likelihood.

However, even if flow vectors are aligned with the epipolar lines, they may belong to moving objects. This is due to the fact that the triangulated point may be located behind one of the two cameras or below the ground surface (for this constraint we make a planar road assumption). Certainly such constellations are only virtually possible, assuming that the point is stationary. In reality such constellations are prohibited by the law of physics. Therefore, such points must be located on moving objects.

In summary, a point is detected as moving if its 3D reconstruction is identified as erroneous. For calculating the distance $d_{\text{valid}}(\mathbf{x})$ between the observed optical flow vector and the closest optical flow vector fulfilling above constraints, we refer to [12] where above verbal descriptions are expressed in mathematical formulations. We calculate the Mahalanobis distance to this closest optical flow vector by weighing the distance with the variance of the optical flow vector, yielding

$$\xi_{\text{motion}}(\mathbf{x}) = \sqrt{d_{\text{valid}}(\mathbf{x})^2 \, \sigma_{u,v}(\mathbf{x})^2}. \tag{16}$$

Note, that due to the coupling in the variational framework, the variances σ_u and σ_v are assumed to be equal.

3.4 Binocular Motion Likelihood

In the stereo setting, the full disparity and scene flow information is available. A point is transformed from the image coordinates (x, y, d) into world coordinates (X, Y, Z) according to $X = (x - x_0)\frac{b}{d}$, $Y = (y - y_0)\frac{b}{d}\frac{f_x}{f_y}$, and $Z = \frac{f_x b}{d}$, where b is the basis length of the stereo camera system, $f_x\,f_y$ are the focal lengths of the camera in pixels, and (x_0, y_0) its principal point. As a simplification, we assume the focal lengths $f_x\,f_y$ to be equal. Transforming the points (x, y, d) and $(x + u, y + v, d + p)$ into world coordinates and compensating the camera rotation $\mathbf{R}$ and translation $\mathbf{T}$ yields the three-dimensional residual translation (or motion) vector $\mathbf{M}$ with

$$\mathbf{M} = \frac{b}{d}\mathbf{R}\begin{bmatrix} x - x_0 \\ y - y_0 \\ f_x \end{bmatrix} - \frac{b}{d + p}\begin{bmatrix} x + u - x_0 \\ y + v - y_0 \\ f_x \end{bmatrix} + \mathbf{T} \tag{17}$$

Using error propagation we calculate the Mahalanobis length of the translation vector. Essentially, this incorporates the variances of the disparity, scene flow estimates, and the camera rotation and translation. Here, we assume the variances of the camera rotation to be negligible. Although this is certainly not true, such procedure is possible because the estimation of the fundamental matrix from the complete optical flow field does yield vanishing variances for the rotational parts. We do however use fixed variances in the camera translation because the translation information from the velocity sensor of the ego-vehicle is rather inaccurate. With the variances σ_u^2, σ_v^2, σ_p^2, and σ_d^2 for the scene flow and $\sigma_{\mathbf{T}}^2$ for the translation this yields the Mahalanobis distance

$$\xi_{\text{motion}}(\mathbf{x}) = \sqrt{\mathbf{M}^{\top}\left(\mathbf{J}^{\top}\text{diag}(\sigma_u, \sigma_v, \sigma_p, \sigma_d, \sigma_{\mathbf{T}})\,\mathbf{J}\right)^{-1}\mathbf{M}}, \tag{18}$$

where $\mathbf{J}$ is the Jacobian of Equation (17).

4 Experimental Results and Discussion

In this section we present results which demonstrate the accurate segmentation of moving objects using scene flow. In the first part, we show that the presented reliability measures greatly improve the segmentation results when compared to a fixed variance for the disparity and scene flow variables. In the second part, we compare the segmentation results using the monocular and binocular motion segmentation approaches.

4.1 Robust Segmentation

Figure 7 illustrates the importance of using the reliability measures to derive individual variances for the scene flow variables. If the propagation of uncertainties is not used at all, the segmentation of moving objects is not possible (top row). Using the same variance for every image pixel the segmentation is more meaningful; but still outliers are present in both, the motion likelihoods and the segmentation results (middle row). Only when the reliability measures are used to derive individual variances for the pixels, is the segmentation accurate and outlier influence is minimized (bottom row).

no error propagation

spatially fixed variances used in error propagation

variances from reliability measures used for error propagation

Fig. 7. Results for different error propagation methods. The *left* images show the motion likelihoods and the *right* images the segmentation results.

PreceedingCar   **HillSide**

Optical Flow Result

Monocular Segmentation of Independently Moving Objects

Binocular Segmentation of Independently Moving Objects

Fig. 8. The figure shows the energy images and the segmentation results for objects moving parallel to the camera movement. This movement cannot be detected monocularly without additional constraints, such as a planar ground assumption. Moreover if this assumption is violated, this yields errors (as in the *HillSide* sequence). In a stereo setting prior knowledge is not needed to solve the segmentation task in these two scenes.

Bushes   **Running**

Optical Flow Result

Monocular Segmentation of Independently Moving Objects

Binocular Segmentation of Independently Moving Objects

Fig. 9. The figure shows the energy images and the segmentation results for objects which move not parallel to the camera motion. In such constallations a monocular as well as a binocular segmentation approach is successfull. However, one can see in the energy images and in the more accurate segmentation results (the head of the person in the *Running* sequence) that stereo is more discriminative. Note, that the also non-rigid independently moving objects are segmented.

4.2 Monocular versus Binocular Segmentation of Independently Moving Object

A binocular camera system will always outperform a monocular system, simply because more information is available. However, in many situations a monocular system is able to detect independent motion and segment the moving objects in the scene. In this section we demonstrate the segmentation of independently moving objects using a monocular and a binocular camera system and discuss the results.

In a monocular setting, motion which is aligned with the epipolar lines cannot be detected without prior knowledge about the scene. Amongst other motion patterns, this includes objects moving parallel to the camera motion. For a camera moving in depth this includes all (directly) preceding objects and (directly) approaching objects. The *PreceedingCar* and *HillSide* sequences in Figure 8 show such constellations.

Using the ground plane assumption in the monocular setting (no virtually triangulated point is allowed to be located below the road surface) facilitates the detection of preceding objects. This can be seen in the *PreceedingCar* experiment, where lower parts of the car become visible. If compared to the stereo settings, which does not use any information about scene structure, the motion likelihood for the lower part of the preceding car is more discriminative. However, if parts of the scene are truly located below the ground plane, as the landscape at the right in the *HillSide* experiment, these will always be detected as moving, too. Additionally, this does not help to detect approaching objects. Both situations are solved using a binocular camera.

If objects do not move parallel to the camera motion, they are essentially *detectable* in the monocular setting (*Bushes* and *Running* sequences in Figure 9). However, the motion likelihood using a binocular system is more discriminative. This is due to the fact that the three-dimensional position of an image point is known from the stereo disparity. Thus, the complete viewing ray for a pixel does not need to be tested for apparent motion in the images, as in the monocular setting. In the unconstrained setting (not considering the ground plane assumption), the stereo motion likelihood therefore is more restrictive than the monocular motion likelihood. Note, that non-rigid objects (as in the *Running* sequence in Figure 9) are detected as well as rigid objects and do not limit the detection and segmentation at any stage.

5 Conclusion

Building up on a recent variational approach to scene flow estimation, we proposed in this paper an energy minimization method to detect and segment independently moving objects filmed in two video cameras installed in a driving car. The central idea is to assign, to each pixel in the image plane, a motion likelihood which specifies whether, based on 3D structure and motion, the point is likely to be part of an independently moving object. Subsequently, these local likelihoods are fused in an MRF framework and a globally optimal spatially coherent labelling is computed using the min cut max flow duality. In challenging real world scenarios where traditional background subtraction techniques would not work (because everything is moving), we are able to accurately localize independently moving objects. The results of our algorithm could directly be employed for automatic driver assistance.

References

1. Sun, J., Zhang, W., Tang, X., Shum, H.: Background Cut. In: Leonardis, A., Bischof, H., Pinz, A. (eds.) ECCV 2006. LNCS, vol. 3952, pp. 628–641. Springer, Heidelberg (2006)
2. Vidal, R., Sastry, S.: Optimal segmentation of dynamic scenes from two perspective views. In: 2003 IEEE Computer Society Conference on Computer Vision and Pattern Recognition, Proceedings., vol. 2 (2003)
3. Brox, T., Rosenhahn, B., Cremers, D., Seidel, H.P.: High accuracy optical flow serves 3-D pose tracking: exploiting contour and flow based constraints. In: Leonardis, A., Bischof, H., Pinz, A. (eds.) ECCV 2006. LNCS, vol. 3952, pp. 98–111. Springer, Heidelberg (2006)
4. Cremers, D., Soatto, S.: Motion competition: A variational framework for piecewise parametric motion segmentation 62(3), 249–265 (2005)
5. Kolmogorov, V., Criminisi, A., Blake, A., Cross, G., Rother, C.: Bi-layer segmentation of binocular stereo video, vol. 2 (2005)
6. Wedel, A., Pock, T., Braun, J., Franke, U., Cremers, D.: Duality TV-L1 flow with fundamental matrix prior. In: Proc. Image and Vision Computing New Zealand, Christchurch, New Zealand (November 2008)
7. Valgaerts, L., Bruhn, A., Weickert, J.: A variational approach for the joint recovery of the optical flow and the fundamental matrix, Munich, Germany, June 2008, pp. 314–324 (2008)
8. Zhang, G., Jia, J., Xiong, W., Wong, T., Heng, P., Bao, H.: Moving object extraction with a hand-held camera, pp. 1–8 (2006)
9. Vaudrey, T., Wedel, A., Rabe, C., Klappstein, J., Klette, R.: Evaluation of moving object segmentation comparing 6D-Vision and monocular motion constraints. In: Proc. Image and Vision Computing New Zealand, Christchurch, New Zealand (November 2008)
10. Boykov, Y., Kolmogorov, V.: An experimental comparison of min-cut/max-flow algorithms for energy minimization in vision, pp. 1124–1137 (2004)
11. Kolmogorov, V., Zabih, R.: What energy functions can be minimized via graph cuts? In: Heyden, A., Sparr, G., Nielsen, M., Johansen, P. (eds.) ECCV 2002. LNCS, vol. 2352, pp. 65–81. Springer, Heidelberg (2002)
12. Klappstein, J., Stein, F., Franke, U.: Detectability of Moving Objects Using Correspondences over Two and Three Frames. In: Hamprecht, F.A., Schnörr, C., Jähne, B. (eds.) DAGM 2007. LNCS, vol. 4713, pp. 112–121. Springer, Heidelberg (2007)
13. Huguet, F., Devernay, F.: A variational method for scene flow estimation from stereo sequences. In: IEEE Eleventh International Conference on Computer Vision, ICCV 2007, Rio de Janeiro, Brazil (October 2007)
14. Wedel, A., Rabe, C., Vaudrey, T., Brox, T., Franke, U., Cremers, D.: Efficient dense scene flow from sparse or dense stereo data. In: Forsyth, D., Torr, P., Zisserman, A. (eds.) ECCV 2008, Part I. LNCS, vol. 5302, pp. 739–751. Springer, Heidelberg (2008)
15. Hirschmüller, H.: Stereo vision in structured environments by consistent semi-global matching, pp. 2386–2393 (2006)
16. Shimizu, M., Okutomi, M.: Precise sub-pixel estimation on area-based matching, pp. 90–97 (2001)
17. Bruhn, A., Weickert, J.: A confidence measure for variational optic flow methods. Geometric Properties for Incomplete Data (March 2006)
18. University of Auckland: enpeda. Image Sequence Analysis Test Site (EISATS) (2008), http://www.mi.auckland.ac.nz/EISATS/

Efficient Global Minimization for the Multiphase Chan-Vese Model of Image Segmentation

Egil Bae[1] and Xue-Cheng Tai[1,2]

[1] Department of Mathematics, University of Bergen, Norway
`Egil.Bae@math.uib.no`
[2] Division of Mathematical Sciences, School of Physical and Mathematical Sciences,
Nanyang Technological University, Singapore
`tai@mi.uib.no`

Abstract. The Mumford-Shah model is an important variational image segmentation model. A popular multiphase level set approach, the Chan-Vese model, was developed for this model by representing the phases by several overlapping level set functions. Recently, exactly the same model was also formulated by using binary level set functions. In both approaches, the gradient descent equations had to be solved numerically, a procedure which is slow and has the potential of getting stuck in a local minima. In this work, we develop an efficient and global minimization method for the binary level set representation of the multiphase Chan-Vese model based on graph cuts. If the average intensity values of the different phases are sufficiently evenly distributed, the discretized energy function becomes submodular. Otherwise, a novel method for minimizing nonsubmodular functions is proposed with particular emphasis on this energy function.

1   Introduction

Multiphase image segmentation is a fundamental problem in image processing. Variational models such as Mumford-Shah [1] are powerful for this task, but efficient numerical computation of the global minimum is a big challenge. The level set method [2,3] is a powerful tool which can used for numerical realization. It was first proposed for the Mumford-Shah model in [4] for two phases and [5] for multiple phases. This approach still has the disadvantage of slow convergence and potential of getting stuck in a local minima.

Graph cuts from combinatorial optimization [6,7,8,9,10,11] is another technique which can perform image segmentation by minimizing certain discrete energy functions. In the recent years, the relationship between graph cuts and continuous variational problems have been much explored [12,13,14,15]. It turns out graph cuts are very similar to the level set method, and can be used for many variational problems with the advantage of a much higher efficiency and ability to find global minima. It can be applied to the 2-phase Mumford-Shah model [16,17], but for multiple phases it is probably not possible to find the exact, global minimum in polynomial time as this is an NP-hard problem. The

D. Cremers et al. (Eds.): EMMCVPR 2009, LNCS 5681, pp. 28–41, 2009.

usual approach to minimization problems with several regions is some heuristic method for finding an approximate, local minimum. Most popular in computer vision are the α-expansion algorithms [7]. Recently, also convex formulations of the continuous multiphase problem have been made in [18,19] by relaxing the integrality constraint. A suboptimal solution is found by converting the real valued relaxed solution to an integral one (e.g. by thresholding).

In this paper we propose a method to globally and efficiently minimize the Mumford-Shah model in the multiphase level set framework of Vese and Chan [5] by using binary level set functions as in [20]. Since the length term is slightly approximated in this framework, global minimization is no longer NP hard. We will construct a graph such that the discrete variational problem can be minimized exactly by finding the minimum cut on the graph. However, the energy function may not be submodular if the average intensity values of the phases are distributed very unevenly. To handle these cases, we have developed a method for minimizing non-submodular functions with particular emphasis on our energy function. The minimization is global if these values are fixed. A local minimization approach for determining these values is also proposed.

Note that in contrast to α-expansion, the approximation is done in the model rather than in the minimization method. Experimental comparison with alpha expansion is out of the scope of this paper. What can be said is that our method is certainly a lot faster. It is also straight forwardly generalizable to non-local measurements of the curve lengths as was done for two phases in [21]. Such a generalization is not obvious for alpha expansion.

In this work we focus on the case of 4 or less phases, but aim to generalize the results to more phases later. Nevertheless, these are important cases since by the four colour theorem, four phases in theory suffices to segment any 2D image.

1.1   The Mumford-Shah Model and Its Level Set Representation

Image segmentation is the task of partitioning the image domain Ω into a set of n meaningful disjoint regions $\{\Omega_i\}_{i=1}^n$. The Mumford-Shah model [1] is an established image segmentation model with a wide range of applications. An energy functional to be minimized is defined over the regions $\{\Omega_i\}_{i=1}^n$, and an approximation image u of the input image u_0. In an especially popular form, u is assumed to be constant within each region Ω_i, in which case the model reads

$$\min_{\{c_i\},\{\Omega_i\}} E(\{c_i\},\{\Omega_i\}) = \sum_{i=1}^n \int_{\Omega_i} |u - c_i|^\beta dx + \sum_{i=1}^n \nu \int_{\partial\Omega_i} ds, \qquad (1)$$

where $\partial\Omega_i$ is the boundary of Ω_i. The power β is usually chosen as $\beta = 2$. As a numerical realization, Chan and Vese [4,5] proposed to represent the above functional with level set functions, and solve the resulting gradient descent equations numerically. By using $m = \log_2(n)$ level set functions, denoted $\phi^1, ..., \phi^m$, n phases could be represented. An important special case is the representation of 4 phases by two level set functions ϕ^1, ϕ^2, as in Table 1. The energy function can then be written

$$\min_{\phi^1, \phi^2, c_1, \ldots, c_4} = \nu \int_\Omega |\nabla H(\phi^1)| + \nu \int_\Omega (|\nabla H(\phi^2)| \tag{2}$$

$$+ \int_\Omega \{H(\phi^1)H(\phi^2)|c_2 - u^0|^\beta + H(\phi^1)(1 - H(\phi^2))|c_1 - u^0|^\beta$$

$$+(1 - H(\phi^1))H(\phi^2)|c_4 - u^0|^\beta + (1 - H(\phi^1))(1 - H(\phi^2))|c_3 - u^0|^\beta\}dx.$$

Note that the length term in (1) is slightly approximated, since some of the boundaries are counted twice. Note also that we have made a small permutation in the interpretation of the phases compared to [5]. The energy is still exactly identical for all feasible solutions. This permutation is crucial for making the corresponding discrete energy function submodular.

The functional in this variational problem is highly non-convex for fixed constant values $c_1, \ldots, c_4$. The traditional minimization approach of solving the gradient descent equations can therefore easily get stuck in a local minima. Furthermore, the numerical solution of the gradient descent PDEs is expensive computationally.

In [20], the same multiphase model was formulated using binary level set functions $\phi^1, \phi^2 \in D = \{\phi \mid \phi : \Omega \mapsto \{0,1\}\}$, representing the phases as in Table 1. This resulted in the energy functional

$$\min_{\phi^1, \phi^2 \in D, c_1, \ldots, c_4} E(\phi^1, \phi^2, c_1, \ldots, c_4) = \nu \int_\Omega |\nabla \phi^1| dx + \nu \int_\Omega |\nabla \phi^2| dx + E^{data}(\phi^1, \phi^2), \tag{3}$$

where

$$E^{data}(\phi^1, \phi^2) = \int_\Omega \{\phi^1 \phi^2 |c_2 - u^0|^\beta + \phi^1(1 - \phi^2)|c_1 - u^0|^\beta$$

$$+(1 - \phi^1)\phi^2 |c_4 - u^0|^\beta + (1 - \phi^1)(1 - \phi^2)|c_3 - u^0|^\beta\}dx.$$

The constraint D was represented by a polynomials in ϕ^1 and ϕ^2. Minimization was carried out by the augmented lagrangian method. Since both the constraint D and the energy functional is non-convex, global minimization could not be

Table 1. Representation of four phases by traditional and binary level set functions

	Traditional level set functions	Binary level set functions
$x \in$ phase 1 iff	$\phi^1(x) > 0, \phi^2(x) < 0$	$\phi^1(x) = 1, \phi^2(x) = 0$
$x \in$ phase 2 iff	$\phi^1(x) > 0, \phi^2(x) > 0$	$\phi^1(x) = 1, \phi^2(x) = 1$
$x \in$ phase 3 iff	$\phi^1(x) < 0, \phi^2(x) < 0$	$\phi^1(x) = 0, \phi^2(x) = 0$
$x \in$ phase 4 iff	$\phi^1(x) < 0, \phi^2(x) > 0$	$\phi^1(x) = 0, \phi^2(x) = 1$

guaranteed. Also, convergence was slow just as in the traditional level set approach. A similar approach could also be used for finding a local minimum with exact curve lengths [22].

Let us mention that a method often referred to as continuous graph cut can be used to globally minimize the Mumford Shah model in case of two phases. The idea, first presented in [23] is to relax the constraint D by the convex constraint $D' = \{\phi \mid \phi : \Omega \mapsto [0, 1]\}$. It was shown that thresholding this solution at almost any threshold in $[0, 1]$ yields the optimal solution within D. The same idea could also be used to minimize (3). The problem is that (3) is not convex, and hence the algorithm may converge to a local minimum.

In general, discrete graph cuts has the disadvantage of some metrication artifacts over continuous graph cuts. However, discrete graph cuts is faster and can elegantly be used for minimization problems with non-local operators. The method we propose can very easily be generalized to minimize non-local measurements of the curve lengths as was done for two phases in [21], by using regularization term

$$\nu \int_\Omega |\nabla_{NL}\phi^1| dx + \nu \int_\Omega |\nabla_{NL}\phi^2| dx.$$

However, that is not the focus of this paper. We will propose a method which globally minimizes (3) for fixed constant values $c_1, ..., c_4$. This new approach, is also shown to be very superior in terms of efficiency compared to gradient descent.

2   Graph Cut Minimization

We will discretize the problem (3) and show that this discrete problem can be minimized globally by finding the minimum cut on a specially designed graph. This is possible when the constant values $c_1, ..., c_4$ are sufficiently evenly distributed. We show that such a distribution makes the discrete energy function sub-modular. The evenness criterion will soon be defined more clearly. We have observed that this criterion makes sense for most practical images. Nevertheless, we later develop an algorithm for minimizing non-submodular functions with particular emphasize on functions of the form (3).

2.1   Brief Overview of Graph Cuts in Computer Vision

Graph cuts were first introduced as a computer vision tool by Greig et. al. [8] in connection with markov random fields [6]

A graph $\mathcal{G} = (\mathcal{V}, \mathcal{E})$ is a set of vertices $\mathcal{V}$ and a set of edges $\mathcal{E}$. We let (a, b) denote the directed edge going from vertex a to vertex b, and let $c(a, b)$ denote the capacity/cost/weight on this edge. In the graph cut scenario there are two distinguished vertices in $\mathcal{V}$, called the source $\{s\}$ and the sink $\{t\}$. A cut on $\mathcal{G}$

is a partitioning of the vertices $\mathcal{V}$ into two disjoint connected sets $(\mathcal{V}_s, \mathcal{V}_t)$ such that $s \in \mathcal{V}_s$ and $t \in \mathcal{V}_t$. The cost of the cut is defined as

$$c(\mathcal{V}_s, \mathcal{V}_t) = \sum_{(i,j) \in \mathcal{E} \text{ s.t. } i \in \mathcal{V}_s, j \in \mathcal{V}_t} c(i,j).$$

A flow f on $\mathcal{G}$ is a function $f : \mathcal{E} \mapsto \mathbb{R}$. For a given flow, the residual capacities are defined as $R(e) = c(e) - f(e) \; \forall e \in \mathcal{E}$. The max flow problem is to find maximum amount of flow that can be pushed from $\{s\}$ to $\{t\}$, under flow conservation constraint at each vertex. A theorem of Ford and Fulkerson [24] says this is the dual to the problem of finding the cut of minimum cost on $\mathcal{G}$, the min-cut problem. Therefore, efficient algorithms for finding max-flow, such as the augmented paths method [24] can be used for finding minimum cuts in graphs. An efficient implementation of this algorithm specialized for image processing problems can be found in [9]. This algorithm, which is available on-line has been used in our experiments.

In computer vision this has been exploited for minimizing energy functions of the form

$$\min_{x \in \{0,1\}^m} E(x) = \sum_i E^i(x_i) + \sum_{i<j} E^{i,j}(x_i, x_j).$$

Typically, $i = 1, ..., m$ denotes the grid points and x contains one binary variable for each grid point. In order to be representable as a cut on a graph, it is required that the energy function is submodular (or regular) [10,6], i.e.

$$E^{i,j}(0,0) + E^{i,j}(1,1) \leq E^{i,j}(0,1) + E^{i,j}(1,0).$$

2.2   Discretization of Energy Functional

Instead of discretizing the Euler-Lagrange equations, we will discretize the variational problem (3). In the next section we show how to minimize the resulting discrete energy function exactly. Let us first mention there are two variants of the total variation term. The isotropic variant, by using 2-norm $TV_2(\phi) = \int_\Omega |\nabla\phi|_2 \, dx = \int_\Omega \sqrt{|\phi_{x_1}|^2 + |\phi_{x_2}|^2} \, dx$, and the anisotropic variant, by using 1-norm $TV_1(\phi) = \int_\Omega |\nabla\phi|_1 \, dx = \int_\Omega |\phi_{x_1}| + |\phi_{x_2}| \, dx$. The anisotropic variant is graph representable and will be considered here. More isotropic variants can be derived by splitting the calculation of $TV_1(\phi)$ between several rotated coordinate systems, see [25].

Let $\mathcal{P} = \{(i,j) \subset \mathbb{Z}^2\}$ denote the set of grid points. For each $p = (i,j) \in \mathcal{P}$, the neighborhood system $\mathcal{N}_p^k \subset \mathcal{P}$ is defined as

$$\mathcal{N}_p^4 = \{(i \pm 1, j), (i, j \pm 1)\} \cap \mathcal{P}$$

$$\mathcal{N}_p^8 = \{(i \pm 1, j), (i, j \pm 1), (i \pm 1, j \pm 1)\} \cap \mathcal{P}.$$

The discrete energy function can be written

$$\min_{\phi^1, \phi^2 \in D, c_1, \ldots, c_4} E_d(\phi^1, \phi^2, c_1, \ldots, c_4) = \nu \sum_{p \in \mathcal{P}} \sum_{q \in \mathcal{N}_p^k} w_{pq} |\phi_p^1 - \phi_q^1| + \nu \sum_{p \in \mathcal{P}} \sum_{q \in \mathcal{N}_p^k} w_{pq} |\phi_p^2 - \phi_q^2|$$

(4)

$$+ \sum_{p \in \mathcal{P}} E_p^{data}(\phi_p^1, \phi_p^2),$$

where

$$E_p^{data}(\phi_p^1, \phi_p^2) = \{\phi_p^1 \phi_p^2 |c_2 - u^0|^\beta + \phi_p^1 (1 - \phi_p^2)|c_1 - u^0|^\beta$$

$$+ (1 - \phi_p^1)\phi_p^2 |c_4 - u^0|^\beta + (1 - \phi_p^1)(1 - \phi_p^2)|c_3 - u^0|^\beta\}.$$

The weights w_{pq} are used to approximate the curve lengths. They can be derived from the continuous functional as in full version [25], or from the Cauchy-Crofton formula as in [12].

2.3   Graph Construction

We will construct a graph $\mathcal{G}$ such that there is a one-to-one correspondence between cuts on $\mathcal{G}$ and the level set functions ϕ^1 and ϕ^2. Furthermore, the minimum cost cut will correspond to the level set functions ϕ^1 and ϕ^2 minimizing the energy (4).

$$\min_{(\mathcal{V}_s, \mathcal{V}_t)} c(\mathcal{V}_s, \mathcal{V}_t) = \min_{\phi^1, \phi^2} E_d(\phi^1, \phi^2, c_1, \ldots, c_4) + \sum_{p \in \mathcal{P}} \sigma_p.$$

(5)

where $\sigma_p \in \mathbb{R}$ for each $p \in \mathcal{P}$. In the graph, two vertices are associated to each grid point $p \in \mathcal{P}$. They are denoted $v_{p,1}$ and $v_{p,2}$, and corresponds to each of the level set functions ϕ^1 and ϕ^2. Hence the set of vertices is formally defined as

$$\mathcal{V} = \{v_{p,i} \mid p \in \mathcal{P}, \ i = 1, 2\} \cup \{s\} \cup \{t\}.$$

(6)

The edges are constructed such that the relationship (5) is satisfied. We begin with the edges constituting the data term of (4). For each grid point $p \in \mathcal{P}$ they are defined as

$$\mathcal{E}_D(p) = (s, v_{p,1}) \cup (s, v_{p,2}) \cup (v_{p,1}, t) \cup (v_{p,2}, t) \cup (v_{p,1}, v_{p,2}) \cup (v_{p,2}, v_{p,1}). \quad (7)$$

The set of all data edges are denoted $\mathcal{E}_D$ and defined as $\cup_{p \in \mathcal{P}} \mathcal{E}_D(p)$. The edges corresponding to the regularization term are defined as

$$\mathcal{E}_R = \{(v_{p,1}, v_{q,1}), (v_{p,2}, v_{q,2}) \ \forall p, q \subset \mathcal{P} \text{ s.t.} q \in \mathcal{N}_p^k\}.$$

(8)

For any cut $(\mathcal{V}_s, \mathcal{V}_t)$, the corresponding level set functions are defined by

$$\phi_p^1 = \begin{cases} 1 & \text{if } v_{p,1} \in \mathcal{V}_s, \\ 0 & \text{if } v_{p,1} \in \mathcal{V}_t, \end{cases} \qquad \phi_p^2 = \begin{cases} 1 & \text{if } v_{p,2} \in \mathcal{V}_s, \\ 0 & \text{if } v_{p,2} \in \mathcal{V}_t. \end{cases}$$

(9)

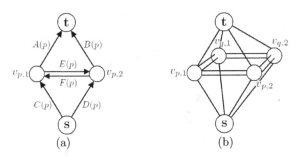

Fig. 1. (a) The graph corresponding to the data term at one grid point p. (b) A sketch of the graph corresponding to the energy function of a 1D signal of two grid points p and q.

Weights are assigned to the edges such that the relationship (5) is satisfied. Weights on the regularization edges are simply given by

$$c(v_{p,1}, v_{q,1}) = c(v_{q,1}, v_{p,1}) = c(v_{p,2}, v_{q,2}) = c(v_{q,2}, v_{p,2}) = \nu w_{pq}, \qquad \forall (p,q) \in \mathcal{N}. \tag{10}$$

We now concentrate on the weights on data edges $\mathcal{E}_D$. For grid point $p \in \mathcal{P}$, let

$$A(p) = c(v_{p,1}, t), \ B(p) = c(v_{p,2}, t), \ C(p) = c(s, v_{p,1}),$$

$$D(p) = c(s, v_{p,1}), E(p) = c(v_{p,1}, v_{p,2}), F(p) = c(v_{p,2}, v_{p,1}).$$

It is clear that these weights must satisfy

$$\begin{cases} A(p) + B(p) & = |c_2 - u_p^0|^\beta + \sigma_p \\ C(p) + D(p) & = |c_3 - u_p^0|^\beta + \sigma_p \\ A(p) + E(p) + D(p) = |c_1 - u_p^0|^\beta + \sigma_p \\ B(p) + F(p) + C(p) = |c_4 - u_p^0|^\beta + \sigma_p \end{cases} \tag{11}$$

This is a non-singular linear system for the weights $A(p), B(p), C(p), D(p),$ $E(p), F(p)$. Negative weights are not allowed. By choosing σ_p large enough there will exist a solution with $A(p), B(p), C(p), D(p) \geq 0$. However, the requirement $E(p), F(p) \geq 0$ implies that

$$|c_1 - u_p^0|^\beta + |c_4 - u_p^0|^\beta = A(p) + B(p) + C(p) + D(p) + E(p) + F(p)$$

$$\geq A(p) + B(p) + C(p) + D(p) = |c_2 - u_p^0|^\beta + |c_3 - u_p^0|^\beta.$$

This condition must hold for all grid points $p \in \mathcal{P}$. Hence, the following condition on the constant values $c_1, ..., c_4$ must be satisfied

$$|c_2 - I|^\beta + |c_3 - I|^\beta \leq |c_1 - I|^\beta + |c_4 - I|^\beta, \qquad \forall I \in [0, L], \tag{12}$$

where L is the maximum intensity value. This condition can be seen in the light of submodular energy functions [10,6]. In fact, the pairwise term $\sum_{p \in \mathcal{P}} E_p^{data}(\phi_p^1, \phi_p^2)$ is submodular if and only if the condition (12) is satisfied.

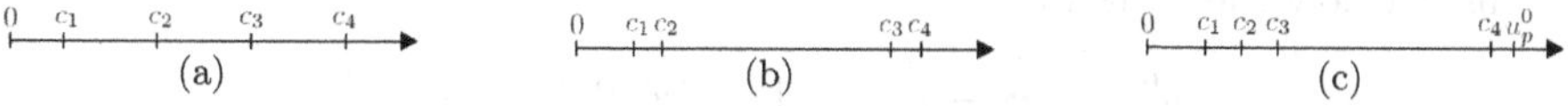

Fig. 2. (a) and (b) distributions of **c** which makes energy function submodular for all β. (c) distribution of **c** which may make energy function nonsubmodular for small β.

Let us analyze this condition further. We assume the constant values are ordered increasingly $0 \leq c_1 < c_2 < c_3 < c_4$. The condition says something about how evenly $\{c_i\}_{i=1}^4$ are distributed. Here is a first observation, the proof of this and the following lemmas can be found in the full version of this work [25]

Lemma 1. *Let $0 \leq c_1 < c_2 < c_3 < c_4$. There exists a $\mathcal{B} \in \mathbb{N}$ such that (12) is satisfied for any $\beta \geq \mathcal{B}$.*

So (12) becomes less strict for larger β. In fact we have observed that for $\beta = 2$, (12) is realistic for most practical images. Here is another observation

Lemma 2. *Let $0 \leq c_1 < c_2 < c_3 < c_4$. (12) is satisfied for all $I \in [c_2, c_3]$.*

The possibility that (12) is not satisfied may happen in two situations: If c_1, c_2, c_3 are very close compared to c_4 and intensity I is close to c_4, or if c_2, c_3, c_4 are very close compared to c_1 and I is close to c_1.

Let us go back to the linear system (11), with restriction $E(p), F(p) \geq 0$. Assuming (12) holds, this has infinitely many solutions.

It was shown in [10] that at most three edges are required for representing a general submodular term of two binary variables. Therefore, it is possible to pick a solution such that at least three of the weights $A(p), B(p), C(p), D(p), E(p), F(p)$ in $\mathcal{E}_D(p)$ becomes zero for each $p \in \mathcal{P}$. The exact construction of the solution can be found in the full version [25]. Hence, at most three edges are required to represent the data term at each grid point. Therefore, by analyzing the complexity of our method in the augmenting paths framework, it is easily seen that the cost of our method is equal to the cost of one single iteration of the alpha expansion method.

2.4 Minimization of Non-submodular Energy Functions

In the last section, we have observed that the energy function (4) is submodular if $c_1, ..., c_4$ satisfies (12). Although this is realistic for most images, we will develop a method for minimizing nonsubmodular functions with particular emphasis on nonsubmodular terms of the kind encountered here. Minimization of nonsubmodular functions via graph cuts has been investigated previously, see [26] for a review. The usual idea is to develop a method for determining most of the variables, while leaving some of the variables undetermined. In our approach, we instead aim to determine all the variables. Even when (12) does not hold, the energy function is "almost submodular", which may explain why the following very efficient algorithms works so well in practice.

Consider now the situation

$$|c_2 - u_p^0|^\beta + |c_3 - u_p^0|^\beta > |c_1 - u_p^0|^\beta + |c_4 - u_p^0|^\beta,$$

for some $p \in \mathcal{P}$. In this case the linear system (11) has a solution only if either $E(p) < 0$ or $F(p) < 0$, in which case one of the edges, $(v_{p,1}, v_{p,2})$ or $(v_{p,2}, v_{p,1})$, will have negative weight. It can be easily seen that if $E(p) < 0$, there exists a solution to the linear system with $F(p) = 0$. Vice versa, if $F(p) < 0$ there exists a solution with $E(p) = 0$. See [25] for the exact construction.

It is difficult interpret physically what is meant by max flow on a graph with negative edge weights. The concept of min-cut, on the other hand, certainly have a meaning even if some of the edges have negative weight. In the extreme case of negative weight on all edges, this becomes equivalent to the max-cut on a graph with negated edge weights. The first step of our procedure finds a good feasible solution, and therefore also a good upper bound on the objective function (4). Very often this feasible solution is in fact the optimal solution. All edges of negative weight will be removed, resulting in a new graph $\overline{\mathcal{G}}$. The motivation is as follows. The previous section discussed the possibility of condition (12) not being satisfied. In this case c_1, c_2, c_3 are close to each other compared to c_4 and I_p at $p \in \mathcal{P}$ is close to c_4. Measured by the data term, the worst assignment of p is to phase 1, which has the cost $|c_1 - u_p^0|^\beta$. By removing the negative edge with $E(p) < 0$, the cost of this assignment becomes even higher $|c_1 - u_p^0|^2 - E(p)$. We therefore expect the minimum cut on $\overline{\mathcal{G}}$ to be almost identical to the minimum cut on $\mathcal{G}$. For easy of notation, we define the sets

$$\mathcal{P}^1 = \{p \in \mathcal{P} \mid E(p) < 0, F(p) \geq 0\}, \quad \mathcal{P}^2 = \{p \in \mathcal{P} \mid F(p) < 0, E(p) \geq 0\}.$$

Assume the maximum flow has been computed on $\overline{\mathcal{G}}$, let $R_A(p), R_B(p), R_C(p), R_D(p)$ denote the residual capacities on the edges $(v_{p,1}, t), (v_{p,2}, t), (s, v_{p,1}), (s, v_{p,2})$ respectively. The following theorem gives a criterion for when the minimum cut on $\overline{\mathcal{G}}$ yields the optimal solution of the original problem.

Theorem 1. *Let $\mathcal{G}$ be a graph as defined in (6)-(8) and (10), with weights $A(p), B(p), C(p), D(p), E(p), F(p)$ satisfying (11). Let $\overline{\mathcal{G}}$ be the graph with weights as in $\mathcal{G}$, with the exception $c(v_{p,1}, v_{p,2}) = 0 \, \forall p \in \mathcal{P}^1$ and $c(v_{p,2}, v_{p,1}) = 0 \, \forall p \in \mathcal{P}^2$.*
Assume the maximum flow has been computed on the graph $\overline{\mathcal{G}}$. If

$$R_A(p) + R_D(p) \geq -E(p), \quad \forall p \in \mathcal{P}^1 \quad and \quad R_B(p) + R_C(p) \geq -F(p), \quad \forall p \in \mathcal{P}^2,$$
$$(13)$$

then min-cut $(\mathcal{G})$ = min-cut $(\overline{\mathcal{G}})$.

Proof. We will create a graph $\underline{\mathcal{G}}$, such that the minimum cut problem on $\underline{\mathcal{G}}$ is a relaxation of the minimum cut problem on $\mathcal{G}$. The graph $\underline{\mathcal{G}}$ is constructed with weights as in $\overline{\mathcal{G}}$ with the following exceptions

$$c(v_{p,1}, t) = A(p) - R_A(p) \quad and \quad c(s, v_{p,2}) = D(p) - R_D(p), \quad \forall p \in \mathcal{P}^1$$

$$c(v_{p,2}, t) = B(p) - R_B(p) \quad and \quad c(s, v_{p,1}) = C(p) - R_C(p), \quad \forall p \in \mathcal{P}^2.$$

Then min-cut($\underline{\mathcal{G}}$) $\leq$ min-cut($\mathcal{G}$) $\leq$ min-cut($\overline{\mathcal{G}}$). The max flow on $\overline{\mathcal{G}}$ is feasible on $\underline{\mathcal{G}}$ and therefore also optimal. Therefore, by duality min-cut($\underline{\mathcal{G}}$) = min-cut($\overline{\mathcal{G}}$) which implies min-cut($\mathcal{G}$) = min-cut($\overline{\mathcal{G}}$).

We have observed that it is often possible to stop at this stage, since (13) is very often satisfied. If not, one could either accept the solution as suboptimal, or make use of the following algorithm, which is designed to handle such cases. The idea is to create a succession of graphs $\{\mathcal{G}_i\}_{i=1}^n$ with only positive edge weights, such that min-cut($\mathcal{G}_i$) $\leq$ min-cut($\overline{\mathcal{G}}$) for all i, min-cut($\mathcal{G}_0$) = min-cut($\overline{\mathcal{G}}$) and min-cut($\mathcal{G}_n$) = min-cut($\mathcal{G}$). For a given flow we define two new sets $\mathcal{P}_0^1 \subseteq \mathcal{P}^1$ and $\mathcal{P}_0^2 \subseteq \mathcal{P}^2$

$$\mathcal{P}_0^1 = \{p \in \mathcal{P}^1 \mid R_A(p) + R_D(p) < -E(p)\}, \quad \mathcal{P}_0^2 = \{p \in \mathcal{P}^2 \mid R_B(p) + R_C(p) < -F(p)\}.$$

The graphs $\mathcal{G}_i$ are constructed such that the minimum cut problems on $\mathcal{G}_i$ are relaxations of the minimum cut problem on $\overline{\mathcal{G}}$. Particularly, for each $p \in \mathcal{P}_0^1$ and each $p \in \mathcal{P}_0^2$, the cost of one of the 4 possible phase assignments is reduced, while the rest of the assignment costs are correct (including the one that was set too high in $\overline{\mathcal{G}}$). The cut on $\mathcal{G}_i$ is feasible if no $p \in \mathcal{P}_0^1 \cup \mathcal{P}_0^2$ is assigned to a phase of reduced cost. The algorithm is iterated until the cut on $\mathcal{G}_i$ becomes feasible.

Algorithm 1:

$\mathcal{G}_0 = \overline{\mathcal{G}}$, $\mathcal{G}_{-1} = \emptyset$, $i = 0$. Find max flow on $\mathcal{G}_0$
while($\mathcal{G}_i \neq \mathcal{G}_{i-1}$ or $i = 0$){
    1. Construct $\mathcal{G}_{i+1}$ as in $\overline{\mathcal{G}}$ except for the following weights

for all $p \in \mathcal{P}_0^1$
   if($v_{p,1} \in V_t$ and $v_{p,2} \in V_t$): set $c(v_{p,1}, t) = A(p) + E(p)$ in $\mathcal{G}_{i+1}$
   if($v_{p,1} \in V_s$ and $v_{p,2} \in V_s$): set $c(s, v_{p,2}) = D(p) + E(p)$ in $\mathcal{G}_{i+1}$
   if($v_{p,1} \in V_s$ and $v_{p,2} \in V_t$): set $c(s, v_{p,1}) = A(p) + E(p)$ in $\mathcal{G}_{i+1}$
   if($v_{p,1} \in V_t$ and $v_{p,2} \in V_s$): set $c(s, v_{p,1}) = D(p) + E(p)$ in $\mathcal{G}_{i+1}$

for all $p \in \mathcal{P}_0^2$
   if($v_{p,1} \in V_t$ and $v_{p,2} \in V_t$): set $c(v_{p,2}, t) = B(p) + F(p)$ in $\mathcal{G}_{i+1}$
   if($v_{p,1} \in V_s$ and $v_{p,2} \in V_s$): set $c(s, v_{p,1}) = C(p) + F(p)$ in $\mathcal{G}_{i+1}$
   if($v_{p,1} \in V_s$ and $v_{p,2} \in V_t$): set $c(s, v_{p,2}) = B(p) + F(p)$ in $\mathcal{G}_{i+1}$
   if($v_{p,1} \in V_t$ and $v_{p,2} \in V_s$): set $c(s, v_{p,2}) = C(p) + F(p)$ in $\mathcal{G}_{i+1}$

    2. Find max-flow on $\mathcal{G}_{i+1}$
    3. Update $\mathcal{P}_0^1$ and $\mathcal{P}_0^2$ by examining residual capacities in graph $\mathcal{G}_{i+1}$
    4. $i \leftarrow i + 1$
}

Theorem 2. *Let $\mathcal{G}_n$ be the graph at termination of Algorithm 1. Then min-cut($\mathcal{G}_n$) = min-cut($\mathcal{G}$).*

Proof. The proof follows some of the same ideas as the proof of theorem 1. We will use $\mathcal{G}_n$ to construct a graph $\underline{\mathcal{G}}$ such that the minimum cut problem on

$\underline{\mathcal{G}}$ is a relaxation of the minimum cut problem on $\mathcal{G}$. Observe first that since $\mathcal{G}_n = \mathcal{G}_{n-1}$, the minimum cut on $\mathcal{G}_n$ is feasible, no edges in the cut have a reduced cost. Therefore, min-cut$(\mathcal{G}_n) \geq$ min-cut$(\mathcal{G})$

The graph $\underline{\mathcal{G}}$ is constructed with weights as in $\mathcal{G}_n$ except (residuals R obtained from flow on $\mathcal{G}_n$)

$$c(v_{p,1}, t) = A(p) - R_A(p) \quad \text{and} \quad c(s, v_{p,2}) = D(p) - R_D(p), \qquad \forall p \in \mathcal{P}^1 \backslash \mathcal{P}_0^1$$

$$c(v_{p,2}, t) = B(p) - R_B(p) \quad \text{and} \quad c(s, v_{p,1}) = C(p) - R_C(p), \qquad \forall p \in \mathcal{P}^2 \backslash \mathcal{P}_0^2.$$

Then min-cut$(\underline{\mathcal{G}}) \leq$ min-cut$(\mathcal{G}) \leq$ min-cut$(\mathcal{G}_n)$. By construction, the max flow on $\mathcal{G}_n$ is feasible on $\underline{\mathcal{G}}$, and therefore also optimal on $\underline{\mathcal{G}}$. Hence, by duality min-cut$(\underline{\mathcal{G}}) =$ min-cut$(\mathcal{G}_n)$ which implies min-cut$(\mathcal{G}) =$ min-cut$(\mathcal{G}_n)$.

Observe that there is a lot of redundancy in this algorithm. It is not necessary to compute the max-flow from scratch in each iteration, especially in the augmenting paths framework. Rather, starting with the max flow in $\mathcal{G}_i$, flow can be pulled back in $s - t$ paths passing through vertices $v_{p,1}, v_{p,2}$ for $p \in \mathcal{P}_0^1 \cup \mathcal{P}_0^2$ until it becomes feasible in graph $\mathcal{G}_{i+1}$. With such an initial flow, only a few augmenting paths are required to find the max flow on $\mathcal{G}_{i+1}$. Since $\mathcal{P}^1$ and $\mathcal{P}^2$ are small subsets of $\mathcal{P}$, and $\mathcal{P}_0^1 \cup \mathcal{P}_0^2$ are small subsets of $\mathcal{P}^1 \cup \mathcal{P}^2$, the cost of this algorithm is negligible.

We are trying to develop a convergence theory for this algorithm. Numerical experiments indicate that convergence is fast and no oscillations occur. We have so far investigated convergence experimentally by applying the algorithm to all images from the segmentation database [27]. We have used both the L^1 and L^2 data fidelity term, and different values on the regularization parameter ν, always resulting in convergence in an average of 3-4 iterations. Let us point out that Algorithm 1 was very rarely needed. However, by setting ν unnaturally high, pathological cases could be created. In order to verify the convergence of the algorithm, we have also successfully tried these extreme choices of ν.

2.5   Local Minimization Algorithm for Estimating c

In order to minimize with respect to both ϕ^1, ϕ^2 and $\mathbf{c}$, we alternate between optimization of ϕ^1, ϕ^2 for fixed $\mathbf{c}$ and optimizing $\mathbf{c}$ for fixed ϕ^1, ϕ^2, as explained in more detail in [25]. This algorithm is shown to be robust and typically only require a few iterations, but can of course not be proven to find a global minimum.

3   Numerical Results

Numerical experiments are made to demonstrate the new minimization methods. We also make comparisons between the PDE approach and combinatorial approach for minimizing (2). In all results, the phases are depicted as bright regions. The values $\mathbf{c}$ used in all experiments are generated from the algorithm in Section 2.5.

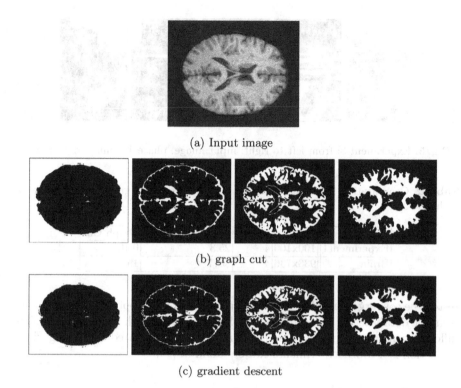

(a) Input image

(b) graph cut

(c) gradient descent

Fig. 3. Experiment 2: From left to right: phase 1 - phase 4

Fig. 4. Experiment 1: L^2 data fidelity

In experiment 1 and 2, Figure (4) and (3), the L^2 norm is used in the data term. The constant values $\{c_i\}_{i=1}^4$ satisfy condition (12) initially and in all iterations until convergence. We next try to use L^1 data fidelity on these images. In this case, condition (12) was not satisfied for all pixels. However, after finding the max flow on $\overline{\mathcal{G}}$ and examining the residual capacities, the criterion (13) was satisfied, and hence the global minimum had been obtained. See Table 2 for computation times.

For the next image, Figure (5), the L^1 norm was used, and for some grid points neither condition (12) nor the criterion (13) was satisfied. Therefore, Algorithm 1 had to be used. For each combination of $\{c_i\}_{i=1}^4$ generated by the algorithm in Section 2.5, it converged in 5-8 iterations. As already mentioned, we have also

Fig. 5. Experiment 3: from left to right: input image, phase 1 - phase 4. L^1 norm

Table 2. Computation times in seconds for gradient descent vs graph cut optimization with $\beta = 2$

	Size	Phases	Gradient descent	Graph Cut
Experiment1	100x100	4	25.3	0.10
Brain	933x736	4	3077	19.4

tested the convergence of Algorithm 1 experimentally by applying it to all images from the database [27]. This includes pathological cases with ν set very high. The different constant values in these experiments were generated by the algorithm in Section 2.5. More experiments can be found in [25].

References

1. Mumford, D., Shah, J.: Optimal approximation by piecewise smooth functions and associated variational problems. Comm. Pure Appl. Math. 42, 577–685 (1989)
2. Dervieux, A., Thomasset, F.: A finite element method for the simulation of a Rayleigh-Taylor instability. In: Approximation methods for Navier-Stokes problems (Proc. Sympos., Univ. Paderborn, Paderborn, 1979). Lecture Notes in Math., vol. 771, pp. 145–158. Springer, Berlin (1980)
3. Osher, S., Sethian, J.: Fronts propagating with curvature dependent speed: algorithms based on hamilton-jacobi formulations. J. Comput. Phys. 79(1), 12–49 (1988)
4. Chan, T., Vese, L.: Active contours without edges. IEEE Image Proc. 10, 266–277 (2001)
5. Vese, L.A., Chan, T.F.: A new multiphase level set framework for image segmentation via the mumford and shah model. International Journal of Computer Vision 50, 271–293 (2002)
6. Geman, S., Geman, D.: Stochastic relaxation, gibbs distributions, and the bayesian restoration of images. In: Readings in uncertain reasoning, pp. 452–472. Morgan Kaufmann Publishers Inc., San Francisco (1990)
7. Boykov, Y., Veksler, O., Zabih, R.: Fast approximate energy minimization via graph cuts. In: ICCV (1), pp. 377–384 (1999)
8. Greig, D.M., Porteous, B.T., Seheult, A.H.: Exact maximum a posteriori estimation for binary images. Journal of the Royal Statistical Society, Series B, 271–279 (1989)
9. Boykov, Y., Kolmogorov, V.: An experimental comparison of min-cut/max-flow algorithms for energy minimization in vision. In: Energy Minimization Methods in Computer Vision and Pattern Recognition, pp. 359–374 (2001)

10. Kolmogorov, V., Zabih, R.: What energy functions can be minimized via graph cuts? IEEE Transactions on Pattern Analysis and Machine Intelligence 26(2), 147–159 (2004)
11. Komodakis, N., Tziritas, G., Paragios, N.: Fast, approximately optimal solutions for single and dynamic mrfs. In: IEEE Conference on Computer Vision and Pattern Recognition, 2007. CVPR 2007, June 17-22, pp. 1–8 (2007)
12. Boykov, Y., Kolmogorov, V.: Computing geodesics and minimal surfaces via graph cuts. In: ICCV 2003: Proceedings of the Ninth IEEE International Conference on Computer Vision, Washington, DC, USA, p. 26. IEEE Computer Society Press, Los Alamitos (2003)
13. Boykov, Y., Kolmogorov, V., Cremers, D., Delong, A.: An integral solution to surface evolution pdes via geo-cuts. In: Leonardis, A., Bischof, H., Pinz, A. (eds.) ECCV 2006. LNCS, vol. 3953, pp. 409–422. Springer, Heidelberg (2006)
14. Darbon, J., Sigelle, M.: Image restoration with discrete constrained total variation part i: Fast and exact optimization. J. Math. Imaging Vis. 26(3), 261–276 (2006)
15. Darbon, J., Sigelle, M.: Image restoration with discrete constrained total variation part ii: Levelable functions, convex priors and non-convex cases. J. Math. Imaging Vis. 26(3), 277–291 (2006)
16. Darbon, J.: A note on the discrete binary mumford-shah model. In: Gagalowicz, A., Philips, W. (eds.) MIRAGE 2007. LNCS, vol. 4418, pp. 283–294. Springer, Heidelberg (2007)
17. Zehiry, N.E., Xu, S., Sahoo, P., Elmaghraby, A.: Graph cut optimization for the mumford-shah model. In: Proceedings of the Seventh IASTED International Conference visualization, imaging and image processing, pp. 182–187. Springer, Heidelberg (2007)
18. Pock, T., Chambolle, A., Bischof, H., Cremers, D.: A convex relaxation approach for computing minimal partitions. In: IEEE Conference on Computer Vision and Pattern Recognition (CVPR), Miami, Florida (to appear, 2009)
19. Lellmann, J., Kappes, J., Yuan, J., Becker, F., Schnorr, C.: Convex multi-class image labeling by simplex-constrained total variation. In: SSVM 2009, pp. 150–162 (2009)
20. Lie, J., Lysaker, M., Tai, X.: A binary level set model and some applications to mumford-shah image segmentation. IEEE Transactions on Image Processing 15(5), 1171–1181 (2006)
21. Bresson, X., Chan, T.: Non-local unsupervised variational image segmentation models (2008)
22. Lie, J., Lysaker, M., Tai, X.: A variant of the level set method and applications to image segmentation. Math. Comp. 75(255), 1155–1174 (2006) (electronic)
23. Nikolova, M., Esedoglu, S., Chan, T.F.: Algorithms for finding global minimizers of image segmentation and denoising models. SIAM Journal on Applied Mathematics 66(5), 1632–1648 (2006)
24. Ford, L., Fulkerson, D.: Flows in networks. Princeton University Press, Princeton (1962)
25. Bae, E., Tai, X.C.: Efficient global optimization for the multiphase chan-vese model of image segmentation by graph cuts. UCLA, Applied Mathematics, CAM-report-09-53 (June 2009)
26. Kolmogorov, V., Rother, C.: Minimizing nonsubmodular functions with graph cuts-a review. IEEE Trans. Pattern Anal. Mach. Intell. 29(7), 1274–1279 (2007)
27. http://www.eecs.berkeley.edu/Research/Projects/CS/vision/grouping/segbench/

Bipartite Graph Matching Computation on GPU

Cristina Nader Vasconcelos[1] and Bodo Rosenhahn[2]

[1] PUC-Rio
crisnv@inf.puc-rio.br
[2] Leibniz Universitaet Hannover
rosenhahn@tnt.uni-hannover.de

Abstract. The *Bipartite Graph Matching Problem* is a well studied topic in Graph Theory. Such matching relates pairs of nodes from two distinct sets by selecting a subset of the graph edges connecting them. Each edge selected has no common node as its end points to any other edge within the subset. When the considered graph has huge sets of nodes and edges the sequential approaches are impractical, specially for applications demanding fast results. In this paper we investigate how to compute such matching on Graphics Processing Units (GPUs) motivated by its increasing processing power made available with decreasing costs. We present a new data-parallel approach for computing bipartite graph matching that is efficiently computed on today's graphics hardware and apply it to solve the correspondence between 3D samples taken over a time interval.

1 Introduction

Graph Matching is one of the fundamental problems in Graph Theory, with a intrinsic combinatorial nature. It can be defined as: given a graph G, its set of edges E and its set of nodes V, a matching M is a set of edges, subset of E, such that no two edges in M are incident to the same node.

Interesting problems in Computer Vision can be formulated as a Graph Matching, specially when an objective function associates weights to the graph edges, semantically related to some benefit or cost of the application. In that case, the weighted graph matching optimization goal is to maximize (or minimize) the sum of the weights of the matched edges.

There exist some parallel algorithms for approximating a graph matching [1,2,3,4]. Usually, the graph is initially distributed over several processors of a parallel computer or a set of computers organized as clusters or distributed systems.

In this paper we are interested in a variant of the general matching proposition, where the considered graph is a bipartite graph. A matching in a bipartite graph is easier to compute than in a general (or non-bipartite) graphs, as the number of possible combinations decreases considerably with the bipartite restriction and that its result can be obtained in a non-approximation way. Such version is usually named *The Assignment Problem* as it can be semantically proposed as: given a set of employees, a set of jobs and some cost (or benefit) function that evaluates the

D. Cremers et al. (Eds.): EMMCVPR 2009, LNCS 5681, pp. 42–55, 2009.
© Springer-Verlag Berlin Heidelberg 2009

employee-job assignment, it is required to designate the people to accomplish the tasks by assigning exactly one employee to each job in such a way that the total cost of the assignment is minimized. Unbalanced versions of the same problem consider that the number of employees and the number of jobs are not equal. In that case, the one in greater number will have some elements unmatched at the final association.

The need for a parallel formulation to *The Assignment Problem* arises from cases considering huge graphs and from applications demanding fast results. Attending to such demand, we propose a GPU friendly formulation motivated by the modern graphics hardware increasing processing power made available with low costs (specially when compared with other high processing power solutions like distributed systems). This paper presents a parallel algorithm developed using the stream processing paradigm. Our algorithm identifies elements that can be arranged within a stream of data and processed independently. It attends to GPU implementation requirements and scales in a transparent way as the number of processors increases.

As an application to the GPU implementation developed, we propose a new formulation for a Computer Vision classical problem: the correspondence problem, here defined of over independent sets of 3D samples taken over a period of time. The quality of the obtained matchings was tested using microscopy data, taken during different time instants, against ground truth results manually obtained. The time efficiency of our solution was tested comparing it against two sequential implementations on CPU and also observing its answer time with a growing set of artificially generated 3D moving points.

The source code containing an implementation of the bipartite graph matching optimization proposed here, coded using CUDA programming language, is available for download from the author's homepage [5].

This paper is organized as follows. The next section presents a formal definition of the matching problem (section 2) and existing algorithms for computing it are presented in section 3. A brief comparison between CPU and GPU computing is presented in section 4. Our stream processing approach is presented in section 5, while results and conclusions are presented in sections 6 and 7.

2   Bipartite Graph Matching Definitions

According to the formal definition, a graph $G = (V, E)$ is bipartite if there exists a partition of its vertexes (or nodes) into two distinct sets, X and Y, such as that three properties are valid: the original set of vertexes V is formed by the union of the generated sets ($V = X \cup Y$); each vertex of the original set of vertexes V belongs exclusively to one of the created sets and not to both of them ($X \cap Y = \emptyset$); and, all the edges of the graph connect a vertex from one of the vertexes sets to the other ($E \subseteq X \times Y$). A Matching is a subset of the edges set ($M \subseteq E$) such that for every $v \in V$ at most one edge in M is incident to v. A vertex v is said to be matched in M if it is an endpoint of an edge in M, otherwise v is said free.

In this paper we are specially interested in weighted bipartite graphs, in which each edge (i, j) is associated to a weight $w(i, j)$. The weight of a matching M is defined as the sum of the weights of edges in M:

$$w(M) = \sum_{e \in M} w(e). \tag{1}$$

There are variants of the weighted bipartite graph matching formulation, including: $w(M)$ maximization or minimization (can be viewed as a maximization problem by just replacing the cost function c with $-c$), perfect matchings and maximum matchings [6].

3   Related Works and Background

In this section we present two approaches for computing a bipartite graph optimal matching in order to evaluate them for a GPU approach. The first one, known as *The Hungarian Algorithm*, is the classical approach from Graphs Theory literature and it is described in Subsection 3.1. In subsection 3.2 we describe a second approach, known as *The Auction Algorithm*, motivated by its distributed formulation.

3.1   The Hungarian Algorithm

The Hungarian Algorithm is a sequential combinatorial optimization algorithm published by Harold Kuhn [7] that solves *The Bipartite Graph Matching Problem* in a polynomial time. It iterates between two phases: the first one is based on the Graph Theory concept of augmenting paths while the second is based on the concept of feasible labellings (dual variable) and equality graphs.

A path in a graph is a sequence of vertexes such that from each of its vertexes there is an edge to the next vertex in the sequence. Both the first and the last vertexes of the sequence are called end or terminal vertexes of the path. Given a matching M and the set of edges E of the graph, a path is alternating if its edges alternate between M and $E - M$ and an alternating path is augmenting if both endpoints are free (see Figure 1).

The property that assures the first phase of *The Hungarian Algorithm* is that an augmenting path has one less edge in M than in $E - M$, thus, it is possible

Fig. 1. A matching M represented with red edges (left); an augmenting path formed by the red (edges from M) and yellow edges (edges from the $E - M$ set) (center); the new path with size incremented by one (right)

to increment size of the matching by replacing the M edges by the $E - M$ ones. The second phase of the algorithm is concerned about dealing with the weights. Considering a vertex labeling as a function $l : V \rightarrow \Re$, a feasible labeling is one such that

$$l(x) + l(y) \geq w(x,y), \forall\ x \in X, y \in Y \qquad (2)$$

The labeling here works as a dual variable in order to solve the problem. Given equation 2, the Equality Graph (with respect to the labeling l) is defined as $G = (V, E_l)$ where E_l is the set of edges satisfying the equation with equality, that is:

$$E_l = (x,y) : l(x) + l(y) = w(x,y) \qquad (3)$$

Figure 2 (left) illustrates an example of weights associated to the vertexes that satisfies equation 2, thus, they compose a feasible labeling. Figure 2 (right) shows an equality graph constructed using the same labeling.

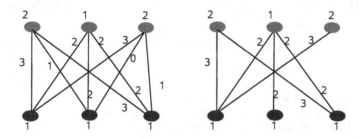

Fig. 2. A feasible labeling (left) and an equality graph (right)

The Kuhn-Munkres Theorem [7] assures that: If l is feasible and M is a perfect matching in E_l then M is a max-weight matching. This theorem transforms the problem from an optimization problem of finding a max-weight matching into a combinatorial one of finding a perfect matching. Its proof assures that for any matching M and any feasible labeling l we have

$$w(M) \leq \sum_{v \in V} L(v) \qquad (4)$$

ie., that the sum of the labels in a feasible labeling defines an upper-bound on the cost of any perfect matching.

Finally, the Hungarian algorithm is described as: start with any feasible labeling l and some matching M in E; while M is not perfect repeat the following:

- find an augmenting path for M in E_l and increase the size of M;
- if M is maximum, then, it has minimum cost among all maximum matchings in G. Stop the search.
- otherwise, if no augmenting path exists, improve l to l' such that $E_l \subset E_{l'}$ (adjust the labels in order to enlarge the Equality Graph). Go back to the first step.

The iteration finishes as in each step of the loop we will either be increasing the size of M or of the set E_l. When it finishes, M is a perfect matching in E_l (if it exists), for some feasible labeling l and, by the Kuhn-Munkres theorem, M is a max-weight matching.

The Hungarian Algorithm Computation. main input is the weights associated with the graph edges that are arranged into a matrix, whose elements (i, j) represent the cost of a matching between the vertexes i and j, thus $w(i, j)$. Its computation is described as:

- Step 1:
  - 1.1 For each row, subtract its smallest element from all its elements.
  - 1.2 For each column, subtract its smallest element from all its elements.
- Step 2:
  - 2.1 Locate a lone zero and assign it.
  - 2.2 Cover the row or column associated with the lone zero depending on where the other zeros are, if any.

Repeat these operations until there are no lone zeros in the matrix. Upon completion go to Step 3.
- Step 3:
  - 3.1 If all the rows have been assigned, then stop.
  - 3.2 If there are uncovered zeros, then go to Step 4.
  - 3.3 Else (if there are no uncovered zeros), go to Step 5.
- Step 4: Assign any one of the uncovered zeros and cover both its row and its column. Then go to Step 2.
- Step 5: The way this step is carried out depends on whether Step 3.2 had been visited since the last time a new zero was created in Step 2. If this step has not been executed since the last time a new zero was created, then new zero(s) are generated by subtracting the smallest uncovered element of the matrix from all the uncovered elements of the matrix. If this step has already been executed, it is necessary first to cover all the zeros of the matrix with the minimum number of lines (rows and/or columns) and then create a new zero. Then go to Step 2.

Some of the presented steps can be easily computed concurrently, but others intrinsically demand the introduction of a non GPU-friendly message pass scheme or of global synchronization points as they provoke ambiguities if computed concurrently or locally. Thus, the definition of independent processing kernels for coding this algorithm as required for a pure GPU computation, is not straightforward.

3.2 The Auction Algorithm

This section presents *The Auction Algorithm* for *The Assignment Problem*. For our purposes, this algorithm has a huge advantage over the Hungarian as it was originally described as a distributed relaxation method [8], very well suited for parallel computation.

The Auction Algorithm is semantically described as a real auction where persons compete for objects by raising their prices through competitive bidding. Suppose that there are n persons and m objects (where $(n \leq m)$). We want to match them in a way that each person should be assigned to a single object and each object should be assigned to at most a single person. The matching should respect the restriction that each person i can only be assigned to object a j if the pair (i,j) belongs to a given set $A(i)$ of possible matching pairs. Analogously, for each object j it is possible to define $B(j)$ as the set of persons that can be matched with j.

There is a benefit a_{ij} for matching a person i with an object j, such that the goal of the auction is to assign persons to objects so as to maximize the total benefit, defined as

$$\sum_{i=0}^{n-1} a_{ij_i} \tag{5}$$

The auction algorithm introduces an economic equilibrium problem that can be seen as a dual problem. It supposes that an object j has a price p_j and that the person who receives the object must pay the price p_j. As each person associates an benefit a_{ij} with each object, then the object j net value of for person i is related with the difference between the corresponding benefit the object price. Each person i would logically want to be assigned to an object j_i with maximal value, that is, with

$$a_{ij_i} - p_{j_i} = max\left\{a_{ij} - p_j\right\}. \tag{6}$$

The Linear Programming Formulation for *The Assignment Problem* associates an assignment A with the set of variables $\{x_{ij}|(i,j) \in A\}$, where $x_{ij} = 1$ if person i is assigned to object j and $x_{ij} = 0$ otherwise. Thus, the value of an assignment is expressed as

$$\sum_{i=0}^{n-1}\sum_{j=0}^{m-1} a_{ij_i}x_{ij} \tag{7}$$

and the restrictions that assure one-to-one mapping are then written as

$$\sum_{j=0}^{m-1} x_{ij} = 1, \text{for every } i \tag{8}$$

and

$$\sum_{i=0}^{n-1} x_{ij} = 1, \text{for every } j \tag{9}$$

As a consequence of the existent duality, it can be assured using Linear Programming Theory that an equilibrium assignment offers the maximum total benefit (and thus solves *The Assignment Problem*), while the corresponding set of prices solves an associated dual problem.

The Auction Algorithm Computation goal is to find an equilibrium assignment and its corresponding price vector. The algorithm iterates between two steps: a bidding phase and an assignment phase.

During the bidding phase, each unassigned person finds an object j which offers maximal value (according to equation 6) and makes a bid for that object offering a bidding increment γ_i calculated as:

$$\gamma_i = v_i - w_i + \epsilon \qquad (10)$$

where (v_i) and (w_i) are respectively the maximal and second maximal net values of objects that the person i is interested in. The inclusion of the positive constant ϵ in equation 10 assures that the bidding increment is not zero, which otherwise would happen in cases where a person has more that one object with the maximum net value, ie., cases where (v_i) is equal to (w_i).

After the bidding phase, the algorithm turns into the assignment phase. Then, each object j, if it was selected as a best object by any nonempty set of people $P(j)$, determines the highest bidder by:

$$i_j = arg\ max_{i \in P(j)}\ \gamma_i \qquad (11)$$

Using the highest bidding increment the object raises its price and gets assigned to the person i, considered as highest bidder i_j. If the object was previously assigned to other person, that person becomes unassigned.

Iterating between those two phases, the algorithm continues until all persons have an assigned object. The termination with a feasible assignment (if it exists) is assured by noting that once an object is assigned to any person, it will never be turned into an unassigned object again. Besides, if an object receives a bid in k iterations, its price must exceed its initial price by at least $k\epsilon$, thus, at some point of the iteration, an assigned object will become expensive enough to be judged less valuable (according to equation 6) than some other object that has not received a bid so far. It follows an object can receive a bid in a limited number of iterations while some other object still has not yet received any bid. On the other hand, once n objects ($n \leq m$) receive at least one bid, the auction terminates.

4   Using CPU *versus* GPU for Computer Vision Tasks

Both microprocessors (CPUs) and graphics hardware (GPUs – Graphics Processing Units) are composed in low level by the same components: the transistors. This means that the CPU and GPU are equally benefited by transistors technology advances [9]. Their performance difference can be illustrated comparing the processing power of models on market, like the NVIDIA GeForce 8800 GTX graphic card (330 GFlops and 80 GB/s bandwidth) and the multi-core processor Intel Core Duo (48 GFlops and 10 GB/s bandwidth).

What defines the main difference between CPUs and GPU being responsible for their efficient disparity is their architecture. CPUs are developed to efficiently attend to a variety class of applications, requiring the disposal of many transistors to offer complex control functionality in hardware (such as branch prediction). The GPUs were originally developed aiming 3D graphics processing which allowed their architecture to concentrate the transistors on computation power, rather than on control chips. Nowadays, GPUs are low-cost stream processors specialized on high arithmetic computation over independent elements of a stream.

Modern GPUs can be seen as fine-grained parallel computers. Its programming model requires to formulate the desired algorithms as what is called data algorithms, that means, to be executed with simultaneous operations across large sets of data. Each set of data is organized in what is called a stream, containing similar elements to be processed. Thus, the algorithm to be computed should be decomposed into similar operations to be applied to each element of a stream of data independently. The computations are defined in a operator called kernel describing the algorithm tasks to be applied over a single element of the stream. In such formulation, the parallelism occurs by processing in parallel the same kernel over different elements of the stream. As the number of processors increases, more elements can be processed in parallel, scaling the algorithm in a transparent way to its developers.

Computer vision tasks are well suited for Graphics Processing Unit (GPUs) hardware as many of their tasks can be seen as similar arithmetically-intensive operations over huge sets of data, exactly the nature of problem to which such hardware is developed for.

Next section presents our formulation to solve the bipartite graph matching problem using graphics hardware by decomposing *The Auction Algorithm* into a data-parallel formulation proper to moderns GPU architectures.

5   The Bipartite Graph Matching on GPU

Our proposal reformulates the auction algorithm for a GPU computation using one kernel for the bidding phase, one kernel for the assignment phase (both processed on the device - GPU) and a loop that iterates between the phases triggering the GPU threads until convergence (controlled by the host - CPU). The iteration cycle is illustrated in Figure 3.

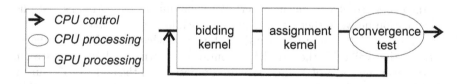

Fig. 3. Our GPU formulation iteration cycle

Before the matching computation begins, data streams have to be created representing (or reserving storage space for) the input, the output and temporary data used by the algorithm.

The initialization task includes the bipartite graph creation, that is, its disjoint sets of nodes and its edges set representation into data streams. The two sets of nodes (X and Y) represent the elements to match. For clarity we will call them here as the set of persons and the set of objects respectively. *The Assignment Problem* looks for a set of one-to-one associations represented as edges in this graph. This means that one should include edges from a vertex $x_i \in X$, to every vertexes $y_j \in Y$, that represents a possible matching in the final association. Consequently, the number of edges created for each vertex is related to the number of possible matchings for each node of the graph.

Any structure used by our algorithm is represented as a 1D or a 2D linear data stream. In our model, the graph nodes, edges and weights are represented within a single two-dimensional stream of constant elements (their values are set during initialization and kept constant during the algorithm), containing the benefit for matching a person i with an object j set as a matrix.

Both requirements that each person i can only be assigned to an object j, or that each object j can only be assigned to a person i, if those pairs belongs to a given set of possible matching pairs (existent edges) are imposed to the algorithm computation in our proposal by setting negative infinity values for the corresponding association of unwanted pairs within the benefit matrix during initialization phase. Thus, the rows and columns of such matrix indirectly represent the existent nodes, while the existent edges are represented with positive values in their corresponding matrix positions.

Other structures used by our algorithm are represented as unidimensional streams. They are dynamic value streams created to represent: the objects prices (initially set as zero); the objects index associated with each person (or a sentinel value if not associated); the person index associated with each object (or a sentinel value if not associated); the bids value suggested by each person in last iteration; the bids target from each person in last iteration (an object index, if any).

Observe that we split a natural two-dimensional data representation for the bids value of each person i to each object j, into two unidimensional streams, by observing that each person can only bid for a single object in each iteration. We create this organization in order to save graphics card memory and to induce faster memory transfers while keeping independent access to stream elements during the kernel computation. With such improvement each person can still write its bidding concurrently to all the others in our unidimensional streams, with no communication between them, but requiring much less storage space.

Once the iteration cycle starts, our algorithm turns what is considered as the input stream that will drive the parallel computation, alternating between a stream composed over the nodes of X and a stream composed over the nodes of Y. During the bidding phase a kernel is coded driven to process a single element of X, while during the assignment phase, a kernel is coded driven to process a single element of Y (see figure 4).

The bidding kernel (bk) is executed by every person concurrently. During its execution, the person decides if he is going to suggest a bid and to which object to bid for, or if he is currently associated to an object (does not ask for another).

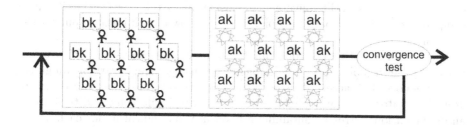

Fig. 4. Independent Parallel Processing

During the next phase of the cycle, every object executes the assignment kernel (ak), each one individually has the task of testing if it has received any bid recently. In such case, the object is responsible for updating its own price, for changing the current bidder (if it has one) and the bid value for the most recently ones. The object is also responsible for setting previous bidder free to let him to start bidding again.

5.1   The Convergence Test

Once that GPU can not trigger its own processes and threads, the convergence decision has to involve the CPU at least to decide if the algorithm cycle stops or if the CPU has to trigger the bidding kernel (bk) and the assignment kernel (ak) once more.

The task of computing the convergence test itself in CPU would require to retrieve data from the GPU streams to CPU memory space. Data transfers are one of the most expensive operations in CPU-GPU programming and such cost is directly proportional to the amount of data retrieved. Based on these facts, our goal is to reduce the amount of data consulted on CPU for the convergence decision.

The convergence criteria evaluates if all persons have already been assigned to an object. A first solution for its computation would involve the transfer of the unidimensional stream containing the objects' index associated with each person and checking in CPU if there is any person associated with the sentinel value. If not, the algorithm has converged.

In a better solution, we observe the algorithm cases when the total number of assigned persons changes and track those cases from the algorithm processing kernels. Supposing that the algorithm starts with no assignments, such number increases only when a free object receives the first bid. In cases when and assigned object changes its corresponding bidder, the total number of assigned persons remains unchanged. This observation is assured by the property of the algorithm that once an object is assigned for any person it never turns to unassigned again, so our counter can never decrease.

With the presented assumption, our algorithm uses a transfer of a single value between the GPU and CPU, representing a counter of the assigned persons total. Using such data, the convergence decision can be taken on CPU as it can be compared with the number of people, which is known as an initial input for the algorithm.

The task of updating the total number of assigned persons on GPU concurrently over several objects can be implemented using a single variable on global space memory, but accessed using atomic operations. In architectures where atomic operators are not available, the kernels can be implemented to write in an unidimensional stream containing a boolean value indicating for each object its status (assigned/unassigned). The total number of assigned objects can be retrieved reducing such boolean-valued stream to a single integer value (the reduction operator can be consulted in [10]).

6   Application and Results

A human being can normally solve visual correspondences quickly and easily, even when the sets of samples observed contain significant amount of noise. Different classes of Computer Vision tasks involve computing correspondences between sets of samples taken from distinct cameras or in distinct time instants.

In this section we use the bipartite graph matching to model and compute the search for the best one-to-one association (according to the metric adopted) that correlates an input data composed by distinct sets of samples. The application developed models a variant of the correspondence problem as a discrete optimization over a bipartite graph. More specifically, we model a correspondence problem between $X = \{x_0, x_1, ..., x_n\}$ and $Y = \{y_0, y_1, ..., y_m\}$, that represent two independent sets of 3D samples to be matched. To attend the algorithm description and convergence criteria presented in section 5, in cases the samples sets have different size we associate X with the smaller one.

As a discrete optimization, we are interested in finding the minimum cost matching between the samples in X and Y. The weights of our graph are defined by a energy function indicating the cost of associating each 3D sample in X to each sample in Y. As we are assuming that the input data does not contain any feature that identifies the individual particles, but only their position in 3D space, the cost function is defined as the Euclidean distance between the time-sampled points.

As our first test set, we explore a data set taken with a microscopy collecting positioning data samples from bacterias moving around 3D space over a certain period of time. The problem is to find the correspondence between each bacteria sampled in a instant of time to the same bacteria in the next set of samples taken in *a posteriori* instant. The single feature taken as input is the sets of 3D positions obtained from the microscopy data. For each pair of samples sets taken during consecutive time instants, we initially consider that any bacteria in the first set of samples can be matched to any bacteria in the second set.

The evaluation of the matching accuracy for those microscopy data reveled that our model has found the right correspondences between all bacterias presented in both samples sets. Wrong matchings appear caused by the fact that a moving bacteria could left the microscopy vision field, while others can enter. That is, wrong cases happened when a particle leaves the vision field from one time instance sampled to the other and at the same time a new particle enters in the field. In those cases, our model do not identify that they are actually two different moving particles and it matches them like if they were the same. The disadvantage of such data is that it does not offer a data set with continuous growing size (neither many different data samples of each size) to evaluate correctly the matching efficiency.

For measuring the timing results, we are interested in creating bipartite graphs with increasing number of nodes and edges. Aiming to produce data sets containing several, increasing size, test samples (with ground truth) we simulated sets of 3D moving points, moving inside a 3D bounding box in a random biased and correlated movement and sampled them in different iteration times. The velocity vector of our particles are composed by two vectors: the correlation vector, as a persistence tendency to keep the particle moving in the same direction, resulting in a correlation between successive steps of the simulation; and the bias vector, a randomly generated vector to disturb regular movement at each time sample. The persistence level in our model is a value between $[0.0, 1.0]$ that indicates a

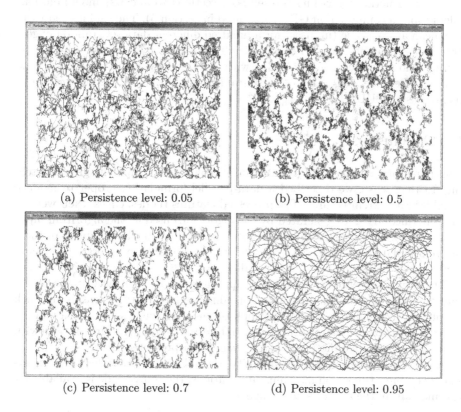

(a) Persistence level: 0.05

(b) Persistence level: 0.5

(c) Persistence level: 0.7

(d) Persistence level: 0.95

Fig. 5. Tracking 500 samples moving in 3D with varying persistence levels

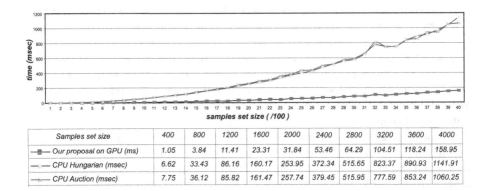

Samples set size	400	800	1200	1600	2000	2400	2800	3200	3600	4000
Our proposal on GPU (ms)	1.05	3.84	11.41	23.31	31.84	53.46	64.29	104.51	118.24	158.95
CPU Hungarian (msec)	6.62	33.43	86.16	160.17	253.95	372.34	515.65	823.37	890.93	1141.91
CPU Auction (msec)	7.75	36.12	85.82	161.47	257.74	379.45	515.95	777.59	853.24	1060.25

Fig. 6. Time comparison between CPU Hungarian, CPU Auction and GPU Auction

linear combination factor between the correlation vector and the bias vector in the composition for the particles movement (Figure 5).

Figure 6 presents the timing results for sequential implementations of the Hungarian and Auction algorithms (presented in sections 3.1 and 3.2) and for our parallel proposal computed on GPU (presented in section 5). The timings represent the mean answer time (in msec) for a hundred different data sets randomly generated given the total number of particles. The tests were performed using a Intel Core 2 Duo processor E6550 2.33Ghz with 2GB of RAM memory processor and a nVidia GeForce 9600 GT (512MB) graphics card. The GPU code was implemented using CUDA [11].

7   Conclusion

This paper presented a GPU formulation for computing a bipartite graph matching. In that sense, we described a data-parallel formulation proper to modern graphics cards architectures. The processed features (nodes, edges, their weights and intermediary data generated by the algorithm) used as the algorithm input and output were reviewed as streams of data, while algorithms applied to them are reformulated as "processing kernels" to be applied over several elements of each stream independently and in parallel.

The results presented show that our approach considerably accelerates the bipartite graph matching computation, opening the possibility of considering such graph and technique as a model in applications using huge sets of data and demanding fast results. In the future we are interested to compare our approach with other algorithms, like the *invisible hand algorithm* presented in [12].

Aiming to let the scientific community to be able to analyze their own experiments on bipartite graph matching applications, the source code is available for download from the author's homepage [5].

References

1. Fischer, T., Goldberg, A.V., Haglin, D.J., Plotkin, S.: Approximating matchings in parallel. Inf. Process. Lett. 46(3), 115–118 (1993)
2. Hougardy, S., Vinkemeier, D.E.: Approximating weighted matchings in parallel. Inf. Process. Lett. 99(3), 119–123 (2006)
3. Lotker, Z., Patt-Shamir, B., Rosen, A.: Distributed approximate matching. In: PODC 2007: Proceedings of the twenty-sixth annual ACM symposium on Principles of distributed computing, pp. 167–174. ACM, New York (2007)
4. Lotker, Z., Patt-Shamir, B., Pettie, S.: Improved distributed approximate matching. In: SPAA 2008: Proceedings of the twentieth annual symposium on Parallelism in algorithms and architectures, pp. 129–136. ACM, New York (2008)
5. Vasconcelos, C.N., Rosenhahn, B.: Bipartite graph matching computation on gpu public code, http://crisnv.googlepages.com/bgm
6. Alexander, H., Saip, B., Lucchesi, C.: Matching algorithms for bipartite graphs. Technical report, DCC-UNICAMP (March 1993)
7. Kuhn, H.W.: The hungarian method for the assignment problem. Naval Research Logistics Quarterly 2, 83–97 (1955)
8. Bertsekas, D.P.: Auction algorithms for network flow problems: A tutorial introduction. Computational Optimization and Applications 1, 7–66 (1992)
9. Fernando, R.: GPU Gems 2 - Programming techniques for High-Performance Graphics and General Purpose Computation. Addison-Wesley Professional, USA (2005)
10. Vasconcelos, C.N., Sá, A., Teixeira, L., Carvalho, P.C., Gattass, M.: Real-time video processing for multi-object chromatic tracking. In: Proceedings of the 12th (BMVC 2008), pp. 113–122 (2008)
11. Cuda programming guide 1.1 (2007), http://developer.download.nvidia.com/
12. Kosowsky, J.J., Yuille, A.L.: The invisible hand algorithm: solving the assignment problem with statistical physics. Neural Netw. 7(3), 477–490 (1994)

Pose-Invariant Face Matching Using MRF Energy Minimization Framework

Shervin Rahimzadeh Arashloo and Josef Kittler

Center for Vision, Speech and Signal Processing
University of Surrey
Guildford GU2 7XH, United Kingdom

Abstract. A pose-invariant face verification system based on an image matching method is presented. The method uses the normalized energy of the established match between images as a measure of goodness-of-match. The method can tolerate moderate global spatial transformations between the gallery and the test images and alleviates the need for geometric and photometric normalization of facial images. It requires no training on non-frontal face images. A number of innovations, such as a dynamic block size and block shape adaptation, as well as label pruning and error prewhitening measures have been introduced to increase the effectiveness of the approach. The experimental evaluation of the method is performed on the rotation shots of the XM2VTS database and promising results are obtained.

1 Introduction

In spite of the impressive progress in face recognition technology, many problems still remain unsolved. The two most challenging requirements for a recognition system to operate in real world conditions are invariance to facial image pose and illumination [23]. A variety of methods have been proposed to deal with the problem of pose changes. One of the earlier attempts to overcome the pose variation problem is the work of Beymer [1] in which images of different views of subjects were stored in a database and every input image was first aligned with the relevant reference images from the database and then a similarity measure was computed for recognition. There are also other works which take advantage of multiple images corresponding to different poses in the gallery *e.g.* the method by Singh et al. [17] which constructs composite images using semi-profile and frontal views. Other work by Pentland *et al.* [13] and Wiskott *et al.* [21] can be considered as relatively robust feature-based methods which could tolerate moderate pose variations. There are also 2D learning-based algorithms that try to synthesize a virtual frontal view in the 2D domain. The active appearance model [3] is a well known example of this category. Other techniques which use 3D methods to construct a novel view of the face image form another category. One of the most successful methods in this category is the 3D morphable model proposed by Blanz and Vetter [2]. In parallel with the methods which try to synthesize a novel view of the face, using either one or multiple images from the

D. Cremers et al. (Eds.): EMMCVPR 2009, LNCS 5681, pp. 56–69, 2009.

gallery, there are other methods which try to learn the most discriminant information between classes across different poses. The work by Kim and Kittler [7] and Kanade and Yamada [6] are some examples of this category. In another work by Kim and Kittler [8] authors have tried to make use of different pose-invariant face recognition experts in a multiple classifier fusion system framework.

In short, pose-invariant face recognition systems fall broadly into three categories. The first group are those which try to synthesize novel views, either in 2D or 3D. The methods, which try to infer the most discriminatory information across different poses between distinct classes, constitute the second category. There are also some methods which make use of multiple images of different poses in the database. The algorithms in the last group fail when only one image per subject is available in the database. The main drawback of the methods which synthesize novel views is the imperfection of the synthesizing process in addition to the requirement for prior labeling of landmarks which is usually carried out manually. An observation regarding the 3D methods is that although these techniques perform slightly better than the 2D alternatives, they still suffer from unresolved problems. The most important one is that in 3D-based geometric normalization methods, the recovered shape and texture are completely determined by the 3D morphable face model fitted to the query 2D face image which has the capacity to reconstruct only the information captured during statistical learning. As a result, these approaches can not recover atypical features that have not been available in the training set. Moreover, the high computational complexity of 3D methods in comparison with 2D algorithms makes them unsuitable for real-time applications.

In this work we propose a face recognition system which operates on 2D images. The images are first matched densely and then a similarity criterion defined as the normalized energy of the match is used to judge the goodness-of-match. The method excludes the need for geometric pre-processing of images by encapsulating a matching stage as part of the method. The underlying idea in this work is not new. Similar approaches have been employed especially in general object recognition systems. In a general object recognition setting, it is commonly believed that in the presence of varying illumination, partial occlusion, change of viewing angle, cluttered background, change of scale *etc.*, graph-based techniques perform better in comparison to other alternatives. In a graph matching approach, the concepts of interest are assumed to be built up from simple neighboring primitives. The primitives are coded as the nodes of a graph while the edges convey the neighborhood structure and contextual dependencies. In the area of face recognition and authentication, this approach has been previously attempted by a number of researchers. In [11] the authors have used a dynamic link architecture to construct model and scene graphs. In a later work [21], the authors extended their previous work in [11] by performing the matching twice. In the first stage, the location and size of the face are estimated. The second matching is performed to find the exact location of fiducial points on the test image. Measuring the similarity of the test image to the models of the database was performed using only the node attributes without taking into account the

structure (distortion) of the underlying graph explicitly. In [20] an extension to
the previous method in [21] was proposed to identify special characteristics of
the unknown facial image. In another similar work [10], a graph matching scheme
was proposed in which instead of Gabor wavelet filter outputs, multi-scale morphological operators were employed as node attributes. In order to take into
account different discriminatory capabilities of nodes, a weighting scheme was
employed. The work presented in [18] is very similar to [10] but the authors here
have tried to estimate the node weights by reformulating Fisher's discriminant
ratio as a quadratic optimization problem which is then solved by combining
statistical pattern recognition methods and support vector machines. There are
also some other approaches in the context of facial image analysis which use
graph theoretic methods to recognize expressions, *e.g.* in [19].

2   Contributions

In this work a method for verification of facial images under varying pose, based
upon the method in [15] is presented. The method takes advantage of an image matching method [16] for establishing correspondences between images and
formulates the similarity criterion between objects as a combination of the normalized distortion energy of the match as well as texture similarities.

The contributions of the present work and modifications to the matching
method in [16] can be outlined as below.

- In order to cope better with matching under different viewing angles, a dynamically deformable block matching method is proposed. In the new generalized block matching scheme, blocks are neither of the same size, nor the
  same shape. Blocks are deformed according to a global projective transformation estimated between the two images. Accordingly, a square block on
  the model image is matched to a patch of pixels in the scene image whose
  shape and area are determined based on the global transformation. The new
  matching scheme allows much denser sampling of the areas of the face which
  have undergone contraction and coarser sampling in the areas of expansion
  as a result of changes in head pose. It has been found that much better
  matches can be established using the new method.
- In the case of a pan movement of the subject's head, only a half of the face
  is used for recognition. It is shown experimentally that the visible half of
  the face contains much more useful shape information and is superior to the
  whole face.
- The data term has been truncated to achieve more robustness against matching of outliers or occlusions.
- Since the matching method should be able to match facial images of different subjects under varying pose, in order to achieve more flexibility in the
  deformation, the binary hard constraints are replaced by quadratic penalty
  functions.
- In order to cope better with illumination changes, the data term has been
  computed using edge maps. The data term is defined as a combination of
  normalized vertical and horizontal edge magnitudes.

– The method in [16] used distance transforms [4] to compute messages in linear time. Here, additional speed gain is achieved by pruning unlikely labels at the node level during optimization.

The paper is organized as follows: In Section 3, the image matching method in [16] is overviewed. In Section 4, a new deformable block matching method is introduced. The method incorporates a label-pruning heuristic to speed up the matching process. A similarity criterion for assessing the quality of a match is presented in Section 5. The results of an experimental evaluation of the method on the rotation shots of XM2VTS database are presented and discussed in Section 6. Section 7 concludes the paper.

3 Image Matching

There are different methods for image matching proposed in the literature. In fact, for recognition purposes, the following properties are desirable in an image matching method:

The method should support large displacements to allow the matching of images taken at different viewing angles and scales. It is also important to achieve good solutions in a reasonable time, *i.e.* the method adopted should be efficient enough so that it can be used for recognition purposes in a large database of images. Both objectives can be realized by taking advantage of recent optimization techniques for MRFs [16].

The efficiency of the image matching technique in [16] is based on the fact that disparities in two directions are modeled by two fields interacting together rather than coding them in a single MRF. Additional efficiency was gained through the application of a fast energy minimization technique [9] and updating messages using distance transforms [4]. In the following we briefly review the method used for image matching followed by an overview of the object recognition method in [15].

3.1 Preliminaries

Many computer vision problems can be formulated in an energy minimization framework where the objective function takes the following form:

$$E(X|\theta) = \sum_{s \in \nu} \theta_s(x_s) + \sum_{(s,t) \in \epsilon} \theta_{st}(x_s, x_t) \tag{1}$$

ν corresponds to sites and ϵ to edges. x_s denotes the label of site $s \in \nu$. θ defines the parameters of the energy: θ_s denotes unary data penalty functions whereas θ_{st} denotes pairwise potentials. It is worth noting that in this formulation only cliques of size up to two are considered.

The minimum energy in equation (1) corresponds to the maximum probability of a Gibbs distribution. According to the Hammersley-Clifford theorem, the configuration of a set of sites with respect to the neighborhood system adopted, is an MRF if and only if it is a Gibbs random field with respect to the same neighborhood system. Thus, the solution on an MRF can be considered as the configuration of a Gibbs distribution with maximum probability or inversely as the configuration with minimum posterior energy.

3.2 Decomposed Model

The method proposed in [16] formulates the image matching as a labeling problem on MRFs with the label set $L_{reg} = \{(x_{s^1}, x_{s^2})|x_{s^1}, x_{s^2} \in L\}$ where x_{s^1} and x_{s^2} denote displacements in horizontal and vertical directions. In fact this technique models the deformation in horizontal and vertical directions by two MRFs interacting together. The edge set of this model is comprised of two separate edge sets(inter-layer and intra-layer edges). The edge potential functions on each of these layers are assumed to be identical (intra-layer edges) while inter-layer edges encode the data term. For the intra-layer edges the following crisp continuity terms are adopted:

$$\theta_{st}(x_s, x_t) = \begin{cases} 0, & x_s = x_t, \\ c_r, & |x_s - x_t| = 1, \\ \infty, & |x_s - x_t| > 1. \end{cases} \tag{2}$$

In order to achieve more flexibility in deformation, hard continuity terms are replaced by quadratic penalty function:

$$\theta_{st}(x_s, x_t) = c(x_s - x_t)^2 \tag{3}$$

where c is a normalizing constant. In our experiments, setting c to 5×10^{-3} was found to give good results. It should be noted that this value depends on the range of input data (normalized to [-1,1] in our case) and also determines the elasticity of the model. In [16], by restricting the neighboring blocks (blocks are of size 4×4) to have relative displacements of no more than one pixel, the scale changes were limited to [.75,1.25] of the model image size whereas by replacing the hard constraints by a quadratic term a much greater range of scales can be accomodated.

The inter-layer edges encode the data term, *i.e.* the cost of assigning label x_{s^1} in layer one and label x_{s^2} in layer two to two isomorphic nodes of the graph. The data term has been constructed using *block model*. In the block model, the pixels are grouped into non-overlapping blocks which correspond to nodes of the graph. The data term for the block model is defined as below:

$$\theta_{s^1 s^2}(x_{s^1}, x_{s^2}) = \frac{1}{\sigma^2} \text{Dis}(I_s^1, I_{s+(x_{s^1}, x_{s^2})}^2), s^1 \in \nu^1, s^2 \in \nu^2 \tag{4}$$

where I_s^1 is a block on image I^1 and the corresponding block on image I^2 is denoted by $I_{s+(x_{s^1}, x_{s^2})}^2$, which is the block with the coordinates $s + (x_{s^1}, x_{s^2})$, where s is the vector pointing to the position of block I_s^1. Dis(.,.) is a dissimilarity measure which is defined as the sum of squared differences over the pixels of

corresponding blocks. Since edge maps are less affected by unwanted illuminations changes, in order to achieve robustness against changes in illumination, we
use horizontal and vertical edge maps instead of grey scale images. Horizontal
and vertical edges are normalized to the range [-1,1] and combined to form the
data term. The data term then becomes:

$$\theta_{s^1 s^2}(x_{s1}, x_{s2}) = \frac{1}{\sigma^2}[\text{Dis}(I_s^{1h}, I_{s+(x_{s1},x_{s2})}^{2h})) + \text{Dis}(I_s^{1v}, I_{s+(x_{s1},x_{s2})}^{2v}))], s^1 \in \nu^1, s^2 \in \nu^2 \quad (5)$$

where I_s^{1h} and $I_{s+(x_{s1},x_{s2})}^{2h}$ denote a block in the horizontal edge map of the first
image and its corresponding block in the horizontal edge map of the target image
respectively. I_s^{1v} and $I_{s+(x_{s1},x_{s2})}^{2v}$ are defined in a similar way.

Since we are interested in comparing configurational arrangements of the entities of the model and scene images, it is desirable to rely more on common
features of the two images and bypass the atypical features which appear only
in one image. This can be achieved by ignoring the weak edges and setting those
below a threshold to zero and also by truncating the data term. By truncating
the data term, the matching becomes more robust to outliers and occlusions.

4   Matching with Deformable Blocks

In [16] it has been assumed that for a block in the model image, there exists a
block with the same size which has undergone some translational motion. This
assumptions ignores any global geometric transformation between the template
and the target which is one of the omnipresent factors when matching objects
viewed from different angles. Obviously, those parts of the object closer to the
sensing device appear larger than the parts further away. In order to handle this
effect it seems appropriate to have much more dense sampling (smaller blocks) in
the areas of contraction while coarser sampling (larger blocks) would be sufficient
in areas of expansion.

In this work the variation in block sizes is controlled by a global projective
transformation. Although in order to estimate a global geometric transformation between two images dense matching is not required and transformation can
be estimated using a variety of techniques, we have used the same matching
scheme [16] and RANSAC to exclude mismatches and estimated a projective
transformation. In the second step of matching, each block on model image is
warped according to the estimated global transformation and then the corresponding patch on the scene image is sought. The advantages of this method are
two fold. First, as mentioned previously, it supports a more realistic sampling of
signals subject to a global transformation. Second, as the global transformation
controls and predicts the relative placement of corresponding blocks, the size of
the neighborhood (search area) that has to be searched for correspondences can
significantly be reduced. This minimizes the computational cost of matching in
the second stage.

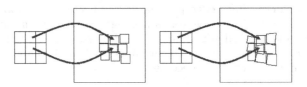

Fig. 1. Left: blocks in [16], Right: blocks in the new deformable block scheme

Considering T:

$$T = \begin{pmatrix} a & b & c \\ d & e & f \\ g & h & 1 \end{pmatrix} \qquad (6)$$

as the estimated projective transformation between the images, the 2D spatial mapping of blocks then can be interpreted as a combination of projective mapping and translational motion:

$$x_{s^1} = \left(\frac{ax + by + c}{gx + hy + 1}\right) + \hat{x}_{s^1}, x_{s^2} = \left(\frac{dx + ey + f}{gx + hy + 1}\right) + \hat{x}_{s^2} \qquad (7)$$

where x_{s^1} and x_{s^2} stand for horizontal and vertical displacements and x and y are coordinates of the block center. $\hat{x}_{s^1}$ and $\hat{x}_{s^2}$ are labels which are inferred in the second stage of matching. Since the projective transformation captures the dominant part of motion, the potential range of $\hat{x}_{s^1}$ and $\hat{x}_{s^2}$ can be reduced during second matching, thus reducing the computational cost. Another advantage of the deformable-block matching method is its enhanced robustness against outliers in matching. In practice, the matching is not perfect and there might be parts of the model image which are not matched correctly to the unknown image. By reducing the search region in the second stage of matching and allowing the estimated global spatial transformation to carry the dominant part of the motion, this shortcoming is partly corrected in the new matching scheme.

4.1   Pruning Unlikely Labels

Inference on the constructed MRF is performed using the sequential tree-reweighted message passing method [9] which is built upon the max-product belief propagation of Pearl [12]. In an ideal case, if the algorithm finds the exact solution, choosing the solution would be based on choosing the label which minimizes the cost at each node. Although the label with the minimum cost at each node might not correspond to the best solution when the number of iterations is limited (because the inference is not exact and because of the existence of multiple minima), it is unlikely for a label with a high cost at a node in an intermediate iteration of the algorithm to correspond to the optimal solution at the end of optimization. Based on this observation, one can prune out labels which are unlikely to be optimal at each node (labels with larger costs) and meet only admissible labels at each node during optimization. Pruning unlikely labels reduces the configurational search space, hence speeds up the method. In practice the following heuristic pruning scheme is found to result in reasonable solutions:

After n_1 iterations, prune out up to n_2 least probable labels at each node based on their corresponding costs ensuring that there are at least n_3 labels left at each node.

The choice of n_1, n_2 and n_3 depends on the difficulty of a specific problem. The easier the problem the smaller n_1 and n_3 and larger n_2. Although the pruning might sometimes lead to better results compared to the original method, in a limited number of iterations, it may sometimes introduce a trade off between speed and accuracy.

Using the matching method described, a model image is matched to the test image. Figure (2) shows an example of warping a gallery image (near frontal) to the test image (non-frontal) using the deformable block matching scheme.

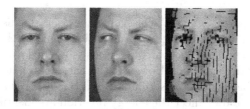

Fig. 2. Left: model image, Middle: scene image, Right: warped model image

Figure (3) shows an example in which the original method in [16] fails to find a correct match especially around the mouth and nose region of the model image. Using the new matching method one can get better matches in the problematic areas. Figure (3) shows the improvement in matching around the mouth and nose of the subject.

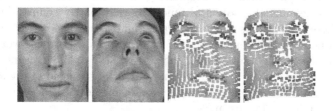

Fig. 3. Left to Right: template, target, result of warping the template using the method in [16], result obtained using deformable-block method

5   Classification

In order to measure the similarity, the two stages are cascaded: first matching the model image image to the unknown image and then computing a similarity/cost function invariant to unwanted global spatial transformation and illumination variations. More explicitly the problem can be described as follows: let I be the image of an ideal subject. Let J be the image of an unknown subject under analysis which depends on its geometrical parameters such as scale s, displacements d_x and d_y, rotation ϕ and perspective effects p, so that $J = J(s, d_x, d_y, \phi, p)$. Let $\mathrm{Dis}(I, J)$ be a dissimilarity function between the ideal image I and unknown image J. The problem is then formulated as calculating:

$$d = \min_{s, d_x, d_y, \phi, p} \mathrm{Dis}(I, J(s, d_x, d_y, \phi, p)) \tag{8}$$

In a hypothesis verification (two class) problem the decision rule is:

Assign J to class ω_r *iff* $\min_{s, d_x, d_y, \phi, p} \mathrm{Dis}(I_r, J(s, d_x, d_y, \phi, p)) < thresh_r$

where I_r is the template for the rth class and $thresh_r$ is the dissimilarity threshold for the rth class. In the context of recognition using MRFs, a cost function corresponding to the unary and pairwise terms is defined and optimized which is then used in the decision rule. However, the energy obtained in this way has been found not to have enough discriminatory capacity for classification. In [15] factors which unfavorably affect the energy function are identified as pose, nonrigidity of the pattern and last but not least statistical dependencies between residual displacements of neighboring sites and the limited cardinality of the potential functions. In order to remove the effect of the rigid motion, the distortion associated with the global spatial transformation was subtracted from distortion vectors thus achieving global spatial transformation invariance. In order to take into account non-rigidity of patterns, a number of different exemplars of each class were matched one to another and the average distortion was considered as a class-specific model of deformations. The problems of inherent correlation between residual displacements of neighboring sites and of the limited cardinality of the cliques defining the potential functions were partly compensated for by modeling these interactions using covariance matrices which convey correlation information between different sites even at a larger range. In this work we estimate covariance matrices for the full face in the case of tilt movement of head and for the half face in the case of pan movement of the head of the subject.

The structural differences between a pair of images is hence formulated in terms of the Mahalanobis distance:

$$D_{Mahalanobis}(I_i, J) = (\bar{e}_v - \bar{\mu}_{iv})^t \textstyle\sum_v^{-1} (\bar{e}_v - \bar{\mu}_{iv}) +$$
$$(\bar{e}_h - \bar{\mu}_{ih})^t \textstyle\sum_h^{-1} (\bar{e}_h - \bar{\mu}_{ih}) \tag{9}$$

where I_i is a template of class i, $\bar{\mu}_{iv}$ and $\bar{\mu}_{ih}$ are the average distortions for this class in vertical and horizontal directions respectively pursued in a raster scan

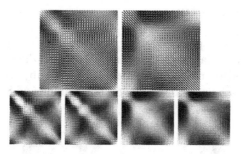

Fig. 4. Covariance matrices of distortions: up row left: full face covariance matrix for vertical direction, up row right: full face covariance matrix for horizontal direction, bottom row from left to right: half face covariance matrices for left half of face in vertical direction, right half of face in vertical direction, left half of face in horizontal direction, right half of face in horizontal direction

fashion. $\bar{e}_v$ and $\bar{e}_h$ are the local distortion vectors obtained after matching I_i to J. In order to obtain the local distortions, after the second stage of matching, another projective transformation is fitted to the set of corresponding points and the effect of rigid motion is subtracted from the distortion field. $\sum_v^{-1}$ and $\sum_h^{-1}$ represent inverse covariance matrices for distortions in vertical and horizontal directions respectively.

5.1   Textural Content

The spatial distortion measure should be complemented by a measure of quality of the match conveyed by the data to refine the cost of match. However, the data term should not be sensitive to unwanted changes in lighting conditions during image capture. Thus, it is essential to use illumination-invariant representation of images for comparison. Local binary patterns have been found to be effective texture descriptors as long as the intensity order of the pixels in a neighborhood is preserved. The textural content of the two images represented as the output of an LBP operator are compared using normalized correlation:

$$NC = \frac{\sum_i \sum_j \sum_h \sum_v (b_{I(i,j)}(h,v) b_{J(i,j)}(h,v))}{\sqrt{\sum_i \sum_j \sum_h \sum_v (b_{I(i,j)}(h,v)^2 b_{J(i,j)}(h,v)^2)}} \quad (10)$$

where $b_{I(i,j)}(h,v)$ is the pixel with horizontal and vertical indices h and v respectively in the block with horizontal and vertical indices i and j in image I. $b_{J(i,j)}(h,v)$ is defined in a similar way.

The final distance measure between a class model (I_i) and the unknown image (J) can be interpreted as a weighted measure of shape and texture distances:

$$D(I_i, J) = w_1(1 - NC) + (1 - w_1)((1 - w_2)D_{Mahal_H} + w_2 D_{Mahal_V}) \quad (11)$$

where D_{Mahal_H} and D_{Mahal_V} correspond to Mahalanobis distance between the two shapes using horizontal and vertical distortions.

6   Experimental Setup

To test our approach to pose-invariant face verification we performed experiments on the XM2VTS data set. This is a multi-modal database which contains color images plus video and sound sequences of 295 subjects. For these experiments we make use of 8 near frontal images for training and 8 rotated head shots per subject, as test images.

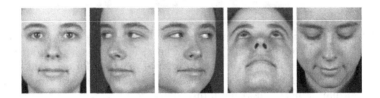

Fig. 5. Example images of frontal and rotated faces in XM2VTS corpus

Some examples of rotated and near frontal images used are shown in Figure 5. The XM2VTS database is divided into a training set of 200 clients, an evaluation set of the same 200 clients plus 25 impostors and a test set of the same 200 clients plus 70 different impostors. In our rotated head shot experiments we used a modified form of the XM2VTS Lausanne test protocol. This retains the same partitioning of subject identities into valid clients and impostors. It differs in that for clients the 2 frontal test images are replaced by 8 rotated head shots (down, left, right, up) and for impostors the 8 frontal images are replaced by 8 rotated head shots. Figure 5 illustrates the severity of the pose deviation from the frontal.

6.1   Results

Two error measures adopted for assessing the performance of the verification system are false acceptance and false rejection rates defined as:

$$FA = EI/I * 100\%, \qquad FR = EC/C * 100\% \tag{12}$$

where I is the number of imposter claims, EI the number of imposter acceptances, C the number of client claims and EC the number of client rejections.

The performance of a verification system is often stated in *Equal Error Rate* (EER) in which the FA and FR are equal and the threshold for the acceptance or rejection of a claimant is set using the true identities of test subjects. In our experiments we use all 8 near frontal images of clients as training images for estimating the covariance matrices. The results of the verification test on images under pan movement of the head using full face and half face shape information are reported in Tables 1 and 2 and the results using texture in Table 4 in which F.F., H.F and Tex. stand for full face, half face and texture respectively.

Table 1. EER for Pan Using Full-Face Shape Information

Horizontal Distortions		Vertical Distortions		Horizontal and Vertical Distortions	
Euc. dist.	Mahal. dist.	Euc. dist.	Mahal. dist.	Euc.n dist.	Mahal. dist.
53.92	31.24	25.13	17.09	25.13	16.86

Table 2. EER for Pan Using Half-Face Shape Information

Horizontal Distortions		Vertical Distortions		Horizontal and Vertical Distortions	
Euc. dist.	Mahal. dist.	Euc. dist.	Mahal. dist.	Euc.n dist.	Mahal. dist.
34.44	31.49	17.44	13.86	17.27	13.77

Table 3. EER for Tilt Using Shape Information

Horizontal Distortions		Vertical Distortions		Horizontal and Vertical Distortions	
Euc. dist.	Mahal. dist.	Euc. dist.	Mahal. dist.	Euc.n dist.	Mahal. dist.
25.80	25.25	42.58	29.37	25.80	24.9

Table 4. EER Using Texture/ Shape and Texture

Pan Movement			Tilt Movement	
F.F. Tex.	H.F. Tex.	F.F. Tex. and H.F. Shape	Tex.	Tex. and Shape
13.25	17.14	9.62	26.5	24.37

The results using the fusion of texture and shape scores are also reported in Table 4.

According to the results in Tables 1, 2 and 4 the following conclusions can be drawn. First, the effect of modeling statistical dependencies in deformation is evident by comparing the results obtained using Euclidean and Mahalanobis distances. Second, in the case of pan movement, distortions in vertical direction are much more discriminative than the distortions in the horizontal direction. Third, using half face image, the configurational information of face images is better represented. This is because the more distant part of the rotated face in the image is partly self-occluded and does not contain useful shape information. The second reason is the inability of a simple projective transformation to model 3D rigid motion of a non-planar object. The results of fusing texture and half face shape are reported in Table 4.

Table 3 reports the results for the tilt movement using shape information. In this case we have used full face matching in our experiments. The effectiveness of the covariance information is again evident. As in the case of pan movement, the distortions perpendicular to the direction of head movement are more discriminative. The result of fusing shape and texture scores is reported in Table 4. From the results it can be concluded that, the recognition of faces subject to pan movement ($EER = 9.62$) is more accurate than that of tilt ($EER = 24.37$). In the case of tilt motion, the self occlusion can not be compensated for by exploiting symmetry. Inevitably this decreases the quality of the match. The results obtained, are considerably better than the state-of-the-art results obtained on the XM2VTS database using AAMs in [5]. In [5] only a subset of the same database was used to test the method whereas here the obtained results are on the whole database. Also the results are significantly better than those obtained in [8] reporting error rates of 30% for the same session and 58% for different sessions on a subset consisting of 125 subjects of the same database. Also, the results compares favorably with the results in [14] which has used only 100 subjects out of 295 in the same database and achieves an error of 23% in recognition.

7 Conclusion

A pose-invariant face recognition system based on an image matching method was presented. The method uses the normalized energy of the established match between images as a criterion for assessing goodness-of-match. The method can tolerate moderate global spatial transformations between the model and the scene object and alleviates the need for geometric normalization of facial images which is commonly required in face recognition. The experimental evaluation of the method was performed on the very challenging rotation shots of the XM2VTS database and promising results were obtained.

References

1. Beymer, D.J.: Face recognition under varying pose. In: Proc. IEEE Conf. Computer Vision and Pattern Recognition, June 1994, pp. 756–761 (1994)
2. Blanz, V., Vetter, T.: Face recognition based on fitting a 3D morphable model. IEEE Trans. Pattern Analysis and Machine Intelligence 25(9), 1063–1074 (2003)
3. Cootes, T.F., Edwards, G.J., Taylor, C.J.: Active appearance models. IEEE Trans. Pattern Analysis and Machine Intelligence 23(6), 681–685 (2001)
4. Felzenszwalb, P.F., Huttenlocher, D.P., Kleinberg, J.M.: Fast algorithms for large-state-space HMMs with applications to web usage analysis. In: NIPS (2003)
5. Guillemaut, J.-Y., Kittler, J., Sadeghi, M., Christmas, W.: General pose face recognition using frontal face model. In: Proceedings of the 11th Iberoamerican Congress in Pattern Recognition, November 2006, pp. 79–98 (2006)
6. Kanade, T., Yamada, A.: Multi-subregion based probabilistic approach toward pose-invariant face recognition. In: Proc. IEEE International Symposium on Computational Intelligence in Robotics and Automation, vol. 2, pp. 954–959 (2003)
7. Kim, T.K., Kittler, J.: Locally linear discriminant analysis for multimodally distributed classes for face recognition with a single model Image. IEEE Trans. Pattern Analysis and Machine Intelligence 27(3), 318–327 (2005)
8. Kim, T.K., Kittler, J.: Design and fusion of pose-invariant face-identification experts. IEEE Trans. Circuits and Systems for Video Technology 16(9), 1096–1106 (2006)

9. Kolmogorov, V.: Convergent tree-reweighted message passing for energy minimization. IEEE Trans. on Pattern Recognition and Machine Intelligence 28(10), 1568–1583 (2006)
10. Kotropoulos, C., Tefas, A., Pitas, I.: Frontal face authentication using morphological elastic graph matching. IEEE Trans. on Image Processing 9(4), 555–560 (2000)
11. Lades, M., Vorbruggen, J.C., Buhmann, J., Lange, J., Von der Malsburg, C., Wurtz, R.P., Konen, W.: Distortion invariant object recognition in the dynamic link architecture. IEEE Trans. on Computers 42(3), 300–311 (1993)
12. Pearl, J.: Probabilistic reasoning in intelligent systems: Networks of plausible inference. Morgan Kaufmann, San Francisco (1988)
13. Pentland, A., Moghaddam, B., Starner, T.: View-based and modular eigenspaces for face recognition. In: Proc. IEEE Conf. Computer Vision and Pattern Recognition, June 1994, pp. 84–91 (1994)
14. Prince, J.D., Elder, J.H., Warrel, J., Felisberti, F.M.: Tie factor analysis for face recognition across large pose differences. IEEE Trans. Pattern Analysis and Machine Intelligence 30(6), 970–984 (2008)
15. Rahimzadeh Arashloo, Sh., Kittler, J.: On matching criteria for non-rigid object recognition, research report, Centre for Vision, Speech and Signal Processing, University of Surrey (July 2008)
16. Shekhovtsov, A., Kovtun, I., Hlavac, V.: Efficient MRF deformation model for non-Rigid image matching. In: Conf. Computer Vision and Pattern Recognition (CVPR), pp. 1–6 (2007)
17. Singh, R., Vatsa, M., Ross, A., Noore, A.: A mosaicing scheme for pose-invariant face recognition. IEEE Trans. Systems, Man, and Cybernetics, Part B: Cybernetics 37(5) (October 2007)
18. Tefas, A., Kotropoulos, C., Pitas, I.: Using support vector machines to enhance the performance of elastic graph matching for frontal face authentication. IEEE Trans. on Pattern Analysis and Machine Intelligence 23(7), 735–746 (2001)
19. Wang, M., Iwai, Y., Yachida, M.: Expression recognition from time-sequential facial images by use of expression change model. In: Proc. Third IEEE International Conf. on Automatic Face and Gesture Recognition, April 1998, pp. 324–329 (1998)
20. Wiskott, L.: Phantom faces for face analysis. In: Proc. International Conf. on image Processing, October 26-29, vol. 3, pp. 308–311 (1997)
21. Wiskott, L., Fellous, J.M., Kuiger, N., Von Der Malsburg, C.: Face recognition by elastic bunch graph matching. IEEE Trans. Pattern Analysis and Machine Intelligence 19(7), 775–779 (1997)
22. Messer, K., Matas, J., Kittler, J., Luettin, J., Maitre, G.: XM2VTSDB: The extended m2vts database. In: Proc. Audio- and Video-based Biometric Person Authentication Conf. (1999)
23. Zhao, W., Chellappa, R., Phillips, P.J.: Face recognition: A literature survey. ACM Computing Surveys 35(4), 399–458 (2003)

Parallel Hidden Hierarchical Fields for Multi-scale Reconstruction

Ying Liu and Paul Fieguth

Department of Systems Design Engineering, University of Waterloo
Waterloo, Ontario, Canada, N2L 3G1
{y30liu,pfieguth}@uwaterloo.ca

Abstract. In any problem involving images having scale-dependent structures, a key issue is the modeling of these multi-scale characteristics. Because multi-scale phenomena frequently possess nonstationary, piece-wise multi-model behaviour, the classic hidden Markov method can not perform well in modeling such complex images. In this paper we provide a new modeling approach to extend previous hierarchical methods, with multiple hidden fields, to perform reconstruction in more complex, nonstationary contexts.

1 Introduction

There are many problems in texture analysis, remote sensing and scientific imaging where the observed image possesses highly scale-dependent structure. Although such structures can, in principle, be represented with sufficiently complex models, the development of such models is a difficult task, and leads to computationally intractable algorithms if executed on a single scale. In this paper, we are interested in efficient hierarchical model structures to reconstruct complex scientific imagery.

Certainly multi-scale image modeling and analysis is common in image processing, given the widespread application of wavelet [1,2,3], Gaussian and Laplacian pyramid methods [4,5], quad-tree based models in the continuous Gaussian case [6,7], and texture synthesis [8]. In addition, multi-fractal analysis is used to characterize the self-similarity property of objects. This technique has been used to study the statistics of natural images [9], been applied to synthesize textures [9], and acted as a prior to regularize reconstruction problems [10]. However, these methods are all applied to continuous-state problems, whereas our interest is in the hierarchical modeling of discrete-state fields, such as a label field underlying an image.

In the discrete-state case, Markov / Gibbs Random Fields have been widely used in image restoration, segmentation, reconstruction [11,12]. However, local Markov / Gibbs Random Fields are limited to describing phenomena at a single scale, and are not naturally suited for multi-scale phenomena.

Instead, hierarchical MRF modeling [13] provides a more natural and efficient way to deal with multi-scale structures. Kato et al [14] proposed a hierarchical

D. Cremers et al. (Eds.): EMMCVPR 2009, LNCS 5681, pp. 70–83, 2009.

 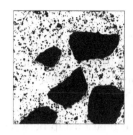 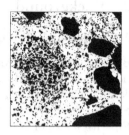

(a) Single-scale example   (b) Two-scale example   (c) Complex, multi-scale, multi-model

Fig. 1. Excerpts from microscopic images of physical porous media. A single-scale structure (a) can be well described by a single hierarchical MRF [17], two-scale nonstationary behaviours in (b) may be described by adding a hidden hierarchical MRF [18], however complex structures (c) with multi-model behaviours pose a modeling challenge.

MRF model with a 3D neighborhood system for modeling label fields, but with considerable computational cost. A MRF model based on a quad-tree structure was discussed by Laferté et al [15], but does not model the interactions within scales. Later, Mignotte et al [16] proposed a model with an inter-scale Markov chain and an intra-scale MRF. However, all of these used simple models at each scale, too limited to capture complex structures, such as those in Fig. 1.

In order to precisely synthesize images of porous media, Alexander and Fieguth [19] proposed a hierarchical model with local MRFs at each scale, but ignoring interrelations between scales. Later, Campaigne [17] proposed a frozen-state hierarchical annealing method, with attractive computational complexity and scale-dependent modeling.

The goal of our research is the extension of [17] to allow hidden fields. Generally, a single hierarchy with a scale-dependent model can capture a stationary structure (Fig. 1(a)), whereas many random fields have some sort of nonstationary piece-wise multi-model behaviour which requires additional hidden fields [20, 21, 22, 23]. Although multiple hidden fields are routinely used in Markov modeling, asserting a hierarchical context creates additional subtleties. Recently, Scarpa et al [24] proposed a hierarchical texture model which represents texture at the region level with a superimposed finite-state hierarchical model. Their approach has some similarities with ours, but focuses on unsupervised model inference, whereas our approach requires more accurate, supervised models, with an emphasis on computational tractability for large problems.

In this paper we build on previous work [18] to perform reconstruction for complex, nonstationary problems. We have chosen to apply our methods to reconstruct scientific images from porous media, such as the one shown in Fig. 1(c), since the images include multiple behaviours, with fractal-like scale dependent structures. The problem reduces to an energy minimization across fields and across scales, with promising results.

2   Background

2.1   Classical Hidden Markov Framework

Based on the hidden Markov field (HMF) [25, 26] framework, image reconstruction can be achieved by estimating a hidden random field X from an observed field Y, where $Y = \{Y_s : s \in S_L\}$ is defined on a Low Resolution (LR) grid space S_L, and $X = \{X_s : s \in S_H\}$ is defined on a High Resolution (HR) grid space S_H with a size of $N \times N$. The relationship between X and Y is expressed by a forward model $Y = g(X) + \nu$, where ν denotes the measurement noise. If ν is i.i.d, the classical HMF is written as

$$p(x|y) \propto \prod_{s \in S_L} p(y_s|x)p(x) \tag{1}$$

where X is assumed to be MRF. However, a single local MRF cannot perform well in modeling a multi-scale nonstationary X; for example Fig. 3(c) shows the failure of the classical HMF method to reconstruct a piece-wise, two-scale image (Fig. 3(a)).

2.2   Single Hierarchical Field

In modeling multi-scale phenomena, the dilemma of a single MRF is that local models cannot strongly assert the presence of nonlocal structures, whereas learning a huge nonlocal model is prohibitive in practice. Very differently, hierarchical modeling defines X via a sequence of fields $\{X^k, k \in K = (0, 1, \cdots, M)\}$, where $k = 0$ defines the finest scale. At each scale k, X^k is defined on site space S^k and results from the downsampling of X^0.

The advantage of hierarchical modeling is that nonlocal large-scale features become local at a sufficiently coarse scale, therefore at each scale a single MRF (X^k) can be used to capture the features local to that scale, inherently allowing scale-dependent structure. In defining a hierarchical model, two issues need emphasizing: the inter-scale context, and the computational complexity.

To model the spatial context, Mignotte et al [16] proposed a Markov chain in scale $p(x^k|x^{K \setminus k}) = p(x^k|x^{k+1})$, whereas the intra-scale relation is a MRF $p(x_s^k|x_{S \setminus s}^k) = p(x_s^k|x_{\wp(s)}^{k+1}, x_{\mathcal{N}_s^k}^k)$, where $\wp(s)$ denotes the parent site of s at the parent scale, and $\mathcal{N}_s^k$ defines a local neighborhood.

To achieve computational efficiency, a Frozen State Hierarchical Field (FSHF) was presented in [17] to synthesize binary images. In their work, a given HR field ($x = x^0$) can be represented by a hierarchical field $\{x^k\}$ (Fig. 2) where $x^k = \Downarrow^k(x^0)$, and $\Downarrow^k(\cdot)$ denotes a downsampling operator. The key idea of the FSHF is that, at each scale, only those sites which are undetermined need to be sampled, with those sites determined by the parent scale fixed (or frozen):

$$p(x_s^k|x_{S \setminus s}^k) = \begin{cases} \delta_{x_s^k, x_{\wp(s)}^{k+1}}, & \text{if } x_{\wp(s)}^{k+1} \in \{0, 1\} \\ p(x_s^k|x_{\mathcal{N}_s^k}), & \text{if } x_{\wp(s)}^{k+1} = \frac{1}{2} \end{cases} \tag{2}$$

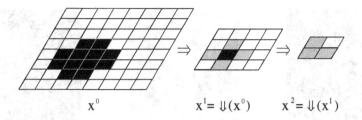

$$\mathbf{x}^0 \qquad \mathbf{x}^1 = \Downarrow(\mathbf{x}^0) \qquad \mathbf{x}^2 = \Downarrow(\mathbf{x}^1)$$

Fig. 2. An example of ternary hierarchical subsampling [17]: a given field x^0 is coarsified by repeated 2×2 subsampling $\Downarrow(\cdot)$. All-white and all-black regions are preserved, with mixtures labeled as uncertain (grey).

where $0, 1$ (black, white) are determined states, and $\frac{1}{2}$ (grey) is undetermined. Since the "grey" interface between black and white regions represents only a small fraction of most images, this approach offers a huge reduction in computational complexity relative to standard, full-sampling hierarchical techniques. The site sampling strategy corresponding to (2) is

$$\hat{x}_s^k = \begin{cases} \hat{x}_{\wp(s)}^{k+1} & \text{if } \hat{x}_{\wp(s)}^{k+1} \in \{0,1\} \\ \text{a sample from } p(x_s^k | x_{\mathcal{N}_s^k}^k) & \text{if } \hat{x}_{\wp(s)}^{k+1} = \frac{1}{2} \end{cases} \qquad (3)$$

where $\hat{x}^k$ is the sampled (estimated) random field at scale k.

2.3 Hidden Hierarchical Fields

A single FSHF works well in modeling stationary scale-dependent structures, however such a model cannot handle X having nonstationary, piece-wise behaviour, because conditioned on X^{k+1}, X^k still has nonstationary features which cannot be captured by a local model. For example, Fig. 3(d) shows that a single hierarchy can not capture the piece-wise two-model behaviour in Fig. 3(a).

To model more general multi-scale cases, we proposed a Hidden Hierarchical Markov Field (HHMF) in [18]. The HHMF has an hidden binary HR field U to capture the model behavior in X. If the nonstationarity in X can be entirely attributed to a single binary behaviour, then conditioned on U becomes X conditional stationary. Assuming X, U to be Markov in scale, the joint relationship $p(X, U) = p(X|U)p(U)$ can be written as

$$p(x, u) \propto \left[\prod p(x^k | x^{k+1}, u^k) \right] \left[\prod p(u^k | u^{k+1}) \right] \qquad (4)$$

We select some coarsest scale as the scale at which decidable state structure appears, such that (4) becomes

$$p(x, u) \propto \left[\prod_{k=0}^{M_x-1} p(x^k | x^{k+1}, u^k) \right] p(x^{M_x} | u^{M_x}) \left[\prod_{k=0}^{M_u-1} p(u^k | u^{k+1}) \right] p(u^{M_u}) \qquad (5)$$

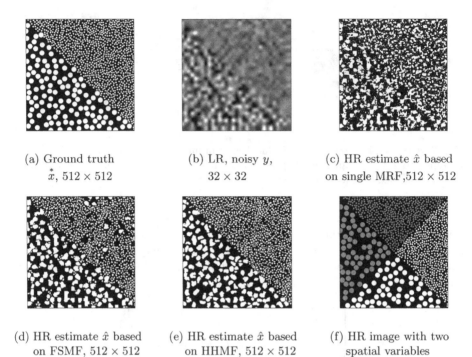

(a) Ground truth
$\overset{*}{x}$, 512 × 512

(b) LR, noisy y,
32 × 32

(c) HR estimate $\hat{x}$ based
on single MRF,512 × 512

(d) HR estimate $\hat{x}$ based
on FSMF, 512 × 512

(e) HR estimate $\hat{x}$ based
on HHMF, 512 × 512

(f) HR image with two
spatial variables

Fig. 3. In this example, we reconstruct a two-scale image (a) from a low resolution measurement (b) with different Hidden Markov Field frameworks. The estimated results are shown in (c)-(e). The clear scale separation of the result from the hidden hierarchy [18] (e) should be compared to the results from single flat MRF based model [27] (c) and single hierarchy based model [17] (d). (f) shows an image with two spatial variables, which creates a modeling challenge for existing models.

where $k = M_x$ and $k = M_u$ denote the coarsest scale of X and U respectively, and where $p(x^k|x^{k+1}, u^k)$ and $p(u^k|u^{k+1})$ are modeled as frozen-state (2), based on the joint, local ternary histogram of [19]. Since U is defined to describe large scale features or model behaviour in X, the decidable state in X is expected to vanish at a finer scale than in U ($M_x < M_u$).

The introduction of a hierarchical hidden field provides a more general, powerful modeling ability, as illustrated in the comparison between Fig. 3(b) and (e), however more complex problems (Fig. 3(f) or Fig. 1(c)) require a more general approach, the subject of this paper.

3   Parallel Hidden Hierarchical Fields

In general, the behaviour of a random field X will be determined by more than one spatial variable ($N_v > 1$), such that X remains nonstationary when

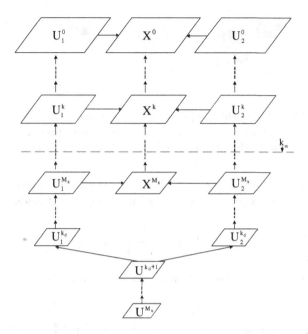

Fig. 4. The proposed Parallel Hidden Hierarchical Markov Field modeling structure, such that the hidden label field U is a joint field only at coarse scales. As the features of different model behaviour become separable at some scale k_d, U is decoupled as multiple parallel hierarchical fields $\{U_1, \cdots, U_{N_v}\}$. At scales coarser than M_x the entire random field X of interest is uncertain, and so only U is represented to scale M_u.

conditioned on a single binary field U. The obvious solution to this problem is to define U as a multi-label field; for example the behaviour of Fig. 3(f) is determined by two variables of scale and shade, with a corresponding quad-label model.

Generalizing the frozen-state annealing algorithm [17] to the non-binary case is a non-trivial task, yet the significant computational benefits of the frozen-state approach motivate us to continue using it. Although multi-label models (e.g., Potts) do exist, the complexity of modeling all pairwise, triplet-wise etc. label interactions at coarser scales makes the problem rather complex.

Our proposed approach is to use multiple, parallel hierarchical hidden fields (PHHF), as illustrated in Fig 4. At finer scales, having many state elements, the hidden fields are decoupled, binary, and simply modeled. The complex, joint hidden structure appears only at very coarse scales, where the small number of state elements allows such a structure to be computationally tractable.

3.1   Parallel Hidden Hierarchical Field

The key idea of PHHF is as follows: instead of using a single hidden field U to model complex multi-model behaviors directly, we introduce multiple binary hidden label fields $U = \{U_i, i \in N_v\}$, such that each field U_i is used to capture

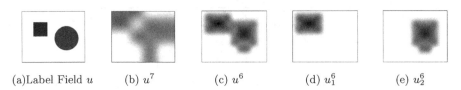

(a)Label Field u (b) u^7 (c) u^6 (d) u_1^6 (e) u_2^6

Fig. 5. For a hidden label field U with nonlocal features (a), at some coarse scale the features of different behaviours interact (b), however at some finer scale the U become separable (c), and can be decoupled to multiple simple fields (d,e)

only a single binary structure or model behavior. With this assumption, the joint field, conditioned on its parent, decouples into independent fields

$$p(u_1^k, \cdots, u_{N_v}^k | u^{k+1}) = \prod_{i=1}^{N_v} p(u_i^k | u^{k+1}) \tag{6}$$

for all scales $k \leq k_d$. The entire PHHF model can then be extended from (5) and written as

$$p(x, u) \propto \prod_{k=0}^{M_x - 1} p(x^k | x^{k+1}, u^k) \cdot p(x^{M_x} | u^k) \cdot \prod_{k=0}^{k_d - 1} \prod_{i=1}^{N_v} p(u_i^k | u_i^{k+1}) \cdot$$

$$\prod_{i=1}^{N_v} p(u_i^{k_d} | u^{k_d + 1}) \prod_{k=k_d+1}^{M_u - 1} p(u^k | u^{k+1}) \cdot p(u^{M_u}) \tag{7}$$

An example of the PHHF with two hidden variables ($N_v = 2$) is shown in Fig. 4. The approach simplifies modeling in three significant ways.

First, the PHHF consists entirely of simple models, both local and stationary. Specifically, although X^k and U^k may have complex, non-local behaviour, the conditional residuals $U^k | U^{k+1}$, $X^k | X^{k+1}, U^k$ are local, by virtue of the fact that all non-local matters have been absorbed into the conditioned (coarser) scale.

Second, the hidden states are primarily decoupled and binary. At coarse scales, where only few pixels exist, it is computationally tolerable to assert a joint model for $U^k, k > k_d$, where the joint model is need to allow the hidden models to interact (Fig. 5(b)). In most problems, empirically, the hidden models become separable at some scale $k \leq k_d$ (Fig. 5(c)), leading to parallel independent fields (Fig. 5(d)(e)).

Third, because $\{X^k\}$ and $\{U_i^k\}$ are modeled using simple, binary models, $\{X^k\}$ and $\{U_i^k\}$ are easily defined as hierarchical frozen states, leading to the computational cost of the PHHF being linear in the number of hidden fields N_v, except at the small, coarse scales.

3.2 Reconstruction

To reconstruct a scale-dependent, near fractal, piece-wise nonstationary image such as the porous medium in Fig. 1(c) is a major modeling challenge. The image in Fig. 1(c) displays three types of behaviour: large-scale pores, regions of high density, and background areas of low density. We therefore propose the ternary hidden field U to be decoupled as two parallel hierarchies, where $\{U_1^k\}$ identifies large pores, and $\{U_2^k\}$ identifies regions of high density.

For image enhancement, the hidden fields are invisible to the measurements, therefore $p(y|x^k, u^k) = p(y|x^k)$, and so the reconstruction model illustrated in Fig. 4 can be written as

$$p(x, u|y) \propto \prod_{s \in S_L} p(y_s|x)p(x, u) \tag{8}$$

where $p(x, u)$ is the prior PHHF, defined as in (7), and where the measurements $p(y_s|x)$ are taken at some scale k_m.

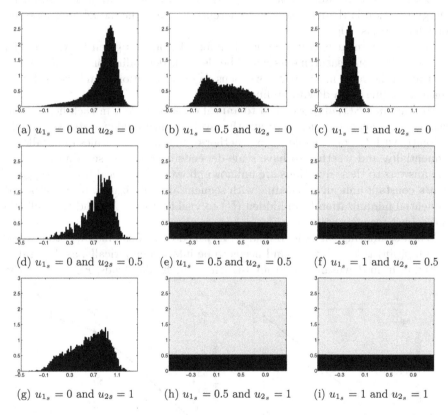

(a) $u_{1_s} = 0$ and $u_{2s} = 0$ (b) $u_{1_s} = 0.5$ and $u_{2_s} = 0$ (c) $u_{1_s} = 1$ and $u_{2_s} = 0$

(d) $u_{1_s} = 0$ and $u_{2s} = 0.5$ (e) $u_{1_s} = 0.5$ and $u_{2s} = 0.5$ (f) $u_{1_s} = 1$ and $u_{2_s} = 0.5$

(g) $u_{1_s} = 0$ and $u_{2s} = 1$ (h) $u_{1_s} = 0.5$ and $u_{2_s} = 1$ (i) $u_{1_s} = 1$ and $u_{2_s} = 1$

Fig. 6. The conditional target histograms of $g(X)$ for decoupling a joint field U^{k_d} into two simple fields $U_1^{k_d}$ and $U_2^{k_d}$; that is, each panel shows the distribution of $g(X)$ for one of nine possibilities on U_1, U_2. Since the hidden fields are asserted to be decoupled, those cases where both fields are asserted (shaded distributions) are never observed, and so are assigned a uniform distribution with low marginal probability. To the extent that the joint state configuration of $(U_{1_s}^{k_d}, U_{2_s}^{k_d})$ relates to distinguishable model behaviour in $g(X)$, we expect the hidden fields to be estimable.

Given measurements Y contaminated by i.i.d. noise, the posterior distribution of (X, U, Y) can be represented as a Gibbs distribution

$$p(x, u|y) \propto exp(-\frac{1}{T}E(x, u|y)) \tag{9}$$

where T is the temperature, such that E is the energy function implying the probability density p. Finding a good estimate $\hat{x}$ therefore corresponds to maximizing $p(x, u|y)$, correspondingly minimizing E, the sum of hidden joint $E_u^k(u^k|u^{k+1})$, decoupled $E_{u_i}^k(u_i^k|u_i^{k+1})$, visible $E_{x|u}^k(x^k|x^{k+1}, u^k)$, and measurement $E_m(y|x)$.

All of the prior models are learned using a nonparametric joint local distribution, the exhaustive joint distribution of a local 3×3 neighbourhood of ternary state elements. The models are learned separately on each scale, based on downsampled training data $\bar{x}^k = \Downarrow^k (\bar{x}^0|\bar{u}^0)$, $\bar{u}_i^k = \Downarrow^k (\bar{u}_i^0)$. The resulting energy function is the least-squares difference between the model and observed joint histograms [19].

The measurement energy function is inferred from the given forward model $g()$. A variety of measurements could be defined, depending on the measuring instrument, however in this paper we focus on reconstruction from low-resolution images, making $g()$ a downsampling operator.

To minimize $E(x, u|y)$, we need to anneal on each scale k in each field X, U_i, with consequent open questions: whether to minimize hidden states separately or jointly with the observable state, whether to minimize the scales in parallel or sequentially, and whether to have scale-dependent annealing schedules. Definitive answers to these questions are unknown, however empirical testing suggests that a constant annealing schedule with sequential minimization over scales and sequential minimization from hidden (U) to visible states (X) lead to a reliable and robust reconstruction.

When estimating the hidden field U, in which case X is unknown, the distribution $p(y|u)$ is less obvious, and needs to be inferred empirically, as illustrated in Fig. 6, from downsampled ground-truth of the hidden fields. Fig. 6 plots the

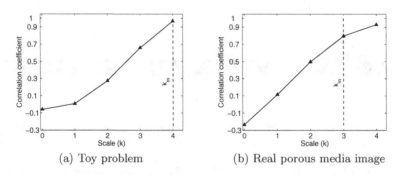

(a) Toy problem   (b) Real porous media image

Fig. 7. Correlation coefficients ρ between the estimates $\hat{x}$ and ground truth $\overset{*}{x}$ as a function of structure scale. For a number of scales below the measured resolution k_m, $\rho(\overset{*}{x}, \hat{x}) > 0$ meaning that some trustable details are reconstructed.

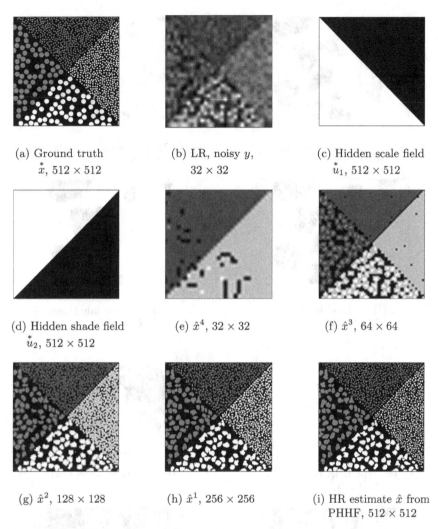

(a) Ground truth
$\overset{*}{x}$, 512 × 512

(b) LR, noisy y,
32 × 32

(c) Hidden scale field
$\overset{*}{u}_1$, 512 × 512

(d) Hidden shade field
$\overset{*}{u}_2$, 512 × 512

(e) $\hat{x}^4$, 32 × 32

(f) $\hat{x}^3$, 64 × 64

(g) $\hat{x}^2$, 128 × 128

(h) $\hat{x}^1$, 256 × 256

(i) HR estimate $\hat{x}$ from
PHHF, 512 × 512

Fig. 8. A toy problem with two spatial variables. For the purpose of this example, we assume the hidden fields for scale and shade to be given, in (c) and (d) respectively. From the low resolution measurements of (b), our estimated results $\{\hat{x}^k, 0 \le k \le 4\}$ are shown in (e)-(i). A clear scale and shade separation are shown in the final result (i), demonstrating the strength of modeling by paralleled hidden fields.

nonparametric histogram in Y as a function of the nine possible joint relationships in U_1 and U_2. Because the hidden fields are decoupled, four of the nine joint relationships are inadmissible (shown as shaded, in the figure), and are modeled as uniform, with a low marginal probability.

The sequential estimation from U to X has the further benefit of determining U in detail, at the pixel level, when estimating X, avoiding the ambiguity of an "undecided" (state $\frac{1}{2}$) hidden state with ambiguous assertions on X.

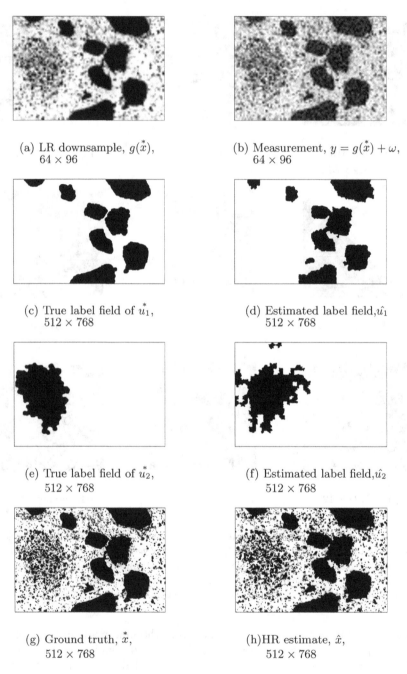

(a) LR downsample, $g(\overset{*}{x})$,
    64×96

(b) Measurement, $y = g(\overset{*}{x}) + \omega$,
    64×96

(c) True label field of $\overset{*}{u_1}$,
    512×768

(d) Estimated label field, $\hat{u_1}$
    512×768

(e) True label field of $\overset{*}{u_2}$,
    512×768

(f) Estimated label field, $\hat{u_2}$
    512×768

(g) Ground truth, $\overset{*}{x}$,
    512×768

(h) HR estimate, $\hat{x}$,
    512×768

Fig. 9. Ground truth and reconstruction results for a real porous media image. Although $\overset{*}{x}$ is not able to fully reconstruct some subtle structures (eg., the connectivities at the interface between the large pores), the improvement in relevant detail of (h) over (b) is stunning. Here, $k_m = 3$, $M_x = 3$, $k_d = 5$, and $M_u = 6$.

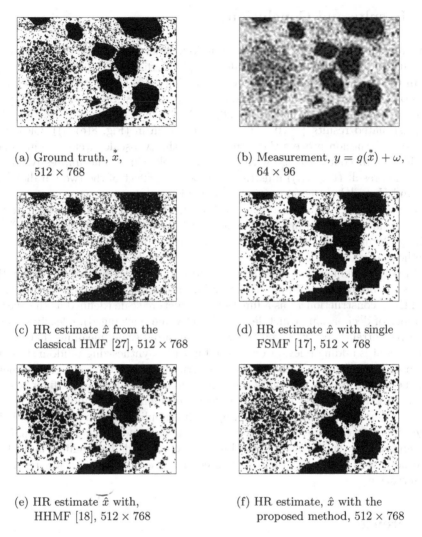

(a) Ground truth, $\overset{*}{x}$,
    512×768

(b) Measurement, $y = g(\overset{*}{x}) + \omega$,
    64×96

(c) HR estimate $\hat{x}$ from the
    classical HMF [27], 512×768

(d) HR estimate $\hat{x}$ with single
    FSMF [17], 512×768

(e) HR estimate $\widetilde{\hat{x}}$ with,
    HHMF [18], 512×768

(f) HR estimate, $\hat{x}$ with the
    proposed method, 512×768

Fig. 10. Reconstruct results for a real porous media image from different frameworks. Compared to the results from other methods (c)-(e), the reconstruction result $\hat{x}$ from our proposed method (f) shows an impressive improvement in modeling multiple spatial nonstationarities. In particular, observe closely the small-scale structures in the high dense regions and the boundaries of large pores.

The computational cost of our proposed approach can be approximated as

$$C_{PHHA} \simeq \sum_{k=0}^{M_u} (N_\nu + 1) \cdot \alpha_k \cdot (4^{-k} N^2) \tag{10}$$

where α_k denotes the fraction of unfrozen pixels at each scale k. Experimentally, C_{PHHA} remains dominated by the cost of the finest scale $O((N_\nu + 1) \cdot \alpha_0 \cdot N^2)$.

4   Results and Conclusions

Our results will be based on a synthetic four-label image (Fig. 3(f)) and a real porous media image with multi-scale, complex structure (Fig. 1(c)).

In the synthetic example we have two hidden fields U_1, U_2, which respectively describe the states of circle-size and shade. For the purpose of estimating X, we will assume U_1, U_2 to be known (Fig. 8(c,d)). Given a LR noisy image (Fig. 8(b)), our estimated results $\{\hat{x}^k, 0 \leq k \leq 4\}$ are shown in (Fig. 8(e)-(i)). Clearly, in the reconstruction process the structures of the two-scale circles are gradually decided from coarse to fine with different label values. The impressive reconstruction result (Fig. 8(i)) illustrates the positive effect of the two parallel fields U_1 and U_2 to label the nonstationary behaviours.

In the porous media example, a much more difficult problem, we estimate first U_1, U_2 and then X, as proposed. The final HR estimate $(\hat{x})$ is shown in Fig. 9. The performance of the proposed PHHF is clear from the comparison to other methods (Fig. 10(c-e)).

Finally, to demonstrate the the reconstruction is actually correctly estimating fine-scale details, Fig. 7 plots the correlation coefficient between ground truth and the reconstruction $\hat{x}^0$ as a function of structure scale (defined as the average number of decimations which leaves a pixel value unchanged). Clearly strongly positive correlations exist well below the measured scale, meaning that the enhancement is adding relevant detail, and not just synthesizing random structure from the prior model. Clearly very tiny structures fail to exert much influence on the measurements, therefore the correlation ρ does decrease at finer scales.

Our research interest is the hierarchical representation of multi-label hidden fields. Although here the proposed approach is applied only to porous media images, it can be extended to more general problems in modeling, analysis, and processing. Our intention is to move this research towards hierarchical multi-label synthesis, and the reconstruction of three-dimensional data from low-resolution observations.

References

1. Mallat, S.G.: A theory for multiresolution signal decomposition. IEEE Trans. Pattern Anal. Mach. Intell. 11, 674–693 (1989)
2. Crouse, M.S., Nowak, R.D., Baraniuk, R.G.: Wavelet-based statistical signal processing using hidden markov models. IEEE Trans. Signal Proc. 46, 886–902 (1998)
3. Choi, H., Baraniuk, R.: Multiscale image segmentation using wavelet-domain hidden markov models. IEEE Trans. Image Processing 10, 1309–1321 (2001)
4. Burt, B., Adelson, E.: The laplacian pyramid as a compact image code. IEEE Trans. Commun. 31, 532–540 (1983)
5. Olkkonen, H., Pesola, P.: Gaussian pyramid wavelet transform for multiresolution analysis of images. Graphical Models and Image Proceesing 58(32), 394–398 (1996)
6. Basseville, M., Benveniste, A., Chou, K.C., Golden, S.A., Nikoukhah, R., Willsky, A.S.: Modeling and estimation of multiresolution stochastic processes. IEEE Trans. Inform. Theory 38, 766–784 (1992)

7. Fieguth, P.: Hierarchical posterior sampling for images and random fields. IEEE Trans. Image Processing (accepted, 2008)
8. Han, C., Risser, E., Ramamoorthi, R., Grinspun, E.: Multiscale texture synthesis. ACM Trans. Graph 27(3), 51:1–51:8 (2008)
9. Chainais, P.: Infinitely divisible cascades to model the statistics of natural images. IEEE Trans. Pattern Anal. Mach. Intell. 29, 2015–2119 (2007)
10. Liao, I.Y., Petrou, M., Zhao, R.: A fractal-based relaxation algorithm for shape from terrain image. Computer Vision and Image Understanding 109, 227–243 (2008)
11. Li, S.: Markov Random Field Modeling in Image Analysis. Springer, Heidelberg (2001)
12. Chellappa, R., Jain, A.: Markov Random Fields: Theory and Applications. Academic Press, London (1993)
13. Graffigne, C., Heitz, F., Perez, P.: Hierchical markov random field models applied to image analysis: a review. In: SPIE, vol. 2568, pp. 2–17 (1994)
14. Kato, Z., Berthod, M., Zerubia, J.: A hierarchical markov random field model and multitemperature annealing for parallel image classificaion. Graphical Models and Image Proceesing 58(1), 18–37 (1996)
15. Laferté, J.M., Pérez, P., Heiz, F.: Discrete markov image modeling and inference on the quadtree. IEEE Trans. Image Processing 9(3), 390–404 (2001)
16. Mignotte, M., Collet, C., Pérez, P., Bouthemy, P.: Sonar image segmentation using an unsupervised hierarchical mrf model. IEEE Trans. Image Processing 9(7), 1216–1231 (2000)
17. Campaigne, W., Fieguth, P., Alexander, S.: Forzen-state hierarchical annealing. In: Campilho, A., Kamel, M.S. (eds.) ICIAR 2006. LNCS, vol. 4141, pp. 41–52. Springer, Heidelberg (2006)
18. Liu, Y., Fieguth, P.: Image resolution enhancement with hierarchical hidden fields. In: ICIAR 2009 (2009)
19. Alexander, S.K., Fieguth, P., Vrscay, E.R.: Hierarchical annealing for random image synthesis. In: Rangarajan, A., Figueiredo, M.A.T., Zerubia, J. (eds.) EMMCVPR 2003. LNCS, vol. 2683. Springer, Heidelberg (2003)
20. Benboudjema, D., Pieczynski, W.: Unsupervised statistical segmentation of non-stationary images using triplet markov fields. IEEE Trans. Pattern Anal. Mach. Intell. 29(8), 1367–1378 (2007)
21. Marroquin, J.L., Santana, E.A., Botello, S.: Hidden markov measure field models for image segmentation. IEEE Trans. Pattern Anal. Mach. Intell. 25(11), 1380–1387 (2003)
22. Rivera, M., Gee, J.C.: Twoclevel mrf models for image restoration and segmentation. In: BMVC 2004, pp. 809–818 (2004)
23. Liu, Y., Mohebi, A., Fieguth, P.: Modeling of multiscale porous media using multiple markov random fields. 4th Biot (June 2009)
24. Scarpa, G., Haindl, M., Zerubia, J.: A hierarchical texture model for unsupervised segmentation of remotely sensed images. In: Ersbøll, B.K., Pedersen, K.S. (eds.) SCIA 2007. LNCS, vol. 4522, pp. 303–312. Springer, Heidelberg (2007)
25. Geman, S., Geman, D.: Stochastic relaxation, gibbs distributions, and the bayesian restoration of images. IEEE Trans. Pattern Anal. Mach. Intell. 6(6), 721–741 (1984)
26. Besag, J.: On the statistical analysis of dirty pictures. J. Roy. Statist. Soc. Ser. B 48(3), 256–302 (1986)
27. Mohebi, A., Fieguth, P.: Posterior sampling of scientific images. In: Campilho, A., Kamel, M.S. (eds.) ICIAR 2006. LNCS, vol. 4141, pp. 339–350. Springer, Heidelberg (2006)

General Search Algorithms for Energy Minimization Problems

Dmitrij Schlesinger

Dresden University of Technology*

Abstract. We describe a scheme for solving Energy Minimization problems, which is based on the A^* algorithm accomplished with appropriately chosen LP-relaxations as heuristic functions. The proposed scheme is quite general and therefore can not be applied directly for real computer vision tasks. It is rather a framework, which allows to study some properties of Energy Minimization tasks and related LP-relaxations. However, it is possible to simplify it in such a way, that it can be used as a stop criterion for LP based iterative algorithms. Its main advantage is that it is exact – i.e. it never produces a discrete solution that is not globally optimal. In practice it is often able to find the optimal discrete solution even if the used LP-solver does not reach the global optimum of the corresponding LP-relaxation. Consequently, for many Energy Minimization problems it is not necessary to solve the corresponding LP-relaxations exactly.

1 Introduction

This work is motivated mainly by the following observations.

At the time there are many algorithms for approximate solutions of Energy Minimization problems, based on LP-relaxation techniques [9,7,6,10,8,5]. However, strongly polynomial algorithms for resulting LP problems are still unknown. Therefore, in practice it is necessary to use approximations, i.e. the results of an iterative procedure, which are reached after finite time. Even if it would be possible to solve LP-relaxations exactly, it is not known in general, whether there is a gap between the optimal relaxed solution and optimal discrete solution of the initial Energy Minimization task. Even if it is known that there is no gap, it is in general not clear, how to obtain the optimal discrete solution, given the optimal relaxed one. To answer the latter two questions it is necessary to solve a Constraint Satisfaction Problem that is NP-complete by itself. Despite of these theoretical drawbacks, the LP-relaxation methods became extremely popular, because in practice it is almost always possible to extract good results from (possibly suboptimal) continuous solutions in an heuristic but reasonable way. Unfortunately, much less papers study relationships between different relaxations of underlying discrete optimization problems as well as relationships between discrete problems and corresponding relaxations. Summarizing, there are many open questions in the scope of LP-relaxation techniques, if they are applied to Energy Minimization problems. Especially the question of strongly polynomial solvability of LP-relaxation is in a certain sense not natural, because this question relates solely to linear programming

---

* Supported by Deutsche Forschungsgemeinschaft, Grant No. FL307/2-1.

D. Cremers et al. (Eds.): EMMCVPR 2009, LNCS 5681, pp. 84–97, 2009.

techniques – i.e. it has in principle nothing in common with the initial discrete optimization problem.

On the other hand there are much less algorithms for Energy Minimization, which have a "discrete nature". A classical example is Dynamic Programming [1]. This algorithm does not relate to any continuous optimization. It just performs a predefined number of operations and gives the solution. Unfortunately, this approach is not suitable for many computer vision problems, because it is applicable only for simple graphs. Another group of algorithms can be characterized as an iterative search – they consider a subset of labelings in each iteration. The simplest method of such kind is Iterated Conditional Modes [2], more elaborated techniques are e.g. α-expansion, α/β-swap [3] and their derivatives. It is noteworthy that these algorithms use MinCut based techniques (which are again indirectly based on LP-relaxations) in order to perform an elementary search step. Unfortunately, here it is often difficult to state, for which tasks these approaches give global optimal solution, what is the precision for general Energy Minimization tasks etc. Summarizing, the branch of research, which is formed by discrete optimization techniques, seems to be neglected (at least in the scope of computer vision problems).

We propose to consider Energy Minimization problems primarily as *discrete* optimization problems and to apply general search techniques to them. In particular, we describe in this paper a scheme, based on the A^* algorithm [4]. Appropriately chosen LP-relaxations are used thereby as heuristic functions. This idea by itself is not new for discrete optimization in general. However, to our knowledge, it was not considered before taking into account special properties of Energy Minimization problems, which result from typical computer vision tasks (like e.g. segmentation, stereo-reconstruction etc.). The main specific here is that it is not possible as a rule to solve the corresponding LP-relaxations exactly due to a very large number of variables and constraints.

The proposed scheme is quite general and therefore can not be applied directly for real computer vision tasks. It is rather a framework, which allows to study some properties of Energy Minimization tasks and related LP-relaxations. However, it is possible to simplify it in such a way, that it can be used as a stop criterion for LP based iterative algorithms (such as Message Passing, the Subgradient method etc.). Its main advantage is that it is exact – i.e. it gives the globally optimal discrete solution for some tasks. For other problems it never stops (e.g. if there is an essential gap between the values of the best discrete solution and the global optimum of the corresponding LP-relaxation). In other words, it never produces a solution that is not optimal. The other important property is that it is often able to find the optimal discrete solution even if the used LP-solver does not reach the global optimum of the corresponding LP-relaxation. To conclude, for many Energy Minimization problems it is not necessary to solve the corresponding LP-relaxations exactly in order to get the globally optimal discrete solution.

2   Notations and Definitions

Problem statement. For simplicity we consider Energy Minimization tasks (MinSum problems) of second order. Let $G = (R, E)$ be a graph (also called the problem graph) with the node set R, a particular node is referred as $r \in R$, the edge set is $E = \{\{r, r'\}\}$.

Let K be a finite label set and $f : R \to K$ be a labeling, i.e. $f(r)$ denotes a label $k \in K$ chosen by the labeling f in a node $r \in R$. Let $q_r : K \to \mathbb{R}$ and $g_{rr'} : K \times K \to \mathbb{R}$ be quality functions, which assign real values to the labels in each node and to the label pairs in each edge respectively. The quality of a labeling is the sum

$$Q(f) = \sum_{r \in R} q_r\big(f(r)\big) + \sum_{\{rr'\} \in E} g_{rr'}\big(f(r), f(r')\big), \tag{1}$$

the task is to find

$$(\arg) \min_f Q(f) =$$

$$= (\arg) \min_f \left[\sum_{r \in R} q_r\big(f(r)\big) + \sum_{\{rr'\} \in E} g_{rr'}\big(f(r), f(r')\big) \right]. \tag{2}$$

In addition we will need the following notations. Let us denote by $G' = (R', E')$ the subgraph induced by a node subset $R' \subset R$, i.e. $E' = \{\{r, r'\} \mid \{r, r'\} \in E, r, r' \in R'\}$. A *partial labeling* $f_i : R' \to K$ is a restriction of f into the subset R'. The index will always depend on the context, determining in each particular case, what subset R' is meant, which labeling on it is considered etc. For a partial labeling $f_i : R' \to K$ and a node $r \in R'$ we will say, that the partial labeling *covers* this node. The quality $Q(f_i)$ of a partial labeling is the same as (1) but summed over the corresponding subgraph G' instead of the whole problem graph G. A *subtask* $\mathcal{A}(f_i)$ induced by a partial labeling f_i is a MinSum problem, that is built from the original one (2) by fixation of the labels in those nodes, which are covered by f_i, that is

$$(\arg) \min_{f_j} \left[Q(f_j) + \sum_{\substack{\{rr'\} \in E \\ r \in R', r' \in R/R'}} g_{rr'}\big(f_i(r), f_j(r')\big) \right], \tag{3}$$

where $f_j : R/R' \to K$ are partial labelings, which cover the complement R/R' (thereby we omit the constant part $Q(f_i)$).

LP-relaxation. A common approach to cope with energy minimization tasks (2) is based on LP-relaxation technique. The first step is to introduce integer weights $\lambda(\cdot) \in \{0, 1\}$ for each pair (r, k) as well as for each triple $(\{rr'\}, k, k')$ and to rewrite (2) in the form[1]

$$\sum_r \sum_k q_r(k)\lambda_r(k) + \sum_{rr'} \sum_{kk'} g_{rr'}(k, k')\lambda_{rr'}(k, k') \to \min_\lambda$$

s.t.

$$\sum_k \lambda_r(k) = 1 \quad \forall r$$

$$\sum_{k'} \lambda_{rr'}(k, k') = \lambda_r(k) \quad \forall r, r', k$$

$$\lambda_r(k), \lambda_{rr'}(k, k') \in \{0, 1\}. \tag{4}$$

---

[1] Sometimes we omit brackets and "$\in$" relations for readability.

Then the last condition is relaxed, i.e. substituted by $\lambda(\cdot) \in [0,1]$, which gives a task of linear programming.

We prefer to consider the task (4) using the notation of equivalent transformations. They are defined as a superpositions of following operations:

$$\hat{q}_r(k) = q_r(k) + \delta, \quad \hat{g}_{rr'}(k,k') = g_{rr'}(k,k') - \delta \;\; \forall k' \tag{5}$$

with an arbitrary finite constant δ. The new functions $\hat{q}_r$ and $\hat{g}_{rr'}$ represent an equivalent task. The application of an equivalent transformation is also called reparametrization of a task. The dual of (4) can be formulated using these notations as follows:

$$\sum_r \min_k \hat{q}_r(k) + \sum_{rr'} \min_{kk'} \hat{g}_{rr'}(k,k') \to \max_{\hat{q},\hat{g}}, \tag{6}$$

where the functions $\hat{q}$ and $\hat{g}$ can be obtained by equivalent transformations from the original functions q and g. For a subtask $\mathcal{A}(f_i)$, we denote by $LP(f_i)$ the solution (the value) of its LP-relaxation (6).

General search algorithms, A^*

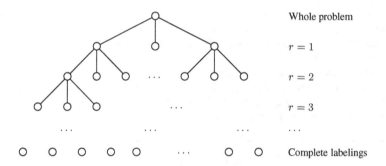

Fig. 1. Search tree

General search algorithms are described very extensive in the literature. Therefore we omit here formal definitions and just consider, how a MinSum task can be formulated as a general search problem. The main idea is to decompose the whole task into subtasks followed by recursive decompositions of these. Let the nodes of our MinSum problem be enumerated. We may decompose the whole problem by fixation of labels e.g. in the first node $r = 1$. In that way we obtain $|K|$ subproblems, each one having one node less than the initial task. Doing such decomposition recursively a search tree is built, which is presented in Fig. 1. Each layer (a particular depth value) corresponds to a node r of the problem graph. Each tree node in a layer r corresponds to a partial labeling $f_i : R' \to K$ with $R' = \{1, 2 \dots r\}$. According to this the subtree with the root f_i represents the subtask $\mathcal{A}(f_i)$. The root of the tree represents the whole problem, the leaves correspond to complete labelings. The quality of a node f_i is the quality $Q(f_i)$ of the corresponding partial labeling. The task is to find a leaf of minimal quality.

We will also need the following notation. A partial labeling f_j is called *successor* of another partial labeling f_i iff

$$f_i : R' \to K, \quad f_j : R' \cup \{r \notin R'\} \to K, \quad f_j(r) = f_i(r) \; \forall r \in R'. \qquad (7)$$

We will say that the partial labeling f_j is obtained from f_i by fixation of a label in a node $r \notin R'$. The set of all successors for a partial labeling f_i, which can be obtained in such a way, is denoted by $succ_r(f_i)$. At the same time it denotes the set of all children of a tree node f_i in the search tree.

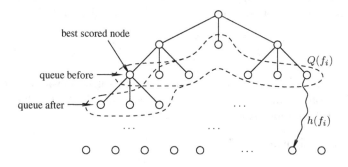

```
Put the root into the queue
Repeat {
    Take the best scored node from the queue
    if it is a leaf
        return it as the solution
    compute all successors
    score them according to Q(fᵢ) + h(fᵢ)
    put them into the queue
}
```

Fig. 2. A^* algorithm

A common approach for general search problems is to use heuristic search techniques to evaluate the search tree in order to reach a leaf of minimal quality. In particular we will use the A^* algorithm. Its functionality is illustrated in Fig. 2. In each step there is a subset of tree nodes (called queue), which are not evaluated so far. At the very beginning it consists only of root of the tree. Each node in the queue is scored according to the *evaluation function*, which is the sum of the node quality and an *heuristic function* (denoted by $h(f_i)$). The latter is an estimator for the difference between the quality of the node and the best leaf in the corresponding subtree. In each step the algorithm takes the node with the best score from the queue. If it is a leaf, the algorithm finishes and returns this node as the solution. Otherwise, the node is *expanded*, i.e. all successors (children of this node in the search tree) are computed. They are scored as described above and put into the queue.

The properties of A^* are determined primarily by the type of the used heuristic function. An heuristic function is called *admissible* (optimistic), if it never overestimates the real difference between the qualities of the node and the best leaf in the corresponding subtree:

$$h(f_i) \leq \underset{f \in subtree(f_i)}{\arg\min} \ Q(f) - Q(f_i) = h^*(f_i). \tag{8}$$

An heuristic is called *monotone* if

$$Q(f_i) + h(f_i) \leq Q(f_j) + h(f_j) \quad f_j \in succ(f_i) \tag{9}$$

holds, i.e. the heuristic is the more precisely (and therefore the less optimistic) the closer to the leaves. The A^* algorithm is exact (always finds a global solution) if the heuristic function is admissible[2].

For the complexity analysis of A^* it is usually assumed that the complexity of computation for $Q(f_i)$ and $h(f_i)$ is a constant. Let n be the maximal number of tree nodes in the queue during the search. The time complexity of A^* is then $O(n \log n)$. Alternatively, the number of expansions can be considered, which are necessary to solve the problem. In this case it is assumed that the time complexity of one expansion step (one loop iteration in Fig. 2) is a constant. In both variants the crucial question is, how the considered numbers (n or the number of expansions) depend on the problem size, for example on the number of nodes $|R|$ in the problem graph. To be precise, we will use the notation "time complexity" having in mind the number of expansions depending on the problem graph size.

The time complexity of the A^* algorithm is exponential in general. However, if the heuristic function is admissible and monotone, then a complexity analysis is possible, that takes into account the precision of the heuristic function. In particular, A^* has linear time complexity, if the heuristic function is exact (denoted by $h^*(f_i)$ in (8)). It means that the A^* algorithm behaves like the depth-first search, i.e. a tree node of maximal depth is always chosen for expansion. In this case the maximal number of nodes in the queue is $O(|R||K|)$ and exactly $|R|$ expansions should be performed in order to reach an optimal leaf.

3 General Scheme

The main idea of our approach is to use LP-relaxation $LP(f_i)$ as the heuristic function $h(f_i)$ for the A^* search. First of all, is should be stated that this is an admissible heuristic, since the LP-relaxation of a discrete optimization task represents a lower bound for its solution. Therefore A^* is exact. Secondly, it can be easily seen that this heuristic is also monotone. The key point to show this is to note that for each pair f_i and $f_j \in succ_r(f_i)$

$$Q(f_j) + LP(f_j) = Q(f_i) + LP'(f_i) \geq Q(f_i) + LP(f_i) \tag{10}$$

holds, where $LP'(f_i)$ is the same as $LP(f_i)$ but with additional constraints for λ (see (4)), which fix the label $k = f_j(r)$ in the node r – i.e. both LP-relaxations have the

---

[2] In general, the heuristic must be both admissible and monotone to guarantee optimality. In our case it is enough that it is admissible only, because the set of leaves is finite.

same objective function but in $LP'(f_i)$ the set of feasible solutions is a proper subset of those in $LP(f_i)$.

We call a discrete task LP-solvable, if there is no gap between the values of its solution and its LP-relaxation. We call a task strictly LP-solvable, if the same holds for all partial labelings, which may be needed during the evaluation of the search tree[3]. It is obvious that in the latter case A^* has linear time complexity, since the heuristic function becomes exact. It is also clear that non strict LP-solvability does not guarantee linear complexity in general.

On the other hand, it is easy to see that strict LP-solvability is only a sufficient condition for linear complexity, but not a necessary one. Even if there is a gap (the task is not LP-solvable), A^* may have linear complexity, if this gap is relatively small – i.e. if it is not big enough to prefer partial labelings of non maximal depth during the search. A weaker sufficient condition can be formulated from A^* itself as follows. The A^* algorithm has linear time complexity if there exist a sequence of partial labelings $f_1, f_2 \ldots f, f_i : \{1 \ldots i\} \to K$, $f_{i+1} \in succ_{i+1}(f_i)$, so that

$$Q(f_i) + LP(f_i) < Q(f_l) + LP(f_l) \quad \forall f_l \in succ(f_j)/f_{j+1}, j < i \qquad (11)$$

holds for each f_i in the sequence. But even this is not a necessary condition. If the above inequality is satisfied non strictly, A^* may behave as depth-first search as well, because it may find the "right" sequence of partial labelings by chance.

4 A^* Based Stop Criterion for LP-Solvers

Let us remember that it is not possible to solve the needed LP-relaxations exactly and efficiently for real tasks. Known LP-solvers are iterative algorithms, that only approach the global optimum. Consequently, the optimal solution is guaranteed only in infinite time. Therefore we are constrained to use approximations, i.e. the results of these algorithms after a finite time. Let us denote these approximations by $\tilde{LP}(f_i)$. In this case we can not guarantee the monotonicity of the heuristic anymore. However, it remains admissible, if methods based on maximization of the dual energy are used (in this case $\tilde{LP}(f_i) \le LP(f_i)$ holds, since these methods try to maximize the lower bound for optimal energy). Therefore A^* remains exact. It makes in principle possible to use these methods with A^* to search for optimal discrete solution.

Let us consider again the behavior of A^* but from a slightly other point of view, taking into account that we use approximations $\tilde{LP}(f_i)$ instead of the exact values $LP(f_i)$. Assume a task for which A^* does not behave like depth-first search, i.e. at some stage it chooses for expansion a tree node f_i, that has non maximal depth – the condition (11) is not satisfied. Obviously, it can not happen, if the task is strictly LP-solvable and LP-relaxations are computed exactly. Consequently, there are only two possible reasons for such a behavior: either the task is not strictly LP-solvable or the approximations $\tilde{LP}(f_i)$ are too far from the corresponding $LP(f_i)$. Suppose, it is known that the task is strictly LP-solvable. Then only the second possibility remains – i.e. for some f_i used during the search the corresponding approximation $\tilde{LP}(f_i)$ was computed too coarsely.

---

[3] For example submodular tasks have this property.

There are basically two ways to proceed further in this situation. The first one is to allow the search to deviate from the depth-first like behavior, i.e. proceed A^* further using current (coarse) approximations $\tilde{LP}(f_i)$. In this case however, the search procedure often starts to massively expand tree nodes of non maximal depth (note that the heuristics $h(f)$ for leaves are zero, i.e. exact per definition). In short, we can not forecast the further behavior of A^* – i.e. it often turns into a complete search.

The second way is to attempt to improve the current approximations $\tilde{LP}(f_i)$ in order to preserve the depth-first like behavior of the search. Unfortunately, this idea can not be used directly, because we can detect the necessity to improve these approximations only *after* the search was already performed (the approximations $\tilde{LP}(f_i)$ were already computed). Nevertheless, we would like to discuss this idea in a little bit more detail. Let us imagine, that we have a hypothetical algorithm at our disposition, which has the following properties. First, it is an iterative procedure that maximizes the lower bound of the energy – i.e. it is something like a usual LP-solver (e.g. Diffusion or Message Passing etc.). Second, besides of the lower bound, it provides some kind of information, that allows to easily compute approximations $\tilde{LP}(f_i)$ for all partial labelings which may be needed during the A^* search – i.e. without to solve the LP-relaxations for each partial labeling separately. Furthermore, let us assume, that the algorithm improves during its work this additional information in a similar way as it improves the lower bound for the energy. In other words, the additional information allows to compute $\tilde{LP}(f_i)$ the more precisely, the more time is spent. Then we could use the idea to detect the necessity of an improvement for $\tilde{LP}(f_i)$ as follows. We run this hypothetical procedure and let it compute all information, that may be needed for estimation of $\tilde{LP}(f_i)$ during the search. Once a while we call the A^* search and check, whether it behaves like the depth-first search. If yes, then we have found the optimal discrete solution. Otherwise we let this hypothetical procedure work further to improve the lower bound as well as the additional information. Summarizing, the depth-first like behavior of the A^* algorithm can serve as a stop criterion for LP-solvers, which have the above mentioned properties.

Unfortunately, we have no such hypothetical LP-solver at our disposition. However, it is possible to to adapt existing algorithms in order to compute the necessary additional information approximately. Let us consider again the task of the maximization of the dual energy (6). In particular, let $\hat{q}$ and $\hat{g}$ be quality functions obtained by an iterative LP-solver from the original q and g at some stage of its work. Note, that we can omit without loss of generality the node terms $\hat{q}_r$, because they can be always set to zero by an equivalent transformation (5) with $\delta = -\hat{q}_r(k)$. Secondly, for a particular instance of $\hat{g}$ in (6) we can always subtract a constant from each $\hat{g}_{rr'}$. In doing so we change both the initial discrete task (qualities of all labelings) and the value of (6) by the same number. When subtracting the minimal value from each $\hat{g}_{rr'}$ the following holds. The current value of the LP-relaxation (the current lower bound) is zero. All functions $\hat{g}_{rr'}$ are nonnegative, all functions $\hat{q}_r$ are zero, the value of the best labeling is therefore nonnegative as well. After such a normalization the numbers

$$marg(r, k) = \sum_{r':\{rr'\}\in E} \min_{k'} \hat{g}_{rr'}(k, k') \qquad (12)$$

can serve as an approximation for min-marginals for a node r and a label k. The current value of the heuristic $h(f_i)$ for a partial labeling $f_i : R' \to K$ can be approximated by

$$h(f_i) = \sum_{\substack{\{rr'\}\in E \\ r\in R', r'\in R/R'}} \min_k \hat{g}_{rr'}(f_i(r), k), \qquad (13)$$

i.e. it is the same as min-marginals (12) but accumulated only over those edges, which connect nodes from the fixed part R' with nodes from its complement. It is easy to see that this heuristic does not represent the needed LP-relaxations for all partial labelings. However, the values computed by (13) are guaranteed not greater and therefore represent an admissible heuristic. Moreover, it can be easily seen that this heuristic is monotone as well.

```
f* is the current best partial labeling,
    set it empty (no labels are fixed)
next_best is the value of the evaluation function for the
    best tree node not expanded so far, despite of f*,
    set next_best = ∞

For each node r {
    compute the set of successors S = {f_i, f_i ∈ succ_r(f*)}
    and their evaluation functions e(f_i) = Q(f_i) + h(f_i);
    extend f* to the best successor: f* = arg min e(f_i);
                                           f_i∈S
    if e(f*) > next_best
        return "solution not found"

    choose f' as the next best successor: f' = arg min    e(f_i);
                                               f_i∈S,f_i≠f*
    if e(f') < next_best
        next_best = e(f')
}
return f* as the best solution
```

Fig. 3. A^* based stop criterion

Summarized, the A^* based stop criterion is presented in Fig. 3. Note, that it is not necessary to store all partial labelings in the queue in order to detect the depth-first like behavior. Only the value of the evaluation function for the best "second pretender" is needed (denoted by $next\_best$ in Fig. 3). It reduces the time complexity of the A^* search from $O(n \log n)$ (as in the general case) to $O(n)$, where n is the number of partial labelings needed during the search, i.e. the maximal number of nodes in the search tree. If the values $\min_{k'} \hat{g}_{rr'}(k, k')$ are precomputed in advance[4] for each $r < r'$ and k, the computation of (13) for a node r can be done in a time, that is proportional to

---

[4] In fact, the solvers compute these numbers in order to perform equivalent transformations. Therefore, they should be only stored for further use by the stop criterion.

the number of edges, which are incident to r. Summarizing, the overall time complexity of the criterion is $O(|E||K|)$.

We would like to stress that this stop criterion does not indicate the exact solution of the corresponding LP-relaxation. It detects only the possibility "to extract" the global optimal discrete solution from the current state of a continuous optimization procedure. In the case the solution was found, it is not known in general, whether the task is strictly LP-solvable or not, whether there is a gap (the task is not LP-solvable), whether the used LP-solver reached the global optimum of the LP-relaxation. But in this case it is not necessary to answer these questions, because the discrete solution was already found. Summarizing, this approach gets rid (to some extent) of the necessity to look for algorithms for exact solution of LP-relaxation.

The interpretation of the non depth-first like behavior of A^* depends on the additional knowledge about the task and/or the used LP-solver. If it is known that the task is strictly LP-solvable, then the non depth-first like behavior indicates that the LP-solver has not reached the optimum so far or the coarsening (13) is too pure. If it is known that the solver is already in optimum, then the task is not strictly LP-solvable or the coarsening (13) is too pure again and so on.

In most practical cases we do not have such additional knowledge. It is then obviously possible that the whole loop (LP-solver with the A^* based stop criterion) never stops. If it is nevertheless necessary to find the exact solution of the initial discrete problem, it is necessary to perform the A^* search in a more general form – i.e. to allow it to have non depth-first like behavior.

5 Experiments

First of all, it is necessary to justify, what we would like to demonstrate by the experiments. The main goal is to show that the A^* based stop criterion gives (as a rule) the exact global discrete solution in the situation that the used LP-solver does not reach the global optimum of the corresponding LP-relaxation. It can be done e.g. by comparison of the qualities for the found discrete solution and the lower bound, found by the used LP-solver – in our case the latter is zero since the task is normalized as described above. The second goal is to examine, whether our stop criterion allows to reduce the time complexity of the used LP-solver compared with a "simple" stop criterion, which does not relate to the best discrete solution, but only detects the optimum of the LP-relaxation – i.e. whether it allows to stop LP-solver essentially earlier. We use the following procedure as such simple stop criterion. First, the approximations for min-marginals (12) are computed. A "seemingly good" labeling is produced by fixation of labels with best min-marginals in each node. If the quality of the labeling produced in such a way is zero, then this labeling is optimal and the algorithm reached the global optimum of the LP-relaxation for the whole task[5].

In our experiments we used the Diffusion algorithm [9,10] and the Subgradient method [7] for maximization of (6). In short, they both are iterative ones, that apply equivalent transformations (each one in its own manner) in order to maximize (6).

---

[5] Obviously, this is a sufficient but not necessary condition for the global optimum.

First, we tested the approach for problems, generated as follows. The problem graph corresponds to a grid with 4-connected neighborhood structure (128×128 for Diffusion and 64×64 for the Subgradient method). The values of the functions q_r are generated uniformly in interval $[0, 1]$. The functions $g_{rr'}$ are chosen as Potts model, i.e. $g_{rr'}(k, k') = \alpha \cdot \mathbb{1}(k \neq k')$. One experiment is organized in the following way. For a given task we call the LP-solver. After each iteration (some portion of work) we check the A^* based stop criterion. If a predefined number of iterations is done and the stop criterion does not produce the solution, we cancel the experiment and state that the solver does not converge for this task. If the criterion was successful (the optimal discrete solution was found), we note the number of iterations made so far. After that we let the solver work further until the simple stop criterion is satisfied. In doing so we note the number of iterations, which was necessary for the solver to converge after the best labeling was found by the A^* based stop criterion. Then we compute ratio of these two numbers of iterations. For certain combinations of the Potts parameter α and the number of labels $|K|$ we performed 10 experiments per combination and average the computed ratio over the successful experiments (i.e. solver converges). These ratios are summarized in Fig. 4. The values given after slash are the numbers of successful experiments for each combination. Empty cells indicate combinations of $|K|$ and α, for which the solvers never converged (either the problems with these combinations have as a rule an essential gap or the solvers were not able to narrow the global optimum of the LP-relaxation good enough in acceptable time).

In addition we would like to note the following. In the majority of the experiments (92.2% for Diffusion and 97.9% for the Subgradient method) the value of the best labeling found by our stop criterion was greater then zero. It confirms our main hypothesis – the criterion finds as a rule the optimal discrete labeling in the situation, that the used LP-solver does not reach the global optimum of the corresponding LP-relaxation. In such cases some additional iterations (together with the normalization) were necessary

$|K| \setminus \alpha$	0.3	0.4	0.5	0.6	0.7	0.8
2	.16/10	.04/10	.01/10	.06/10	.89/10	.94/10
3			.02/8	.03/10	.85/10	.71/10
4				.09/9	.20/9	.46/10
5				.07/8	.16/10	.34/10

(a) Diffusion (ratios are multiplied by 100)

$|K| \setminus \alpha$	0.3	0.4	0.5	0.6	0.7	0.8
2	.42/10	.46/10	.21/10	.07/10	.10/10	.05/10
3	.07/1	.75/6	.16/9	.04/10	.03/9	.03/10
4		.49/3	.36/9	.07/9	.18/10	.03/10
5		.44/1	.27/5	.05/9	.02/10	.02/9

(b) Subgradient method

Fig. 4. Average ratios for generated problems

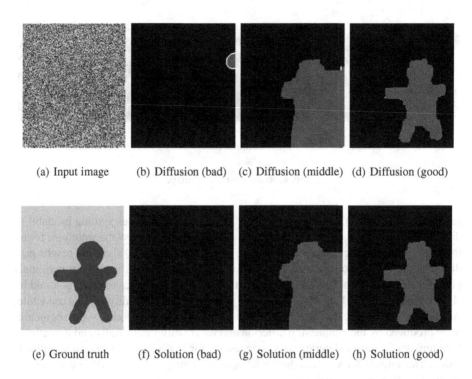

(a) Input image   (b) Diffusion (bad)   (c) Diffusion (middle)   (d) Diffusion (good)

(e) Ground truth   (f) Solution (bad)   (g) Solution (middle)   (h) Solution (good)

Fig. 5. Results for an artificial image

for simple stop criterion to find the optimal discrete solution of zero quality. There were also situation (6 times for Diffusion and 18 times for Subgradient), that the LP-solver did not converge in acceptable time after the best labeling was already found by the A^* based stop criterion. The typical situation for Diffusion was, that the value of the dual energy did not change at all for a very long time[6]. For the Subgradient method it was always the case that the dual energy changed, but so slowly that we were not able to wait. Sometimes the dual energy even decreased after the solution was found[7]. Comparing Diffusion and Subgradient, we observed that the A^* stop criterion has much more essential impact for Subgradient (sometimes more as 75% additional iterations were necessary) as for Diffusion (about a fraction of a percent).

The next experiment was made for an artificial image shown in Fig. 5. The input image in Fig. 5(a) was produced from the ground truth (Fig. 5(e)) by generation a gray-value in each pixel independently according to segment specific probability distributions. The resulting MinSum problem was a Potts model with negated logarithms of corresponding probability values for q_r. We performed experiments with Diffusion as the LP-solver for a "sequence" of models, starting from one with "bad" functions q_r (the corresponding probability distributions are almost the same and uniform for both

---

[6] Note, that Diffusion does not converge in general to the global optimum of the LP-relaxation.

[7] The Subgradient method converges to the global optimum of the LP-relaxation but not monotonously.

(a) Input image   (b) Solution (bad)   (c) Solution (good)

Fig. 6. Results for a real image

segments – Fig. 5(b,f)) and finishing with "good" one (the corresponding probability distributions are the true ones – Fig. 5(d,h)). Figures 5(f,g,h) show the solutions, found by the A^* based stop criterion in each stage. The figures 5(b,c,d) show the results produced by the simple stop criterion (labels with the best approximation of min-marginals (12) were chosen in each pixel independently) just after the exact solution is found by A^*. Those edges for which the chosen state pair has non zero quality are marked white. A more or less significant effect was observed mainly for bad models – the segmentations produced by the simple stop criterion and by A^* differ essentially and a relatively big number of additional iterations was necessary for Diffusion to find optimal segmentation. This impact decreases as the model becomes better – in Fig. 5(d) there are almost no edges with non zero qualities.

Finally, we tested our approach on real images. One example is presented in Fig. 6. The experiment was organized in the same manner, as for the previous one. Mixtures of multivariate Gaussians were chosen as the probability distributions of colors for segments. This example just demonstrates that the A^* based stop criterion is able to work with real images as well.

6   Conclusion

In this work we discussed, how general search techniques can be applied for Energy Minimization problems. We presented a scheme, which is based on the A^* algorithm accomplished with appropriate chosen LP-relaxations as heuristic functions. Based on it we derived a stop criterion for iterative LP-solvers, which is often able to find the global optimal discrete solution even if the used LP-solver does not reach the global optimum of the corresponding LP-relaxation. Summarizing, for many Energy Minimization problems it is not necessary to solve the corresponding LP-relaxations exactly.

This work is a first trial at most. Consequently, there are many open questions. Here we would like to mention only some of them. Obviously, the general search scheme can be built in different ways. In this paper we used a simple enumeration of nodes – i.e. the search tree is fixed in advance. Of course, other variants are possible as well. A related topic is that our construction does not take into account the structure of the problem graph. Especially for computer vision tasks it would be profitable to account for it, because here graphs are often very sparse. In the paper we considered an "hypothetical LP-solver" that has certain properties, which allow to use it with the A^* based stop criterion. As we do not really have such a solver, it was necessary to coarsen the needed estimations. Even with this coarsening the algorithm performs well. However, a real LP-solver with the necessary properties would obviously further improve it.

References

1. Bellman, R.E., Dreyfus, S.E.: Applied Dynamic Programming. Princeton University Press, Princeton (1962)
2. Besag, J.: On the statistical analysis of dirty pictures (with discussion). Journal of the Royal Statistical Society, Series B 48(3), 259–302 (1986)
3. Boykov, Y., Veksler, O., Zabih, R.: Fast approximate energy minimization via graph cuts. In: ICCV, pp. 377–384 (1999)
4. Hart, P.E., Nilsson, N.J., Raphael, B.: A formal basis for the heuristic determination of minimum cost paths. IEEE Transactions on Systems Science and Cybernetics, 100–107 (1968)
5. Kohli, P., Shekhovtsov, A., Rother, C., Kolmogorov, V., Torr, P.: On partial optimality in multi-label MRFs. In: McCallum, A., Roweis, S. (eds.) Proceedings of the 25th Annual International Conference on Machine Learning (ICML 2008), pp. 480–487. Omnipress (2008)
6. Kolmogorov, V.: Convergent tree-reweighted message passing for energy minimization. IEEE Transactions on Pattern Analysis and Machine Intelligence (PAMI) 28(10), 1568–1583 (2006)
7. Schlesinger, M.I., Giginyak, V.V.: Solution to structural recognition (max,+)-problems by their equivalent transformations. Control Systems and Computers (1,2) (2007)
8. Komodakis, N., Paragios, N., Tziritas, G.: Mrf optimization via dual decomposition: Message-passing revisited. In: ICCV, pp. 1–8 (2007)
9. Schlesinger, M.I.: Mathematical Methods of Image Processing. Naukova Dumka, Kiev (1989)
10. Werner, T.: A linear programming approach to max-sum problem: A review. Technical Report CTU–CMP–2005–25, Center for Machine Perception, K13133 FEE Czech Technical University (December 2005)

Complex Diffusion on Scalar and Vector Valued Image Graphs*

Dohyung Seo[1] and Baba C. Vemuri[2]

[1] Department of Electrical and Computer Engineering (ECE)
University of Florida, USA
dhseo@ufl.edu
[2] Department of Computer and Information Science and Engineering (CISE)
University of Florida, USA
vemuri@cise.ufl.edu

Abstract. Complex diffusion was introduced in the image processing literature as a means to achieve simultaneous denoising and enhancement of scalar valued images. In this paper, we present a novel geometric framework to achieve complex diffusion for color images represented by image graphs. In this framework, we develop a novel variational formulation that involves a modified harmonic map functional and is quite distinct from the Polyakov action described by Sochen et al. Our formulation provides a novel framework for simultaneous feature preserving denoising and enhancement. We also develop a quaternionic diffusion that can be applied to color image data represented by a quaternion in the image graph framework. In this framework, the real and imaginary parts can be interpreted as low and high-pass filtered data respectively. Finally, we suggest novel ways to use the imaginary part of complex diffusion toward image reconstruction. We present results of comparison between the complex diffusion, quaternionic diffusion and the well known Beltrami flow in the image graph framework.

1   Introduction

Image denoising is a quintessential component of most image analysis tasks and there are numerous denoising methods reported in the literature. In the past few decades, methods based on partial differential equations (PDEs) have become very popular. Some of the PDE-based methods are derived from minimization principles while others are not. The general mathematical form of a feature preserving anisotropic diffusion is given by

$$\frac{\partial I}{\partial t} = Div(g(|\nabla u|)\nabla u)$$

Here, $u(x, y; t)|_{t=0} = I(x, y)$ is the function being smoothed. The choice of $g(|\nabla u|)$ in the above leads to various types of diffusion flows.

---

* This research was supported in part by the NIH grant NS46812.

D. Cremers et al. (Eds.): EMMCVPR 2009, LNCS 5681, pp. 98–111, 2009.

Alternatively, one may represent the 2D image as a graph by embedding it in R^3, as a surface Σ with local coordinates (σ^1, σ^2). The embedding map X is given by, $X : (\sigma^1, \sigma^2) \rightarrow (x, y, I(x,y))$. This provides a geometric interpretation to the PDEs as those that modify some geometric property such as area of the 2D manifold representing the image surface. In the case of vector-valued images, the embedding map X is given by, $X : (\sigma^1, \sigma^2) \rightarrow (x, y, I^i(\sigma^1, \sigma^2))$, where, $I^i(x,y)$ are the channels of the given vector-valued image, and the $2+i$ dimensional manifold, $(x, y, I^i(\sigma^1, \sigma^2))$ is refered to as the space-feature manifold, M [2]. This graph representation also provides a geometric way to handle the interaction between the components (channels) of the vector-valued images. Kimmel et al., [1,2,3] pioneered the use of image graph representation to perform image smoothing in scalar and vector-valued image data sets. They introduced the Polyakov action [4] to derive various flows such as the Beltrami, mean curvature, and the Perona-Malik flows. One of the benefits of this approach is that the channels in multi-channel (vector-valued) images such as color images can be correlated in a geometrical way. Diffusing the RGB channels in a color image while retaining their correlation is essential. If we perform isotropic or anisotropic diffusion of each channel independently, all correlations are ignored and the solution would be erroneous.

Alternatively, one may simply extend the traditional diffusion to the complex-domain, which was pioneered by Gilboa et al. [5,6,7]. In complex diffusion, an image, $I(x,y)$, which is a real-valued function in general, is extended to the complex domain, i.e., $I(x,y) = I_R(x,y) + iI_M(x,y)$. Then, the isotropic diffusion equation is generalized to, $I = C\triangle I$ where, C is a complex number with unit norm $e^{i\theta}$, and $\triangle$ is defined as usual by $\frac{\partial^2}{\partial x^2} + \frac{\partial^2}{\partial y^2}$.

More generally, diffusion equations are given by

$$\frac{\partial I}{\partial t} = H(t)I \tag{1}$$

where $H(t)$ is a diffusion operator which can be either isotropic or anisotropic and can produce linear or nonlinear scale-spaces respectively. In the case of complex diffusion, $H(t)$ is a complex operator and can be rewritten as follows:

$$H(t) = H_R(t) + iH_M(t) = e^{i\theta}h(t) \tag{2}$$

where $h(t)$ is a real-valued operator. Then, the diffusion equations for the real and imaginary parts are given by

$$\frac{\partial I_R}{\partial t} = \cos(\theta)h(t)I_R - \sin(\theta)h(t)I_M \tag{3}$$

$$\frac{\partial I_M}{\partial t} = \sin(\theta)h(t)I_R + \cos(\theta)h(t)I_M. \tag{4}$$

The processed input image is considered as a solution to Eq.(3). In the case of isotropic diffusion, $h(t)$ becomes the $\triangle$ operator. Gilboa et al. showed that small positive values of θ lead to approximating the real part of I by the regular isotropic diffusion and the imaginary part by the smoothed second derivative of the real part. This allows one to achieve denoising and enhancement simultaneously. Therefore, regular (non-complex) diffusions discussed in the previous paragraphs can be seen as special cases of complex diffusions. Because the imaginary part represents the smoothed second derivative of the real part, the imaginary part contains the edge information of the real part. They applied this aspect of the imaginary part to denoise and enhance the images. For the task of denoising, they introduced a new anisotropic diffusion by replacing $|\nabla u|$ in $g(|\nabla u|)$ of Perona-Malik diffusion with that of the imaginary part as follows:

$$\frac{\partial I}{\partial t} = \nabla \cdot \left(\frac{e^{i\theta}}{1 + \left(\frac{I_M}{k\theta}\right)^2} \nabla I \right) \tag{5}$$

where I_M is the imaginary part of the complex image I and k is a threshold parameter. This allowed the diffusion flow to avoid the stair-casing effect. They also introduced a shock filter which used the imaginary part as edge information. However, they did not apply the complex diffusion model to multi-channel images and did not suggest a method to account for the coupling of the channels. *In this paper, we present a novel model for simultaneous smoothing and enhancement by mapping the real and complex channels to $\boldsymbol{C}^n$, introducing an image-surface metric and constructing an action functional distinct from the Polyakov action in [1].* In our approach, the correlation between the color channels (R, G and B) is introduced via the metric on the image graph manifold. Additionally, we applied our action functional to the quaternion representation of a color image in the graph representation. Liu et al.[8] have suggested a way to treat the color channels as a quaternion assuming that the R,G, and B channels were correlated through the quaternion algebra. In this approach, the R,G, and B were mapped to the pure quaternion parts with one extra dimension, which was the real part of the quaternion representation.

We present several experimental results depicting the performance of our model in comparison to the complex diffusion model of Gilboa et al. [5], for the scalar image denoising and enhancement case as well as with the Beltrami flow [2] for color image denoising. The rest of this paper is organized as follows. In Section 2, we present a novel metric for the complex image manifold and a novel functional whose minimization yields the desired flow equation. This is followed by a description of the quaternion representation for color images, a novel formulation of the functional and the accompanying flow equation for color image denoising and enhancement. In Section 3, we present experimental results for our model applied to color images along with comparisons to other models. In Section 3.2, we describe techniques to reduce computational time by considering the diffusion of the real part as a low-pass filter and the imaginary part as high-pass filter, and adding these two parts for the denoised reconstruction. In Section 4, we demonstrate that this reconstruction method can be applied to achieve high quality reconstruction. We draw conclusions in Section 5.

2 Action Formalism for Complex Diffusion

In this section, we introduce a metric for the complex image manifold for multi-channel images, and construct an action functional that is minimized to derive the complex diffusion equation. In addition, we applied the metric and the action functional to the quaternion representation of RGB images.

2.1 The Image Metric

The general idea of complex diffusion has been investigated in [5]. However, their primary focus was on gray level images. There was no description of extensions to vector-valued data sets. Since we deal with processing of multi channel images here, the key challenges involve processing the vector-valued data and capturing the correlation between the channels. In [1,2], a norm functional called the Polyakov action [4] and an embedding map $\mathbf{X} : \Sigma \to \mathbf{R}^n$ were introduced, where Σ is a 2-D manifold. They were used to capture the interaction between the multiple channels, and minimize the norm functional to obtain specific flows that smooth images in different ways. In this paper, we suggest an alternative to the Polyakov action, where the image manifold, Σ, is mapped to an n-dimensional complex manifold by $\mathbf{Z} : \Sigma \to \mathbf{C}^n$. Upon denoting the local coordinates on the 2-D manifold Σ by (σ_1, σ_2), the map $\mathbf{Z}$ is given by $[Z^1(\sigma^1, \sigma^2), Z^2(\sigma^1, \sigma^2), ..., Z^n(\sigma^1, \sigma^2)]$, where all the Z's are complex-valued. For example, a color (RGB) image can be mapped by $\mathbf{Z}$ as follows:

$$\mathbf{Z} : (\sigma^1, \sigma^2) \to [z^1, \bar{z}^1, Z^l = I^l(\sigma^1, \sigma^2), \bar{Z}^l] \tag{6}$$

where $z = \sigma^1 + i\sigma^2$, $\bar{z}$ is the complex conjugate of z, I^l is a complex-valued channel, $I_R^l(\sigma^1, \sigma^2) + iI_M^l(\sigma^1, \sigma^2)$, $\bar{Z}^l$ is the complex conjugate of Z^l and the index l runs over R,G, and B.

Let M, the space-feature manifold denote the embedding manifold of the complex image graph, with the map $\mathbf{Z} : \Sigma \to M$. Let $g_{\mu\nu}$ be the metric on the image manifold, Σ, and h_{ij} be the metric on M. Here, h_{ij} is defined such that $h_{ij}dZ^idZ^j$ gives a length element on M, and this metric makes the manifold M a Riemannian manifold with $(n \times 2) + 2$ dimensions ,where n is the number of channels and the local spatial coordinates are represented by two additional dimensions. For a gray level image, h_{ij} becomes

$$h = \begin{pmatrix} 0 & \frac{1}{2} & 0 & 0 \\ \frac{1}{2} & 0 & 0 & 0 \\ 0 & 0 & 0 & \frac{1}{2} \\ 0 & 0 & \frac{1}{2} & 0 \end{pmatrix} \tag{7}$$

so that the length element is $dzd\bar{z} + dId\bar{I} = (d\sigma^1)^2 + (d\sigma^2)^2 + dI_R^2 + dI_M^2$. Then, the image metric, $g_{\mu\nu}$ is given explicitly as follow:

$$g_{\mu\nu}(\sigma^1, \sigma^2) = h_{ij}(\mathbf{Z})\partial_\mu Z^i \partial_\nu Z^j \tag{8}$$

where, $\partial_\mu Z^i = \partial Z^i / \partial \sigma_\mu$. The image metric for the n-channel case is given explicitly by,

$$
g_{\mu\nu} = \begin{pmatrix} 1 + \sum_{l=1}^{n} I_x^l \bar{I}_x^l & \frac{1}{2} \sum_{l=1}^{n} (I_x^l \bar{I}_y^l + I_y^l \bar{I}_x^l) \\ \frac{1}{2} \sum_{l=1}^{n} (I_x^l \bar{I}_y^l + I_y^l \bar{I}_x^l) & 1 + \sum_{l=1}^{n} I_y^l \bar{I}_y^l \end{pmatrix} \tag{9}
$$

where x and y are the spatial coordinates. We are now ready to present the action formalism.

2.2   The Action Formalism

Images in computer vision are usually real-valued. Therefore, it is natural to pose them as a real-valued graph with a real-valued metric. However, in this paper we seek an action appropriate for complex-valued functions and one that is distinct from the Polyakov action presented in [2]. We would like the gradient descent (flow) equation of the new action to match the complex diffusion introduced in [5] under a special geometry and depict edge-preserving flows on a graph. We propose a specific action for n-channel images satisfying the conditions above, given by:

$$
S = \int \int \mathbf{F}(z, \bar{z}, I_x^l, I_y^l \bar{I}_x^l, \bar{I}_y^l) \sqrt{g} dx dy \tag{10}
$$

$$
\mathbf{F} = \frac{1}{2} \sum_{l=1}^{n} (\nabla I^l \cdot \nabla I^l e^{l\theta_l} + \nabla \bar{I}^l \cdot \nabla \bar{I}^l e^{-l\theta_l}). \tag{11}
$$

Here, x and y are local coordinates, and g is the determinant of the image metric $g_{\mu\nu}$, Eq.(9). In Eq.(10) and Eq.(11), I is complex image, $I_R + iI_M$ and $\bar{I}$ is its complex conjugate. In Eq.(11), generally, we can assign different phase θ_l to each channel. Setting g equal to the identity matrix and minimizing Eq.(10) by applying calculus of variation to Eq.(10), we can derive the isotropic complex diffusion equation introduced in [5] and details are given in following paragraphs.

We derive the gradient descent of Eq. (10) by evaluating the Euler-Lagrange equation with respect to the embedding. For this, we fix the x and y coordinates or z and $\bar{z}$, and vary the action with respect to I [2]. Then, the flow equation for I^l is given by:

$$
\frac{\partial I^l}{\partial t} = \frac{1}{g^\beta} \left[\frac{d}{dx} \left(\frac{P^l}{\sqrt{g}} \right) + \frac{d}{dy} \left(\frac{Q^l}{\sqrt{g}} \right) \right] \tag{12}
$$

where, P^l and Q^l are defined as:

$$
P^l = g \frac{\partial F}{\partial I_x^l}, \qquad Q^l = g \frac{\partial F}{\partial I_y^l}. \tag{13}
$$

In Eq.(12), we multiply the right hand side of the equation by a positive function, $1/(g^\beta)$, that will produce nonlinear scale-space and keep the flow geometrical as suggested in [2]. The exponent β will be discussed subsequently. When β is large, the flow becomes more sensitive to edges. Eq.(12) can now be rewritten as follows:

$$\frac{\partial I^l}{\partial t} = \frac{1}{g^{(\beta+0.5)}} \left[P_x^l + Q_x^l - \frac{1}{2g}(g_x P^l + g_y Q^l) \right]. \tag{14}$$

Here, I^l, P^l and Q^l are complex valued defined as: $I^l(x,y) = I_R^l(x,y) + iI_M^l(x,y)$, $P^l(x,y) = P_R^l(x,y) + iP_M^l(x,y)$, and $Q^l(x,y) = Q_R^l(x,y) + iQ_M^l(x,y)$, where l is a channel index. As a special case, we can easily obtain the isotropic complex diffusion equation introduced in [5], by applying Eq. (10) to gray scale images and setting the metric $g_{\mu\nu}$ to be the identity matrix. Then, g is equal to 1, $I(x,y) = I_R(x,y) + iI_M(x,y)$, and the Eq. (11) becomes

$$\boldsymbol{F} = \cos\theta(|\nabla I_R|^2 - |\nabla I_M|^2) - 2\sin\theta(I_{Rx}I_{Mx} + I_{Ry}I_{My}) \tag{15}$$

The gradient descent of Eq.(14) results in the following flow equations:

$$\frac{\partial I_R}{\partial t} = cos(\theta)\triangle I_R - sin(\theta)\triangle I_M \tag{16}$$

$$\frac{\partial I_M}{\partial t} = sin(\theta)\triangle I_R + cos(\theta)\triangle I_M. \tag{17}$$

Here, we recover the complex isotropic diffusion introduced in [5]. There is no imaginary part in the initial condition of complex image I, and the target image is assigned to the real part of the initial condition. However, we can create an imaginary part from a non-zero theta via the time iteration of Eq.(14).

Another special case of Eq.(10) is obtained by setting θ equal to zero in Eq.(15) with same g and initial conditions as before. In this case, $\boldsymbol{F}$ reduces to $|\nabla I_R|^2$ and the gradient descent of Eq.(14) recovers the ordinary isotropic diffusion equation: $\frac{\partial I}{\partial t} = \triangle I$, $(I = I_R)$.

In Fig.1, we have compared the results of anisotropic diffusion using Eq.(14) with isotropic diffusion obtained using Eq.(16) and Eq.(17) and anisotropic diffusion from Eq.(5) when applied to a gray-level image. Fig.1(a) is the given input image and also the real part of the initial complex image. We can observe that Fig.1(h) has no blurring across edges compared to Fig.1(f) and is smoother than Fig.1(g). The real part of Eq.(5) is less smooth than others for θ larger than 5 degrees.

2.3   Quaternion Representation for Color Images

The geometric coupling of channels in the RGB image via the image metric term is not the only way to achieve the coupling. Labunets [9] suggested applying hypercomplex techniques to multi-channel images. He considered R,G, and B color channels as a triplet number. In his framework, color space is identified with the so-called triplet algebra. Instead of the triplet representation of color, Liu et al. [8] employed quaternion to represent the color channels. They considered the diffusion of quaternion images as an extension to the diffusion of

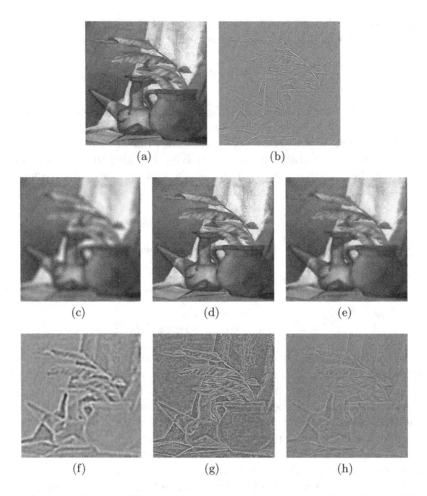

Fig. 1. (a) original image. (b) after one iteration of isotropic complex diffusion. (c) and (f) real and imaginary parts of (a) obtained by isotropic complex diffusion. (d) and (g) real and imaginary parts of (a) obtained using Eq.(5) with $k = 2$. (e) and (h) real and imaginary parts of (a) using Eq.(14) with $\beta = 5/6$. 100 time iterations have been processed with step size, 0.1 and $\theta = \pi/3$. All imaginary parts have been rescaled to 8-bit images for display.

complex images, and discussed the isotropic and anisotropic diffusion of quaternion valued RGB image. One of the choices of mapping RGB channels to a quaternion is to map R, G and B channels to pure quaternion parts, introducing an extra dimension which corresponds to real parts of quaternion, Then the quaternion of RGB channels, Q is represented by $Q = Q_0 + \mathbf{i}R + \mathbf{j}G + \mathbf{k}B$. Here, we introduce a novel geometric approach to achieve smoothing and enhancement of color images using the quaternions based representation of RGB, $[q, \bar{q}, Q, \bar{Q}]$, where $q = \sigma_1 + \mathbf{i}\sigma_2 + \mathbf{j}0 + \mathbf{k}0$. *We emphasize that this representation has never been used earlier and is indeed novel.* The second and third components of the pure quaternion parts of q are fixed to zero. For the action formulation, Eq. (10), we choose Eq. (9) as the image metric after replacing z and I with q and Q respectively,

$$g_{\mu\nu} = \begin{pmatrix} 1 + Q_x\bar{Q}_x & \frac{1}{2}(Q_x\bar{Q}_y + Q_y\bar{Q}_x) \\ \frac{1}{2}(Q_x\bar{Q}_y + Q_y\bar{Q}_x) & 1 + Q_y\bar{Q}_y \end{pmatrix}, \tag{18}$$

The functional $\boldsymbol{F}$ in Eq. (11) must now be rewritten using the quaternion algebra as follows:

$$\boldsymbol{F} = \frac{1}{2}((\nabla Q \cdot \nabla Q)C + \bar{C}(\nabla\bar{Q} \cdot \nabla\bar{Q})) \tag{19}$$

Here, C is a quaternion coefficient defined as $e^{\mathbf{e}_\phi\theta_\phi} = cos\theta_\phi + \mathbf{e}_\phi sin\theta_\phi$, and $\mathbf{e}_\phi = \mathbf{i}C_R + \mathbf{j}C_G + \mathbf{k}C_B$, where $C_R^2 + C_G^2 + C_B^2 = 1$ [10], and $\bar{C}$ is the quaternion conjugate of C. Then, we can have the flow equation, Eq.(14) for quaternion RGB by replacing Eq.(13) with

$$P^i = g\frac{\partial F}{\partial Q_x^i}, \qquad Q^i = g\frac{\partial F}{\partial Q_y^i} \tag{20}$$

where $Q^i \in \{I_0, R, G, B\}$. The correlation between the channels are introduced via a quaternion multiplication between $Q(\bar{Q})$ and $C(\bar{C})$ [8] as well as the metric on the image (graph) manifold. When we set $g_{\mu\nu}$ to the identity metric as we have done previously, and $C_R = C_G = C_B = 1/\sqrt{3}$, we have the isotropic diffusion of the color image in the quaternion framework presented in [8]:

$$\frac{dI_0}{dt} = \cos\theta_\phi\triangle I_0 - \sin\theta_\phi\frac{1}{\sqrt{3}}\triangle(R + G + B), \tag{21}$$

$$\frac{dQ^i}{dt} = \cos\theta_\phi\triangle Q^i + \sin\theta_\phi\frac{1}{\sqrt{3}}\triangle(I_0 + Q^j - Q^k), \tag{22}$$

where i,j and k follow the cyclic permutation of R,G and B. When θ_ϕ is negative, Eq.(21) will have a form similar to that of Eq.(17). This implies that the scalar part of the quaternion diffusion will capture the smoothed second order of $(R+G+B)/\sqrt{3}$ [8]. Additionally, recalling that correlation between channels is introduced only by the image metric, Eq.(9) in the case of complex diffusion, we can recognize that the quaternion algebra introduces alternative type of correlation between channels in Eq.(22).

3   Denoising and Edge Enhancement Experiments

In this paper, we apply our method to noisy color images using an image graph representation. There are two parameters in our model: the exponent β in Eq.(14) and θ in Eq.(11). In [5], large values of phase, θ, made edges represented by the imaginary part thicken with increasing iterations, and small θ less than 5 degrees was recommended for isotropic and anisotropic diffusion to retain sharp edges. In contrast, in our work here, large phase values increase the magnitude of the imaginary part and slow down diffusion speed near edges, which prevent thick edges due to large θ. The exponent, β of the non-linear scale multiplier influences the diffusion flows geometrically. For example, the diffusion equations from Polyakov action with different β values results in different flows like the Beltrami flows, Panora-Malik flows, Mean curvature flows and others [2,11]. The main purpose of this multiplicative factor is to achieve edge-preserving denoising. In this paper, we choose β from the interval $[0.5, 1]$. These two free parameters are chosen empirically based on the amount of noise in the data.

3.1 Denoising Experiments

The results of denoising depend on parameters, θ and β, similar to the earlier approaches [2,5]. The optimal parameter values depend on the amount of noise. The larger phase angles, θ and βs, lead to diffusions that are more sensitive to edges. We have applied the complex RGB flow and the quaternion flow to color images with added Gaussian noise, and compared the results with Beltrami flow. Our test image had an additive Gaussian noise of 25.3dB. Fig.2(a) and Fig.2(b) show original image and the noisy version respectively. We used the peak SNR (PSNR) as the stopping criteria for iterations. We stopped the iterations when the denoised images reached the maximum PSNR. Fig.2(c) and Fig.2(d) show denoised images obtained using the complex (RGB) flow with $\theta = 7\pi/30$ and $\beta = 5/6$, and the quaternionic flow with $\theta_\phi = -7\pi/30$,$\beta = 5/6$ and $C_R = C_G = C_B = 1/\sqrt{3}$. All the experiments reported here were implemented in Matlab 2007a, on an Intel Core Duo 2.16GHz CPU. The step size of time iteration is 0.1. We achieved the denoising using the complex (RGB) flow with a maximum PSNR of 26.6 dB in 38.6 seconds. Similarly, for the quaternionic flow the maximum PSNR is 26.5 dB and the processing time is 27.8 seconds. Fig.2(e) shows a denoised image using Beltrami flow with maximum PSNR of 25.4 dB and a processing time of 13.8 seconds (76 iterations). The result of the complex flow depicts higher degree of smoothing than that due to the Beltrami flow. When the noise is in the image detail, Beltrami flow tends to confuse the noise as detail, and this effect slows down diffusion velocity locally. Fig.2(f) shows the denoised image using Beltrami flow after a processing time of 89.8 secs. (500 iterations). The result is still noisy even after several iterations compared with the results from complex diffusion. The complex diffusion and quaternion diffusion yield results comparable to each other in quality, and are better than the Beltrami flow. However, the quaternion diffusion required less processing time compared to the complex diffusion case. This is due to the fact that the quaternion representation is 6-dimensional when using an RGB color image graph, where as the complex diffusion of the RGB image graph is 8-dimensional. Fig.3 shows imaginary parts of Figures 2(a)-2(d). Fig.3(a) and Fig.3(b) have been achieved after just one iteration on Fig.2(a) and Fig.2(b) respectively. We can see that the imaginary parts are also smoothed along with their corresponding real parts, which we consider as the processed images of the target image.

(a)   (b)   (c)

(d)   (e)   (f)

Fig. 2. (a) and (b) an original image and the image with the Gaussian noise of peak SNR 25.3dB respectively. (c) denoised image using complex RGB flows. (d) denoised image using quaternionic flows. The parameters are $\theta = 7\pi/30$ and $\beta = 5/6$ and the three pure quaternion components are $1/\sqrt{3}$. (e) and (f) images obtained using Beltrami flows with different processing times.

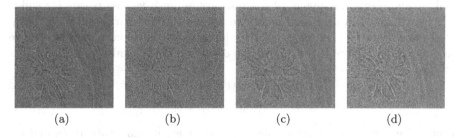

(a)   (b)   (c)   (d)

Fig. 3. (a) Imaginary part of Fig.2(a). (b) Imaginary part of Fig.2(b). (c) and (d) imaginary parts corresponding to Fig.2(c) and Fig.2(d) respectively (a) and (b) have been achieved after one iteration. All images are rescaled to 8-bit images for display.

3.2   Image Reconstruction

Recall that the imaginary part of complex diffusion corresponds to the smooth second order derivative and the real part corresponds to smoothed image. Hence, it is very natural to consider the imaginary part as a high-pass filter and the real

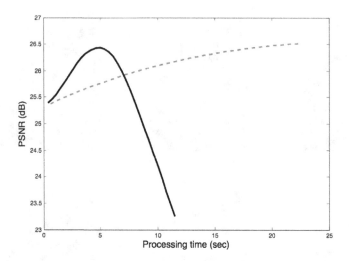

Fig. 4. Gray dashed line: PSNR of a denoised image by Eq. (14) without the reconstruction. Black solid line: PSNR with the reconstruction. The maxima of the black and gray lines are 26.44 dB and 26.57 dB respectively.

part as a low-pass filter, and think of addition of these two parts to recover original image which is an enhanced version of the original and contains smoothed edges. This process is similar to the image reconstruction via wavelet transformation, in which we add the lowest resolution version of low-pass filtered image with a sequence of high pass filtered images from the lowest resolution up to the desired resolution. However, we add the real part (low pass-filtered version) to the imaginary part (high pass-filtered version) so as to recover the smoothed and enhanced original image. (θ must always be positive, the reason for which will be explained in next section.) If we update the real parts by this addition after every iteration, the image can be denoised by diffusion as well as achieve reconstruction. To see this in detail, we discretize Eq.(3) and Eq.(4) in time, as follows:

$$I_R^{i+1} = I_R^i + \Delta t(\cos \theta h^i I_R^i - \sin \theta h^i I_M^i) \tag{23}$$

$$I_M^{i+1} = I_M^i + \Delta t(\sin \theta h^i I_R^i + \cos \theta h^i I_M^i) \tag{24}$$

Here, Δt is the time-step size of the iteration. Then we can evaluate I_R and I_M after i iterations, by a recursive relation :

$$I_R^{i+1} = I_R^0 + \Delta t \sum_{j=0}^{i} (\cos \theta h^j I_R^j - \sin \theta h^j I_M^j), \tag{25}$$

$$I_M^{i+1} = I_M^0 + \Delta t \sum_{j=0}^{i} (\sin \theta h^j I_R^j + \cos \theta h^j I_M^j). \tag{26}$$

Here, I_{M0} is set to be zero. However, if we reset the real part to a sum of the imaginary and real part so as to obtain a reconstruction ($I_R^{i+1} \to I_R^{i+1} + I_M^{i+1}$) after each iteration (and before next iteration), Eq.(23) can be rewritten as,

$$I_R^{i+1} = I_R^i + I_M^i + \Delta t((\cos\theta + \sin\theta)h^i I_R^i + (\cos\theta - \sin\theta)h^i I_M^i), \qquad (27)$$

Rewriting Eq.(25) using Eq.(27) gives us the following recursive relationship:

$$I_R^{i+1} = I_R^0 + \sum_{j=0}^{i} I_M^j + \Delta t \sum_{j=0}^{i}((\cos\theta + \sin\theta)h^j I_R^j + (\cos\theta - \sin\theta)h^j I_M^j). \quad (28)$$

noise with 25.3dB PSNR. The gray dashed line in Fig.4 represents the PSNR results obtained by applying Eq. (14) without the reconstruction, and the black solid line in Fig.4 represents PSNR results with the reconstruction. The maxima of the black and gray line are 26.44 dB and 26.57 dB respectively. We can see that the reconstruction at each iteration is improves the smoothing process. This test was done with the parameter values: $\beta = \frac{5}{6}$ and $\theta = \frac{\pi}{6}$.

(a)   (b)   (c)

(d)   (e)   (f)

Fig. 5. (a) Original image. (b) and (c) Enhanced images with $\beta = 1/2$, $\theta = -\pi/6$ and $\theta = -\pi/3$ respectively after 16 iterations. (d) and (e) Enhanced images with $\beta = 5/6$, $\theta = -\pi/6$ and $\theta = -\pi/3$ respectively after 20 iterations. (f) Enhanced image with same parameters after 25 iterations.

4   Image Enhancement

It has been shown in [5] that the imaginary part of the isotropic complex diffusion can be applied to shock filter since the imaginary part contains the edge information of the real part. We consider the real part as the processed image of the target image. This characteristic of the imaginary part allows us to apply the imaginary part to edge-preserving smoothing as well as image enhancement. In the previous section, we have introduced the idea of image reconstruction. In the case of smoothing, θ has been set to be positive. According to Eq.(26), the diffusion of imaginary part behaves as edge smoothing due to the first term, where $\sin\theta$ is positive. On the other hand, negative θ makes Eq.(26) perform edge enhancement. Therefore, in order to enhance images, we reconstruct the image after every iteration with negative θ as was done in section 3.2.

Fig.5 shows the enhanced results with various parameters. Fig.5(a) is the original image, and Fig.5(b) and Fig.5(c) are the enhanced images with the parameters vales: $\beta = \frac{1}{2}$, and $\theta = -\frac{\pi}{6}$ and $\theta = -\frac{\pi}{3}$ respectively after 16 iterations. We can see that the edges are over-enhanced in Fig.5(c) due to the larger $\sin\theta$ of Eq.(26) than those in Fig.5(b). Also, the small value of β makes the diffusion flow less sensitive to edges and produces thick edges. Images in the bottom row of Fig.5 show enhancement results with $\beta = \frac{5}{6}$. The larger value of β results in sharper edges and more details.

5   Conclusion and Discussion

In this paper, we presented a novel formulation of complex diffusion for simultaneous image smoothing and edge enhancement. The formulation involved the use of an image graph representation as an embedded manifold, a novel image metric and a novel action functional yielding a new complex diffusion. Additionally, we developed a new quaternionic diffusion using this geometric framework for color images and demonstrated improved performance over the Beltrami flow. Comparisons were reported on data with added noise, using PSNR as a quantitative measure. Finally, we presented a "wavelet-like" interpretation of the complex diffusion. We interpreted the real and imaginary part of the complex diffusion as a low pass and high pass filter respectively and applied this concept to image reconstruction and enhancement. When performing the image reconstruction iteratively, we achieved faster convergence to the PSNR with positive θ and image enhancement with negative θ. Our future work will involve application of the proposed model to the complex-valued MRI data.

References

1. Kimmel, R., Sochen, N., Malladi, R.: From high energy physics to low level vision. In: ter Haar Romeny, B.M., Florack, L.M.J., Viergever, M.A. (eds.) Scale-Space 1997. LNCS, vol. 1252, pp. 236–247. Springer, Heidelberg (1997)

2. Sochen, N., Kimmel, R., Malladi, R.: A general framework for low level vision. IEEE Transaction on Image Processing, Special Issue on PDE based Image Processing 7(3), 310–318 (1998)
3. Kimmel, R., Malladi, R., Sochen, N.: Image as embedded maps and minimal surface: movies, color, texture, and volumetric medical images. International Journal of Computer Vision 39(2), 111–129 (2000)
4. Polyakov, A.M.: Quantum geometry of bosonic strings. Physics Letters 103B, 207 (1981)
5. Gilboa, G., Sochen, N., Zeevi, Y.Y.: Image Enhancement and Denoising by Complex Diffusion Process. IEEE Transaction on pattern analysis and machine intelligence (PAMI) 25(8), 1020–1036 (2004)
6. Gilboa, G., Zeevi, Y.Y., Sochen, N.A.: Complex Diffusion Process for image filtering. In: Kerckhove, M. (ed.) Scale-Space 2001. LNCS, vol. 2106, pp. 299–307. Springer, Heidelberg (2001)
7. Gilboa, G., Sochen, N.A., Zeevi, Y.Y.: Regularized Shock Filter and Complex Diffusion. In: Heyden, A., Sparr, G., Nielsen, M., Johansen, P. (eds.) ECCV 2002. LNCS, vol. 2350, pp. 399–413. Springer, Heidelberg (2002)
8. Liu, Z.-X., Lian, S.-G., Ren, Z.: Quaternion diffusion for color image filtering. J. comput. Sci. Technol. 21(1), 126–136 (2006)
9. Labunets, V.: Clifford Algebras as Unified Language for Image Processing and Pattern Recognition. NATO Advanced Study Institute, Computational Noncommutative Algebra and Applications, July 6-19 (2003), http://www.prometheus-inc.com/asi/algebra2003/papers/labunets2.pdf
10. Adler, S.L.: Quaternionic Quantum Mechanics and Quantum Fields. Oxford University Press, Oxford (1995)
11. El-Fallah, A.I., Ford, G.E., Algazi, V.R., Estes, R.R.: The invariance of edges and corners under mean curvature diffusions of images. In: Proc. Processing III SPIE, vol. 2421, pp. 2–14 (1994)
12. Perona, P., Malik, J.: Scale-space and edge detection usion ansiotropic diffusion. IEEE Trans. Pattern Anal. Machine Intell 12, 629–639 (1990)

A PDE Approach to Coupled Super-Resolution with Non-parametric Motion

Mehran Ebrahimi and Anne L. Martel

Department of Medical Biophysics, University of Toronto
Imaging Research, Sunnybrook Health Sciences Centre
Toronto, Ontario, Canada
mehran.ebrahimi@sri.utoronto.ca, anne.martel@sri.utoronto.ca

Abstract. The problem of recovering a high-resolution image from a set of distorted (e.g., warped, blurred, noisy) and low-resolution images is known as *super-resolution*. Accurate motion estimation among the low-resolution measurements is a fundamental challenge of the super-resolution problem. Some recent promising advances in this area have been focused on coupling or combing the super-resolution reconstruction and the motion estimation. However, the existing approach is limited to parametric motion models, e.g., affine. In this paper, we shall address the coupled super-resolution problem with a *non-parametric* motion model. We address the problem in a variational formulation and propose a PDE-approach to yield a numerical scheme. In this approach, we use diffusion regularizations for both the motion and the super-resolved image. However, the approach is flexible and other suitable regularization schemes may be employed in the proposed formulation.

1 Introduction

Naturally, there is always a demand for higher quality and higher resolution images. The level of image detail is crucial for the performance of many computer vision algorithms [2,3,7,8,9,12,14,16,20].

Many of the current imaging devices typically consist of arrays of light detectors. A detector determines pixel intensity values depending upon the amount of light detected from its assigned area in the scene. The spatial resolution of images produced is proportional to the density of the detector array: the greater the number of pixels in the image, the higher the spatial resolution [16]. In many applications, however, the imaging sensors have poor resolution output. When resolution can not be improved by replacing sensors, either because of cost or hardware physical limits, one can resort to resolution enhancement algorithms. Even when superior equipment is available, such algorithms provide an inexpensive alternative. The problem of recovering a high-resolution (HR) image from a set of distorted (e.g., warped, blurred, noisy) and low-resolution (LR) images is known as *super-resolution* [2,7,8,12,14,16,20].

Fusion of the information from the observations is a fundamental challenge in the recovery process. With just one imaging device and under the same lighting conditions, we require some relative motions from frame to frame. Each LR

D. Cremers et al. (Eds.): EMMCVPR 2009, LNCS 5681, pp. 112–125, 2009.

frame should provide a different look at the same scene. Motion and nonredundant information obtained from different frames are what make super-resolution feasible [16].

1.1 A Brief History

The super-resolution literature has significantly expanded in the past 20 years. A rather recent and comprehensive survey of super-resolution techniques is given in [8]. Historically, Irani and Peleg [14] proposed an iterative back-projection method to address the super-resolution problem. Sauer and Allebach [18], and Tekalp, Ozkan and Sezan [23] modelled super-resolution as an interpolation problem with nonuniformly sampled data and used a projection onto convex sets algorithm to reconstruct the image. Ur and Gross [27] considered Papoulis' generalized multichannel sampling theorem [17] for interpolating values on a higher resolution grid. Shekarforoush and Chellappa [22] extended Papoulis' theorem for merging the nonuniform samples of multiple channels into HR data. Aizawa et al. [1] also modelled super-resolution as an interpolation problem with nonuniform sampling and used a formula related to Shannon's sampling theorem [21] to estimate values on a HR grid. Tsai and Huang [26] were among the first to superresolve a HR image from several sampled LR frames. Hardie et al. [12] proposed a joint MAP registration and restoration algorithm using a *Gibbs image prior*. Schultz and Stevenson [20] used a Markov random field model with Gibbs prior to better represent image discontinuities, such as transitions across sharp edges. More recently Farsiu et al. [9] proposed an alternative data fidelity, or regularization term based on the ℓ^1 norm which has been shown to be robust to data outliers. They proposed a novel regularization term called Bilateral-TV which provides robust performance while preserving the edge content common to real image sequences.

1.2 Coupled Motion Estimation

Accurate motion estimation has been a very important aspect of super-resolution schemes. In many existing super-resolution approaches, the motion is computed directly from the LR frames, while many other super-resolution algorithms unrealistically assume that motion parameters are precisely known. In general, however, accurate motion estimation of subpixel accuracy remains a fundamental challenge in super-resolution reconstruction algorithms.

In a recent work [6], it has been suggested that the motion can be relaxed from a strict grid mapping to a multi-pixel-pair intensity relation. In this view, pixel-pairs in different frames may be relevant to each other with some measured probability of confidence. In the method proposed in [6], instead of estimating the motion vectors explicitly, a framework is provided in which such confidence measures are evaluated and employed in the HR image reconstruction. However, the algorithm is computationally intensive.

In general, it is believed that a combined super-resolution reconstruction and motion estimation may be the key to address the super-resolution problem.

A novel method towards this direction is proposed in [4]. Although the authors of [4] appreciate the importance of considering non-parametric motion models, their proposed method is restricted to the parametric affine motion model. The fact that authors of [4] have preferred to work with a parametric motion model rather than a non-parametric one can be associated to the complexity of formulations of the non-parametric approaches as discussed in [4].

1.3   The Agenda

In this work, we propose the coupled multi-frame super-resolution problem with a *non-parametric* motion model. In Section 2, we will introduce the problem as a minimization and present its corresponding variational formulation. For consistency, we adopt our notations from [15]. In Section 3, we derive a PDE with a steady-state solution that corresponds to the solution of the described problem. The discretization and derivation of a numerical scheme for the PDE is followed in Section 4. Finally, we will present various computational experiments and concluding remarks in Sections 5 and 6.

2   Mathematical Formulation

Throughout, images are d-dimensional and are assumed as compactly supported elements of $\mathcal{L}^2(\Omega)$, $\Omega \subset \mathbb{R}^d$, unless otherwise stated.

Forward Model. Assume that m low-resolution measurement images y_1, $y_2, \ldots y_m$ of an ideal image f are given. For every $i = 1, \ldots, m$, y_i is a noisy, low-resolution realization of deformed copies of f via a d-dimensional vector field $u_i = (u_{i,1}, \ldots, u_{i,d})$. Namely,

$$y_i := \mathcal{H}f_{u_i} + n_i, \quad i = 1, \ldots, m, \tag{1}$$

where n_i is the additive noise, and f_{u_i} denotes the deformed image f via u_i, i.e., $f_{u_i}(x) = f(x - u_i(x))$. Throughout, we may also use the alternative notation of

$$\mathcal{S}_i f := f_{u_i}.$$

Note that the operator $\mathcal{S}_i$ is linear with respect to f although f_{u_i} is a nonlinear expression with respect to u_i. The operator $\mathcal{H} : \mathcal{L}^2(\Omega) \to \mathcal{L}^2(\Omega)$ is assumed to be a known linear degradation operator modeled as a composition of a spatially invariant blur $\mathcal{K}$ followed by a down-sampling operator $\mathcal{D}$, i.e., $\mathcal{H} := \mathcal{D} \circ \mathcal{K}$. Here, $\mathcal{D}$ is an impulse train constructed using the sum of uniformly spaced Dirac functions [16,8,9,3,7]. To proceed, we formulate the corresponding super-resolution problem. As opposed to what is typically common in the literature, we assume that both the deformations and the high-resolution image are unknown and try to recover both simultaneously.

Problem 1. *Given a set of m low-resolution measured images represented by*
$y := \{y_1, y_2, \ldots y_m\}$ *and a degradation operator* $\mathcal{H}$, *find a corresponding set of*
deformations $u := \{u_1, u_2, \ldots, u_m\}$ *and a high-resolution image* f *that minimizes*

$$\mathcal{J}[u, f] := \mathcal{C}[y; (u, f)] + \mathcal{R}[u, f]$$

in which $\mathcal{C}$ *measures the consistency of the measurements* y *with the high-*
resolution image f, *and* $\mathcal{R}$ *is a regularization expression on* $[u, f]$. *Here, we*
use the sum of squares of intensity differences for the consistency measure

$$\mathcal{C}[y; (u, f)] := \frac{1}{2} \sum_{i=1}^{m} \|y_i - \mathcal{H}f_{u_i}\|_{\mathcal{L}^2(\Omega)}^2, \tag{2}$$

and the regularization is defined by

$$\mathcal{R}[u, f] := \sum_{i=1}^{m} \alpha_i \mathcal{P}[u_i] + \beta \mathcal{Q}[f], \tag{3}$$

in which $\alpha_1, \ldots, \alpha_m, \beta \in \mathbb{R}^+$ *are positive regularizing parameters. Hence, the*
objective is to minimize

$$\mathcal{J}[u, f] = \frac{1}{2} \sum_{i=1}^{m} \|y_i - \mathcal{H}f_{u_i}\|_{\mathcal{L}^2(\Omega)}^2 + \sum_{i=1}^{m} \alpha_i \mathcal{P}[u_i] + \beta \mathcal{Q}[f]. \tag{4}$$

We shall present a mathematical formulation to solve Problem 1. Briefly speaking, we seek necessary conditions for optimality of $[u, f]$ by finding the Gâteaux derivatives of the components of $\mathcal{J}$ with respect to $[u, f]$. This shall provide us with the corresponding Euler-Lagrange equations that will be used to form a PDE which will be solved numerically.

Theorem 1. *Let* $d \in \mathbb{N}$, *and* $f, y_1, y_2, \ldots y_m$ *are* d-*dimensional real-valued im-*
ages, i.e., functions from $\Omega \subset \mathbb{R}^d \to \mathbb{R}$, $f \in C^2(\mathbb{R}^d)$, $u_1, \ldots, u_m : \mathbb{R}^d \to \mathbb{R}^d$,
$v : \mathbb{R}^d \to \mathbb{R}^{md+1}$, $\Omega :=]0, n[^d$. *The Gâteaux derivative of* $\mathcal{C}[y; (u, f)]$ *is given by*

$$d\mathcal{C}[y; (u, f); v] = -\int_{\Omega} \langle \Phi(x, u(x), f(x)), v(x) \rangle_{\mathbb{R}^{md+1}} dx,$$

in which $\Phi : \mathbb{R}^d \times \mathbb{R}^{md} \times \mathbb{R} \to \mathbb{R}^{md+1}$,

$$\Phi(x, u(x), f(x)) = [p_1(x), \ldots, p_m(x), q(x)],$$

where

$$p_i(x) := \mathcal{H}^*[\mathcal{H}\mathcal{S}_i f(x) - y_i(x)]\nabla \mathcal{S}_i f(x), \quad i = 1, \ldots, m,$$

$$q(x) := -\sum_{i=1}^{m} \mathcal{S}_i^* \mathcal{H}^*[\mathcal{H}\mathcal{S}_i f(x) - y_i(x)],$$

in which $\mathcal{H}^*$ *and* $\mathcal{S}_i^*$ *represent the adjoint of operators* $\mathcal{H}$ *and* $\mathcal{S}_i$ *respectively.*
[see the proof in Appendix 1. Cf. [15] pp. 80.]

Here, we focus on the special case where $\mathcal{P}$ and $\mathcal{Q}$ are diffusion regularization expressions [15,11,10,13,24,25].

Theorem 2. *Assume $\mathcal{P}$ and $\mathcal{Q}$ are diffusion regularization expressions and the functionals $\mathcal{P}_i^e$ and $\mathcal{Q}^e$ are respectively trivial extensions of $\mathcal{P}$ and $\mathcal{Q}$, i.e.,*

$$\mathcal{P}_i^e[(u,f)] := \mathcal{P}[u_i] := \frac{1}{2}\sum_{j=1}^d \int_\Omega \langle \nabla u_{i,j}, \nabla u_{i,j}\rangle \, dx, \quad i = 1,\ldots,m, \qquad (5)$$

$$\mathcal{Q}^e[(u,f)] := \mathcal{Q}[f] := \frac{1}{2}\int_\Omega \langle \nabla f, \nabla f\rangle \, dx. \qquad (6)$$

Also, assume that Neumann boundary conditions are imposed, i.e.,

$$\langle \nabla f(x), \overrightarrow{n}(x)\rangle_{\mathbb{R}^d} = \langle \nabla u_{i,j}(x), \overrightarrow{n}(x)\rangle_{\mathbb{R}^d} = 0 \quad for \quad x \in \partial\Omega \ \ and \ \ j = 1,\ldots,d,$$

in which $\overrightarrow{n}$ denotes the outer normal unit vector of $\partial\Omega$ (boundary of Ω). The Gâteaux derivative of $\mathcal{P}_i^e[(u,f);v]$ and $\mathcal{Q}^e[(u,f);v]$ are respectively

$$d\mathcal{P}_i^e[(u,f);v] = -\int_\Omega \langle \mathcal{A}_i[u](x), v(x)\rangle_{\mathbb{R}^{d+1}} \, dx, \quad i = 1,\ldots,m,$$

$$d\mathcal{Q}^e[(u,f);v] = -\int_\Omega \langle \mathcal{B}[f](x), v(x)\rangle_{\mathbb{R}^{d+1}} \, dx$$

where,
$$\mathcal{A}_i[u](x) = (\underbrace{0_{\mathbb{R}^d},\ldots,0_{\mathbb{R}^d}}_{i-1 \ times}, \Delta u_{i,1}(x),\ldots,\Delta u_{i,d}(x),\underbrace{0_{\mathbb{R}^d},\ldots,0_{\mathbb{R}^d}}_{m-i \ times},0)$$

$$= (\underbrace{0_{\mathbb{R}^d},\ldots,0_{\mathbb{R}^d}}_{i-1 \ times}, \Delta u_i(x),\underbrace{0_{\mathbb{R}^d},\ldots,0_{\mathbb{R}^d}}_{m-i \ times},0),$$

$$\mathcal{B}[f](x) = (\underbrace{0_{\mathbb{R}^d},\ldots,0_{\mathbb{R}^d}}_{m \ times}, \Delta f(x)) = (0_{\mathbb{R}^{md}}, \Delta f(x)).$$

Proof. The result yields applying the Green's formula similar to [15] pp. 138.

Theorem 3. *The Euler-Lagrange equations corresponding to the objective expression $\mathcal{J} = \mathcal{C} + \sum_{i=1}^m \alpha_i \mathcal{P}_i^e + \beta \mathcal{Q}^e$ identical to Equation (4) where $\mathcal{C}$ is defined by Equation (2) and $\mathcal{P}_i^e$, $\mathcal{Q}^e$ are defined by Equations (5,6) respectively are*

$$\Phi(x, u(x), f(x)) + \sum_{i=1}^m \alpha_i \mathcal{A}_i[u](x) + \beta \mathcal{B}[f](x) = 0, \qquad x \in \Omega, \qquad (7)$$

with Neumann boundary conditions. These can also be written as

$$\mathcal{H}^*[\mathcal{H}\mathcal{S}_i f(x) - y_i(x)]\nabla \mathcal{S}_i f(x) + \alpha_i \Delta u_i(x) = 0_{\mathbb{R}^d},$$

$$i = 1,\ldots,m, \quad x \in \Omega,$$

$$-\sum_{i=1}^{m} \mathcal{S}_i^* \mathcal{H}^* [\mathcal{HS}_i f(x) - y_i(x)] + \beta \Delta f(x) = 0, \qquad x \in \Omega,$$

$$\langle \nabla f(x), \overrightarrow{n}(x) \rangle_{\mathbb{R}^d} = \langle \nabla u_{i,j}(x), \overrightarrow{n}(x) \rangle_{\mathbb{R}^d} = 0,$$

$$j = 1, \dots, d, \quad i = 1, \dots, m, \quad x \in \partial \Omega.$$

Proof. The result yields from Theorem 1 and substitution (Cf. [15] pp. 138.) .

3 A Corresponding PDE

There exist various ways to solve Equation (7). A possibility that we pursue here is to formulate the solution as the steady-state solution of a corresponding PDE similar to [15]. We propose

$$\partial_t(u(x,t), s\ f(x,t)) = \Phi(x, u(x,t), f(x,t)) + \sum_{i=1}^{m} \alpha_i \mathcal{A}_i[u](x) + \beta \mathcal{B}[f](x) \quad x \in \Omega, \ t \ge 0,$$

where s is a scale factor. Assuming $\Phi = (p_1, \dots, p_m, q)$ the PDE can be written as

$$\partial_t u^{(i)}(x,t) = p_i(x, u(x,t), f(x,t)) + \alpha_i \Delta u_i(x,t), \tag{8}$$

$$i = 1 \dots, m, \quad x \in \Omega, \ t \ge 0,$$

$$s\ \partial_t f(x,t) = q(x, u(x,t), f(x,t)) + \beta \Delta f(x,t), \qquad x \in \Omega, \ t \ge 0, \tag{9}$$

where,

$$p_i(x, u, f) := \mathcal{H}^*[\mathcal{HS}_i f(x) - y_i(x)] \nabla \mathcal{S}_i f(x), \qquad i = 1 \dots, m,$$

$$q(x, u, f) := -\sum_{i=1}^{m} \mathcal{S}_i^* \mathcal{H}^*[\mathcal{HS}_i f(x) - y_i(x)].$$

4 Discretization and Numerical Scheme

To numerically solve the derived PDE in Equations (8,9), we evaluate expressions at discrete time variable $\{t_{k+1}\}$

$$\partial_t u_i(x, t_{k+1}) = p_i(x, u(x, t_k), f(x, t_k)) + \alpha_i \Delta u_i(x, t_{k+1}), \tag{10}$$

$$i = 1, \dots, m, \quad x \in \Omega,$$

$$s\ \partial_t f(x, t_{k+1}) = q(x, u(x, t_{k+1}), f(x, t_{k+1})) + \beta \Delta f(x, t_{k+1}), \quad x \in \Omega. \tag{11}$$

Notice that due to the nonlinearity of p_i with respect to u_i, f is evaluated at t_k instead of t_{k+1} in Equation (10) [cf. [15] pp. 80] which translates to applying a fixed-point iteration scheme. However, q is linear with respect to f and t_{k+1} is used consistently in Equation (11). Using a spatial discretization X of Ω that includes n^d voxels (pixels) corresponding to a unit space step in every dimension due to the definition of $\Omega :=]0, n[^d$, and a time step of τ_1, we define for $j = 1, \ldots, d$, and $k = 0, 1, 2, \ldots$

$$U_{i,j}^k(X) := u_{i,j}(X, \tau_1 k) := Discretized(u_j(x, t_k)),$$
$$F^k(X) := f(X, \tau_1 k) := Discretized(f(x, t_k)),$$
$$Y_i(X) := y_i(X) := Discretized(y_i(x)).$$

Furthermore, $\quad \mathbf{A} U_{i,j}^k := \Delta u_{i,j}(X, \tau_1 k) := Discretized(\Delta u_{i,j}(x, t_k)),$

$$\mathbf{A} F^k := \Delta f(X, \tau_1 k) := Discretized(\Delta f(x, t_k)),$$

in which $\mathbf{A} \in \mathbb{R}^{n^d \times n^d}$ is defined such that

$$\mathbf{A} U_{i,j}^k \approx \sum_{l=1}^d \partial_{x_l, x_l} u_{i,j}(X, \tau_1 k) \quad \text{and} \quad \mathbf{A} F^k \approx \sum_{l=1}^d \partial_{x_l, x_l} f(X, \tau_1 k).$$

[See Appendix 2 for the precise definition of $\mathbf{A}$.] Also, assume that $\mathbf{H}$ is a matrix that represents $\mathcal{H}$, and its transpose $\mathbf{H}^T$ represents $\mathcal{H}^*$ in the discretization. Appendix 3 gives the precise definition of $\mathbf{H}$, which is assumed as the matrix product of the local averaging blur $\mathbf{K}$ by a zooming factor of $z \in \mathbb{N}$ in every direction, multiplied by down-sampling matrix of factor z in every direction represented by $\mathbf{D}$. Furthermore, $\mathbf{S}_i^k$ is the sparse matrix constructed using U_i^k. This matrix provides a discrete approximation of the operator $\mathcal{S}_i$ for $i = 1, \ldots, m$. [See [4] Equations (3,4) for the precise construction of such matrix using linear interpolation for 2-dimensional images i.e., $d = 2$.]

Substituting the discretization in the PDEs of Equations (10,11) leads that for $j = 1, \ldots, d$, $k = 0, 1, 2, \ldots$

$$\frac{U_{i,j}^{k+1} - U_{i,j}^k}{\tau_1} = \mathbf{H}^T [\mathbf{H} \mathbf{S}_i^k F^k - Y_i] \cdot \partial_j \mathbf{S}_i^k F^k + \alpha_i \mathbf{A} U_{i,j}^{k+1},$$

$$s \frac{F^{k+1} - F^k}{\tau_1} = -\sum_{i=1}^m (\mathbf{S}_i^{k+1})^T \mathbf{H}^T [\mathbf{H} \mathbf{S}_i^{k+1} F^{k+1} - Y_i] + \beta \mathbf{A} F^{k+1}.$$

Defining $\tau_2 := \tau_1 / s$ gives

$$\left(\mathbf{I} - \tau_1 \alpha_i \mathbf{A} \right) U_{i,j}^{k+1} = U_{i,j}^k + \tau_1 \mathbf{H}^T [\mathbf{H} \mathbf{S}_i^k F^k - Y_i] \cdot \partial_j \mathbf{S}_i^k F^k,$$

$$\left(\mathbf{I} - \tau_2 \beta \mathbf{A} + \tau_2 \sum_{i=1}^m (\mathbf{S}_i^{k+1})^T \mathbf{H}^T \mathbf{H} \mathbf{S}_i^{k+1} \right) F^{k+1} = F^k + \tau_2 \sum_{i=1}^m (\mathbf{S}_i^{k+1})^T \mathbf{H}^T Y_i,$$

where $\mathbf{I} \in \mathbb{R}^{n^d \times n^d}$ is the identity matrix.

This yields

$$U_{i,j}^{k+1} = \left(\mathbf{I} - \tau_1 \alpha_i \mathbf{A} \right)^{-1} \left[U_{i,j}^k + \tau_1 \mathbf{H}^T [\mathbf{H} \mathbf{S}_i^k F^k - Y_i] \cdot \partial_j \mathbf{S}_i^k F^k \right],$$

$$F^{k+1} = \left(\mathbf{I} - \tau_2 \beta \mathbf{A} + \tau_2 \sum_{i=1}^m (\mathbf{S}_i^{k+1})^T \mathbf{H}^T \mathbf{H} \mathbf{S}_i^{k+1} \right)^{-1} \left[F^k + \tau_2 \sum_{i=1}^m (\mathbf{S}_i^{k+1})^T \mathbf{H}^T Y_i \right].$$

Finally, we use the initialization vectors $F^0 = U_{i,j}^0 = 0_{\mathbb{R}^{n^d}}$, for $i = 1, \ldots, m$, $j = 1, \ldots, d$.

5   Computational Experiments

In this Section we present a few computational examples to verify the derived numerical scheme. Figures 1 and 2 show the results of evaluating the proposed super-resolution algorithm on image sequences taken from the data-set library of MDSP at U. California Santa Cruz
(http://www.soe.ucsc.edu/ milanfar/software/sr-datasets.html).

In Figure 1, the first $m = 30$ frames, of an 8-bits text sequence, of size 48×48 is used and independent additive white Gaussian noise of standard deviation $\sigma = 5$ is added to the LR frames. The results are shown along with the parameters described in the caption of the Figure.

In Figure 2, the same kind of experiment is performed over a total of $m = 60$ frames, of an 8-bits surveillance sequence, of size 32×32 and again independent additive white Gaussian noise of standard deviation $\sigma = 5$ is added to the LR frames. The results are shown along with the parameters described in the Figure's caption.

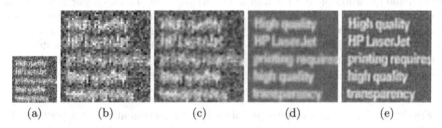

(a)   (b)   (c)   (d)   (e)

Fig. 1. (a) LR frame #1. (b) Nearest neighbor interpolation on the LR frame #1. (c) Bilinear interpolation on the LR frame #1. (d) Average of the bilinear interpolation of all 30 LR frames. (e) Super resolution result of frame #1 in $d = 2$ dimensions of $n = 48$ pixels in every dimension, zooming factor $z = 2$, total number of LR frames $m = 30$, regularization parameters $\alpha_i = 4000$, $i = 1, \ldots, 30$, $\beta = 1$, and time steps $\tau_1 = 0.01$, $\tau_2 = 10^{10}$, where 10 iterations are applied.

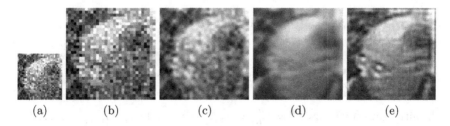

(a)   (b)   (c)   (d)   (e)

Fig. 2. (a) LR frame #1. (b) Nearest neighbor interpolation on the LR frame #1. (c) Bilinear interpolation on the LR frame #1. (d) Average of the bilinear interpolation of all 60 LR frames. (e) Super resolution result of frame #1 in $d = 2$ dimensions of $n = 32$ pixels in every dimension, zooming factor $z = 2$, total number of LR frames $m = 60$, regularization parameters $\alpha_i = 4000$, $i = 1, \ldots, 60$, $\beta = 1$, and time steps $\tau_1 = 0.01$, $\tau_2 = 10^{10}$, where 10 iterations are applied.

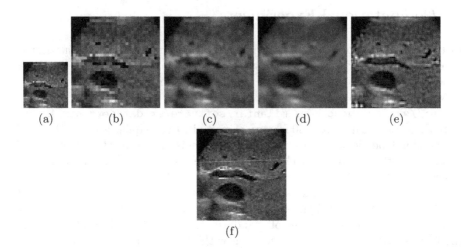

(a)   (b)   (c)   (d)   (e)

(f)

Fig. 3. (a) LR frame #1. (b) Nearest neighbor interpolation on the LR frame #1. (c) Bilinear interpolation on the LR frame #1. (d) Average of the bilinear interpolation of all 40 LR frames. (e) Super resolution result of frame #1 in $d = 2$ dimensions of $n = 32$ pixels in every dimension, zooming factor $z = 2$, total number of LR frames $m = 40$, regularization parameters $\alpha_i = 4000$, $i = 1, \ldots, 40$, $\beta = 0.1$, and time steps $\tau_1 = 0.01$, $\tau_2 = 10^{10}$, where 10 iterations are applied. (f) The HR ground truth relating to frame #1.

Finally, in Figure 3, we have performed the experiments over a portion of a locally-averaged and down-sampled ultrasound sequence of $m = 60$ frames of size 32×32, for which the original HR image relating to frame #1 is given in Figure 3(f). The efficiency of the technique can be simply observed by comparing the image (e) of each Figure to the other images in the figure.

Note that in all of the experiments, we have assumed $\tau_2 = 10^{10}$. In general, if we assume $\tau_2 = \infty$, [i.e., solving Equation (11) for f assuming $s = 0$] at each iteration yields

$$U_{i,j}^{k+1} = \left(\mathbf{I} - \tau_1 \alpha_i \mathbf{A}\right)^{-1} \left[U_{i,j}^k + \tau_1 \mathbf{H}^T[\mathbf{H}\mathbf{S}_i^k F^k - Y_i] \cdot \partial_j \, \mathbf{S}_i^k F^k\right],$$

$$F^{k+1} = \left(-\beta\mathbf{A} + \sum_{i=1}^{m}(\mathbf{S}_i^{k+1})^T\mathbf{H}^T\mathbf{H}\mathbf{S}_i^{k+1}\right)^{-1}\left[\sum_{i=1}^{m}(\mathbf{S}_i^{k+1})^T\mathbf{H}^T Y_i\right],$$

with initialization vectors $F^0 = U_{i,j}^0 = 0_{\mathbb{R}^{nd}}$, for $i = 1, \ldots, m$, $j = 1, \ldots, d$.

6   Concluding Remarks

Accurate motion estimation is a fundamental challenge of the super-resolution problem. Some recent promising advances in this area have been focused on

coupling or combing the super-resolution reconstruction and the motion estimation. However, the existing approach is limited to parametric motion models,
e.g., affine [4]. In this paper, we addressed the coupled super-resolution problem
with a *non-parametric* motion model. We addressed the problem in a variational
formulation and proposed a PDE-approach to yield a numerical scheme. In this
approach, we used diffusion regularizations for both the motion and the super-
resolved image. However, the approach is flexible and other suitable regularization schemes (e.g., total variation) may be employed in the proposed formulation.
Furthermore, multi-scale implementations of the approach seem feasible and can
improve the convergence of the numerical scheme towards a global minimizer.
Finally, computational validation on other image sequences and addressing the
problem of automatic parameter selection are natural steps in extending the
presented theory.

Acknowledgements

This research was supported in part by the Natural Sciences and Engineering
Research Council of Canada (NSERC) in the form of a Post-Doctoral Fellowship
for Mehran Ebrahimi. This work was also supported by the Terry Fox Foundation
for Cancer Research. The authors would like to thank Dr. Peter Burns and Ross
Williams of the Sunnybrook Health Sciences Centre for providing the ultrasound
data.

References

1. Aizawa, K., Komatsu, T., Saito, T.: Acquisition of very high-resolution images
   using stereo cameras. In: Proceedings of SPIE Visual Communications and Image
   Processing, Boston, MA, November 1991, vol. 1605, pp. 318–328 (1991)
2. Borman, S.: Topics in Multiframe Superresolution Restoration. PhD thesis, Graduate Program in Electrical Engineering, University of Notre Dame, Indiana (April
   2004)
3. Chaudhuri, S.: Super-Resolution Imaging. Kluwer, Boston (2001)
4. Chung, J., Haber, E., Nagy, J.: Numerical methods for coupled super-resolution.
   Inverse Problems 22, 1261–1272 (2006)
5. Ebrahimi, M.: Inverse Problems and Self-Similarity in Imaging. PhD thesis, University of Waterloo (2008)
6. Ebrahimi, M., Vrscay, E.R.: Multi-frame super-resolution with no explicit motion
   estimation. In: Proceedings of The 2008 International Conference on Image Processing, Computer Vision, and Pattern Recognition, IPCV, Las Vegas, Nevada,
   USA, pp. 455–459 (2008)
7. Elad, M., Feuer, A.: Restoration of a single superresolution image from several
   blurred, noisy, and undersampled measured images. IEEE Transactions on Image
   Processing 6(12), 1646–1658 (1997)
8. Farsiu, S., Robinson, D., Elad, M., Milanfar, P.: Advances and challenges in super-
   resolution. International Journal of Imaging Systems and Technology 14(2), 47–57
   (2004)

9. Farsiu, S., Robinson, D., Elad, M., Milanfar, P.: Fast and robust multi-frame super-resolution. IEEE Transactions on Image Processing 13(10), 1327–1344 (2004)

10. Fischer, B., Modersitzki, J.: Fast image registration: a variational approach. Applied Numerical Analysis and Computational Mathematics 1(1-2), A69–A74 (2004); (NACoM-2003), Cambridge, UK., May 23-26, vol. 1, pp. A69–A74. Wiley-VCH, Germany (2004)

11. Fischer, B., Modersitzki, J.: Ill-posed medicine-an introduction to image registration. Inverse Problems 24(3), 034008, 16 (2008)

12. Hardie, R., Barnard, K., Armstrong, E.: Joint MAP registration and high-resolution image estimation using a sequence of undersampled images. IEEE Transactions on Image Processing 6(12), 1621–1633 (1997)

13. Horn, B.K.P., Schunck, B.G.: Determining optical flow. Artificial Intelligence 17(1-3), 185–203 (1981)

14. Irani, M., Peleg, S.: Improving resolution by image registration. CVGIP: Graphical Model and Image Processing 53, 324–335 (1993)

15. Modersitzki, J.: Numerical methods for image registration. Oxford University Press, Oxford (2004)

16. Nguyen, N.X.: Numerical Algorithms for Image Superresolution. PhD thesis, Graduate Program in Scientific Computation and Computational Mathematics, Stanford University (July 2000)

17. Papoulis, A.: Generalized sampling expansion. IEEE Transactions on Circuits and Systems 24, 652–654 (1997)

18. Sauer, K., Allebach, J.: Iterative reconstruction of band-limited images from non-uniformly spaced samples. IEEE Transactions on Circuits and Systems, 1497–1505 (1987)

19. Schoenemann, T., Cremers, D.: High resolution motion layer decomposition using dual-space graph cuts. In: CVPR. IEEE Computer Society, Los Alamitos (2008)

20. Schultz, R., Stevenson, R.: Extraction of high-resolution frames from video sequences. IEEE Transactions on Image Processing 5, 996–1011 (1996)

21. Shannon, C.E.: Communication in the presence of noise. Proceedings of the IRE 37, 10–21 (1949)

22. Shekarforoush, H., Chellappa, R.: Data-driven multi-channel super-resolution with application to video sequences. Journal of the Optical Society of America A 16(3), 481–492 (1999)

23. Tekalp, A., Ozkan, M., Sezan, M.: High-resolution image econstruction from lower-resolution image sequences and space-varying image restoration. In: Proceedings of the IEEE International Conference on Acoustics, Speech and Signal Processing, San Francisco, CA, vol. 3, pp. 169–172 (1992)

24. Thirion, J.P.: Fast non-rigid matching of 3d medical images. In: Medical Robotics and Computer Aided Surgery (MRCAS), Baltimore, pp. 47–54 (1995)

25. Thirion, J.P.: Image matching as a diffusion process: an analogy with maxwell's demons. Medical Image Analysis 2(3), 243–260 (1998)

26. Tsai, R., Huang, T.: Multi-frame image restoration and registration. In: Advances in Computer Vision and Image Processing, Greenwich, CT, vol. 1 (1984)

27. Ur, H., Gross, D.: Improved resolution from subpixel shifted pictures. CVGIP: Graphical Models and Image Processing 54(2), 181–186 (1992)

Appendix 1: Proof of Theorem 1

Proof. Split the variation $v(x)$ as $v(x) := \big(v_1(x), \ldots, v_m(x), g(x)\big)$ in which $v_i(x) \in \mathbb{R}^d$, $i = 1, \ldots, m$ and $g(x) \in \mathbb{R}$. Also define,

$$p_i(x) := \mathcal{H}^*[\mathcal{H}\mathcal{S}_i f(x) - y_i(x)]\nabla \mathcal{S}_i f(x), \quad i = 1, \ldots, m,$$

$$q(x) := -\sum_{i=1}^{m} \mathcal{S}_i^* \mathcal{H}^*[\mathcal{H}\mathcal{S}_i f(x) - y_i(x)].$$

Using the Taylor expansion of $f_{u_i+hv_i}(x)$ and $g_{u_i+hv_i}(x)$ with respect to h at the point $x - u(x)$,

$$f_{u_i+hv_i}(x) = f(x - u_i(x) - hv_i(x))) = f_{u_i}(x) - h\langle \nabla f_{u_i}(x), v_i(x)\rangle_{\mathbb{R}^d} + \mathcal{O}(h^2),$$

$$g_{u_i+hv_i}(x) = g(x - u_i(x) - hv_i(x))) = g_{u_i}(x) - h\langle \nabla g_{u_i}(x), v_i(x)\rangle_{\mathbb{R}^d} + \mathcal{O}(h^2),$$

for every $i = 1, \ldots, m$. Directly using the definitions and the linearity of $\mathcal{H}$

$$
\begin{aligned}
&d\,\mathcal{C}\,[y; (u, f); v] \\
&= \lim_{h\to 0} \frac{1}{h}(\mathcal{C}[y; (u, f) + hv] - \mathcal{C}[y; (u, f)]) \\
&= \lim_{h\to 0} \frac{1}{h}(\mathcal{C}[y; (u_1 + hv_1, \ldots, u_m + hv_m, f + hg)] - \mathcal{C}[y; (u_1, \ldots, u_m, f)]) \\
&= \lim_{h\to 0} \frac{1}{2h} \sum_{i=1}^{m} \int_{\Omega} [y_i(x) - \mathcal{H}(f + hg)_{u_i+hv_i}(x)]^2 - [y_i(x) - \mathcal{H}f_{u_i}(x)]^2 \, dx \\
&= \lim_{h\to 0} \frac{1}{2h} \sum_{i=1}^{m} \int_{\Omega} [y_i(x) - \mathcal{H}f_{u_i+hv_i}(x) - h\mathcal{H}g_{u_i+hv_i}(x)]^2 - [y_i(x) - \mathcal{H}f_{u_i}(x)]^2 \, dx \\
&= \lim_{h\to 0} \frac{1}{2h} \sum_{i=1}^{m} \int_{\Omega} [y_i(x) - \mathcal{H}(f_{u_i}(x) - h\langle \nabla f_{u_i}(x), p(x)\rangle_{\mathbb{R}^d} + \mathcal{O}(h^2)) - h\mathcal{H}g_{u_i}(x) + \mathcal{O}(h^2)]^2 \\
&\quad - [y_i(x) - \mathcal{H}f_{u_i}(x)]^2 \, dx \\
&= \lim_{h\to 0} \frac{1}{2h} \sum_{i=1}^{m} \int_{\Omega} 2[y_i(x) - \mathcal{H}f_{u_i}(x)][h]\mathcal{H}(\langle \nabla f_{u_i}(x), v_i(x)\rangle_{\mathbb{R}^d} - g_{u_i}(x)) + \mathcal{O}(h^2)dx \\
&= -\sum_{i=1}^{m} \int_{\Omega} [\mathcal{H}f_{u_i}(x) - y_i(x)]\mathcal{H}(\langle \nabla f_{u_i}(x), v_i(x)\rangle_{\mathbb{R}^d} - g_{u_i}(x))dx \\
&= -\sum_{i=1}^{m} \int_{\Omega} \left[[\mathcal{H}f_{u_i}(x) - y_i(x)]\mathcal{H}(\langle \nabla f_{u_i}(x), v_i(x)\rangle_{\mathbb{R}^d})\right]dx + \sum_{i=1}^{m} \int_{\Omega} [\mathcal{H}f_{u_i}(x) - y_i(x)]\mathcal{H}g_{u_i}(x)dx.
\end{aligned}
$$

Hence,

$$
d\,\mathcal{C}\,[y;(u,f);v]
$$

$$
= -\sum_{i=1}^{m}\Big\langle [\mathcal{H}f_{u_i}(x) - y_i(x)], \mathcal{H}(\langle \nabla f_{u_i}(x), v_i(x)\rangle_{\mathbb{R}^d})\Big\rangle_{\mathcal{L}^2(\Omega)}
$$

$$
+\sum_{i=1}^{m}\Big\langle [\mathcal{H}f_{u_i}(x) - y_i(x)], \mathcal{H}g_{u_i}(x)\Big\rangle_{\mathcal{L}^2(\Omega)}
$$

$$
= -\sum_{i=1}^{m}\Big\langle \mathcal{H}^*[\mathcal{H}f_{u_i}(x) - y_i(x)], (\langle \nabla f_{u_i}(x), v_i(x)\rangle_{\mathbb{R}^d})\Big\rangle_{\mathcal{L}^2(\Omega)}
$$

$$
+\sum_{i=1}^{m}\Big\langle \mathcal{H}^*[\mathcal{H}f_{u_i}(x) - y_i(x)], g_{u_i}(x)\Big\rangle_{\mathcal{L}^2(\Omega)}
$$

$$
= -\sum_{i=1}^{m}\Big\langle \mathcal{H}^*[\mathcal{H}f_{u_i}(x) - y_i(x)], (\langle \nabla f_{u_i}(x), v_i(x)\rangle_{\mathbb{R}^d})\Big\rangle_{\mathcal{L}^2(\Omega)}
$$

$$
+\sum_{i=1}^{m}\Big\langle \mathcal{H}^*[\mathcal{H}f_{u_i}(x) - y_i(x)], \mathcal{S}_i g(x)\Big\rangle_{\mathcal{L}^2(\Omega)}
$$

$$
= -\sum_{i=1}^{m}\Big\langle \mathcal{H}^*[\mathcal{H}f_{u_i}(x) - y_i(x)], (\langle \nabla f_{u_i}(x), v_i(x)\rangle_{\mathbb{R}^d})\Big\rangle_{\mathcal{L}^2(\Omega)}
$$

$$
+\sum_{i=1}^{m}\Big\langle \mathcal{S}_i^*\mathcal{H}^*[\mathcal{H}f_{u_i}(x) - y_i(x)], g(x)\Big\rangle_{\mathcal{L}^2(\Omega)}
$$

$$
= -\sum_{i=1}^{m}\int_{\Omega} \langle p_i(x), v_i(x)\rangle_{\mathbb{R}^d}\,dx + \sum_{i=1}^{m}\int_{\Omega} \mathcal{S}_i^*\mathcal{H}^*[\mathcal{H}f_{u_i}(x) - y_i(x)]g(x)\,dx
$$

$$
= -\int_{\Omega}\Big[\sum_{i=1}^{m}\langle p_i(x), v_i(x)\rangle_{\mathbb{R}^d} - \sum_{i=1}^{m}\mathcal{S}_i^*\mathcal{H}^*[\mathcal{H}f_{u_i}(x) - y_i(x)]g(x)\Big]dx
$$

$$
= -\int_{\Omega}\Big[\sum_{i=1}^{m}\langle p_i(x), v_i(x)\rangle_{\mathbb{R}^d} + q(x)g(x)\Big]dx
$$

$$
= -\int_{\Omega}\Big[\langle p_1(x), v_1(x)\rangle_{\mathbb{R}^d} + \cdots + \langle p_m(x), v_m(x)\rangle_{\mathbb{R}^d} + \langle q(x), g(x)\rangle_{\mathbb{R}}\Big]dx
$$

$$
= -\int_{\Omega}\Big\langle \big(p_1(x),\ldots,p_m(x),q(x)\big), \big(v_1(x),\ldots,v_m(x),g(x)\big)\Big\rangle_{\mathbb{R}^{md+1}}dx
$$

$$
= -\int_{\Omega}\langle \Phi(x, u(x), f(x)), v(x)\rangle_{\mathbb{R}^{md+1}}\,dx.
$$

Appendix 2: Definition of A

$\mathbf{A} \in \mathbb{R}^{n^d \times n^d}$ is defined as $\mathbf{A} := \sum_{l=1}^{d} \mathbf{A}_l$ where $\mathbf{A}_l = \underbrace{\mathbf{I} \otimes \cdots \otimes \mathbf{I}}_{l-1 \text{ times}} \otimes \mathbf{B} \otimes \underbrace{\mathbf{I} \otimes \cdots \otimes \mathbf{I}}_{d-l \text{ times}}$,

in which $\mathbf{I} \in \mathbb{R}^{n \times n}$ is identity matrix and $\otimes$ denotes the Kronecker product of matrices. The lth factor $\mathbf{B} \in \mathbb{R}^{n \times n}$ is an approximation of the second order derivative in only one spatial direction. More precisely, it can be defined as the tridiagonal matrix

$$
\mathbf{B} = \begin{pmatrix}
-2 & 1 & 0 & \ldots & 0 \\
1 & -2 & 1 & \ldots & 0 \\
\vdots & \ddots & \ddots & \ddots & \vdots \\
0 & \ldots & 1 & -2 & 1 \\
0 & \ldots\ldots & & 1 & -2
\end{pmatrix}.
$$

Appendix 3: Construction of Matrices K, D, and H

Given the zooming factor $z \in \mathbb{N}$, we define $\mathbf{K} \in \mathbb{R}^{(n-z+1)^d \times n^d}$ as $\mathbf{K} := \underbrace{\mathbf{K}_1 \otimes \cdots \otimes \mathbf{K}_1}_{d \text{ times}}$

in which $\mathbf{K}_1 \in \mathbb{R}^{(n-z+1) \times n}$ is

$$
\mathbf{K}_1 = \frac{1}{z}
\begin{pmatrix}
\underbrace{1\ldots 1}_{z \text{ times}}\underbrace{0\ldots\ldots\ldots\ldots}_{n-z \text{ times}} 0 \\
0\underbrace{1\ldots 1}_{z \text{ times}}\underbrace{0\ldots\ldots\ldots\ldots}_{n-z-1 \text{ times}}0 \\
0\,0\underbrace{1\ldots 1}_{z \text{ times}}\underbrace{0\ldots\ldots\ldots}_{n-z-2 \text{ times}}0 \\
\vdots \quad \vdots \quad \vdots \quad \vdots \\
0\underbrace{\ldots\ldots\ldots\ldots}_{n-z \text{ times}}0\underbrace{1\ldots 1}_{z \text{ times}}
\end{pmatrix}.
$$

Also, $\mathbf{D} \in \mathbb{R}^{\lfloor n/z \rfloor^d \times (n-z+1)^d}$ is defined as $\mathbf{D} := \underbrace{\mathbf{D}_1 \otimes \cdots \otimes \mathbf{D}_1}_{d \text{ times}}$, where

$\mathbf{D}_1 \in \mathbb{R}^{\lfloor n/z \rfloor \times (n-z+1)}$ is

$$
\mathbf{D}_1 =
\begin{pmatrix}
1 & \underbrace{0\ldots\ldots\ldots\ldots\ldots\ldots\ldots\ldots\ldots\ldots\ldots}_{n-z \text{ times}}0 \\
\underbrace{0\ldots\ldots 0}_{z \text{ times}} 1 \underbrace{0\ldots\ldots\ldots\ldots\ldots\ldots}_{n-2z \text{ times}}0 \\
\underbrace{0\ldots\ldots\ldots\ldots}_{2z \text{ times}}0\ 1\ \underbrace{0\ldots\ldots\ldots\ldots}_{n-3z}0 \\
\vdots \quad \vdots \quad \vdots \quad \vdots \\
\underbrace{0\ldots\ldots\ldots\ldots\ldots\ldots\ldots}_{(\lfloor n/z \rfloor-1)z \text{ times}}0\ 1\ \underbrace{0\ldots\ldots 0}_{n-(\lfloor n/z \rfloor)z \text{ times}}
\end{pmatrix}.
$$

Finally, $\mathbf{H} \in \mathbb{R}^{\lfloor n/z \rfloor^d \times n^d}$ is defined as the matrix multiplication of $\mathbf{H} := \mathbf{D} \times \mathbf{K}$.

On a Decomposition Model for Optical Flow

Jochen Abhau[1], Zakaria Belhachmi[2], and Otmar Scherzer[3,4]

[1] Department of Mathematics, University of Innsbruck,
Technikerstrasse 21a, A-6020 Innsbruck, Austria
Jochen.Abhau@uibk.ac.at
http://infmath.uibk.ac.at
[2] Laboratoire de Mathématiques LMAM, UMR 7122,
Université UPV-Metz, ISGMP,
Batiment A, Ile du Saulcy, 57045 Metz, France
belhach@univ-metz.fr
[3] Department of Mathematics, University of Vienna,
Nordbergstrasse 15, A-1090 Wien, Austria
[4] Radon Institute of Computational and Applied Mathematics,
Altenberger Strasse 69, A-4040 Linz, Austria
otmar.scherzer@oeaw.ac.at

Abstract. In this paper we present a variational method for determining cartoon and texture components of the optical flow of a noisy image sequence. The method is realized by reformulating the optical flow problem first as a variational denoising problem for multi-channel data and then by applying decomposition methods. Thanks to the general formulation, several norms can be used for the decomposition. We study a decomposition for the optical flow into bounded variation and oscillating component in greater detail. Numerical examples demonstrate the capabilities of the proposed approach.

1   Introduction

Let be given $\Omega \subseteq \mathbb{R}^2$, a rectangular domain, and let the spatial-temporal function $f : \Omega \times [0, \infty) \to \mathbb{R}$ be a representation of a continuous image sequence. The goal of this work is to apply *image* decomposition methods to separate the optical flow belonging to f in texture and cartoon parts.

For this purpose we first review on optical flow estimation and decomposition methods.

Optical flow estimation
Optical flow estimation is used to determine the motion in an image sequence by tracking pixels of constant intensity. For an excellent overview on optical flow estimation we refer to [11]. The standard optical flow model is differential and based on a Taylor series expansion, requiring that $f \in C^1(\Omega \times [0, \infty); \mathbb{R}^2)$. The optical flow is a characteristics $\boldsymbol{w} = (w_1, w_2)^T$, $w_1 = w_1(x_1, x_2, t)$, $w_2 = w_2(x_1, x_2, t)$ of the differential equation

$$f_t + f_{x_1} w_1 + f_{x_2} w_2 = 0 \text{ for } (x_1, x_2) \in \Omega, t \in [0, \infty) . \tag{1}$$

D. Cremers et al. (Eds.): EMMCVPR 2009, LNCS 5681, pp. 126–139, 2009.
© Springer-Verlag Berlin Heidelberg 2009

In mathematical terms *characteristics* are the paths of constant intensity. Variational optical flow methods are based on least squares formulations, consisting in minimization of the functional

$$\boldsymbol{w} \to \mathcal{S}(\boldsymbol{w}) := \frac{1}{2} \|f_t + f_{x_1} w_1 + f_{x_2} w_2\|_{L^2(\Omega)}^2 \ . \tag{2}$$

The minimization problem is ill-posed, which is usually overcome by adding a convex regularization term $\mathcal{R}$ to $\mathcal{S}$. For $\lambda > 0$, the regularized optical flow problem consists in minimization of

$$\boldsymbol{w} \to \frac{1}{\lambda} \mathcal{S}(\boldsymbol{w}) + \mathcal{R}(\boldsymbol{w}) . \tag{3}$$

Optical flow methods have been pioneered in [12]. There, the squared L^2-norm of the gradient is used for regularization and therefore the method consists in minimization of

$$\boldsymbol{w} \to \frac{1}{\lambda} \mathcal{S}(\boldsymbol{w}) + \frac{1}{2} \|\nabla \boldsymbol{w}\|_{L^2(\Omega;\mathbb{R}^2)}^2 \ . \tag{4}$$

This regularization approach has the drawback that the computed optical flow field $\boldsymbol{w}$ is not aligned with edges in f_1 and f_2. To overcome this drawback generalized regularization functionals $\mathcal{R}$ have been considered in the literature. See for instance [9,10,19,6], to name but a few. An extensive survey on variational methods in optical flow estimation is given in [18]

Image decomposition models.
Recently, decomposition models of gray-value images into structural and textural components have been studied [14,16,15]. Generally speaking, for an image I, these models consist in minimizing the functional

$$(u, v) \mapsto \frac{1}{2\lambda} \|u + v - I\|^2 + N_U(u) + \gamma N_V(v) \text{ over } u \in U, \ v \in V \tag{5}$$

The minimizer (u, v) of (5) is considered the structural and textural component of I. In [4], various spaces U (such as $BV(\Omega)$, the Sobolev spaces $W_0^{1,p}(\Omega)$, and the homogeneous Besov space $U = \dot{B}_{1,1}^1$) with associated seminorms N_U and duals $V = U^*$, $N_V = (N_U)^*$ have been examined. As it is reported there, the choice $U = BV(\Omega)$ and N_U the total variation seminorm is very interesting, since the dual of the total variation seminorm approximates Meyer's G-norm [14]. The G-norm is suitable to model texture, because it takes small values on oscillating functions.

Optical flow decomposition models
Quite recently, there have been established decomposition models for optical flow models. In particular, for analyzing experimental fluid flow data, decomposition into solenoidal components (div $\boldsymbol{w}$) and vortices (curl $\boldsymbol{w}$) of the flow $\boldsymbol{w}$ are calculated (see [13,21,20]). There, minimizers of functionals of the form

$$\mathcal{S}(\boldsymbol{w}) + \lambda_d \int_\Omega |\nabla \text{div } \boldsymbol{w}|^{p_d} dxdy + \lambda_c \int_\Omega |\nabla \text{curl } \boldsymbol{w}|^{p_c} dxdy + \gamma \int_{\partial\Omega} (\partial_n \boldsymbol{w})^2 ds \tag{6}$$

with $p_d, p_c \in \{1, 2\}$ are used for optical flow decomposition.

A duality based model for optical flow estimation is proposed in [22]. Functionals of the form

$$E_\theta(\boldsymbol{u}, \boldsymbol{v}) = \int_\Omega \left\{ \sum_d |\nabla u_d| + \frac{1}{2\theta} \sum_d v_d^2 + \lambda |\rho(\boldsymbol{u} + \boldsymbol{v})| \right\} dx \qquad (7)$$

are minimized. Here, $\boldsymbol{u} = (u_d)_d$ and $\boldsymbol{v} = (v_d)_d$ are flow fields, $|\rho|$ is a data fidelity function, being small if $\boldsymbol{u} + \boldsymbol{v}$ solves the optical flow equation, and $\lambda > 0$, $\theta > 0$ are weighting parameters. The arising optical flow $\boldsymbol{w} = \boldsymbol{u} + \boldsymbol{v}$ is implicitly decomposed into a component $\boldsymbol{u}$ of small total variation and a component $\boldsymbol{v}$ of small L^2-norm.

Outline of the paper

In this paper we apply the variational decomposition models of [4] to optical flow problems. To do so, we first reformulate the optical flow problem as an image denoising problem for vector valued data (cf. Section 2). In section 3 we recall recent methods for image decomposition of color data [5] and decomposition models for gray valued data [3] and adopt them for optical flow decomposition. We present a general formulation of variational optical flow decomposition which allows for utilizing various spaces and seminorms. In section 4, we particularly focus on the total variation seminorm and Meyer's G-norm. Moreover, a variant of Chambolle's algorithm (originally used for total variation denoising) is used to compute numerical examples in section 5, which demonstrate the feasibility of the proposed method.

2   Reformulation as a Denoising Problem and Optical Flow Decomposition

The matrix $A_0 := \nabla f (\nabla f)^T$ has rank one, is positive semi-definite with nontrivial kernel, which consists of all vector valued functions, which are orthogonal to ∇f. Moreover, $\langle \boldsymbol{u}, \boldsymbol{v} \rangle_{A_0} = \int_\Omega \boldsymbol{u}^T A_0 \boldsymbol{v}$ is an inner product and by $|\boldsymbol{u}|_{A_0}^2 := \langle \boldsymbol{u}, \boldsymbol{u} \rangle_{A_0} = \int_\Omega \boldsymbol{u}^T A_0 \boldsymbol{u}$ a seminorm is given. For further rewriting the optical flow least-squares functional $\mathcal{S}$, defined in (2), we use a full rank approximation of A_0, which is derived in two steps. We first regularize A_0 by setting $\tilde{A} := ((A_0)^T A_0 + \epsilon \text{Id})^{\frac{1}{2}}$. Here, Id denotes the identity matrix and $\epsilon > 0$ is a small regularization parameter. This way, $\tilde{A}$ is positive definite. Second, we apply an anisotropic evolution to the matrix $\tilde{A}$ to enhance the structure of the underlying image data. We solve

$$\partial_t a_{ij} = \text{div}\, g(|\nabla A \nabla A^T|) \nabla a_{ij} \qquad (8)$$

$$a_{ij}(0) = \tilde{a}_{ij} \qquad (9)$$

It can be checked easily, that the matrix A is positive definite. Moreover, in [7,17], it is reported that this preprocessing is very appropriate for optical flow estimation in noisy data. As a consequence, $\langle u, v \rangle_A = \int_\Omega u^T A v$ is a scalar product (which we call A-scalar product) on the weighted L^2-space

$$L^2(\Omega; A) = \left\{ u : \|u\|_A := \sqrt{\langle u, u \rangle_A} < \infty \right\} . \tag{10}$$

The optical flow least squares functional $\mathcal{S}$, defined in (2), can now be approximated by the squared of the A-norm of the optical flow residual. To see this let

$$\hat{f} := \frac{1}{|\nabla f|} (-f_t f_x, -f_t f_y)^T \tag{11}$$

Note that $A_0^{1/2}$, defined by spectral decomposition, equals $\frac{1}{|\nabla f|} A_0$. Then,

$$\left\| A_0^{1/2} \cdot w - \hat{f} \right\|_{L^2(\Omega; \mathbb{R}^2)}^2$$

$$= \int_\Omega [(A_0^{1/2} \cdot w)_1 - \hat{f}_1]^2 + [(A_0^{1/2} \cdot w)_2 - \hat{f}_2]^2$$

$$= \int_\Omega \left[\frac{f_x^2 w_1 + f_x f_y w_2}{|\nabla f|} + \frac{f_x f_t}{|\nabla f|} \right]^2 + \left[\frac{f_x f_y w_1 + f_y^2 w_2}{|\nabla f|} + \frac{f_y f_t}{|\nabla f|} \right]^2 \tag{12}$$

$$= \int_\Omega \frac{f_x^2 + f_y^2}{|\nabla f|^2} (f_x w_1 + f_y w_2 + f_t)^2$$

$$= \|\nabla f \cdot w + f_t\|_{L^2(\Omega)}^2 ,$$

Using the notation

$$\tilde{f} = A^{-\frac{1}{2}} \hat{f}, \tag{13}$$

we find that

$$\|\nabla f \cdot w + f_t\|_{L^2(\Omega)} = \left\| A_0^{1/2} \cdot w - \hat{f} \right\|_{L^2(\Omega; \mathbb{R}^2)} \approx \left\| w - \tilde{f} \right\|_A . \tag{14}$$

This relation shows that the optical flow least squares functional $\mathcal{S}$ defined in (2) can be approximated, and in fact replaced, by the squared norm of the weighted L^2-space defined in (10).

3 Decomposition Models for Optical Flow

From now on, taking into account (14), we regard the optical flow problem as an imaging problem. The actual difference to standard image decomposition [3] is that here the function to be filtered, $\tilde{f}$, is vector valued and that weighted norms are used in the fit-to-data functional.

Inspired by the work on variational decomposition of color data in [5] we consider minimizing

$$\mathcal{L}(u, v) := \frac{1}{2\lambda} \left\| (u + v) - \tilde{f} \right\|_A^2 + N_U(u) + \gamma N_V(v) ; \tag{15}$$

over $\boldsymbol{u} \in U$ and $\boldsymbol{v} \in V$. Thus, optical flow $\boldsymbol{w}$ is decomposed into $\boldsymbol{w} = \boldsymbol{u} + \boldsymbol{v}$. Several spaces and seminorms can be considered for U, V, N_U, N_V to model structure and texture component of optical flow, compare [4].

Total variation model
Here, we take $U = \mathrm{BV}(\Omega, \mathbb{R}^2)$, which is defined as the space of functions $\boldsymbol{f} \in L^\infty(\Omega, \mathbb{R}^2)$, with the property

$$J(\boldsymbol{f}) := \sup\left\{ \int_\Omega \boldsymbol{f} \cdot \mathrm{div}(\boldsymbol{p}) \ : \ \boldsymbol{p} \in C_c^1(\Omega, \mathbb{R}^2 \times \mathbb{R}^2), \ \|\boldsymbol{p}\|_{L^\infty(\Omega, \mathbb{R}^2 \times \mathbb{R}^2)} \leq 1 \right\} < \infty, \tag{16}$$

where $C_c^1(\Omega, \mathbb{R}^2 \times \mathbb{R}^2)$ is the space of differentiable functions with compact support in Ω; moreover, $\|\boldsymbol{p}\|_{L^\infty(\Omega, \mathbb{R}^2 \times \mathbb{R}^2)}$ is the L^∞-norm of the Frobenius-norm of $\boldsymbol{p}$.

We take $J(\boldsymbol{f})$, which is called total variation seminorm of $\boldsymbol{f}$, as N_U. The space V consists of $\mathbb{R}^2$- valued distributions $\boldsymbol{g}$, which can be written as $\boldsymbol{g} = \mathrm{div}(\boldsymbol{p})$, with $\boldsymbol{p} \in L^\infty(\Omega, \mathbb{R}^2 \times \mathbb{R}^2)$. The divergence operator is a distributional derivative here. On V we use for N_V the G-norm [14], which is defined as

$$\|\boldsymbol{g}\|_G := \inf\{\|\boldsymbol{p}\|_{L^\infty(\Omega, \mathbb{R}^2 \times \mathbb{R}^2)} \ : \ \boldsymbol{p} \in L^\infty(\Omega, \mathbb{R}^2 \times \mathbb{R}^2), \boldsymbol{g} = \mathrm{div}(\boldsymbol{p})\} \tag{17}$$

The G-norm is suitable to model texture, because it takes small values on oscillating functions.

Sobolev space model
In this model, we choose U as the Sobolev space $W_0^{1,p}(\Omega, \mathbb{R}^2)$ and N_U as the standard Sobolev seminorm (see [2] for a reference of general Sobolev spaces). Moreover, we use for V the dual space of U, which is $V = W^{-1,q}(\Omega, \mathbb{R}^2)$, where q is the conjugate of p, and N_V is the corresponding Sobolev-Norm.

Besov space model
Here $U = \dot{B}_{1,1}^1(\Omega, \mathbb{R}^2)$ is the homogenous Besov space with norm N_U (see also [2]) and V is its dual $\dot{B}_{-1,\infty}^\infty(\Omega, \mathbb{R}^2)$

For gray valued images, as stated in [4], the total variation model gives the most meaningful decomposition results of the three mentioned above. This is why we concentrate on this case in detail and apply it for the decomposition of optical flow.

4   Numerical Implementation

In the following section, we discretize the total variation model and derive numerical algorithms for its minimization. From now on we only consider a discrete and finite-dimensional setting.

Discrete one-channel images are matrices $U = (u_{i,j})$ of size $M \times N$, representing continuous images on Ω. Analogously, multi-channel images are matrices of size $M \times N$ with vectorial entries $\boldsymbol{u}_{i,j} = (u_{i,j}^1, u_{i,j}^2)^T$. We denote by $X := (\mathbb{R}^2)^{M \times N}$ the space of multi-channel matrices.

For a discrete matrix $H = (h_{ij})$ we define the discrete gradient $\nabla H = (\nabla_x H, \nabla_y H)^T$ as

$$\nabla_x h_{i,j} := \begin{cases} h_{i+1,j} - h_{i,j} & \text{if } i < M \\ 0 & \text{if } i = M \end{cases} \tag{18}$$

and

$$\nabla_y h_{i,j} := \begin{cases} h_{i,j+1} - h_{i,j} & \text{if } j < N \\ 0 & \text{if } j = N \end{cases}. \tag{19}$$

The discrete total variation of a vector field $\boldsymbol{u}$ is defined by

$$J(\boldsymbol{u}) = \sum_{i,j} \sqrt{(\nabla_x u_{i,j}^1)^2 + (\nabla_y u_{i,j}^1)^2 + (\nabla_x u_{i,j}^2)^2 + (\nabla_y u_{i,j}^2)^2}. \tag{20}$$

Moreover, the discrete divergence operator of the tensor $\boldsymbol{u}$ is defined by

$$[\text{div}(\boldsymbol{u})]_{i,j} = \begin{cases} u_{i,j}^1 - u_{i-1,j}^1 & \text{if } 1 < i < M \\ u_{i,j}^1 & \text{if } i = 1 \\ -u_{i-1,j}^1 & \text{if } i = M \end{cases} + \begin{cases} u_{i,j}^2 - u_{i,j-1}^2 & \text{if } 1 < j < N \\ u_{i,j}^2 & \text{if } j = 1 \\ -u_{i-1,j}^2 & \text{if } j = N \end{cases}.$$

The discrete divergence operator, as in the continuous setting again denoted by div, of $\boldsymbol{u}$ is defined as the discrete divergences of the components. For definition of the discrete time derivative, we fix a small constant $\delta t > 0$. Given two subsequent frames $U_k, U_{k+1} \in \mathbb{R}^{M \times N}$ of discrete one-channel images, we define

$$\nabla_t U = \frac{U_{k+1} - U_k}{\delta t}. \tag{21}$$

The discrete formulas in (18), (19) and (21) give approximations of $\tilde{\boldsymbol{f}}$ as defined in (13) and hence also of A_0 and its regularization A. The A-scalar product in $L^2(\Omega; A)$ as defined in (10) is then approximated by

$$\langle \boldsymbol{u}, \boldsymbol{v} \rangle_X = \sum_{i,j} \boldsymbol{u}_{i,j}^T A_{i,j} \boldsymbol{v}_{i,j}, \tag{22}$$

where $A_{i,j} \in \mathbb{R}^{2 \times 2}$.

Approximation of the discrete G-Norm by the dual J^* of J.
For definition of the G-norm, we set

$$K = \{ v \in X : \text{ there exists } \boldsymbol{p} = (p_1, p_2) \in X \times X \text{ such that } v = \text{div}(\boldsymbol{p}) \} \tag{23}$$

For $v \in K$, the discrete G-norm $\|\cdot\|$ is then given by

$$\|v\|_G = \inf \{\|\boldsymbol{p}\|_\infty : \boldsymbol{p} \in X \times X, v = \operatorname{div}(\boldsymbol{p})\}. \tag{24}$$

Here $\|\boldsymbol{p}\|_\infty$ denotes the l^∞-norm of the Frobenius-norm of the matrix $\boldsymbol{p}$. This definition of the discrete G-norm is difficult to implement numerically. For numerical purposes it is convenient, as proposed in [4], to use the Fenchel dual

$$J^*(v) := \sup\{\langle v, u \rangle - J(u) : (u, v) \in X \times X\} \tag{25}$$

of J. Since J is a seminorm, an elementary calculation shows that there exists a convex, closed set $K_1 \subseteq X$ such that

$$J^*(v) = \begin{cases} \infty & \text{if } v \in K_1 \\ 0 & \text{otherwise} \end{cases}. \tag{26}$$

Similar to [4], one can show that K_1 in (26) is given by

$$K_1 = \{\operatorname{div}(\boldsymbol{p}) : \boldsymbol{p} \in X \times X \text{ and } \|\boldsymbol{p}\|_\infty \leq 1\} \tag{27}$$

Because of this characterization, J^* is much easier to compute than the G-norm. The close relationship between G-norm and J^* is revealed by the following theorem:

Theorem 1. *Let α, λ, $\gamma > 0$. Consider the following minimization problems over $X \times X$:*

(A) $\min_{u+v=\tilde{f}} J(u) + \alpha \|v\|_G$
(B) $\min_{u+v=\tilde{f}} J(u) + J^*(\frac{v}{\gamma})$
(C)

$$\min_{(u,v)} H_{\lambda,\gamma}(u, v) := \frac{1}{2\lambda} \left\| u + v - \tilde{f} \right\|_X^2 + J(u) + J^*(\frac{v}{\gamma}) \tag{28}$$

Then minimizers for all three problems exist, and for (C) it is unique. Moreover, there exists a relation between α and γ, such that a minimizer of (A) is a minimizer of (B) and vice versa. Moreover, as $\lambda \downarrow 0$, the minimizers of (C) converge to a minimizer of (B).

Proof. Here, we only prove existence and uniqueness of a minimizer of (C), following the proof in [3] for gray valued images. The other statements can then be proven analogously and are therefore omitted.

Existence of a minimizer of $H_{\lambda,\gamma}$: The set $X \times \gamma K$ is closed in the finite dimensional space $X \times X$ and the restriction of $H_{\lambda,\gamma}$ to $X \times \gamma K$ is continuous. Therefore lower semicontinuity of $H_{\gamma,\lambda}$ holds, that is, for every sequence $(u_k, v_k) \in X \times X$

$$H_{\gamma,\lambda}(u, v) \leq \liminf_{(u_k,v_k) \to (u,v)} H_{\gamma,\lambda}(u_k, v_k).$$

Next we show coercivity for $H_{\lambda,\gamma}$ on $X \times X$, i.e. we prove that

$$H_{\lambda,\gamma}(\boldsymbol{u}, \boldsymbol{v}) \to \infty \text{ if } \|(\boldsymbol{u}, \boldsymbol{v})\|_{X \times X} \to \infty \, .$$

Let $(\boldsymbol{u}, \boldsymbol{v}) \in X \times \gamma K$. By the definition of γK there exists $\boldsymbol{p} \in X \times X$ such that $\text{div}(\boldsymbol{p}) = \boldsymbol{v}$ and $\|\boldsymbol{p}\|_\infty \leq \gamma$. Therefore we have

$$\|\boldsymbol{v}\|_X^2 = \sum_{i,j} \text{div}(\boldsymbol{p})_{i,j}^T A_{i,j} \text{div}(\boldsymbol{p})_{i,j} \, ,$$

Since $\boldsymbol{p}$ is uniformly bounded with respect to $\boldsymbol{v}$, we see that $\|\cdot\|_X$ is bounded on γK. Therefore, if $\|(\boldsymbol{u}, \boldsymbol{v})\|_{X \times X} \to \infty$ with $\boldsymbol{v} \in \gamma K$, it follows that $\|\boldsymbol{u}\|_X \to \infty$, hence $H_{\lambda,\gamma}(\boldsymbol{u}, \boldsymbol{v}) \geq \left\|\boldsymbol{u} + \boldsymbol{v} - \tilde{\boldsymbol{f}}\right\|_X \to \infty$, which gives coercivity. In summary, since $H_{\lambda,\gamma}$ is lower semi-continuous and coercive in $X \times X$, a minimizer for (15) exists.

Uniqueness of a minimizer of $H_{\lambda,\gamma}$: The functional $H_{\lambda,\gamma}(\boldsymbol{u}, \boldsymbol{v})$ is strictly convex on $X \times \gamma K$ up to direction $(\boldsymbol{u}, -\boldsymbol{u})$. So it might happen for some $t > 0$, that both $(\boldsymbol{u}, \boldsymbol{v})$ and $(\boldsymbol{u}+t\boldsymbol{u}, \boldsymbol{v}-t\boldsymbol{u})$ are (global) minimizers of (15). We show that in this case, $\boldsymbol{u} = 0$, which means that the two minimizers coincide. Indeed, from $H_{\lambda,\gamma}(\boldsymbol{u} + t\boldsymbol{u}, \boldsymbol{v} - t\boldsymbol{u}) = H_{\lambda,\gamma}(\boldsymbol{u}, \boldsymbol{v}) + tJ(\boldsymbol{u})$ and the assumption that both are global minimizers, we conclude that $J(\boldsymbol{u}) = 0$. From the definition of the discrete total variation it follows that $\boldsymbol{u}$ is a constant. Moreover, since $\boldsymbol{u} \in \gamma K$, there exists $\boldsymbol{p} \in X \times X$ such that $\boldsymbol{u} = \text{div}(\boldsymbol{p})$. Therefore

$$\sum_{i,j} \boldsymbol{u}_{i,j} = \sum_{i,j} \text{div}(\boldsymbol{p})_{i,j} = 0.$$

The last equality holds, since each summand occurs exactly four times in $\sum_{i,j} \text{div}(\boldsymbol{p})_{i,j}$, twice with a plus, twice with a minus sign.

Theorem (1) enables us to discretize the G-norm in functional(15) by J^* and minimize $H_{\lambda,\gamma}$ numerically. Using Equation (26), $H_{\lambda,\gamma}(\boldsymbol{u}, \boldsymbol{v})$ can be expressed in the more convenient form

$$H_{\lambda,\gamma}(\boldsymbol{u}, \boldsymbol{v}) = \begin{cases} \frac{1}{2\lambda}\left\|\boldsymbol{u} + \boldsymbol{v} - \tilde{\boldsymbol{f}}\right\|_X^2 + J(\boldsymbol{u}) & \text{if } \boldsymbol{v} \in \gamma K_1 \\ \infty & \text{otherwise} \end{cases}. \tag{29}$$

In the following we investigate an *alternating direction* algorithm for minimization of $H_{\lambda,\gamma}$. It consists of the following two steps:

1. Choose $\boldsymbol{v}^{(0)} \in \gamma K_1$
2. For $k = 0, 1, 2, \ldots$
    − Calculate a minimizer $\boldsymbol{u}^{(k)} \in \gamma K$ of

$$\frac{1}{2\lambda}\left\|\boldsymbol{u} + \boldsymbol{v}^{(k)} - \tilde{\boldsymbol{f}}\right\|_X^2 + J(\boldsymbol{u}) \, . \tag{30}$$

– Calculate a minimizer $v^{(k+1)}$ of

$$\left\| u^{(k)} + v - \tilde{f} \right\|_X^2 . \tag{31}$$

– Continue

We state convergence of the iteration process. The proof is omitted here, since it is again along the lines in [3].

Theorem 2. *The sequence* $(u^{(k)}, v^{(k)})$ *converges to the unique solution* $(\hat{u}, \hat{v})$ *of* (28).

We stress that $\tilde{v}$ solves (31) if and only if $\tilde{w} = \tilde{f} - u - \tilde{v}$ is a minimizer of the functional

$$w \to \frac{1}{2\gamma} \left\| u + w - \tilde{f} \right\|_X^2 + J(w) . \tag{32}$$

This formulation actually shows that also minimization with respect to v can be realized by total variation denoising. Therefore both iteration steps of the alternating direction algorithm can be realized by total variation denoising, which can be implemented with a variant of Chambolle's projection algorithm [8]. Therefore, in the sequel, we only consider total variation denoising for vector valued data u^δ, which consists in minimization of

$$\frac{1}{2\lambda} \left\| u - u^\delta \right\|_X^2 + J(u) . \tag{33}$$

Following Chambolle's algorithm [8], we derive an iterative procedure for minimizing (33).

Let

$$p(0) := (p^1(0), p^2(0)) := \begin{pmatrix} p^{1,1}(0) & p^{1,2}(0) \\ p^{2,1}(0) & p^{2,2}(0) \end{pmatrix} = 0 .$$

For $k = 0, 1, 2, \ldots$ set

$$q(k) = \left[\mathrm{div}(p(k)) - u^\delta/\lambda \right]$$

Each entry $A_{i,j}$ of the matrix A is a 2×2-matrix, which is positive definite. We set

$$S_{i,j} := A_{i,j}^{1/2} ,$$

the root of A. Then componentwise

$$p^1(k+1) = \frac{p^1(k) + \tau \left[\nabla \left(S_{1,1} q^1(k) + S_{1,2} q^2(k) \right) \right]}{1 + \tau \left| \nabla \left(S_{1,1} q^1(k) + S_{1,2} q^2(k) + S_{2,1} q^1(k) + S_{2,2} q^2(k) \right) \right|} \tag{34}$$

and

$$p^2(k+1) = \frac{p^2(k) + \tau \left[\nabla \left(S_{2,1} q^1(k) + S_{2,2} q^2(k) \right) \right]}{1 + \tau \left| \nabla \left(S_{1,1} q^1(k) + S_{1,2} q^2(k) + S_{2,1} q^1(k) + S_{2,2} q^2(k) \right) \right|} . \tag{35}$$

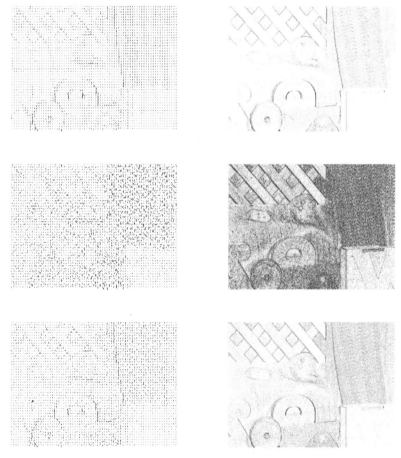

Fig. 1. (a) ON TOP: Frame 10 of the rubber whale sequence (584 × 388 pixels). (b) TOP LEFT: Cartoon part u of the flow between frame 10 and frame 11. (c) TOP RIGHT: Norm of cartoon part u. (d) MIDDLE LEFT: Texture part v. (e) MIDDLE RIGHT: Norm of texture part v. (f) BOTTOM LEFT: The flow field $w = u + v$. (g) BOTTOM RIGHT: Norm of the flow field $w = u + v$.

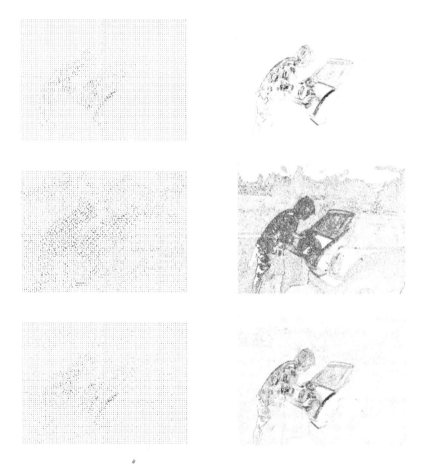

Fig. 2. (a) ON TOP: Frame 10 of the Mini Cooper sequence (640 × 480 pixels). (b) TOP LEFT: Cartoon part u of the flow between frame 10 and frame 11. (c) TOP RIGHT: Norm of cartoon part u. (d) MIDDLE LEFT: Texture part v. (e) MIDDLE RIGHT: Norm of texture part v. (f) BOTTOM LEFT: The flow field $w = u + v$. (g) BOTTOM RIGHT: Norm of the flow field $w = u + v$.

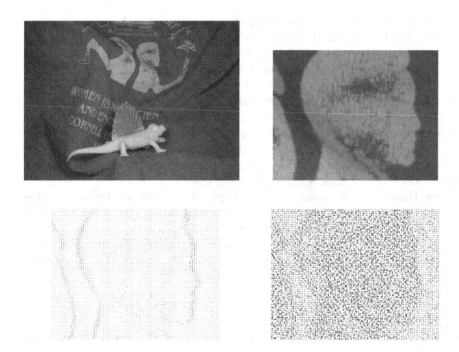

Fig. 3. (a) TOP LEFT: Frame 10 of the dimetrodon sequence (584 × 388 pixels). (b) TOP RIGHT: A detailed view of the head in the upper middle of (a). (c) BOTTOM LEFT: Detailed view of cartoon part u of the flow field between frame 10 and frame 11. (d) BOTTOM RIGHT: Detailed view of texture part v.

Theorem 3. *Let $\pi_{\lambda K}$ the orthogonal projector onto λK, where K is as in (23). If τ, as in (34) and (35), is chosen sufficiently small, then div $\boldsymbol{p}(k)$ converges to $\pi_{\lambda K}(\boldsymbol{u}^\delta)$. The solution of (33) is given by*

$$\boldsymbol{u} = \boldsymbol{u}^\delta - \pi_{\lambda K}(\boldsymbol{u}^\delta) \,. \tag{36}$$

Proof. The proof is along the lines of [8] and thus it is omitted here.

From Theorem 3 it follows that the minimizer u_α of (33) satisfies

$$u_\alpha = \boldsymbol{u}^\delta - \lambda \lim_{k \to \infty} \operatorname{div} \boldsymbol{p}(k) \,.$$

This theorem completes the numerical analysis of the functional $H_{\lambda,\gamma}$.

5   Results

We performed numerical experiments on several image sequences, which are publicly available at [1]. We decomposed the optical flow in the *rubber whale* sequence (see Figure 1). The curtain and the two round pieces at the bottom move to the left, while the other pieces move to the right, see (a). The moving large structures, such as the fence and the two round objects, are captured in the cartoon component (b), while for instance the curtain contains texture movement in (d). In order to improve visibility, the (scaled) magnitude of the flow u and v is shown in (c) and (e), respectively. The whole optical flow field is shown in (f) and (g). In the *Mini Cooper* sequence shown in Figure 2, a man is closing the trunk of a car (a). The cartoon part u of the optical flow consists of the moving tailgate and the man (b). The magnitude of flow u is shown in (c). The slight motion of the trees in the background is completely contained in the texture component v of the optical flow, shown in (d). The (upscaled) magnitude of v is given in (e).

These two examples show the advantages of our method quite well, separating different kinds of movements in an image sequence.

In the Dimetrodon sequence shown in Figure 3, we examine directions of the computed flow field and its decomposition in greater detail. The main direction of movement of the head can be seen in flow field u in (c), where neighboring flow vectors are often parallel. Oscillating patterns are captured in v shown in (d), where the directions of the flow vectors can differ very much between neighboring pixels.

6   Conclusion

We presented a general approach for decomposition of the optical flow of an image sequence into structural and textural components. A variational framework has been established, which allows to use various functionals for different kinds of texture extraction. Numerical examples demonstrate the effectivity of total variation norm, respectively G-norm, decomposition. In the future, we plan to study other possible seminorms in (15) as well and plan to compare the outcome of the different methods.

Acknowledgement

This work has been supported by the Austrian Science Fund (FWF) within the national research networks Industrial Geometry, project 9203-N12, and Photoacoustic Imaging in Biology and Medicine, project S10505-N20. The research stay of Dr. Zakaria Belhachmi in Innsbruck has been funded by the "'Frankreich Schwerpunkt"' of the University of Innsbruck. The work of the first author has been partly supported by the Austrian Ministry for Economy and Labour and by the Government of Upper Austria within the framework "Industrial Competence Centers".

We also want to thank the referees for their useful comments.

References

1. http://vision.middlebury.edu/flow/
2. Amann, H.: Compact embeddings of sobolev and besov spaces. Glasnik Matematicki 35(55), 161–177 (2000)
3. Aujol, J.-F., Aubert, G., Blanc-Feraud, L., Chambolle, A.: Image decomposition into a bounded variation component and an oscillating component. Journal of Mathematical Imaging and Vision 22(1), 71–88 (2005)

4. Aujol, J.-F., Chambolle, A.: Dual norms and image decomposition models. International Journal of Computer Vision 63(1), 85–104 (2005)
5. Aujol, J.-F., Kang, S.-H.: Color image decomposition and restoration. Journal of Visual Communication and Image Representation 17(4), 916–928 (2006)
6. Brox, T., Bruhn, A., Papenberg, N., Weickert, J.: High accuracy optical flow estimation based on a theory for warping. In: Pajdla, T., Matas, J(G.) (eds.) ECCV 2004. LNCS, vol. 3024, pp. 25–36. Springer, Heidelberg (2004)
7. Bruhn, A.: Variational optic flow computation: Accurate modeling and efficient numerics. PhD thesis, Saarland University, Germany (2006)
8. Chambolle, A.: An algorithm for total variation minimization and applications. Journal of Mathematical Imaging and Vision 20(1-2), 89–97 (2004)
9. Cohen, I.: Nonlinear variational method for optical flow computation. In: Proc. Eigth Scandinavian Conference on Image Analysis, Tromso, Norway, May 1993, vol. 1, pp. 523–530 (1993)
10. Deriche, R., Kornprobst, P., Aubert, G.: Optical flow estimation while preserving its discontinuities. In: Proc. Second Asian Conference on Computer Vision, Singapore, December 1995, vol. 2, pp. 290–295 (1995)
11. Fleet, D.J., Weiss, Y.: Optical Flow Estimation. In: Mathematical Models of Computer Vision: The Handbook, ch. 15, pp. 239–258. Springer, Heidelberg (2005)
12. Horn, B.K.P., Schunck, B.G.: Determining optical flow. Artificial Intelligence 17, 185–203 (2004)
13. Kohlberger, T., Memin, E., Schnoerr, C.: Variational dense motion estimation using helmholtz decomposition. In: Griffin, L.D., Lillholm, M. (eds.) Scale-Space 2003. LNCS, vol. 2695, pp. 432–448. Springer, Heidelberg (2003)
14. Meyer, Y.: Oscillating patterns in image processing and in some nonlinear evolution equations. In: The Fifteenth Dean Jaqueline B. Lewis Memorial Lectures (March 2001)
15. Starck, J.L., Elad, M., Donoho, D.L.: Image decomposition via the combination of sparse representations and a variational approach. IEEE Transactions on Image Processing 14(10), 1570–1582 (2005)
16. Vese, L.A., Osher, S.: Modeling textures with total variation minimization and oscillating patterns in image processing. Journal of Scientific Computing 19, 553–572 (2003)
17. Weickert, J., Bruhn, A., Brox, T., Papenberg, N.: A Survey on Variational Optic Flow Methods for Small Displacements. In: Mathematical Models for Registration and Applications to Medical Imaging, ch. 1, pp. 103–136. Springer, Heidelberg (2006)
18. Weickert, J., Schnoerr, C.: A theoretical framework for convex regularizers in pde-based computation of image motion. International Journal of Computer Vision 45(3), 245–264 (2001)
19. Weickert, J., Schnoerr, C.: Variational optic flow computation with a spatio-temporal smoothness constraint. Journal of Mathematical Imaging and Vision 14(3), 245–255 (2001)
20. Yuan, J., Schnoerr, C., Steidl, G.: Simultaneous higher-order optical flow estimation and decomposition. SIAM J. Scientfic Computing 29(6), 2283–2304 (2007)
21. Yuan, J., Schnörr, C., Steidl, G., Becker, F.: A study of non-smooth convex flow decomposition. In: Paragios, N., Faugeras, O., Chan, T., Schnörr, C. (eds.) VLSM 2005. LNCS, vol. 3752, pp. 1–12. Springer, Heidelberg (2005)
22. Zach, C., Pock, T., Bischof, H.: A duality based approach for realtime TV-shape L^1 optical flow. In: Hamprecht, F.A., Schnörr, C., Jähne, B. (eds.) DAGM 2007. LNCS, vol. 4713, pp. 214–223. Springer, Heidelberg (2007)

A Schrödinger Wave Equation Approach to the Eikonal Equation: Application to Image Analysis

Anand Rangarajan and Karthik S. Gurumoorthy[*]

Department of Computer and Information Science and Engineering
University of Florida, Gainesville, FL, USA
{anand,ksg}@cise.ufl.edu

Abstract. As Planck's constant $\hbar$ (treated as a free parameter) tends to zero, the solution to the eikonal equation $|\nabla S(X)| = f(X)$ can be increasingly closely approximated by the solution to the corresponding Schrödinger equation. When the forcing function $f(X)$ is set to one, we get the Euclidean distance function problem. We show that the corresponding Schrödinger equation has a closed form solution which can be expressed as a discrete convolution and efficiently computed using a Fast Fourier Transform (FFT). The eikonal equation has several applications in image analysis, viz. signed distance functions for shape silhouettes, surface reconstruction from point clouds and image segmentation being a few. We show that the sign of the distance function, its gradients and curvature can all be written in closed form, expressed as discrete convolutions and efficiently computed using FFTs. Of note here is that the sign of the distance function in 2D is expressed as a winding number computation. For the general eikonal problem, we present a perturbation series approach which results in a sequence of discrete convolutions once again efficiently computed using FFTs. We compare the results of our approach with those obtained using the fast sweeping method, closed-form solutions (when available) and Dijkstra's shortest path algorithm.

1 Introduction

While image analysis borrows liberally from classical mechanics—with variational principles [1], Euler-Lagrange equations, Hamiltonians [2] and Hamilton-Jacobi theory [3] all in widespread use at the present time—other than a few pioneering works [4,5], there isn't a concomitant borrowing from quantum mechanics. Given the very close relationship between Hamilton-Jacobi theory and Schrödinger wave mechanics, this is somewhat surprising. In this work, we begin with a brief overview of the classical mechanics sequence of i) variational principles and Euler-Lagrange equations, ii) Legendre transformations leading to the Hamiltonian [6], iii) canonical transformation of the Hamiltonian which yields the Hamilton-Jacobi theory [6], and finally iv) first quantization to obtain the Schrödinger wave equation [7]. This well known path of development needs an image analysis payoff which we next describe.

---

[*] Corresponding Author.

D. Cremers et al. (Eds.): EMMCVPR 2009, LNCS 5681, pp. 140–153, 2009.

It has been a decade since EMMCVPR 1999 at which event we saw [2] the advent of Hamiltonian mechanics to solve the eikonal equation. More specifically, the Hamiltonian approach was also used to analyze the Euclidean distance function problem—an important special case of the eikonal problem wherein $|\nabla S(X)| = 1$ and X a regular grid. The Euclidean distance function problem in its image analysis incarnation can be stated as follows: Given a set of shape silhouettes whose boundaries are parameterized as piecewise smooth curves, compute the *signed* distance at every location on a grid w.r.t. the boundary points. Furthermore, we often seek the gradient, divergence, curvature and medial axes of the signed distance function which are not easy to obtain by other approaches such as the fast marching [8] and fast sweeping methods [9] due to the lack of differentiability of the signed distance function. In sharp contrast, we show—using our previous work on this topic [10]—that the Schrödinger wave equation approach to the eikonal results in a closed-form solution which can be expressed as a discrete convolution and computed in $O(N \log N)$ time using a Fast Fourier Transform (FFT) [11] where N is the number of grid points. While the fast marching method is also $O(N \log N)$ (and even $O(N)$ with cleverly chosen data structures [8]), these methods are based on spatial discretizations of the derivative operator (in $|\nabla S(X)| = 1$) whereas the Schrödinger approach does not require derivative discretization. A caveat is that our Euclidean distance function is an approximation since it is obtained for a small but non-zero value of Planck's constant $\hbar$.

The Schrödinger equation approach to the eikonal gives us an unsigned distance function. We complement this by independently finding the sign of the distance function in $O(N \log N)$ time on a regular grid in 2D. We achieve this by efficiently computing the *winding number* for each location in the 2D grid. The winding number is the number of times a closed curve winds around a point. We show that just as in the case of the Schrödinger equation, the winding number can also be written in closed-form, expressed as a discrete convolution and efficiently computed using an FFT. The fact that the winding number can be expressed as a discrete convolution for every location in a 2D grid appears to be a new contribution.

We also leverage the closed-form solution for the unsigned distance function obtained from Schrödinger. Since our distance function is differentiable everywhere, we can once again write down closed-form expressions for the gradients and curvature, express them as discrete convolutions and efficiently compute these quantities in $O(N \log N)$ using FFTs. We visualize the gradients and the maximum curvature using 2D shape silhouettes as the source. The maximum curvature—despite its fundamental drawback of being extrinsic—has a haunting similarity to the medial axis. To our knowledge, the fast computation of the derivatives of the distance function on a regular grid using discrete convolutions is new.

Next, we present a general eikonal solver using a Born expansion-based perturbation method [12]. We cannot solve for $S(X)$ in closed form here, and instead we express the solution as a sequence of discrete convolutions, with each term efficiently computed using an FFT. We apply this method to image segmentation by first seeding a set of points on the interior and the exterior of the segmentation regions and then solving the eikonal using a forcing function $f(X)$ derived from the image gradients. These results are compared to those obtained using fast sweeping and Dijkstra's shortest path algorithm [13] (since the ground truth is not available). Since all results are obtained for very low values of $\hbar$, some numerical instability issues arise in the FFT-based convolutions. Consequently, higher precision numerical support for fast, discrete convolution is a fundamental requirement and one that we plan to address in future work.

2   A Schrödinger Equation for the Eikonal Problem

In this section, we briefly review the Schrödinger equation approach to (unsigned) Euclidean distance functions. We begin with a Lagrangian variational principle, derive the Hamiltonian via a Legendre transformation, use a canonical transformation to obtain the Hamilton-Jacobi equation and finally quantize Hamilton-Jacobi to obtain the Schrödinger equation.

The Lagrangian variational principle for Euclidean distance functions is an objective function whose solution is the shortest distance between two points in $\mathbb{R}^d$—the Euclidean distance. While we use $d = 2$ for illustration purposes, the approach is general and not restricted to a particular choice of dimension:

$$I[q] = \int_{t_0}^{t_1} L(q_1, q_2, \dot{q}_1, \dot{q}_2, t) dt \tag{1}$$

where $q(t) = \{q_1(t), q_2(t)\}$ is a C^2 path between two points in time t_0 and t_1 with

$$L(q_1, q_2, \dot{q}_1, \dot{q}_2, t) = \frac{1}{2} \left(\dot{q}_1^2 + \dot{q}_2^2 \right). \tag{2}$$

The corresponding Euler-Lagrange equations are

$$\ddot{q}_1(t) = 0, \text{ and } \ddot{q}_2(t) = 0 \tag{3}$$

which are tantamount to a straight line in 2D. This choice of L actually yields the *squared* Euclidean distance between two points $q(t_0)$ and $q(t_1)$. If we used the square root of this quantity in the Lagrangian, it becomes homogeneous of degree one in $(\dot{q}_1, \dot{q}_2)$ [as in $L(q_1, q_2, \lambda\dot{q}_1, \lambda\dot{q}_2, t) = \lambda L(q_1, q_2, \dot{q}_1, \dot{q}_2, t)$] and this creates problems for the Legendre transform. Note that the Lagrangian is independent of time t and this fact will later allow us to derive a static Schrödinger equation.

The Hamiltonian is obtained via a Legendre transform [6] applied to the Lagrangian:

$$H(q_1, q_2, p_1, p_2, t) = \sum_{i=1}^{2} p_i \dot{q}_i(p_1, p_2, t) - L(q_1, q_2, \dot{q}_1(p_1, p_2), \dot{q}_2(p_1, p_2), t) \tag{4}$$

where the momenta p_1, p_2 are defined as

$$p_i \equiv \frac{\partial L}{\partial \dot{q}_i}, \; i = 1, 2. \tag{5}$$

Equation (5) can be inverted to obtain $\dot{q}_i = \dot{q}_i(p_1, p_2, t)$ [and this fails if the Lagrangian is homogeneous of degree one in $(\dot{q}_1, \dot{q}_2)$].

The Hamilton-Jacobi equation is obtained via a canonical transformation [6] of the Hamiltonian. In classical mechanics, a canonical transformation is defined as a change of variables which leaves the form of the Hamilonian unchanged. For a type 2 canonical transformation, we have

$$\sum_{i=1}^{2} p_i \dot{q}_i - H(q_1, q_2, p_1, p_2, t) = \sum_{i=1}^{2} P_i \dot{Q}_i - K(Q_1, Q_2, P_1, P_2, t) + \frac{dF}{dt} \quad (6)$$

where $F \equiv -\sum_{i=1}^{2} Q_i P_i + F_2(q, P, t)$ which gives

$$\frac{dF}{dt} = -\sum_{i=1}^{2} \left(\dot{Q}_i P_i + Q_i \dot{P}_i \right) + \frac{\partial F_2}{\partial t} + \sum_{i=1}^{2} \left(\frac{\partial F_2}{\partial q_i} \dot{q}_i + \frac{\partial F_2}{\partial P_i} \dot{P}_i \right). \quad (7)$$

When we pick a *particular* type 2 canonical transformation wherein $\dot{P}_i = 0$, $i = 1, 2$ *and* $K(Q_1, Q_2, P_1, P_2, t) = 0$, we get

$$\frac{\partial F_2}{\partial t} + H(q_1, q_2, \frac{\partial F_2}{\partial q_1}, \frac{\partial F_2}{\partial q_2}, t) = 0 \quad (8)$$

where we are forced to make the identification $p_i = \frac{\partial F_2}{\partial q_i}$, $i = 1, 2$. Note that the new momenta P_i are constants of the motion (usually denoted by α_i, $i = 1, 2$). Changing F_2 to S as in common practice, we have the standard Hamilton-Jacobi equation for the function $S(q_1, q_2, \alpha_1, \alpha_2, t)$. To complete the circle back to the Lagrangian, we take the total time derivative of the Hamilton-Jacobi function S to get

$$\frac{dS(q_1, q_2, \alpha_1, \alpha_2, t)}{dt} = \sum_{i=1}^{2} \frac{\partial S}{\partial q_i} \dot{q}_i + \frac{\partial S}{\partial t}$$

$$= \sum_{i=1}^{2} p_i \dot{q}_i - H(q_1, q_2, \frac{\partial S}{\partial q_1}, \frac{\partial S}{\partial q_2}, t) = L(q_1, q_2, \dot{q}_1, \dot{q}_2, t). (9)$$

Consequently $S(q_1, q_2, \alpha_1, \alpha_2, t) = \int_{t_0}^{t} L dt$ and the constants $\{\alpha_1, \alpha_2\}$ can now be interpreted as integration constants.

For the Euclidean distance function problem, following (4) and (8), we get $H(q_1, q_2, p_1, p_2, t) = \frac{1}{2} (p_1^2 + p_2^2)$ and the Hamilton-Jacobi equation is

$$\frac{\partial S}{\partial t} + \frac{1}{2} \left[\left(\frac{\partial S}{\partial q_1} \right)^2 + \left(\frac{\partial S}{\partial q_2} \right)^2 \right] = 0. \quad (10)$$

The Schrödinger equation can sometimes be "derived" using a Feynman path integral approach. The more common approach—termed first quantization[1]—is

---

[1] First quantization is still mysterious. For an informal but illuminating treatment, please see http://math.ucr.edu/home/baez/categories.html

to convert the relation $p_i = \frac{\partial S}{\partial q_i}$, $i = 1, 2$ into an operator relation $p_i = i\hbar\frac{\partial}{\partial q_i}$, $i = 1, 2$. In a similar fashion, the time operator is $i\hbar\frac{\partial}{\partial t}$. When we quantize the Euclidean distance function problem, we get

$$i\hbar\frac{\partial\psi}{\partial t} + \frac{\hbar^2}{2}\left(\frac{\partial^2\psi}{\partial x^2} + \frac{\partial^2\psi}{\partial y^2}\right) = 0. \tag{11}$$

At first glance, there appear to be some similarities between the Hamilton-Jacobi equation in (10) and the Schrödinger equation in (11). Due to first quantization, the squared first derivatives w.r.t. space in the former have morphed into second derivative operators in the latter. Both equations have first derivatives w.r.t. time.

We now show that the time independence of the Lagrangian in (2) allows us to simplify the former into the static Hamilton-Jacobi equation and the latter into the static Schrödinger equation.

If the Lagrangian is not an explicit function of time, we can seek solutions for the Hamilton-Jacobi equation that are time independent. Setting $S(q_1, q_2, \alpha_1, \alpha_2, t) = S^*(q_1, q_2, \alpha_1, \alpha_2) - Et$ where $E = \frac{1}{2}$ is the total energy for the Euclidean distance function problem, we get

$$\left(\frac{\partial S^*}{\partial q_1}\right)^2 + \left(\frac{\partial S^*}{\partial q_2}\right)^2 = 1 \tag{12}$$

which is the eikonal equation with the forcing term set to one—a nonlinear, first-order differential equation. In a similar fashion, when we set $\psi(x, t) = \phi(x)\exp\left(\frac{it}{2\hbar}\right)$ and use $E = \frac{1}{2}$, we see that $\phi(x)$ satisfies the screened Poisson equation

$$\hbar^2\left(\frac{\partial^2\phi}{\partial x^2} + \frac{\partial^2\phi}{\partial y^2}\right) = \phi \tag{13}$$

which is a linear, second-order differential equation. A close relationship between ϕ and S^* can be shown by setting $\phi(x) = \exp\left\{-\frac{\hat{S}(x)}{\hbar}\right\}$ and rewriting (13) to get

$$\left(\frac{\partial\hat{S}}{\partial x_1}\right)^2 + \left(\frac{\partial\hat{S}}{\partial x_2}\right)^2 - \hbar\left(\frac{\partial^2\hat{S}}{\partial x_1^2} + \frac{\partial^2\hat{S}}{\partial x_2^2}\right) = 1 \tag{14}$$

which is strikingly similar to the eikonal equation in (12) with the important difference being a viscosity regularization term [14] modulated by the free parameter $\hbar$. [Note that the viscosity term arises naturally from (13)—an intriguing result.] As $\hbar \to 0$, $\hat{S} \to S^*$ which implies that we can solve the static Schrödinger equation in (13) instead of the eikonal equation in (12). In the next section, we describe fast algorithms for solving (14) and also present fast methods for computing the signed distance function and the derivatives of the Euclidean distance function.

3 Fast Computation of the Signed Euclidean Distance Function and Its Derivatives

In our previous work [10], we showed that the static Schrödinger equation in (13) can be efficiently solved using a Fast Fourier Transform (FFT) approach in arbitrary dimensions. The complexity of the FFT is $O(N \log N)$ where N is the total number of grid points (in any dimension). We briefly summarize the FFT-based Euclidean distance function algorithm.

3.1 Unsigned Euclidean Distance Functions

In the Euclidean distance function problem, we begin by considering the forced version of (13) in 2D:

$$- \hbar^2 \nabla^2 \phi + \phi = \sum_{k=1}^{K} \delta(X - Y_k). \tag{15}$$

The points $Y_k, k \in \{1, \ldots, K\}$ are a set of seed locations at which $S^*(Y_k) = 0, \forall Y_k, k \in \{1, \ldots, K\}$ with the set X being the locations at which we wish to compute the Euclidean distance function. A Green's function approach can be pursued since the above differential equation is homogeneous except at the seed locations Y. The Green's functions [15] (for an unbounded domain with Dirichlet boundary conditions) are

$$G_{1D}(X - Y) = \frac{1}{2\hbar} \exp\left(\frac{-|X - Y|}{\hbar}\right), \tag{16}$$

$$G_{2D}(X - Y) = = \frac{1}{2\pi\hbar^2} K_0\left(\frac{\|X - Y\|}{\hbar}\right), \tag{17}$$

and

$$G_{3D}(X - Y) = \frac{1}{4\pi\hbar^2} \frac{\exp\left(\frac{-\|X-Y\|}{\hbar}\right)}{\|X - Y\|} \tag{18}$$

in 1D, 2D and 3D respectively where $K_0(r)$ is the modified Bessel function of the second kind. We avoid the singularity at the origin in 2D and 3D by replacing their Green's functions with the exponential function (similar to the 1D Green's function). This is a very good approximation as $\hbar \to 0$ since the 2D and 3D Green's functions converge *uniformly* to the exponential function everywhere away from the origin. With this in place, we write the solution for $\phi(X)$ as

$$\phi(X) = \sum_{k=1}^{K} G(X) * \delta(X - Y_k) = \sum_{k=1}^{K} G(X - Y_k) \tag{19}$$

and the corresponding approximate solution to the eikonal equation (after removing terms independent of X) is

$$\hat{S}(X) = -\hbar \log \sum_{k=1}^{K} \exp\left\{-\frac{\|X - Y_k\|}{\hbar}\right\} \tag{20}$$

with the caveat being that we are using an approximate, unbounded domain Green's function $G(X)$ here. We have shown that an approximate solution for the eikonal (with the forcing term set to one) can be obtained in closed-form as in (20) and efficiently computed using an FFT since equation (19) expresses a *discrete* convolution [11] between the functions

$$G(X) = \exp\left\{-\frac{\|X\|}{\hbar}\right\} \tag{21}$$

and

$$Y_{\mathrm{kron}}(X) \equiv \sum_{k=1}^{K} \delta_{\mathrm{kron}}(X - Y_k). \tag{22}$$

(Here $\delta_{\mathrm{kron}}(X)$ is a Kronecker delta function.) This is a significant result since the time complexity of the discrete convolution is $O(N \log N)$ and the expression $\hat{S}(X)$ in (20) for the Euclidean distance function is continuous and differentiable everywhere (except in 1D at the seed locations).

3.2   Winding Numbers for the Signed Distance Function in 2D

The solution for the approximate Euclidean distance function in (20) is lacking in one respect: there is no information on the sign of the distance. This is to be expected since the distance function was obtained only from a set of *points* Y and not a curve or surface. We now describe a new method for computing the signed distance in 2D using winding numbers [16]. (The equivalent concept in 3D and higher dimensions is the topological degree which appears to be a straightforward extension but with possible unexpected pitfalls.)

Assume that we have a closed, parametric curve $\{x^{(1)}(t), x^{(2)}(t)\}$, $t \in [0, 1]$. We seek to determine if a grid location in the set $\{X_i \in \mathbb{R}^2, i \in \{1, \ldots, N\}\}$ is inside the closed curve. The winding number is the number of times the curve winds around the point X_i (if at all) and if the curve is oriented, counterclockwise turns are counted as positive and clockwise turns as negative. If a point is inside the curve, the winding number is a non-zero integer. If the point is outside the curve, the winding number is zero. If we can efficiently compute the winding number for all points on a grid w.r.t. to a curve, then we would have the sign information (inside/outside) for all the points. We now describe a fast algorithm to achieve this goal.

If the curve is C^1, then the angle $\theta(t)$ of the curve is continuous and differentiable and $d\theta(t) = \left(\frac{x^{(1)}\dot{x}^{(2)} - x^{(2)}\dot{x}^{(1)}}{r^2}\right) dt$ where $r(t) = \sqrt{[x^{(1)}]^2 + [x^{(2)}]^2}$. Since we need to determine whether the curve winds around each of the points $X_i, i \in \{1, \ldots, N\}$, define $(\hat{x}_i^{(1)}, \hat{x}_i^{(2)}) \equiv (x^{(1)} - X_i^{(1)}, x^{(2)} - X_i^{(2)})$, $\forall i$. Then the winding numbers for all grid points in the set X are

$$\mu_i = \frac{1}{2\pi} \oint_C \left(\frac{\hat{x}_i^{(1)}\dot{\hat{x}}_i^{(2)} - \hat{x}_i^{(2)}\dot{\hat{x}}_i^{(2)}}{\left[\hat{x}_i^{(1)}\right]^2 + \left[\hat{x}_i^{(2)}\right]^2}\right) dt, \; \forall i \in \{1, \ldots, N\}. \tag{23}$$

As it stands, we cannot actually compute the winding numbers without performing the integral in (23). To this end, we discretize the curve and produce a sequence of points $\{Y_k \in \mathbb{R}^2, k \in \{1, \ldots, K\}\}$ with the understanding that the curve is closed and therefore the "next" point after Y_K is Y_1. (The winding number property holds for piecewise continuous curves as well.) The integral in (23) becomes a discrete summation and we get

$$
\mu_i = \frac{1}{2\pi} \sum_{k=1}^{K} \left(\frac{\left[Y_k^{(1)} - X_i^{(1)}\right]\left[Y_{k\oplus1}^{(2)} - Y_k^{(2)}\right] - \left[Y_k^{(2)} - X_i^{(2)}\right]\left[Y_{k\oplus1}^{(1)} - Y_k^{(1)}\right]}{\left[\left(Y_k^{(1)} - X_i^{(1)}\right)^2 + \left(Y_k^{(2)} - X_i^{(2)}\right)^2\right]} \right)
\tag{24}
$$

$\forall i \in \{1, \ldots, N\}$, where the notation $Y_{k\oplus1}^{(\cdot)}$ denotes that $Y_{k\oplus1}^{(\cdot)} = Y_{k+1}^{(\cdot)}$ for $k \in \{1, \ldots, K-1\}$ and $Y_{K\oplus1}^{(\cdot)} = Y_1^{(\cdot)}$. We can simplify the notation in (24) (and obtain a measure of conceptual clarity as well) by defining the "tangent" vector $\{Z_k, k = \{1, \ldots, K\}\}$ as $Z_k^{(\cdot)} = Y_{k\oplus1}^{(\cdot)} - Y_k^{(\cdot)}, k \in \{1, \ldots, K\}$ with the $(\cdot)$ symbol indicating either coordinate. Using the tangent vector Z, we rewrite (24) as

$$
\mu_i = \frac{1}{2\pi} \sum_{k=1}^{K} \left(\frac{\left[Y_k^{(1)} - X_i^{(1)}\right] Z_k^{(2)} - \left[Y_k^{(2)} - X_i^{(2)}\right] Z_k^{(1)}}{\left[\left(Y_k^{(1)} - X_i^{(1)}\right)^2 + \left(Y_k^{(2)} - X_i^{(2)}\right)^2\right]} \right), \forall i \in \{1, \ldots, N\}
\tag{25}
$$

We now make the somewhat surprising observation (to us at any rate) that μ in (25) is a sum of two discrete convolutions. The first convolution is between two functions $f_{cr}(X) \equiv f_c(X)f_r(X)$ and $g_2(X) = \sum_{k=1}^{K} Z_k^{(2)} \delta_{\text{kron}}(X - Y_k)$ where the Kronecker delta function is a product of two Kronecker delta functions, one for each coordinate. The second convolution is between two functions $f_{sr}(X) \equiv f_s(X)f_r(X)$ and $g_1(X) \equiv \sum_{k=1}^{K} Z_k^{(1)} \delta_{\text{kron}}(X - Y_k)$. The functions $f_c(X), f_s(X)$ and $f_r(X)$ are defined as

$$
f_c(X) \equiv \frac{X^{(1)}}{\sqrt{\left[X^{(1)}\right]^2 + \left[X^{(2)}\right]^2}}, \quad f_s(X) \equiv \frac{X^{(2)}}{\sqrt{\left[X^{(1)}\right]^2 + \left[X^{(2)}\right]^2}}, \quad \text{and} \tag{26}
$$

$$
f_r(X) \equiv \frac{1}{\sqrt{\left[X^{(1)}\right]^2 + \left[X^{(2)}\right]^2}}. \tag{27}
$$

where we have abused notation somewhat and let $X^{(1)}$ ($X^{(2)}$) denote the x (y)-coordinate of all the points in the grid set X. Armed with these relationships, we rewrite (25) to get

$$
\mu(X) = \frac{1}{2\pi} \left[-f_{cr}(X) * g_2(X) + f_{sr}(X) * g_1(X)\right] \tag{28}
$$

which can be computed in $O(N \log N)$ time using two FFTs. We have shown that the sign component of the Euclidean distance function can be separately computed (without knowledge of the distance) in parallel in $O(N \log N)$ on a regular 2D grid.

3.3   Fast Computation of the Derivatives of the Distance Function

Just as the approximate Euclidean distance function $\hat{S}(X)$ can be efficiently computed in $O(N \log N)$, so can the derivatives. This is important because fast computation of the derivatives of $\hat{S}(X)$ on a regular grid can be very useful in medial axes and curvature computations. Below, we detail how this can be achieved. We begin with the gradients and for illustration purposes, the derivations are performed in 2D:

$$\hat{S}_x(X) = \frac{\sum_{k=1}^{K} \frac{\left(X^{(1)}-Y_k^{(1)}\right)}{\sqrt{\left(X^{(1)}-Y_k^{(1)}\right)^2+\left(X^{(2)}-Y_k^{(2)}\right)^2}} \exp\left\{-\frac{\|X-Y_k\|}{\hbar}\right\}}{\sum_{k=1}^{K} \exp\left\{-\frac{\|X-Y_k\|}{\hbar}\right\}}, \tag{29}$$

A similar expression can be obtained for $\hat{S}_y(X)$. These first derivatives can be rewritten as discrete convolutions:

$$\hat{S}_x(X) = \frac{f_c(X)\exp\left\{-\frac{X}{\hbar}\right\} * Y_{\mathrm{kron}}(X)}{\hat{S}(X)}, \ \hat{S}_y(X) = \frac{f_s(X)\exp\left\{-\frac{X}{\hbar}\right\} * Y_{\mathrm{kron}}(X)}{\hat{S}(X)}, \tag{30}$$

where $f_c(X)$ and $f_s(X)$ are as defined in (26) and $Y_{\mathrm{kron}}(X)$ is as defined in (22).

The second derivative formulae are somewhat involved. Rather than hammer out the algebra in a turgid manner, we merely present the final expressions—all discrete convolutions—for the three second derivatives in 2D:

$$\hat{S}_{xx}(X) = -(1+\frac{1}{\hbar})\frac{f_c^2(X)\exp\left\{-\frac{X}{\hbar}\right\} * Y_{\mathrm{kron}}(X)}{\hat{S}(X)} + \frac{1}{\hbar}\hat{S}_x^2(X)$$
$$+\frac{f_r(X)\exp\left\{-\frac{X}{\hbar}\right\} * Y_{\mathrm{kron}}(X)}{\hat{S}(X)}, \tag{31}$$

$$\hat{S}_{yy}(X) = -(1+\frac{1}{\hbar})\frac{f_s^2(X)\exp\left\{-\frac{X}{\hbar}\right\} * Y_{\mathrm{kron}}(X)}{\hat{S}(X)} + \frac{1}{\hbar}\hat{S}_y^2(X)$$
$$+\frac{f_r(X)\exp\left\{-\frac{X}{\hbar}\right\} * Y_{\mathrm{kron}}(X)}{\hat{S}(X)}, \text{ and} \tag{32}$$

$$\hat{S}_{xy}(X) = -(1+\frac{1}{\hbar})\frac{f_c(X)f_s(X)\exp\left\{-\frac{X}{\hbar}\right\} * Y_{\mathrm{kron}}(X)}{\hat{S}(X)} + \frac{1}{\hbar}\hat{S}_x(X)\hat{S}_y(X) \tag{33}$$

where $f_r(X)$ is as defined in (27). We also see that

$$\hat{S}_x^2(X)+\hat{S}_y^2(X)-\hbar\left[\hat{S}_{xx}(X) + \hat{S}_{yy}(X)\right] = (1+\hbar)-2\hbar\frac{f_r(X)\exp\left\{-\frac{X}{\hbar}\right\} * Y_{\mathrm{kron}}(X)}{\hat{S}(X)} \tag{34}$$

[since $f_c^2(X) + f_s^2(X) = 1$] with the right side going to one as $\hbar \to 0$ for points in X away from points in the seed point-set Y. This is in accordance with (14) and vindicates our choice of the replacement Green's function in (21).

Since we can efficiently compute the first and second derivatives of the approximate Euclidean distance function $\hat{S}(X)$ everywhere on a regular grid, we can also compute derived quantities such as curvature (Gaussian, mean and principal curvatures for a two-dimensional surface). In the next section, we visualize the derivatives and maximum curvature for shape silhouettes.

4   Euclidean Distance Function Experiments

We executed the Schrödinger Euclidean distance function algorithm on a set of 2D shape silhouettes[2]. The grid size is $-20 \leq x \leq 20$ and $-20 \leq y \leq 20$ with a grid spacing of 0.25 and $\hbar = 0.3$. The winding number discrete convolution algorithm is used to mark points as either inside or outside each shape. We visualize the vector fields $(\hat{S}_x, \hat{S}_y)$ in Figure 1 for the 8 shapes and the maximum curvature for a subset of the shapes in Figure 2. We chose the maximum curvature (defined as $H + \sqrt{H^2 - K}$ where H and K are the mean and Gaussian curvatures respectively of the Monge patch given by $\left\{ x, y, \hat{S}(x,y) \right\}$) as the vehicle to visualize the medial axes of each shape after first considering the divergence of the unit gradient $\left[\nabla \cdot g = \nabla \cdot \left(\frac{\nabla \hat{S}(X)}{|\nabla \hat{S}(X)|} \right) \right]$ and the entropy $\left(-\frac{\partial \hat{S}(X)}{\partial \hbar} \right)$. The divergence is a good choice for the medial axes provided we update an adaptive grid whereas the entropy requires very high precision numerical computation (which we plan to pursue in the future).

Next, we ran a comparison of the Schrödinger Euclidean distance function algorithm with the fast sweeping method [9] and the exact Euclidean distance. We used a "Dragon" point-set obtained from the Stanford 3D Scanning Repository[3] in 3D and executed the three approaches to construct isosurfaces which are visualized in Figure 3. The common grid was $-2 \leq x \leq 2$, $-2 \leq y \leq 2$ and $-2 \leq z \leq 2$ with a grid spacing of 0.125. Numerical underflow errors in the FFT forced us to run the Schrödinger Euclidean distance function algorithm at four values of $\hbar$, namely, 0.025, 0.045, 0.06, and 0.08. We used the following decision criterion for $\hat{S}(X)$: $\hat{S} = \hat{S}|_{\hbar=0.08}$ if $\hat{S} \geq 2$, $\hat{S} = \hat{S}|_{\hbar=0.06}$ if $1.5 \leq \hat{S} < 2$, $\hat{S} = \hat{S}|_{\hbar=0.045}$ if $0.75 \leq \hat{S} < 1.5$ and $\hat{S} = \hat{S}|_{\hbar=0.025}$ if $\hat{S} < 0.75$. The initial conditions $\hat{S}(Y_k) = 0, \forall k \in \{1, \ldots, K\}$ were used to translate the $\hat{S}$ values (upwards or downwards) such that the minimum value was zero. The average percentage error in the Schrödinger approach was 3.89% whereas the average percentage error in the fast sweeping method (where the Gauss-Seidel iterates were run until convergence) was 6.35%. Our FFT-based approach does not begin by discretizing the spatial differential operator as is the case with the fast marching and fast sweeping methods and this could help account for the increased accuracy.

---

[2] We thank Kaleem Siddiqi for providing us the set of 2D shape silhouettes used in this paper.

[3] This dataset is available at http://graphics.stanford.edu/data/3Dscanrep/

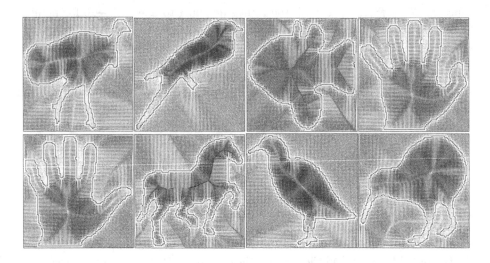

Fig. 1. A quiver plot of $\nabla \hat{S} = (\hat{S}_x, \hat{S}_y)$ for a set of silhouette shapes

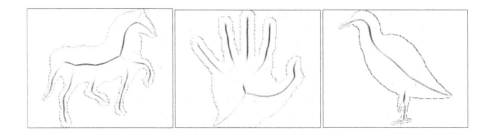

Fig. 2. Maximum curvature plots: i) Horse, ii) Hand, and iii) Bird

Fig. 3. Dragon surface reconstructed using i) Schrödinger, ii) Exact Euclidean distance and iii) Fast sweeping

5   A Perturbation Approach for the General Eikonal Problem

In this section, we briefly summarize our perturbation approach (using the well known Born expansion) [12,10] for the general eikonal equation (with forcing functions $f(X)$ bounded away from zero). We consider the static Schrödinger equation (in 2D) with a forcing function $f(X)$:

$$-\hbar^2 \bigtriangledown^2 \phi + f^2\phi = \sum_{k=1}^{K} \delta(X - Y_k). \tag{35}$$

Equation (35) can be rewritten as

$$(-\hbar^2 \bigtriangledown^2 + \tilde{f}^2)\left[1 + (-\hbar^2 \bigtriangledown^2 + \tilde{f}^2)^{-1} \circ (f^2 - \tilde{f}^2)\right]\phi = \sum_{k=1}^{K} \delta(X - Y_k) \tag{36}$$

with $\tilde{f}(X)$ a constant forcing function. Now, defining the operator A as $A \equiv (-\hbar^2 \bigtriangledown^2 + \tilde{f}^2)^{-1} \circ (f^2 - \tilde{f}^2)$ and ϕ_0 as $\phi_0 \equiv (1 + A)\phi$, we see that ϕ_0 satisfies

$$(-\hbar^2 \bigtriangledown^2 + \tilde{f}^2)\phi_0 = \sum_{k=1}^{K} \delta(X - Y_k) \tag{37}$$

and

$$\phi = (1 + A)^{-1}\phi_0. \tag{38}$$

Using a geometric series approximation for $(1 + A)^{-1}$, we obtain the solution for ϕ as

$$\phi \approx \phi_0 - \phi_1 + \phi_2 - \phi_3 + \ldots + (-1)^T \phi_T \tag{39}$$

where ϕ_i satisfies the recurrence relation

$$(-\hbar^2 \bigtriangledown^2 + \tilde{f}^2)\phi_i = (f^2 - \tilde{f}^2)\phi_{i-1}, \forall i \in \{1, 2, \ldots, T\}. \tag{40}$$

The solutions for ϕ_i can then be obtained by convolution

$$\phi_0(X) = \sum_{k=1}^{K} G(X) * \delta(X - Y_k) = \sum_{k=1}^{K} G(X - Y_k), \tag{41}$$

$$\phi_i(X) = G(X) * \left[(f^2 - \tilde{f}^2)\phi_{i-1}\right], \forall i \in \{1, 2, \ldots, T\} \tag{42}$$

and an approximate solution to the eikonal equation can be obtained from $\hat{S}(X) = -\hbar \log \phi(X)$. The discrete convolutions in (41) and in (42) can be efficiently computed via FFTs. The number of terms (T) used in the geometric series approximation of $(1 + A)^{-1}$ is *independent* of the grid size N. We set $\tilde{f} = \sqrt{\frac{[\min_X f(X)]^2 + [\max_X f(X)]^2}{2}}$ which turns out to be the optimal value [10].

6   Image Segmentation Results Using the Eikonal Solver

To test the eikonal solver, we obtained two images (3096 and 101085) from the Berkeley Segmentation Dataset and Benchmark[4]. After first smoothing them using a 5×5 Gaussian with standard deviation 0.25, we used the following sets of parameters for the two images. For image 3096: we ran the perturbation method for $T = 5$ iterations at $\hbar$ values of 0.05, 0.15, 0.25 and 0.35 with thresholds of 1.5, 2.5 and 4, grid size $-8 \leq x \leq 8$, $-6 \leq y \leq 6$ with grid spacing 0.0313 and the forcing function $f(X) = \frac{|\nabla I(X)|}{\max_X |\nabla I(X)|} + 0.5$. For image 101085, the only changes were: grid size $-5 \leq x \leq 5$, $-8 \leq y \leq 8$ with grid spacing 0.0313 and the forcing function $f(X) = \frac{|\nabla I(X)|}{\max_X |\nabla I(X)|} + 0.01$. Both images were seeded with a set of interior/exterior points, the eikonal algorithms were run twice and we displayed the boundaries in Figure 4 using the eikonal "winner"—the one with the smaller distance. The same approach was used for the fast sweeping method and Dijkstra's shortest path algorithm [13] (since closed form solutions are not available). The results are obviously anecdotal but serve to illustrate the correspondence between the Schrödinger (quantum) and the fast sweeping (classical) approaches. Note the larger scale boundaries in the Schrödinger segmentation of image 101085 which we attribute to the viscosity term in (14).

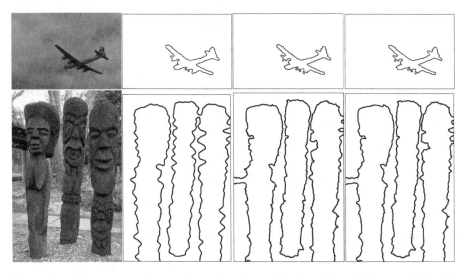

Fig. 4. Image segmentation results. Top from left to right: i) Image 3096, ii) Schrödinger, iii) Dijkstra, iv) Fast sweeping. Bottom from left to right: i) Image 101085, ii) Schrödinger, iii) Dijkstra, iv) Fast sweeping.

7   Discussion

While energy minimization methods have permeated image analysis in the past two decades, one overarching generalization we can make is that the formulations

---
[4] This dataset is available at
    http://www.eecs.berkeley.edu/Research/Projects/CS/vision/grouping/segbench/

are inspired by classical and not quantum mechanics. Despite the fact that a close correspondence exists between Hamilton-Jacobi theory and Schrödinger wave mechanics, we have not seen image analysis leverage this relationship. When we solve the Schrödinger equation at small values of $\hbar$, we obtain closed-form solutions for the Euclidean distance function problem that can be efficiently computed in $O(N \log N)$. This has immediate applications in image analysis as the sign of the distance function, its gradients and curvature can all be written in closed-form and efficiently computed via FFTs. When applied to shape silhouettes, the gradients and curvature information can aid in medical axes computation. We also show that a perturbation series approach leads to a fast eikonal solver which can be used in image segmentation.

References

1. Horn, B.: Robot Vision. MIT Press, Cambridge (1986)
2. Siddiqi, K., Tannenbaum, A., Zucker, S.: A Hamiltonian approach to the eikonal equation. In: Hancock, E.R., Pelillo, M. (eds.) EMMCVPR 1999. LNCS, vol. 1654, pp. 1–13. Springer, Heidelberg (1999)
3. Osher, S., Fedkiw, R.: Level Set Methods and Dynamic Implicit Surfaces. Springer, Heidelberg (2002)
4. Gilboa, G., Sochen, N., Zeevi, Y.: Image enhancement and denoising by complex diffusion processes. IEEE PAMI 26(8), 1020–1036 (2004)
5. Emms, D., Severini, S., Wilson, R., Hancock, E.: Coined quantum walks lift the cospectrality of graphs and trees. In: Rangarajan, A., Vemuri, B.C., Yuille, A.L. (eds.) EMMCVPR 2005. LNCS, vol. 3757, pp. 332–345. Springer, Heidelberg (2005)
6. Goldstein, H., Poole, C., Safko, J.: Classical Mechanics. Addison-Wesley, Reading (2001)
7. Griffiths, D.: Introduction to Quantum Mechanics. Benjamin-Cummings (2004)
8. Yatziv, L., Bartesaghi, A., Sapiro, G.: O(N) implementation of the fast marching algorithm. Journal of Computational Physics 212(2), 393–399 (2006)
9. Zhao, H.: A fast sweeping method for eikonal equations. Mathematics of Computation, 603–627 (2005)
10. Gurumoorthy, K., Rangarajan, A.: A fast eikonal equation solver using the Schrödinger wave equation. Technical Report CVGMI-09-05, Center for Computer Vision, Graphics and Medical Imaging (CVGMI), University of Florida (2009)
11. Bracewell, R.: The Fourier Transform and its Applications. 3rd edn. McGraw-Hill Science and Engineering (1999)
12. Fernandez, F.: Introduction to Perturbation Theory in Quantum Mechanics. CRC Press, Boca Raton (2000)
13. Cormen, T., Leiserson, C., Rivest, R., Stein, C.: Introduction to Algorithms, 2nd edn. MIT Press, Cambridge (2001)
14. Crandall, M., Ishii, H., Lions, P.: User's guide to viscosity solutions of second order partial differential equations. Bulletin of the AMS 27(1), 1–67 (1992)
15. Abramowitz, M., Stegun, I.: Handbook of Mathematical Functions with Formulas, Graphs, and Mathematical Tables. Government Printing Office, USA (1964)
16. Aberth, O.: Precise Numerical Methods Using C++. Academic Press, London (1998)

Computing the Local Continuity Order of Optical Flow Using Fractional Variational Method

K. Kashu[1], Y. Kameda[1], A. Imiya[2], T. Sakai[2], and Y. Mochizuki[1]

[1] School of Advanced Integration Science, Chiba University
{kashu,yu-kameda,motchy}@graduate.chiba-u.jp
[2] Institute of Media and Information Technology, Chiba University
Yayoi-cho 1-33, Inage-ku, Chiba, 263-8522, Japan
{imiya,tsakai}@faculty.chiba-u.jp

Abstract. We introduce variational optical flow computation involving priors with fractional order differentiations. Fractional order differentiations are typical tools in signal processing and image analysis. The zero-crossing of a fractional order Laplacian yields better performance for edge detection than the zero-crossing of the usual Laplacian. The order of the differentiation of the prior controls the continuity class of the solution. Therefore, using the square norm of the fractional order differentiation of optical flow field as the prior, we develop a method to estimate the local continuity order of the optical flow field at each point. The method detects the optimal continuity order of optical flow and corresponding optical flow vector at each point. Numerical results show that the Horn-Schunck type prior involving the $n + \varepsilon$ order differentiation for $0 < \varepsilon < 1$ and an integer n is suitable for accurate optical flow computation.

1 Introduction

In this paper, we introduce the prior involving fractional order differentiations [3,9,10,11] for variational optical flow computation [1]. The order of the differentiations in the prior controls the continuity class of the solution. Therefore, using the fractional order differentiations, we can estimate the order of local continuities of optical flow vectors. Furthermore, we can obtain the optical flow flow vector with the optical continuity at each point.

Total variational (TV) regularisation [2] is a successful method of optical flow computation of an image with a discontinuity of the gray values and the optical flow field. TV regularisation uses the total variation of optical flow field as the prior, although the classical Horn-Schunck method [12,13] uses the L_2 norm of the gradient of flow field. TV regularisation optical flow computation [1] derives a nonlinear elliptic partial differential equation as the Euler-Lagrange equation of the energy functional of the problem.

The generalisation of the order of differentiation in a Horn-Schunck type prior is another modification of the original Horn-Schunck regularisation, since this

D. Cremers et al. (Eds.): EMMCVPR 2009, LNCS 5681, pp. 154–167, 2009.

generalisation yields a linear Euler-Lagrange equation. There are two types of generalisation of the differentiations in priors; the first one is to deal with higher-order differentiations, and the second one is to deal with fractional order differentiations. We focus on the second type of generalisation, that is, we deal with the variational optical flow computation whose prior term involves a fractional order differentiation of optical flow vectors.

Recently, fractional partial differential equations [9,10,11] has been widely used in various areas in science and engineering, because fractional differential equations describe diffusion and wave propagations in inhomogeneous media and fractal structures [19,20,21], and viscoelastic flow and deformation [9,10,11]. Fractional order differentiations are typical tools in signal and image processing [5,6,7] and is applied to the edge detection of images [8]. In edge detection, a zero-crossing set of a fractional order Laplacian derives good performance [8]. As a sequel of edge detection using fractional order differentiations, we propose variational optical flow computation involving the prior with fractional order differentiations on optical flow vectors.

Since fractional differentiations are linear operations [9,10,11], the fractional order regularisation for optical flow computation [1] derives a linear fractional order elliptic partial differential equation as the Euler-Lagrange equation of the energy functional. Therefore, we can numerically solve the problem using the same strategy that is used to solve the Horn-Schunck method. Furthermore, since the order of the differentiation in the prior controls the continuity class of the solution, by selecting the order of differentiations in the prior, we can estimate an appropriate order of continuity of the optical flow vector at each point. This is a mathematical advantage to use the fractional differentiations in the priors.

The Riemann-Liouville fractional differentiation [16] involves the Cauchy integral formula, which is unstable to numerical implementation, since the formula involves a singular integral. A numerically stable fractional differentiation is computed using the Grünwald-Letnikow [16,17,18] definition, which describes fractional differentiation as a finite series. The other definition of fractional differentiation is based on the Fourier transform of differential operations, which is easily implemented using FFT and the filter theory [5,6]. We solve the spatially fractional partial differential equation using the Fourier transform method to compute fractional derivatives.

The image-driven [14,15] and flow driven [1] diffusions are two modifications of the Horn-Schunck method. We introduce the operator-driven method as the third method. The operator-driven method selects the optimal differential operator involved in the prior of the variational energy for optical flow computation. Using the fractional-order differentiations, we obtain a two-path algorithm. The algorithm first estimates the continuity order of each point simultaneously solving a collection of variational problems whose priors involve various order differentiations. Then, the algorithm detects the continuity order of the optical flow field which establish the minimum of the residual. Secondly, using the estimated local continuity order, the algorithm recomputes the optical flow vector of each point. This paper evaluates the performances of the first step, that is, we focus on a local continuous-order estimation method using a variational optimisation problem.

2 Variational Optical Flow Computation

For a spatiotemporal image $f(\boldsymbol{x}, t)$, $\boldsymbol{x} = (x, y)^\top$, the optical flow vector $\boldsymbol{u} = \dot{\boldsymbol{x}} = (\dot{x}, \dot{y})^\top$, for $\dot{x} = u = u(x, y)$ and $\dot{y} = v = v(x, y)$, of each point $\boldsymbol{x} = (x, y)^\top$ is the solution of the singular equation

$$f_x u + f_y v + f_t = \nabla f^\top \boldsymbol{u} + \partial_t f = 0. \tag{1}$$

To solve this equation, a regularisation method [12,13] which minimises the criterion

$$J(\boldsymbol{u}) = \int_{\mathbf{R}^2} \left\{ (\nabla f^\top \boldsymbol{u} + \partial_t f)^2 + \kappa tr \nabla \boldsymbol{u} \nabla \boldsymbol{u}^\top \right\} d\boldsymbol{x} \tag{2}$$

is employed. The Euler-Lagrange equation of the energy function defined by eq. (2) and the associated diffusion equation of the Euler-Lagrange equation are

$$\Delta \boldsymbol{u} = \frac{1}{\kappa}(\nabla f^\top \boldsymbol{u} + f_t)\nabla f, \quad \frac{\partial \boldsymbol{u}}{\partial t} = \Delta \boldsymbol{u} - \frac{1}{\kappa}(\nabla f^\top \boldsymbol{u} + f_t)\nabla f, \tag{3}$$

with the boundary condition $\frac{\partial \boldsymbol{u}}{\partial \boldsymbol{n}} = 0$ for the unit normal $\boldsymbol{n}$ on the boundary.

For the sampled function $f_{mn} = f(hm, hn)$ and vector field $\boldsymbol{u}_{mn} = (u_{mn}, v_{mn})^\top$, $u_{mn} = u(hm, hn)$ and $v_{mn} = v(hm, hn)$, where h is the unit sample interval, the diffusion equation of eq. (3) derives the discrete equation

$$\frac{\boldsymbol{u}_{mn}^{(l+1)} - \boldsymbol{u}^{(i)}}{\Delta \tau} = (\Delta \boldsymbol{u}^{(l)})_{mn} - \frac{1}{\kappa}((\nabla f)_{mn}^\top \boldsymbol{u}_{mn}^{(l+1)} + (\partial_t f)_{mn})(\nabla f)_{mn} \tag{4}$$

and the associated iteration equation

$$(\boldsymbol{I} + \frac{\Delta \tau}{\kappa} \boldsymbol{S}_{mn})\boldsymbol{u}_{mn}^{(l+1)} = \boldsymbol{u}_{mn}^{(l)} + \Delta \tau \sum_{ij} l_{ij} \boldsymbol{u}_{m-in-j}^{(l)} - \frac{\Delta \tau}{\kappa} \boldsymbol{c}_{mn}, \ l \geq 0, \tag{5}$$

where $\boldsymbol{S}_{mn} = (\nabla f)_{mn}(\nabla f)_{mn}^\top$ and $\boldsymbol{c}_{mn} = (\partial_t f)_{mn}(\nabla f)_{mn}$, and the discrete Laplacian operation

$$\sum_{ij} l_{ij} \boldsymbol{u}_{m-in-j} = \begin{pmatrix} u_{i+1j} + u_{i-1j} + u_{ij+1} + u_{ij-1} - 4u_{ij} \\ v_{i+1j} + v_{i-1j} + v_{ij+1} + v_{ij-1} - 4v_{ij} \end{pmatrix}, \tag{6}$$

using the operator splitting [4]. The eigenvalue analysis of the discrete Laplacian implies the following proposition

Proposition 1. *Equation (5) generates a sequence which converges to the solution of the problem, if the relation $|1 - \Delta \tau 2 \frac{2^2}{h^2}| < 1$, where h is the sampling interval, is satisfied.*

3   Fractional Order Variational Optical Flow Computation

3.1   Some Properties of the Fractional Order Differentiations

Using the Fourier transform pair

$$F(\xi, \eta) = \frac{1}{2\pi} \int_{-\infty}^{\infty} \int_{-\infty}^{\infty} f(x, y) e^{-i(x\xi + y\eta)} \, dx \, dy, \tag{7}$$

$$f(x, y) = \frac{1}{2\pi} \int_{-\infty}^{\infty} \int_{-\infty}^{\infty} F(\xi, \eta) e^{i(x\xi + y\eta)} \, d\xi \, d\eta, \tag{8}$$

we define the operations

$$\partial_x^\alpha f(x, y) = \frac{1}{2\pi} \int_{-\infty}^{\infty} \int_{-\infty}^{\infty} (i\xi)^\alpha F(\xi, \eta) e^{i(x\xi + y\eta)} \, d\xi \, d\eta, \tag{9}$$

$$\partial_y^\alpha f(x, y) = \frac{1}{2\pi} \int_{-\infty}^{\infty} \int_{-\infty}^{\infty} (i\eta)^\alpha F(\xi, \eta) e^{i(x\xi + y\eta)} \, d\xi \, d\eta. \tag{10}$$

Setting the operator Λ to be

$$\Lambda f(x, y) = \frac{1}{2\pi} \int_{-\infty}^{\infty} \int_{-\infty}^{\infty} (\sqrt{\xi^2 + \eta^2}) F(\xi, \eta) e^{i(x\xi + y\eta)} \, d\xi \, d\eta \tag{11}$$

from eqs. (9) and (10), we obtain the relation

$$\Lambda^{2\alpha} = (-\Delta)^n (\Lambda)^{2\varepsilon} = (-\Delta)^n (-\Delta)^\varepsilon, \tag{12}$$

for $\alpha = n + \varepsilon$ where n is an integer and $0 < 1 < \varepsilon$. Furthermore, we have the equality[1]

$$\int_{-\infty}^{\infty} \int_{-\infty}^{\infty} |\nabla f|^2 \, dx \, dy = \int_{-\infty}^{\infty} \int_{-\infty}^{\infty} |\Lambda f|^2 \, dx \, dy, \tag{13}$$

since

$$\int_{-\infty}^{\infty} \int_{-\infty}^{\infty} |f|^2 \, dx \, dy = \int_{-\infty}^{\infty} \int_{-\infty}^{\infty} |F|^2 \, d\xi \, d\eta. \tag{14}$$

3.2 α Optical Flow Computation

For the positive integer $n \geq 1$, setting the operator D^n to be

$$D^{n+1} f = \begin{pmatrix} \partial_x D^n f \\ \partial_y D^n f \end{pmatrix}, \quad D f = \nabla f = \begin{pmatrix} \partial_x f \\ \partial_y f \end{pmatrix}, \tag{15}$$

1

$$\int_{-\infty}^{\infty} \int_{-\infty}^{\infty} (|f_x|^2 + |f_y|^2) \, dx \, dy = \int_{-\infty}^{\infty} \int_{-\infty}^{\infty} (|i\xi F|^2 + |i\eta F|^2) \, d\xi \, d\eta$$

$$= \int_{-\infty}^{\infty} \int_{-\infty}^{\infty} (\xi^2 + \eta^2) |F|^2 \, d\xi \, d\eta = \int_{-\infty}^{\infty} \int_{-\infty}^{\infty} |\Lambda f|^2 \, dx \, dy.$$

we define the operation

$$|T^\alpha f|^2 = \begin{cases} |D^\alpha f|^2, & \text{if } \alpha \text{ is an integer,} \\ |\Lambda^\alpha f|^2, & \text{otherwise.} \end{cases} \tag{16}$$

The mathematical properties of the operator Λ allow us to focus on variational optical flow computation which minimises the functional

$$J_\alpha(\boldsymbol{u}) = \int_{\mathbf{R}^2} F(\boldsymbol{u}; \alpha, \kappa) dx dy, \tag{17}$$

$$F(\boldsymbol{u}; \alpha, \kappa) = (\nabla f^\top \boldsymbol{u} + \partial_t f)^2 + \kappa(|T^\alpha u|^2 + |T^\alpha v|^2), \tag{18}$$

for $\alpha \geq 0$ and $\kappa \geq 0$ as an extension of eq. (2), that is, we have the relation $J(\boldsymbol{u}) = J_1(\boldsymbol{u})$.

Definition 1. *We call the solution of eq. (17) the alpha optical flow* [2].

Since $\Lambda = \Lambda^*$, we obtain the Euler-Lagrange equation [3] from eq. (17) as

$$\frac{1}{\kappa} \Lambda^{2\alpha} \boldsymbol{u} + (\nabla f^\top \boldsymbol{u} + \partial_t f) \nabla f = 0. \tag{19}$$

3.3   Numerical Computation for Fractional Differentiation

Equation (19) derives the iteration form

$$(\boldsymbol{I} + \frac{\Delta\tau}{\kappa} \boldsymbol{S}_{mn}) \boldsymbol{u}_{mn}^{(l+1)} = \{(\boldsymbol{u}_{mn}^{(l)} + \Delta\tau(-\Lambda^{2\alpha})\} \boldsymbol{u}_{mn}^{(l)} - \frac{\Delta\tau}{\kappa} \boldsymbol{c}_{mn}, \ l \geq 0, \tag{20}$$

for the numerical computation of α-optical flow, where the numerical Fourier transform achieves the operation $(-\Lambda^{2\alpha} \boldsymbol{u})_{mn}^{(l)}$. To use the FFT (Fast Fourier Transform) under the Neumann condition $\frac{\partial \boldsymbol{u}}{\partial n} = 0$, the function $f(x, y)$ defined in $0 \leq x, y \leq L$ is expanded using the relations $f(L + x, L + y) = f(L - x, L - y)$ and $f(x, y) = (x + 2mL, y + 2nL)$ for integers m and n.

---

[2] Multiresolution image analysis using the diffusion equation $f_\tau = \Delta f$ is called the scale space method. The generalisation using the equation $f_\tau = (-\Delta)^{\frac{\alpha}{2}} f$ is called the α-scale space method [22].

[3] First,

$$\frac{\delta}{\delta u} J_\alpha(\boldsymbol{u}) = 2(\nabla f^\top \boldsymbol{u} + \partial_t f) \nabla f + \kappa \frac{\delta}{\delta u}(|\Lambda^\alpha u|^2 + |\Lambda^\alpha v|^2).$$

Furthermore,

$$\frac{\delta}{\delta u} \int_{-\infty}^{\infty} \int_{-\infty}^{\infty} (|\Lambda^\alpha u|^2 + |\Lambda^\alpha v|^2) dx dy = \frac{\delta}{\delta u} \int_{-\infty}^{\infty} \int_{-\infty}^{\infty} (\Lambda^\alpha u)(\Lambda^\alpha u) dx dy$$

$$= \frac{\delta}{\delta u} \int_{-\infty}^{\infty} \int_{-\infty}^{\infty} ((\Lambda^\alpha)^* \Lambda^\alpha u) u dx dy = \frac{\delta}{\delta u} \int_{-\infty}^{\infty} \int_{-\infty}^{\infty} (\Lambda^{2\alpha} u) u dx dy = \Lambda^{2\alpha} u$$

since $\Lambda^* = \Lambda$. Moreover, in the same manner, we have $\frac{\delta}{\delta v}(|\Lambda^\alpha u|^2 + |\Lambda^\alpha v|^2) = 2\Lambda^{2\alpha} v$, Then, we have eq. (19), using the notation $\Lambda \boldsymbol{u} = (\Lambda u, \Lambda v)^\top$.

For $\{f_{mn}\}_{m,n=0}^{N-1}$ such that $N = 2^k$, setting

$$g_{mn} = \begin{cases} f_{mn} & , m = 0, 1, \cdots, \frac{N}{2} - 1, \, n = 0, 1, \cdots, \frac{N}{2} - 1 \\ f_{m\,N-1+n} & , m = 0, 1, \cdots, \frac{N}{2} - 1, \, , n = -1, -2, \cdots, -\frac{N}{2} \\ f_{N-1+m\,n} & , m = -1, -2, \cdots, -\frac{N}{2}, \, n = 0, 1, \cdots, \frac{N}{2} - 1 \\ f_{N-1+m\,N-1+n} & , m = -1, -2, \cdots, -\frac{N}{2}, \, , n = -1, -2, \cdots, -\frac{N}{2} \end{cases}, \quad (21)$$

the DFT pair is expressed as

$$G_{pq} = F_2 g_{mn} = \frac{1}{N} \sum_{m\,n=-\frac{N}{2}}^{\frac{N}{2}-1} g_{mn} \exp\left(-2\pi i \frac{mp + nq}{N}\right), \quad (22)$$

$$g_{mn} = F_2^{-1} G_{pq} = \frac{1}{N} \sum_{p\,q=-\frac{N}{2}}^{\frac{N}{2}-1} G_{pq} \exp\left(2\pi i \frac{mp + nq}{N}\right). \quad (23)$$

Then, we have

$$\frac{d}{dx} g(x, y) \approx g_{m+\frac{1}{2}\,n} - g_{m-\frac{1}{2}\,n} = F_2^{-1} 2i \sin\left(\frac{\pi p}{N}\right) G_{pq} \quad (24)$$

$$\frac{d}{dy} g(x, y) \approx g_{m\,n+\frac{1}{2}\,n} - g_{n\,m-\frac{1}{2}} = F_2^{-1} 2i \sin\left(\frac{\pi q}{N}\right) G_{pq} \quad (25)$$

The first term of the iteration of eq. (20) is computed using the following filtering operation

$$\Lambda^{2\alpha} \boldsymbol{u}_{mn} = (-\Delta)^{\alpha} \boldsymbol{u}_{mn} = F_2^{-1} \left\{ 4\sin^2\left(\frac{\pi p}{N}\right) + 4\sin^2\left(\frac{\pi q}{N}\right) \right\}^{\alpha} \boldsymbol{U}_{pq} \quad (26)$$

where $\boldsymbol{U}_{pq} = (U_{pa}, V_{pq})^{\top} = F_2 \boldsymbol{u}_{mn} = (F_2 u_{mn}, F_2 v_{mn})^{\top}$.

3.4 Convergence Analysis

Using the vectorisation of the array

$$\boldsymbol{u} = \mathrm{vec}\left(\mathrm{vec}\begin{pmatrix} u_{11} & u_{12} & \cdots & u_{1N} \\ u_{21} & u_{22} & \cdots & u_{2N} \\ \vdots & \vdots & \ddots & \vdots \\ u_{M1} & u_{M2} & \cdots & u_{MN} \end{pmatrix}, \mathrm{vec}\begin{pmatrix} v_{11} & v_{12} & \cdots & v_{1N} \\ v_{21} & v_{22} & \cdots & v_{2N} \\ \vdots & \vdots & \ddots & \vdots \\ v_{M1} & v_{M2} & \cdots & v_{MN} \end{pmatrix}\right), \quad (27)$$

eq. (20) is expressed as

$$A\boldsymbol{u}^{(l+1)} = \boldsymbol{P}^{\top} \boldsymbol{B} \boldsymbol{P} \boldsymbol{u}^{(l)} + \boldsymbol{c}, \quad (28)$$

for

$$\boldsymbol{B} = \boldsymbol{I} + \Delta\tau(-\boldsymbol{L})^{\alpha} \quad (29)$$

where L is the Laplacian matrix for the vector u, that is,

$$L = \begin{pmatrix} D \otimes D & O \\ O & D \otimes D \end{pmatrix}, \quad D = \begin{pmatrix} -1 & 1 & 0 & \cdots & 0 & 0 \\ 1 & -2 & 1 & 0 \cdots & & \\ \vdots & \vdots & & \ddots & & \vdots \\ 0 & 0 & \cdots & 0 & 1 & -1 \end{pmatrix}, \tag{30}$$

and the permutation matrix P is

$$P = \begin{pmatrix} Q & O \\ O & Q \end{pmatrix}, \quad Q(\text{vec}U, \text{vec}V) = (\text{vec}U^\top, \text{vec}V^\top). \tag{31}$$

Setting Φ and Σ to be the DCT matrix and its eigenmatrix, the matrix L^α for is rewritten [4] as

$$L^\alpha = \begin{pmatrix} \Phi \otimes \Phi & O \\ O & \Phi \otimes \Phi \end{pmatrix} \begin{pmatrix} \Sigma^\alpha \otimes \Sigma^\alpha & O \\ O & \Sigma^\alpha \otimes \Sigma^\alpha \end{pmatrix} \begin{pmatrix} \Phi \otimes \Phi & O \\ O & \Phi \otimes \Phi \end{pmatrix}^*. \tag{32}$$

Furthermore, since the matrix A is expressed as

$$A = Diag(I + \frac{\Delta\tau}{\kappa} S_{mn}), \tag{33}$$

$\rho(A) > 1$. Therefore, if $\rho(B) < 1$, eq. (28) generates a sequence which converges to the solution. Since the eigenvalues of the matrix D are $\lambda_i = (2 - 2\cos\frac{2i\pi}{N})$ and $\rho(D) = 2$, the eigenvalues of L^α is $\sigma_{ij} = \lambda_i^\alpha + \lambda_j^\alpha$. These mathematical properties derive the following theorem.

Theorem 1. *If $|1 - \Delta\tau 2\left(\frac{2}{h}\right)^\alpha| < 1$, the spectrum of B is smaller than 1 and eq. (20) satisfies the convergence condition.*

3.5   Selection of Order of Prior

The solution involving the kth-order prior is

$$u(x, y) = \left(\sum_{i,j=0}^{k-1} a_{ij} x^i y^j, \sum_{i,j=0}^{k-1} b_{ij} x^i y^j \right)^\top \tag{34}$$

for nonnegative integers k, that is, the solution is locally a $(k-1)$th-order polynomial of x and y. This property implies that the priors involving the first- and second- order differentiations derive a piecewise linear and affine optical flow, respectively.

Let $u(x, y, t; \alpha)$ be the optical flow vector computed for a fixed α. For each point x, we select

$$u(x, y, t; \alpha^*) = \arg\min_\alpha F(u; \alpha, \kappa), \quad \alpha^*(x, y, t) = \arg\min F(u; \alpha, \kappa) \tag{35}$$

---

[4] For a positive definite matrix A and a real number α, the eigenvalues of A^α is λ^α for $Au = \lambda u$, where u is the eigenvector of A.

for a predetermined positive parameter κ as the solution of the optical flow vector of the point $\boldsymbol{x}$. Equation (35) estimates the local continuity order of the optical flow vector, that is, the point $\boldsymbol{x}$ with the optical flow vector $\boldsymbol{u}(x, y, t; \alpha)$ is the class $(\alpha - 1)$ function of the $\boldsymbol{x}$. We call $\alpha^* = \alpha(x, y, t)$, which establishes the minimum of eq. (35) , the α-map of the optical flow field (α-map in abbreviation.).

4 Numerical Examples

4.1 Performance of Fractional Order Differentiations

Figure 1 shows the computational results of optical flow for $\kappa = \frac{1}{4}\max|\nabla f|_{\max}$ and $\kappa = \max|\nabla f|_{\max}$, for $\alpha = 1.5$. In these results, optical flow vectors are expressed using colour charts in the left rows, that is, a colour of a point in optical flow field images represents the direction and length of the optical flow vector at the point. These results show that if $\alpha = 1 + \varepsilon$ for $0 < \varepsilon < 1$, we can detect the motion boundary with motion discontinuity. Figure 1 shows that the minimums of these measures are not established for $\alpha = 1$ or $\alpha = 2$.

We have evaluated the following measures.

- $AE(\boldsymbol{u}) = \arccos \hat{\boldsymbol{u}}_{computer}^{\top} \hat{\boldsymbol{u}}_{grandtruth}$ for $\hat{\boldsymbol{u}} = \frac{1}{\sqrt{u^2 + v^2 + 1}}(u, v, 1)^{\top}$, where $\boldsymbol{u} = (u, v)^{\top}$.
- $OFCE(\boldsymbol{u}) = |f_x u + f_y + f_t|$
- $AAE(\boldsymbol{u})$ the average of $AE(\boldsymbol{u})$ on an image.
- $AOFCE(\boldsymbol{u})$ the average of $OFCE(\boldsymbol{u})$ on an image.
- $VAE(\boldsymbol{u})$ the variance of $AE(\boldsymbol{u})$ on an image.
- $VOFCE(\boldsymbol{u})$ the variance of $VOFCE(\boldsymbol{u})$ on an image.

Figure 2 shows graph of these parameters for image sequences **Yosemite, New Marbled Block**, and **Rotating Sphere**. The results suggest that the local continuity order of the optical flow field is fractional. Furthermore, the order depends on the point.

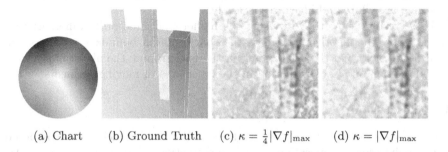

  (a) Chart   (b) Ground Truth   (c) $\kappa = \frac{1}{4}|\nabla f|_{\max}$ (d) $\kappa = |\nabla f|_{\max}$

Fig. 1. Computational results Form the left to right the colour charts for flow vectors, The ground truth of the motion field, For $\kappa = \frac{1}{4}|\nabla f|_{\max}$ and $\kappa = |\nabla f|_{\max}$ with $\alpha = 1.5$. These results for $\alpha = 1.5$ detects the motion boundary with motion discontinuity.

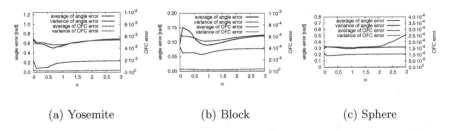

(a) Yosemite   (b) Block   (c) Sphere

Fig. 2. Computational results. Evaluation parameters of image sequences **Yosemite**, **New Marbled Block**, and **Rotating Sphere**. These graph curves suggest that the minimums of evaluation measures are established for $\alpha > 1$.

4.2   Selection of Local Order

Next, we show the results on the selection of $0 \leq \alpha \leq 3$ for each point using eq. (35).

Figure 3 shows the results of rotating sphere. In the first row from left to right, figures show motion field, residual curve, α-map, and results. In the second row, from left to right, figures shows the results for $\alpha = 1.0, 1.5, 2.0$, and 2.5, where $\kappa = 0.1$. In the third row, from left to right, figures shows the results $\alpha = 1.0, 1.5, 2.0$, and 2.5, where $\kappa = 1.0$. Figure 2(d) is the result by our method using the α-map of Fig. 2(c). Fig. 2(c) shows that the boundary of the sphere on the image is sharply extracted, since on the boundary both the gray-value of the image and the optical flow process discontinuity in the gray-value topography and motion filed, respectively. This discontinuity is extracted by the fractional order optical flow. Figure 4 shows the results for the new-marbled-block sequence. The proposed method clearly extracts the objects from the background using the difference in motion continuity order on the boundaries of the segments. Figure 5 shows the application of our method to the motion analysis of a real image sequence. This result shows that the proposed method extracts the moving cars using the difference of motion continuity order on the boundaries of the segments. In these results the maximum and minimum values of α-map are white and black, respectively.

These results show that by using the higher fractional order differentiation in the prior, the algorithm extracts the sharp motion boundary with a sharp optical flow field. On the other hand, in the background, the algorithm detects null motion and stationary motion in the inner area of moving segments.

4.3   Discussion

For $\alpha = n + \varepsilon$ such that $0 < \varepsilon < 1$, a fractional order Laplacian is decomposed into a polyharmonic operation $(-\Delta)^n$ and fractional Laplacian $\Lambda^{2\varepsilon} = (-\Delta)^\varepsilon$ as eq. (12). This decomposition can be read that $\Lambda^{2\alpha} f$ is achieved by applying the polyharmonic operation $(-\Delta)^n$ to $g = (-\Delta)^\varepsilon f$. The numerical filtering of the

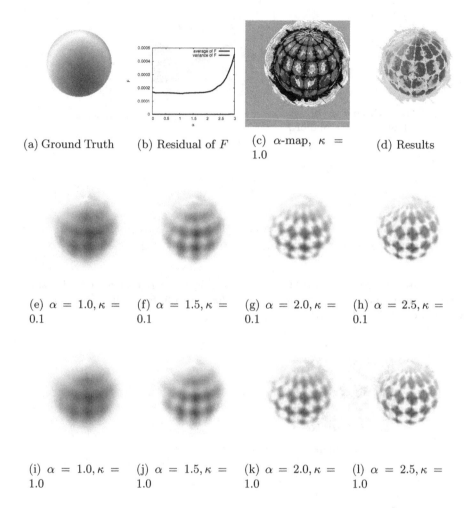

(a) Ground Truth

(b) Residual of F

(c) α-map, $\kappa = 1.0$

(d) Results

(e) $\alpha = 1.0, \kappa = 0.1$

(f) $\alpha = 1.5, \kappa = 0.1$

(g) $\alpha = 2.0, \kappa = 0.1$

(h) $\alpha = 2.5, \kappa = 0.1$

(i) $\alpha = 1.0, \kappa = 1.0$

(j) $\alpha = 1.5, \kappa = 1.0$

(k) $\alpha = 2.0, \kappa = 1.0$

(l) $\alpha = 2.5, \kappa = 1.0$

Fig. 3. Results 1: Rotating sphere. First row: from left to right, motion field, residual curve, α-map, and the results, Second row: for $\kappa = 0.1$, from left to right $\alpha = 1.0$, $\alpha = 1.5$, $\alpha = 2.0$, and $\alpha = 2.5$. Third row: for $\kappa = 1.0$, from left to right $\alpha = 1.0$, $\alpha = 1.5$, $\alpha = 2.0$, and $\alpha = 2.5$.

operation $(-\Delta)^\varepsilon$ derived in eq. (26) possesses a smoothing effect to the optical flow field $\boldsymbol{u}^{(l)}$ in each iteration step[5]. Therefore, our numerical scheme derived in the previous section generates a smoothed optical flow before applying the polyharmonic operation, which is the main part of the prior for the selection of model in optical flow computation. This presmoothing property of the numerical scheme yields better performance for $\alpha = n + \varepsilon$ such that $0 < \varepsilon < 1$.

From the results, we observe the following properties on the boundary motion.

---

[5] The operation $\Lambda^{2\varepsilon} f$ is achieved by convolution between the image f and the Riesz potential $g_{2\varepsilon}(x, y) = \dfrac{c}{(\sqrt{x^2+y^2})^{1+2\varepsilon}}$ for a positive constant c.

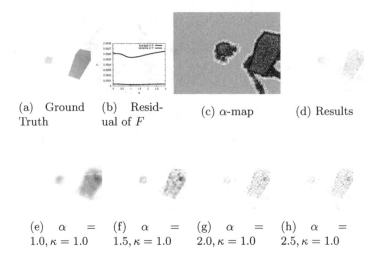

(a) Ground Truth

(b) Residual of F

(c) α-map

(d) Results

(e) $\alpha = 1.0, \kappa = 1.0$

(f) $\alpha = 1.5, \kappa = 1.0$

(g) $\alpha = 2.0, \kappa = 1.0$

(h) $\alpha = 2.5, \kappa = 1.0$

Fig. 4. Results 2: New marbled-block First row: from left to right, motion field, residual Curve, α-map, and the results. Second row: for $\kappa = 1.0$, from left to right $\alpha = 1.0$, $\alpha = 1.5$, $\alpha = 2.0$, and $\alpha = 2.5$.

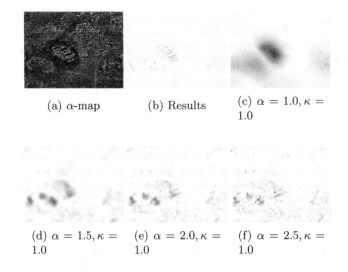

(a) α-map

(b) Results

(c) $\alpha = 1.0, \kappa = 1.0$

(d) $\alpha = 1.5, \kappa = 1.0$

(e) $\alpha = 2.0, \kappa = 1.0$

(f) $\alpha = 2.5, \kappa = 1.0$

Fig. 5. Results 3: Hamburg taxi. From left to right α-map, results, and results of $\alpha = 1.0, 1.5, 2.0, 2.5$ for $\kappa = 1.0$.

Observation 1. *If the motion of points in the neighbourhood of the boundary is locally stationary, for instance, the motion is pure translation in a region, the projection of ridges boundary moves elastically on the image. Therefore, the constraint*

$$E(\boldsymbol{u}) = \int\int_{\mathbf{R}^2}(|D^2u|^2 + |D^2v|^2)dxdy, \tag{36}$$

is suitable to detect moving boundaries.

Observation 2. *If the motion of points in the neighbourhood of the boundary on the image is nonstationary because of motion delay in the neighbourhood, for instance, the delay in the global translation caused by local rotation, the projection of ridges boundary moves viscoelastically on the image. Therefore, the constraint*

$$V(\boldsymbol{u}) = \int\int_{\mathbf{R}^2}(|\Lambda^{1+\varepsilon}u|^2 + |\Lambda^{1+\varepsilon}v|^2)dxdy \tag{37}$$

for $0 < \varepsilon < 1$ is suitable to detect moving boundary.

Our method estimates the continuity order of optical flow field at each point. Furthermore, as shown in the results, our method also extracts the higher order optical flow if the gray-value distribution of an image is discontinuous. The results mathematically provides a method to estimate the local continuity order of optical flow field, and theoretically shows that for motion boundary extraction and tracking, the prior with higher order differentiation is preferable. For the tracking of the image of an elastic boundary of a ridged object in space, the order of the differentiation is two. If the optimal order of the points is between 1 and 2, the points are viscoelastically moving [16,17] on an image.

Although TV regularisation accurately and stably computes an optical flow field and extracts moving segments from the background, the operation is non-linear. The results leads to the conclusion that using the local continuity order, it is possible to extract the motion boundary and separate moving segments from the background. The first step of our method is achieved by a collection of linear operation. This is the most important advantage of our method over TV regularisation.

5   Concluding Remarks

We introduced a method to compute optical flow by selecting the optimal local continuity order of the optical flow field using the variational principle. Our variational method derives a collection of linear partial differential equations. The linear equations can be numerically solved using the same numerical scheme, since the method is described by a differential operation with a parameter. The detection of discontinuity part by our method is compatible with TV regularisation, which derives a nonlinear iteration form as a numerical scheme.

The Nagel-Enkelmann [14,15] and TV regularisation [1] methods are image and flow driven, respectively. On the other hand, our method is an operator driven, since the order of the differentiation is selected at each point. In the Nagel-Enkelmann method, the structure tensor, which is the local moment of the gradient image, controls the local coordinate of optical flow computation, although the numerical scheme is linear. In TV regularisation, the direction of the flow vector itself controls the direction to compute local average and the numerical scheme is nonlinear. Although the local continuity order of an optical flow field is shift-variant, our method first solves many shift-invariant operations and select the optimal solution using the residual of the solution.

For accurate and stable numerical computation, the presmoothing of data is a typical procedure. In our iterative method to solve numerically a fractional partial differential equation, the DFT-based computation of the fractional part of higher order derivative acts as intermediate smoothing in each step of the iteration. Usually, presmoothing operations are heuristically selected. However, in the computation of fractional differentiations in our problem, the operations automatically defines a class of smoothing operations. This is the second advantage of the method.

References

1. Papenberg, N., Bruhn, A., Brox, T., Didas, S., Weickert, J.: Highly accurate optic flow computation with theoretically justified warping. IJCV 67, 141–158 (2006)
2. Yin, W., Goldfarb, D., Osher, S.: A comparison of three total variation based texture extraction models. J. Visual Communication and Image Representation 18, 240–252 (2007)
3. Tadjeran, C., Meerschaert, M.M.: A second-order accurate numerical method for the two-dimensional fractional diffusion equation. J. of Computational Physics 220, 813–823 (2007)
4. Eckstein, J., Bertsekas, D.P.: On the Douglas-Rachford splitting method and the proximal point algorithm for maximal monotone operators. Mathematical Programming 55, 293–318 (1992)
5. Davis, J.A., Smith, D.A., McNamara, D.E., Cottrell, D.M., Campos, J.: Fractional derivatives-analysis and experimental implementation. Applied Optics 32, 5943–5948 (2001)
6. Tseng, C.-C., Pei, S.-C., Hsia, S.-C.: Computation of fractional derivatives using Fourier transform and digital FIR differentiator. Signal Processing 80, 151–159 (2000)
7. Zhang, J., Wei, Z.-H.: Fractional variational model and algorithm for image denoising. In: Proceedings of 4th International Conference on Natural Computation, vol. 5, pp. 524–528 (2008)
8. Mathieu, B., Melchior, P., Oustaloup, A., Ceyral, Cn.: Fractional differentiation for edge detection. Signal Processing 83, 2421–2432 (2003)
9. Sabatier, J., Agrawel, O.P., Tenreiro Machado, I.A.: Advances in Fractional Calculus: Theoretical Development and Applications in Physics and Engineering. Springer, Netherlands (2007)
10. Oldham, K.B., Spanier, J.: The Fractional Calculus: Theory And Applications of Differentiation And Integration to Arbitrary Order (Dover Books on Mathematics). Dover (2004)
11. Podlubny, I.: Fractional Differential Equations. An Introduction to Fractional Derivatives, Fractional Differential Equations, Some Methods of Their Solution and Some of Their Applications. Academic Press, London (1999)
12. Horn, B.K.P., Schunck, B.G.: Determining optical flow. Artificial Intelligence 17, 185–204 (1981)

13. Beauchemin, S.S., Barron, J.L.: The computation of optical flow. ACM Computer Surveys 26, 433–467 (1995)
14. Nagel, H.-H., Enkelmann, W.: An investigation of smoothness constraint for the estimation of displacement vector fields from image sequences. IEEE Trans. on PAMI 8, 565–593 (1986)
15. Nagel, H.-H.: On the estimation of optical flow:Relations between different approaches and some new results. Artificial Intelligence 33, 299–324 (1987)
16. Momani, S., Odibat, Z.: Numerical comparison of methods for solving linear differential equations of fractional order. Chaos, Solitons and Fractals 31, 1248–1255 (2007)
17. Murio, D.A.: Stable numerical evaluation of Grünwald-Letnikov fractional derivatives applied to a fractional IHCP. Inverse Problems in Science and Engineering 17, 229–243 (2009)
18. Gorenfloa, R., Abdel-Rehimb, E.A.: Convergence of the Grünwald-Letnikov scheme for time-fractional diffusion. J. of Computational and Applied Mathematics 205, 871–881 (2007)
19. Debbi. L., Explicit solutions of some fractional partial differential equations via stable subordinators. J. of Applied Mathematics and Stochastic Analysis, Article ID 93502, 1–18 (2006)
20. Debbi, L.: On some properties of a high order fractional differential operator which is not in general selfadjoint. Applied Mathematical Sciences 1, 1325–1339 (2007)
21. Chechkin, A.V., Gorenflo, R., Sokolov, I.M.: Fractional diffusion in inhomogeneous media. J. Phys. A: Math. Gen. 38, L679–L684 (2005)
22. Duits, R., Felsberg, M., Florack, L.M.J., Platel, B.: α scale spaces on a bounded domain. In: Griffin, L.D., Lillholm, M. (eds.) Scale-Space 2003. LNCS, vol. 2695, pp. 502–518. Springer, Heidelberg (2003)

A Local Normal-Based Region Term for Active Contours

Julien Mille and Laurent D. Cohen

CEREMADE, CNRS UMR 7534, Université Paris Dauphine
Place du Maréchal de Lattre de Tassigny, 75775 Paris, France
{mille,cohen}@ceremade.dauphine.fr

Abstract. Global region-based active contours, like the Chan-Vese model, often make strong assumptions on the intensity distributions of the searched object and background, preventing their use in natural images. We introduce a more flexible local region energy achieving a trade-off between local features of gradient-like terms and global region features[1]. Relying on the theory of parallel curves, we define our region term using constant length lines normal to the contour. Mathematical derivations are performed on an explicit curve, leading to a form allowing efficient implementation on a parametric snake. However, we provide implementations on both explicit and implicit contours.

1 Introduction

Active contours, whether parametric [1] or level-set based [2], were initially attached to data by means of edge-based terms. The increasing use of region terms inspired by the Mumford-Shah functional [3][4] has proven to overcome limitations of gradient-based only models, especially when dealing with data sets suffering from noise and lack of contrast between neighboring structures. Early work including the mixed model of Cohen *et al* [5] and the active region model by Ivins and Porrill [6] introduced the use of region terms in the evolution of parametric snakes. On the other hand, many papers have dealt with region-based approaches using the level set framework, including the active contours without edges by Chan and Vese [7], the deformable regions by Jehan-Besson *et al* [8] and the geodesic active regions by Paragios and Deriche [9], benefiting of adaptive topology at the expense of computational cost. Classical region-based deformable models segment images according to statistical data computed over the object of interest and the background. Image partitions should be uniform in terms of pixel intensities or higher level features like texture descriptors [9]. Considering for instance the Chan-Vese model [7], the region term penalizes the curve splitting the image into heterogeneous regions, using intensity variances. It is devoted by essence to the segmentation of uniform objects and backgrounds.

Such an ideal case is rarely encountered in most of computer vision applications, as the background usually contains various structures, which differ in their

---

[1] This work was partially supported by ANR grant MESANGE ANR-08-BLAN-0198.

overall intensities or textures. In this context, the multiphase approach [10] allows to partition the image into more than two regions, provided that the number of partitions is known. When one wishes to extract a particular object from the background without any prior knowledge about the number of actual regions, strict homogeneity is not desirable property for the background. In order to account for spatially varying intensity, local statistics in region-based segmentation have emerged recently [11][12][13][14]. Basically, these methods express the data term as a sum of local region energies computed over neighborhoods of pixels inside and outside the evolving curve. We believe these approaches have the drawback of not formulating the region energy fully explicitly in terms of the curve, which only leads to a level set implementation. However, many applications benefit from explicit implementations of active contours, including low computational cost and topological control. This justifies the use of an explicit mathematical framework.

We introduce a local normal-based region energy handling configurations in which the outer neighborhood of the object is piecewise uniform. Unlike other region terms, whether local or global, this new type of combination allows to handle the common case where one seeks for a uniform object in a heterogeneous background. We formulate it as the intensity variance over the inner region and finite length lines along outward normals to the curve. The theory of parallel curves [15][16] leads to an explicit formulation of our energy, which is suitable for mathematical derivation and implementation on parametric contours. In order to allow gradient descent afterwards, we determine the variational derivative of the region energy thanks to calculus of variations. Then, we deal with numerical implementation on both parametric snakes and level sets. Finally, experiments are carried out on medical data and natural color images. The tests present the advantages of our new data term over an edge term, a global region term as well as a recent local region-based approach [12].

2   Local Normal-Based Region Energy

2.1   Active Contour Model

Given a simple closed curve Γ with position vector $\mathbf{c}(u) = [x(u) \; y(u)]^T$ with $u \in \Omega = [0, 1]$, segmentation is performed by finding the curve minimizing a weighted sum of smoothness term and our local normal-based region (LNBR) energy:

$$E[\Gamma] = \omega E_{\text{smooth}}[\Gamma] + (1 - \omega)E_{\text{LNBR}}[\Gamma] \tag{1}$$

where the user-provided ω weights the significance of the smoothness term, which can be classicaly written with squared magnitudes of first and second order derivatives. Curve Γ splits the image domain $\mathcal{D}$ into an inner region R_I and an outer region R_O. Instead of formulating our data term on R_I and R_O, we use the narrow band principle, which has proven its efficiency in the evolution of level sets [2]. Hence, in addition to the inner region R_I, instead of dealing

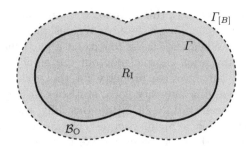

Fig. 1. Inner region and outer band for LNBR energy

with the entire image domain, we consider an outer band $\mathcal{B}_O$ in the vicinity of Γ, as depicted in fig. 1.

The purpose of the LNBR energy is to handle cases where the inner region is homogeneous and the background is locally homogeneous in the outer band. For now, we express the outer term using a local descriptor depending on current position $\mathbf{x}$:

$$E_{\text{LNBR}}[\Gamma] = \iint\limits_{R_I} (I(\mathbf{x})-\mu_I)^2 d\mathbf{x} + \iint\limits_{\mathcal{B}_O} (I(\mathbf{x})-\mu(\mathcal{B}_O,\mathbf{x}))^2 d\mathbf{x} \qquad (2)$$

In what follows, we explain how E_{LNBR} can be explicitly formulated in terms of curve Γ.

2.2 Parallel Curve

Let B be the band thickness, constant along Γ. The theoretical background of our narrow band framework is based on parallel curves [15][16]. The curve $\Gamma_{[B]}$ is called a parallel curve of Γ, as its position vector $\mathbf{c}_{[B]}$ is defined by:

$$\mathbf{c}_{[B]}(u) = \mathbf{c}(u) - B\mathbf{n}(u) \qquad (3)$$

where $\mathbf{n}$ is the unit inward normal. Hereafter, we will use the index $[B]$ to denote all quantities related to the parallel curve. The definition in eq. (3) is suitable to our narrow band formulation, in the sense that $\mathcal{B}_O$ is bounded by Γ and $\Gamma_{[B]}$. Afterwards, we denote $R_{I[B]}$ the dilated inner region bounded by $\Gamma_{[B]}$.

Given length element $\|\mathbf{c}_u\|$ and curvature κ, an important property resulting from the definition in eq. (3) is that the velocity vector of the parallel curve can be expressed as a function of the velocity vector of Γ, as well as its curvature and normal. Using the identity $\mathbf{n}_u = -\kappa\mathbf{c}_u$, we have:

$$\mathbf{c}_{[B]_u} = \mathbf{c}_u - B\mathbf{n}_u = (1 + B\kappa)\mathbf{c}_u \qquad (4)$$

and the corresponding length element is $\|\mathbf{c}_{[B]_u}\| = |1+B\kappa| \, \|\mathbf{c}_u\|$, which implies a constraint on the maximal curvature of curve Γ. We should assume that Γ

is smooth enough so that $\kappa(u) > -1/B$, $\forall u \in \Omega$, so that curve $\Gamma_{[B]}$ does not exhibit singularities. This has an impact on explicit numerical implementation, which is discussed in section 3.1.

We rely on the principle of parallel curve to transform region integrals over $\mathcal{B}_O$. Introducing a variable thickness b and using Green's theorem to convert region integrals into boundary integrals, it can be shown that:

$$\iint_{\mathcal{B}_O} f(\mathbf{x})d\mathbf{x} = \int_\Omega \int_0^B f(\mathbf{c} - b\mathbf{n}) \|\mathbf{c}_u\| (1 + b\kappa)db\,du \tag{5}$$

2.3   Transformation and Derivation of LNBR Energy

We provide the final expression of the LNBR term as it is implemented, in contrast with the temporary form of eq. (2). We now assume that piecewise uniformity over the outer band is verified if intensity is uniform along line segments in the direction normal to the object boundary. We first calculate the average intensity along the outward local normal line of length B at a given contour point. We use the same curvature-dependent weighting than in eq. (5), leading to:

$$\mu_{\text{LN}}(u) = \frac{2}{B(2 + B\kappa)} \int_0^B I(\mathbf{c} - b\mathbf{n})(1 + b\kappa)db \tag{6}$$

where $I(\mathbf{x}) \in [0, 1]$ is the image intensity. The LNBR energy should penalize non-uniformity over the whole inner region and over all normal lines. Thus, we write:

$$E_{\text{LNBR}}[\Gamma] = \iint_{R_I} (I(\mathbf{x}) - \mu_I)^2 d\mathbf{x} + \int_\Omega \|\mathbf{c}_u\| \int_0^B (I(\mathbf{c} - b\mathbf{n}) - \mu_{\text{LN}}(u))^2 (1 + b\kappa)db\,du \tag{7}$$

where μ_I is the average intensity of inner region R_I. To some extent, the outer band $\mathcal{B}_O$ is split into infinitesimal trapezoids with parallel sides $\|\mathbf{c}_u\|$ and $\|\mathbf{c}_u\| (1 + B\kappa)$. This principle is represented on the discretized curve in fig. 2. The LNBR energy has the following first variation (details of derivation are provided in the appendix):

$$\frac{\delta E_{\text{LNBR}}}{\delta \Gamma} \approx \|\mathbf{c}_u\| \left[-(I(\mathbf{c}) - \mu_I)^2 - (1 + B\kappa)(I(\mathbf{c}_{[B]}) - \mu_{\text{LN}})^2 + (I(\mathbf{c}) - \mu_{\text{LN}})^2 \right] \mathbf{n} \tag{8}$$

The derivative holds the term $(I(\mathbf{c}) - \mu_{\text{LN}})^2 - (I(\mathbf{c}) - \mu_I)^2$, which is clearly in accordance with the region-based segmentation principle. Indeed, the sign of the above quantity depends on the likeness of the current point's intensity with respect to μ_I or μ_{LN}. If $I(\mathbf{c})$ is closer to μ_I than μ_{LN}, the contour will locally expand, as it would be the case with a region growing approach. The derivative holds an additional curvature-dependent term which effect is discussed in section 3.1.

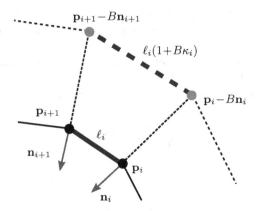

Fig. 2. Neighboring vertices with corresponding points on the discrete parallel curve

3   Numerical Implementation

3.1   Explicit Representation

Implementation on an explicit curve is pertinent when speed and topology preservation is a major concern. The contour is discretized as a closed polygonal line made up of a set of n vertices, denoted $\mathbf{p}_i = [x_i \; y_i]^T$. Their coordinates are iteratively modified using gradient descent of eq. (1):

$$\mathbf{p}_i^{(t+1)} = \mathbf{p}_i^{(t)} + \Delta t \mathbf{f}(\mathbf{p}_i) \tag{9}$$

where $\mathbf{f}(\mathbf{p}_i)$ is the force vector depending on the discretization of the energy derivative at a given vertex $\mathbf{p}_i$. In addition to the squared differences between $I(\mathbf{c})$ and the average intensities, the variational derivative in eq. (8) also contains a curvature-based term depending on the intensity at point $\mathbf{c}_{[B]}$. Actually, this term turns out to go against the region growing or shrinking principle, as it opposes the other terms depending on $I(\mathbf{c})$. As stated in [17], the usual energy gradient may not be systematically the best direction to take, which justifies our choice to remove side effect terms. The region force is:

$$\mathbf{f}_{\mathrm{LNBR}}(\mathbf{p}_i) = \left[(I(\mathbf{p}_i)-\mu_{\mathrm{I}})^2 - (I(\mathbf{p}_i)-\mu_{\mathrm{LN}}(\mathbf{p}_i))^2 \right] \mathbf{n}_i \tag{10}$$

Given ℓ_i, $\mathbf{n}_i$ and κ_i the finite differences discretizations of length element, normal and curvature at vertex $\mathbf{p}_i$, the average intensity along the normal line is implemented as:

$$\mu_{\mathrm{LN}}(\mathbf{p}_i) = B \left(1 + \frac{\kappa_i(B+1)}{2} \right) \sum_{b=1}^{b=B} (1+b\kappa_i) I(\mathbf{p}_i - b\mathbf{n}_i)$$

Fig. 2 depicts two neighboring vertices on a locally convex polygon, with corresponding length elements and points on the parallel polygon. There are two complementary techniques to address the regularity condition $\kappa_i > -1/B$. The first one is to prevent vertices from making sharp angles with their neighbors, so that κ_i is well bounded. Moreover, the case of a negative length element can be handled. Hence, $\ell_i(1+b\kappa_i)$ is actually computed as $\max(0, \ell_i(1+b\kappa_i))$.

In a particular case, the formulation of $\mathbf{f}_{\text{LNBR}}$ presents a shortcoming. Indeed, the magnitude of $\mathbf{f}_{\text{LNBR}}$ is low when μ_I and μ_{LN} are similar. This situation also arises in local region-based methods [12][13] when the curve, including the outer neighborhood, is initialized inside a uniform area. However, we expect the contour to grow if the intensity at the current vertex matches the inner region features, whatever the value of μ_{LN}. Thus, we introduce a bias acting like a balloon force [18] which expands the boundary in the normal direction:

$$\mathbf{f}_{\text{bias}}(\mathbf{p}_i) = -\alpha(1 - (\mu_I - \mu_{\text{LN}}(\mathbf{p}_i))^2)\mathbf{n}_i \tag{11}$$

with $\alpha \in [0, 1]$. Forces $\mathbf{f}_{\text{LNBR}}$ and $\mathbf{f}_{\text{bias}}$ are summed up, so that the bias is predominant when mean intensities are close. Consequently, we do not loose the convergence ability of global region-based active contours. The region bias guarantees the contour has a similar capture range as global region-based contours.

Let us give a note on the implementation of Green's theorem. Our experiments include a comparison between the LNBR energy and a global region energy, similar to the data term of the Chan-Vese model. The implementation of the latter on the explicit polygon raises the difficulty of computing region integrals. A naive solution consists in using region filling algorithms to determine inner pixels [6] which would be computationally expensive if performed after each deformation step. Another solution, which we chose, is based on a discretization of Green-Riemann theorem. We compute and store the summed intensities P and Q in the respective directions x and y only once, before deformation is performed. This reduces the algorithmic complexity to $O(n)$, whereas the LNBR term induces a $O(nB)$ complexity.

3.2   Implicit Representation

On the other hand, we provide an implicit implementation of the LNBR energy. In this case, the contour is the zero level set of $\psi : \mathbb{R}^2 \to \mathbb{R}$. We define the region enclosed by the contour as $R_I = \{\mathbf{x}|\psi(\mathbf{x}) \leq 0\}$. Function ψ evolves according to the following PDE:

$$\frac{\partial\psi}{\partial t} = F(\mathbf{x})\,\|\nabla\psi(\mathbf{x})\| \;\; \forall \mathbf{x} \in \mathbb{R}^2$$

where speed function F is to some extent the level set-equivalent of the explicit energy in eq. (1), i.e. a weighted sum of smoothness and region terms:

$$F(\mathbf{x}) = \omega F_{\text{smooth}}(\mathbf{x}) + (1 - \omega)F_{\text{LNBR}}(\mathbf{x})$$

where the smoothness term is expressed as usual using curvature. Areas and average intensities on the outer band are easily computed on the level set implementation, since a circular window of radius B may be considered around each pixel located on the front.

$$\mathcal{B}_O = \{\mathbf{x}|\psi(\mathbf{x}) \geq 0 \text{ and } \exists \mathbf{y} \in \mathcal{W}_B(\mathbf{x}) \text{ s.t. } \psi(\mathbf{y}) = 0\}$$

Considering the sign of ψ, pixels belonging to $\mathcal{B}_O$ are easily determined by dilating the front with circular window $\mathcal{W}_B$. As regards the average intensity along outward normal lines, we rely on the curvature-based formulation of the explicit curve. In the level set framework, it gives:

$$\mu_{\mathrm{LN}}(\mathbf{x}) = \frac{2}{B(2 + B\kappa_\psi(\mathbf{x}))} \int_0^B I(\mathbf{x} + b\mathbf{n}_\psi(\mathbf{x}))(1 + b\kappa_\psi(\mathbf{x}))db$$

with unit outward normal $\mathbf{n}_\psi$ and curvature κ_ψ:

$$\mathbf{n}_\psi(\mathbf{x}) = \frac{\nabla\psi(\mathbf{x})}{\|\nabla\psi(\mathbf{x})\|} \qquad \kappa_\psi(\mathbf{x}) = \mathrm{div}\left(\frac{\nabla\psi(\mathbf{x})}{\|\nabla\psi(\mathbf{x})\|}\right)$$

Computed as is, in order for $\mathbf{n}_\psi$ to be actually normal to the front, ψ should remain a distance function. This implies to update ψ as a signed Euclidean distance in the neighborhood of the front before estimating normal vectors. From eq. (7), we write the level set formulation of the LNBR term:

$$E_{\mathrm{LNBR}}[\psi] = \iint_{\mathcal{D}} (1 - H(\psi(\mathbf{x})))(I(\mathbf{x}) - \mu_{\mathrm{I}})^2 d\mathbf{x}$$
$$+ \iint_{\mathcal{D}} \delta(\psi(\mathbf{x})) \int_0^B (I(\mathbf{x} + b\mathbf{n}_\psi(\mathbf{x})) - \mu_{\mathrm{LN}}(\mathbf{x}))^2 (1 + b\kappa_\psi(\mathbf{x}))dbd\mathbf{x}$$

where H and δ are the Heaviside step and Dirac impulse functions. For a point $\mathbf{x}$ located on the front, the corresponding speed is approximated from eq. (10):

$$F_{\mathrm{LNBR}}(\mathbf{x}) = (I(\mathbf{x}) - \mu_{\mathrm{LN}}(\mathbf{x}))^2 - (I(\mathbf{x}) - \mu_{\mathrm{I}})^2$$

Eventually, the reader may note that an equivalent bias technique as the one used in the explicit implementation (see eq. (11)) is applied in the level set model. The level set function ψ evolves according to the narrow band technique [2], so that only pixels located on the front are updated.

4   Results and Discussion

4.1   Concurrent Methods

We compare the behavior of explicit and implicit active contours endowed with different data terms: an edge term [19], a global region term similar to one of the Chan-Vese model [7] and the uniform modeling energy of Lankton-Tannenbaum [12]. The goal of our experiments is not to compare explicit and implicit implementations, since it is well accepted that both exhibit their own advantages. We intend to show the interest of the LNBR energy whatever implementation is used. In the edge-based model, the region force is replaced by an edge force resulting from the differentiation of the image gradient magnitude:

$$\mathbf{f}_{\mathrm{edge}}(\mathbf{p}_i) = \nabla \|\nabla G_\sigma * I(\mathbf{p}_i)\| - \alpha\mathbf{n}_i$$

where α weights an additional balloon force [18] increasing the capture range and consequently allowing the snake to be initialized far from the target boundaries. The gradient magnitude is computed on data convolved with first-order derivative of gaussian G_σ, where scale σ is empirically chosen to yield the most significant edges. A similar speed term F_{edge} is implemented in the level set contour. As stated by their authors, the Chan-Vese (CV) and Lankton-Tannenbaum (LT) models directly rely on an implicit formulation of the curve. We give their corresponding region speed terms:

$$F_{\text{global}}(\mathbf{x}) = (I(\mathbf{x}) - \mu_I)^2 - (I(\mathbf{x}) - \mu_O)^2$$

$$F_{\text{LT}}(\mathbf{x}) = \iint_{\mathcal{W}_B(\mathbf{x})} \delta(\psi(\mathbf{y}))(I(\mathbf{y}) - \mu_I(\mathbf{x}))^2 - (I(\mathbf{y}) - \mu_O(\mathbf{x}))^2 d\mathbf{y}$$

where $\mu_I(\mathbf{x})$ is the local inner average intensity over the ball of radius B centered at $\mathbf{x}$, and similarly for the local outer average intensity $\mu_O(\mathbf{x})$. One may note that in the initial paper by Chan and Vese, the region term is asymmetric, as inner and outer terms are independently weighted, so that the variance minimization may be favoured inside or outside. However, we chose to use a symmetric term, as it is commonly the case with region-based active contours. Incidentally, future experiments could be done using asymmetry on all compared region terms. Moreover, the localized term of Lankton-Tannenbaum suffers from a weak capture range, since the front cannot evolve if inner and outer local means are similar. Thus, we also embedded into this energy the bias force of eq. (11). We used the same curvature-based regularization term for all tested approaches.

For all datasets, the model was initialized as a small circle fully or partially inside the area of interest, far from the target boundaries. Results are shown in fig. 3. Explicit contours are drawn in red whereas implicit ones appear in blue. For all experiments, the regularization weight ω was set to 0.5. On noisy data, we found that contours with lower ω were prone to boundary leaking. In addition, insufficient regularization makes level set implementations leave spurious isolated pixels inside and outside the inner region. Conversely, values above 0.8 turn out to prevent the surface from propagating into narrow structures. Experiments are carried on grayscale and color images as well. For the latter ones, we should point out that the minimal variance principle is easily extended to vector quantities. Let us consider the vector-valued image $\mathbf{I}$ and average intensities $\mathbf{m}_I$ and $\mathbf{m}_{\text{LN}}$. In the inner term, the integrand becomes $\|\mathbf{I} - \mathbf{m}_I\|^2$ and similarly for the outer term. The synthetic image in row 3, made up of color ellipses corrupted with gaussian noise, was segmented using RGB values. The natural images depicted in rows 4 to 7 hold nearly color-uniform objects. They were segmented using the ab components of the perceptually uniform CIE Lab color space. Neglecting the brightness L makes color statistics insensitive to illumination changes in visually uniform regions, allowing to handle highlights and shadows properly.

4.2   A Note on the Choice of the Band Thickness

The band thickness B is an important parameter of our method and should be discussed. Apart from its impact on the algorithmic complexity - computing average intensities along normal lines takes at least $O(nB)$ operations - it controls the trade-off between local and global features around the object. If $B = 1$, the region energy is as local as an edge term. The main image property having an effect on the minimal band thickness is the edges sharpness. Indeed, the deformable curve needs a larger band as the boundaries of the target object are fuzzy. To put this phenomenon into evidence, we applied the active contour on an increasingly blurred image. Bands thinner than the minimal one caused the contour to flow into neighboring structures. The original image was segmented with $B = 2$. For subsequent images, increasing the band turned out to be necessary. As the blur level of the last image in the sequence is rarely encountered in the applications we aim at, $B = 10$ was a suitable value in our experiments.

4.3   Segmentation Results

Since we are looking for perceptually homogeneous objects, segmentation quality can be assessed visually. One can reasonably admit that the target object corresponds to the area containing the major part of the initial region. Gradient-based deformable models fail on images where noise and low contrast between neighboring objects prevent the extraction of reliable edges. Except for the last image in fig. 3, we could not find a suitable balloon weight preventing the contour from being trapped in spurious noisy edges inside the shape while stopping on the actual boundaries. Indeed, the edge-based energy is inefficient when the sharpness of boundaries decreases, as the contour may pass through the actual edges and stop on false ones simultaneously. In order to keep a critical eye on our approach, we draw the attention on the equivalence between the global and local region energies in particular images.

Row 2 and 4 depict typical configurations where there is no particular benefit in using localized region energies. In the MRI short-axis view of the human heart, the background is not uniform but still significantly darker than the bright left ventricle. Thus, the global region speed manages to make the front stabilize on the actual boundaries. The background of image 4 is obviously color-uniform as well. However, in other images containing various objects surrounding the structure of interest, the global region term captures all areas considered as different from the background. By definition, any two-phase segmentation model may fail at recovering a particular object when it is surrounded by many different objects, except in very particular cases such as a bright object surrounded by several dark objects. This phenomenon is well illustrated in row 3. Due to the averaging performed over the outer region, the global region approach turns out to split the image with respect to the blue component, since it is the dominant color in the background and it is absent of all areas in the inner region. Row 1 is a particular case in which the contour endowed with the global region-based active contour does not manage to grow, as inner and outer average intensities are not sufficiently different.

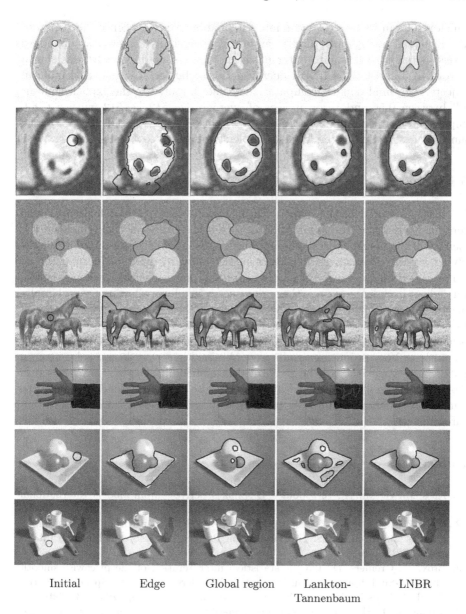

Initial   Edge   Global region   Lankton-   LNBR
                                                 Tannenbaum

Fig. 3. Segmentation results on medical and natural color images. Starting from common initializations shown in column 1, the LNBR term is compared with three other energies (edge energy, global region energy and local uniform modeling energy of Lankton-Tannenbaum). The image in row 4 was taken from the Berkeley Segmentation Dataset [20].

In the extent of our experiments, the Lankton-Tannenbaum method turns out to somewhat more sensitive to initialization than the LNBR active contour. In row 2, an inner dark papillary muscle is partially included in the initial region, which results in its incorporation into the final inner region. Since the Lankton-Tannenbaum energy only implies uniformity over balls centered at boundary pixels, it tends to flow into outer parts and leave some inner parts, as shown in rows 4, 5 and 6. The LNBR energy performs better at segmenting uniform objects. As a final remark, computational times imputed to the explicit contour fall between 0.5s and 1s on images of average size 512×512, with a C++ implementation running on an Intel Core 2 Duo 2GHz with 1Gb RAM. On the same images, we found the level-set implementations 3 to 4 times slower.

5   Conclusion

We have presented in this paper a local region-based method for deformable contours, relying on the assumption of a piecewise uniform background in the vicinity of the target object. The approach is based on a novel region term implying average intensities along lines in the outward direction normal to the curve. Based on the theory of parallel curves, a mathematical development was carried out in order to express the region energy in a form allowing natural implementation on explicit models. The local normal-based region energy managed to overcome the drawbacks of deformable models relying exclusively on edge terms or global region terms. We provided explicit and level-set based implementations. Very promising results were obtained in grayscale and color images. Further investigations will be performed in embedding local region terms into more geometrically constrained models. We also plan to extend the model to temporal segmentation, in order to track evolving objects in videos, and to textured images.

References

1. Kass, M., Witkin, A., Terzopoulos, D.: Snakes: active contour models. International Journal of Computer Vision 1(4), 321–331 (1988)
2. Malladi, R., Sethian, J., Vemuri, B.: Shape modeling with front propagation: a level set approach. IEEE Transactions on Pattern Analysis and Machine Intelligence 17(2), 158–175 (1995)
3. Brox, T., Cremers, D.: On the statistical interpretation of the piecewise smooth Mumford-Shah functional. In: International Conference on Scale Space and Variational Methods in Computer Vision (SSVM), Ischia, Italy, pp. 203–213 (2007)
4. Mumford, D., Shah, J.: Optimal approximation by piecewise smooth functions and associated variational problems. Communications on Pure and Applied Mathematics 42(5), 577–685 (1989)
5. Cohen, L., Bardinet, E., Ayache, N.: Surface reconstruction using active contour models. In: SPIE Conference on Geometric Methods in Computer Vision, San Diego, CA, USA (1993)
6. Ivins, J., Porrill, J.: Active region models for segmenting textures and colours. Image and Vision Computing 13(5), 431–438 (1995)
7. Chan, T., Vese, L.: Active contours without edges. IEEE Transactions on Image Processing 10(2), 266–277 (2001)

8. Jehan-Besson, S., Barlaud, M., Aubert, G.: DREAM²S: Deformable regions driven by an eulerian accurate minimization method for image and video segmentation. International Journal of Computer Vision 53(1), 45–70 (2003)
9. Paragios, N., Deriche, R.: Geodesic active regions and level set methods for supervised texture segmentation. International Journal of Computer Vision 46(3), 223–247 (2002)
10. Vese, L., Chan, T.: A multiphase level set framework for image segmentation using the Mumford and Shah model. International Journal of Computer Vision 50(3), 271–293 (2002)
11. Alemán-Flores, M., Alvarez, L., Caselles, V.: Texture-oriented anisotropic filtering and geodesic active contours in breast tumor ultrasound segmentation. Journal of Mathematical Imaging and Vision 28(1), 81–97 (2007)
12. Lankton, S., Tannenbaum, A.: Localizing region-based active contours. IEEE Transactions on Image Processing 17(11), 2029–2039 (2008)
13. Li, C., Kao, C., Gore, J., Ding, Z.: Implicit active contours driven by local binary fitting energy. In: IEEE Computer Vision and Pattern Recognition (CVPR), Minneapolis, Minnesota, USA, pp. 17–22 (2007)
14. Piovano, J., Papadopoulo, T.: Local statistic based region segmentation with automatic scale selection. In: Forsyth, D., Torr, P., Zisserman, A. (eds.) ECCV 2008, Part II. LNCS, vol. 5303, pp. 486–499. Springer, Heidelberg (2008)
15. Elber, G., In-Kwon, L., Myung-Soo, K.: Comparing offset curve approximation methods. IEEE Computer Graphics and Applications 17(3), 62–71 (1997)
16. Pressley, A.: Elementary differential geometry. Springer, Heidelberg (2002)
17. Charpiat, G., Maurel, P., Pons, J.P., Keriven, R., Faugeras, O.: Generalized gradients: priors on minimization flows. International Journal of Computer Vision 73(3), 325–344 (2007)
18. Cohen, L.: On active contour models and balloons. Computer Vision, Graphics, and Image Processing: Image Understanding 53(2), 211–218 (1991)
19. Caselles, V., Kimmel, R., Sapiro, G.: Geodesic active contours. International Journal of Computer Vision 22(1), 61–79 (1997)
20. Martin, D., Fowlkes, C., Tal, D., Malik, J.: A database of human segmented natural images and its application to evaluating segmentation algorithms and measuring ecological statistics. In: IEEE International Conference on Computer Visison (ICCV), Vacouver, Canada, vol. 2, pp. 416–423 (2001)
21. Zhu, S., Yuille, A.: Region competition: unifying snakes, region growing, Bayes/MDL for multiband image segmentation. IEEE Transactions on Pattern Analysis and Machine Intelligence 18(9), 884–900 (1996)

A Calculus of Variations

A.1 Derivative of Parallel Curve-Based Term

In classical active contours, curve Γ is a local minimizer of the functional

$$\mathcal{E}[\Gamma] = \int_\Omega \mathcal{L}(\mathbf{c}, \mathbf{c}_u, \mathbf{c}_{uu})\, du$$

when the following variational derivative vanishes:

$$\frac{\delta\mathcal{E}[\Gamma]}{\delta\Gamma} = \frac{\partial\mathcal{L}}{\partial\mathbf{c}} - \frac{d}{du}\left\{\frac{\partial\mathcal{L}}{\partial\mathbf{c}_u}\right\} + \frac{d^2}{du^2}\left\{\frac{\partial\mathcal{L}}{\partial\mathbf{c}_{uu}}\right\} \tag{12}$$

Considering now a functional expressed on the parallel curve,

$$\mathcal{E}'[\Gamma_{[B]}] = \int_{\Omega} \mathcal{L}'(\mathbf{c}_{[B]}, \mathbf{c}_{[B]_u}) du,$$

determining directly the derivative of $\mathcal{E}'[\Gamma_{[B]}]$ with respect to Γ leads to tedious calculations. Instead, we find more practical to determine $\delta\mathcal{E}'[\Gamma_{[B]}]/\delta\Gamma_{[B]}$ first, and then relate it to $\delta\mathcal{E}'[\Gamma_{[B]}]/\delta\Gamma$ using the following general expression:

$$
\begin{aligned}
\frac{\delta\mathcal{E}'}{\delta\Gamma} = {} & (1+B\kappa)\left\langle \frac{\delta\mathcal{E}'}{\delta\Gamma_{[B]}}, \mathbf{t} \right\rangle \mathbf{t} + (1-B\kappa)\left\langle \frac{\delta\mathcal{E}'}{\delta\Gamma_{[B]}}, \mathbf{n} \right\rangle \mathbf{n} \\
& + \frac{B\|\mathbf{c}_u\|_u}{\|\mathbf{c}_u\|^2}\left\langle \frac{\delta\mathcal{E}'}{\delta\Gamma_{[B]}}, \mathbf{t} \right\rangle \mathbf{n} - \frac{B}{\|\mathbf{c}_u\|}\left\langle \frac{d}{du}\left\{\frac{\delta\mathcal{E}'}{\delta\Gamma_{[B]}}\right\}, \mathbf{t} \right\rangle \mathbf{n}
\end{aligned}
\tag{13}
$$

where $\langle , \rangle$ is the L^2 inner product and $\mathbf{t}$ is the unit tangent vector. To some extent, we designed the expression in eq. (13) as a chain rule for parallel curve-based energies. Hereafter, we use it to determine the derivative of the LNBR term.

A.2 Derivative of Region Terms

We now need to express derivatives of general region terms over R_I and $\mathcal{B}_\mathrm{O}$. Region terms are transformed into boundary integrals using Green's theorem. In this way, inners terms are differentiated according to the following template formula (the detailed derivation may be found for example in the appendix of [21]):

$$\frac{\delta}{\delta\Gamma}\left\{\iint_{R_\mathrm{I}} f(\mathbf{x})d\mathbf{x}\right\} = -\|\mathbf{c}_u\| f(\mathbf{c})\mathbf{n} \tag{14}$$

Integrals over $\mathcal{B}_\mathrm{O}$ are more conveniently differentiated when expressed with integrals over R_I and its dilated counterpart $R_{\mathrm{I}[B]}$. Since $\mathcal{B}_\mathrm{O} = R_{\mathrm{I}[B]}\backslash R_\mathrm{I}$, we have:

$$\frac{\delta}{\delta\Gamma}\left\{\iint_{\mathcal{B}_\mathrm{O}} f(\mathbf{x})d\mathbf{x}\right\} = \frac{\delta}{\delta\Gamma}\left\{\iint_{R_{\mathrm{I}[B]}} f(\mathbf{x})d\mathbf{x}\right\} - \frac{\delta}{\delta\Gamma}\left\{\iint_{R_\mathrm{I}} f(\mathbf{x})d\mathbf{x}\right\} \tag{15}$$

From eq. (14), we have:

$$\frac{\delta}{\delta\Gamma_{[B]}}\left\{\iint_{R_{\mathrm{I}[B]}} f(\mathbf{x})d\mathbf{x}\right\} = -\|\mathbf{c}_u\|(1+B\kappa)f(\mathbf{c}_{[B]})\mathbf{n}$$

which is intuitively obtained by substituting Γ with $\Gamma_{[B]}$. In eq. (13), we replace $\delta\mathcal{E}_{[B]}/\delta\Gamma_{[B]}$ with the previous result. Since $\langle \mathbf{n}, \mathbf{t} \rangle = 0$, the derivative eventually reduces to:

$$\frac{\delta}{\delta\Gamma}\left\{\iint_{R_{\mathrm{I}[B]}} f(\mathbf{x})d\mathbf{x}\right\} = -\|\mathbf{c}_u\|(1+B\kappa)f(\mathbf{c}_{[B]})\mathbf{n} \tag{16}$$

A.3   Derivative of LNBR Energy

To determine the derivative of E_{LNBR}, we consider eqs (14), (15) and (16) and instantiate f with $(I - \mu_{\text{I}})^2$ or $(I - \mu_{\text{LN}})^2$ where appropriate. We approximate the derivative of the outer term of the LNBR energy from the derivative of the general integral $J(f, \mathcal{B}_{\text{O}})$. Doing this, average intensities μ_{I} and μ_{LN} are assumed to be curve-independent. This is actually a shortcut since they do obviously depend on Γ (see eq. (6)). However, one may note that a similar derivation is made in the work by Chan-Vese [7], where average intensities μ_{I} and μ_{O} are initially formulated as variables and, by means of gradient descent, are actually assigned to average intensities. Finally, we have:

$$\frac{\delta E_{\text{LNBR}}}{\delta \Gamma} \approx \|\mathbf{c}_u\| \left[-(I(\mathbf{c}) - \mu_{\text{I}})^2 - (1 + B\kappa)(I(\mathbf{c}_{[B]}) - \mu_{\text{LN}})^2 + (I(\mathbf{c}) - \mu_{\text{LN}})^2 \right] \mathbf{n}$$

$$(17)$$

Hierarchical Pairwise Segmentation Using Dominant Sets and Anisotropic Diffusion Kernels

Andrea Torsello and Marcello Pelillo

Dipartimento di Informatica
Università Ca' Foscari di Venezia
{torsello,pelillo}@dsi.unive.it

Abstract. Pairwise data clustering techniques are gaining increasing popularity over traditional, feature-based central grouping techniques. These approaches have proved very powerful when applied to image-segmentation problems. However, they are mainly focused on extracting flat partitions of the data, thus missing out on the advantages of the inclusion constraints typical of hierarchical coarse-to-fine segmentations approaches very common when working directly on the image lattice. In this paper we present a pairwise hierarchical segmentation approach based on dominant sets [12] where an anisotropic diffusion kernel allows for a scale variation for the extraction of the segments, thus enforcing separations on strong boundaries at a high level of the hierarchy. Experimental results on the standard Berkeley database [9] show the effectiveness of the approach.

1   Introduction

Proximity-based, or pairwise, data clustering techniques are gaining increasing popularity over traditional central grouping techniques, which are centered around the notion of "feature" (see, e.g., [5,14,15,13]). In many application domains, in fact, the objects to be clustered are not naturally representable in terms of a vector of features. On the other hand, quite often it is possible to obtain a measure of the similarity/dissimilarity between objects. Although such a representation lacks geometric notions such as scatter and centroid, it is attractive as similarity information arising from sources of very different nature can be incorporated very easily, often not requiring more than adding together distances or multiplying similarities calculated from different sources. In contrast, integrating information of different nature within the central clustering framework requires an integrated feature model capable of simultaneously characterizing all information at the feature level.

These approaches have proven very powerful when applied to image segmentation problems [15,8,5,2]. Here, the possibility of easily integrating different

D. Cremers et al. (Eds.): EMMCVPR 2009, LNCS 5681, pp. 182–192, 2009.

sources of information has been used to incorporate color, texture, and proximity information between pair of pixels. Conversely, feature based segmentation algorithms must explicitly integrate all types of information into a single geometrical model which requires a stronger characterization of the geometry of the image.

Despite the promise of ease of integration of inhomogeneous information, most actual implementations of pairwise segmentation only integrate local appearance based information, with color- and texture-based similarities taking the lion's share over all pairwise measures found in the literature. With few exceptions, little work has been done to integrate locality and boundary information in a pairwise setting. Among these we note Malik and coworkers' proposal to incorporate boundary information in the normalized-cut framework by looking for an *intervening contour* [8]. However, their approach only looks for detected edges in the straight line joining two pixels; hence, it is strongly dependant on the quality of the edge extractor and tends to separate pixels belonging to a single region if this is not convex. The normalized cut framework is relatively forgiving about this problem, but it is particularly severe when using pairwise clustering algorithms that favor "compact" globular clusters such as the dominant sets framework [12]. Furthermore, intervening contour information alone is not able to separate regions with fuzzy or unclear boundaries such as regions delimited by relatively smooth gradients. An alternative is to use the minimal boundary separation along all possible paths [16], however the selection of the optimal path renders this approach not robust with respect to the misdetection of a single boundary point. A more robust path-based segmentation can be achieved using random walks on the image lattice. For example, Grady uses random walks to extract a semi-supervised segmentation [4] where a pixel belongs to the class of whose label is expected to find first on a random walk.

Further, pairwise segmentation algorithms are generally concerned with flat partitions, thus missing out on the advantages of the inclusion constraints typical of hierarchical coarse-to-fine segmentations approaches very common when working directly on the image lattice.

In this paper, we propose a coarse-to-fine segmentation algorithm based on a hierarchical variant of the dominant sets framework [10]. Here, however, the regularizer term is substituted with a heat diffusion kernel [7], which enforces locality and boundary separation based on a limited time random walk on the image lattice. At the beginning of the diffusion process the effects are local and hence the long-range similarity is dominated by the color and texture appearance, while as the time increases the range of the kernel expands thus enforcing a coarser segmentation. To this end we start by segmenting at a high time value, and then we iteratively reduce the time to obtain finer-grain separations. The anisotropic diffusion properties of the heat kernel have been used in the computer graphics and vision communities to perform controlled smoothing [6,1]. Here, however, we are using it to define an explicit scale space on which to base a recursive hierarchical partitioning scheme.

2 Hierarchical Dominant Sets

The dominant set framework [12] is a pairwise clustering approach based on a recursive characterization of the weight $w_S(i)$ of element i with respect to a set S of elements, and characterizes a group as a *dominant set*, i.e., a set that satisfies:

1. $w_S(i) > 0$, for all $i \in S$
2. $w_{S \cup \{i\}}(i) < 0$, for all $i \notin S$.

These conditions correspond to the two main properties of a cluster: the first regards internal homogeneity, whereas the second regards external inhomogeneity.

The main result presented in [12] provides a one-to-one relation between dominant sets and strict local maximizers of the following quadratic program

$$\begin{aligned} \text{maximize} \quad & \mathbf{x}'A\mathbf{x} \\ \text{subject to} \quad & \mathbf{x} \in \Delta \end{aligned} \tag{1}$$

where $A = (a_{ij})$ is the matrix of similarities of the n elements to be grouped,

$$\Delta = \{\mathbf{x} \in \mathbb{R}^n \ : \ x_i \geq 0 \text{ for all } i = 1, \ldots, n \text{ and } \mathbf{1}'\mathbf{x} = 1\}$$

is the standard simplex of $\mathbb{R}^n$, and $\mathbf{1}$ is a vector of appropriate length consisting of unit entries.

Specifically, in [12] it is proven that if $\mathbf{x}$ is a strict local solution of program (1) then its support $S = \sigma(\mathbf{x})$ is a dominant set. Here, the *support* of a vector $\mathbf{x} \in \Delta$ is the set of indices corresponding to its positive components. The local maxima of (1) is found using the replicator equations, a dynamical systems mutuated from game-theory. The approach has proven to be a very effective and robust pairwise clustering approach that has in its speed one of its major selling points.

In [10] a hierarchical approach was presented by taking into consideration the regularized quadratic program

$$\begin{aligned} \text{maximize} \quad & \mathbf{x}'(A - \alpha I)\mathbf{x} \\ \text{subject to} \quad & \mathbf{x} \in \Delta \end{aligned} \tag{2}$$

where α is a scale parameter that defines the hierarchy. In [10] was shown that for sufficiently large values of α all elements where clustered into a single group and a recursive divisive algorithm was applied to the data as α was reduced. However, no indication of how to select the relevant values of α was provided.

3 Anisotropic Diffusion Kernel

Let $\mathcal{M} = (V_M, E_M)$ be the regular mesh defined over the image by connecting each pixel to its 4-neighbors. Further, let $\gamma : V_M \times V_M \to \mathbb{R}_+$ be an edge weight function which reflects how similar two neighboring pixels are. In our boundary-based segmentation approach we set

$$\gamma(i,j) = \begin{cases} e^{-k\frac{\nabla I_i + \nabla I_j}{2}} & \text{if } (i,j) \in E_M \\ 0 & \text{otherwise,} \end{cases}$$

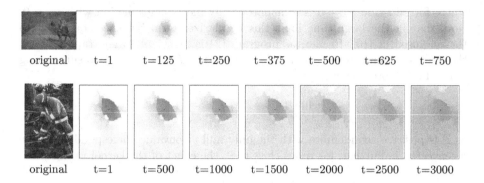

| original | t=1 | t=125 | t=250 | t=375 | t=500 | t=625 | t=750 |

| original | t=1 | t=500 | t=1000 | t=1500 | t=2000 | t=2500 | t=3000 |

Fig. 1. Location distribution of a walker at time t. the black dot marks the starting position.

where ∇I_i is the image gradient at pixel i. We define $M = (m_{ij})$ as the weighted adjacency matrix of $\mathcal{M}$, where we set $m_{ij} = \gamma(i, j)$.

An anisotropic diffusion on $\mathcal{M}$ is a lazy random walk from a pixel in the image lattice to other pixels along the mesh connections, where the transition probabilities are proportional to the value of the edge weight function γ. The walk is lazy as at each time step the walker has a non-null probability $1/Z_i$ of remaining at location i and a probability m_{ij}/Z_i of moving to location j, where $Z_i = 1 + \sum_j m_{ij}$ is a normalizing factor. The use of a lazy walk forces the walker to diffuse rapidly on flat image regions, when Z_i is high, while slowing down when there is a complex edge structure around i, i.e., when Z_i is close to 1.

The expected position at time t of a random walker starting from position i is governed by the heat diffusion kernel $e^{-\mathcal{L}t}$, where $\mathcal{L}$ is the Laplacian of $\mathcal{M}$, with $\mathcal{L} = D - M$ where D is the diagonal matrix with elements $d_{ii} = \sum_j m_{ij}$. At $t = 0$, the diffusion kernel is an identity matrix; as t increases the kernel assigns non-zero values to elements in the vicinity of position i spreading more rapidly along flat locations and stopping on boundaries, while as t becomes very large the support of the kernel is very diffused and far-reaching. Figure 1 shows two images and the location probability of a walker starting on the pixel marked with the black dot. As it can be seen, for small times the support is mostly restricted to a local segment, bound by strong boundaries in the image, while as time increases the support becomes more diffused.

4 Diffusion Regularizers

In the definition of our hierarchical coarse-to-fine segmentation approach we substitute the identity matrix I in the regularizer term of program (2) with the kernel $I - e^{-\mathcal{L}t}$, obtaining the following regularized quadratic program

$$\text{maximize} \quad f_t(\mathbf{x}) = \mathbf{x}' \left[A - \alpha_t \left(I - e^{-\mathcal{L}t} \right) \right] \mathbf{x}$$
$$\text{subject to} \quad \mathbf{x} \in \Delta \tag{3}$$

where for all $t > 0$ $\alpha_t \geq 0$ is a monotonically increasing function with $\alpha_0 = 0$. Note that in the large time limit this regularizer term becomes equivalent to the one used in [10]. In fact, as program (2) is invariant to constant shifts in the matrix A, subtracting the regularizing term αI is equivalent to subtracting $\alpha(I - \mathbf{1}\mathbf{1}')$. Moreover, we have

$$\lim_{t \to \infty} \left(I - e^{-\mathcal{L}t}\right) = I - \mathbf{1}\mathbf{1}'$$

as $\mathcal{L}$ is positive semidefinite with the only null eigenvalue corresponding to the eigenvector $\mathbf{1}$. Further, for times close to 0, the effect of the kernel vanishes as

$$\lim_{t \to \infty} \left(I - e^{-\mathcal{L}t}\right) = \mathbf{0}.$$

Intuitively we are substituting a regularizing term that increases the support equivalently to all elements, with one that increases the support to neighboring pixels first.

We can now prove that the time parameter indeed produces a scale-space as cluster hierarchy collapses to the full image for sufficiently large times.

Proposition 1. *Let* $\lambda_1(A), \lambda_2(A), \ldots, \lambda_n(A)$ *represent the largest, second largest,..., smallest eigenvalue of matrix* A. *If* $\alpha_t > \frac{\lambda_1(A)}{1-e^{-\lambda_{n-1}(\mathcal{L})t}}$, *then* f_t *is a strictly concave function in* Δ. *Further, if* $\alpha_t > \frac{n\lambda_1(A)}{1-e^{-\lambda_{n-1}(\mathcal{L})t}}$ *the only solution of (2) belongs to the interior of* Δ.

Proof. Note that $\mathcal{L}\mathbf{1} = 0$, which implies for all t $e^{-\mathcal{L}t}\mathbf{1} = \mathbf{1}$. Further, all other eigenvalues are in the open interval $(0, 1)$. The function f_t is strictly concave in Δ if for all $\mathbf{y} \in \mathbb{R}^n$ with $\mathbf{y}'\mathbf{1} = 0$, we have $\mathbf{y}' \left[A - \alpha_t \left(I - e^{-\mathcal{L}t}\right)\right] \mathbf{y} < 0$. However,

$$\mathbf{y}' \left(A - \alpha_t \left(I - e^{-\mathcal{L}t}\right)\right) \mathbf{y} \leq \lambda_1(A)\mathbf{y}'\mathbf{y} - \alpha_t \left(\mathbf{y}'\mathbf{y} - \mathbf{y}'e^{-\mathcal{L}t}\mathbf{y}\right) \leq$$
$$\lambda_1(A)\mathbf{y}'\mathbf{y} - \alpha_t \left(\mathbf{y}'\mathbf{y} - \lambda_2 \left(e^{-\mathcal{L}t}\right)\mathbf{y}'\mathbf{y}\right) = \mathbf{y}'\mathbf{y} \left(\lambda_1(A) - \alpha_t \left(1 - e^{-\lambda_{n-1}(\mathcal{L})t}\right)\right) < 0.$$
$$(4)$$

Since f_t is strictly concave in Δ, program (2), which is a concave with convex constraints, has a unique solution. To prove the second result, suppose by contraddiction that this solution $\mathbf{x}$ lies on the boundary of Δ, then we have $x_i = 0$ for some index i. There is a unique $\mathbf{y} \in \mathbb{R}^n$ with $\mathbf{y}'\mathbf{1} = 0$ such that $\mathbf{x} = \frac{1}{n}\mathbf{1} + \mathbf{y}$. Further, since by hypothesis we have $x_i = 0$, then $\mathbf{y}'\mathbf{y} \geq \frac{1}{n(n-1)}$. With this we have

$$\mathbf{x}'A\mathbf{x} - \alpha_t \left(\mathbf{x}'\mathbf{x} - \mathbf{x}'e^{-\mathcal{L}t}\mathbf{x}\right) \leq \lambda_1(A)(\mathbf{y}'\mathbf{y} + \frac{1}{n}) - \mathbf{y}'\mathbf{y}\alpha_t \left(1 - e^{-\lambda_{n-1}(\mathcal{L})t}\right) =$$
$$\mathbf{y}'\mathbf{y} \left(\lambda_1(A) \left(1 + \frac{1}{n\mathbf{y}'\mathbf{y}}\right) - \alpha_t \left(1 - e^{-\lambda_{n-1}(\mathcal{L})t}\right)\right) \leq$$
$$\mathbf{y}'\mathbf{y} \left(\lambda_1(A)n - \alpha_t \left(1 - e^{-\lambda_{n-1}(\mathcal{L})t}\right)\right) < 0.$$

Recall that a point $\mathbf{x} \in \Delta$ satisfies the Karush-Kuhn-Tucker (KKT) conditions for problem (2) if

$$
\begin{aligned}
(A\mathbf{x})_i - \alpha_t \left(x_i - e^{-\mathcal{L}t}\mathbf{x} \right)_i &= \mathbf{x}'A\mathbf{x} - \alpha_t \left(\mathbf{x}'\mathbf{x} - \mathbf{x}'e^{-\mathcal{L}t}\mathbf{x} \right) \\
&\quad \text{if } i \in \sigma(\mathbf{x}) \\
(A\mathbf{x})_i + \alpha_t \left(e^{-\mathcal{L}t}\mathbf{x} \right)_i &\le \mathbf{x}'A\mathbf{x} - \alpha_t \left(\mathbf{x}'\mathbf{x} - \mathbf{x}'e^{-\mathcal{L}t}\mathbf{x} \right) \\
&\quad \text{otherwise.}
\end{aligned}
\tag{5}
$$

Then, we have

$$
(A\mathbf{x})_i + \alpha_t \left(e^{-\mathcal{L}t}\mathbf{x} \right)_i \le \mathbf{x}'A\mathbf{x} - \alpha_t \left(\mathbf{x}'\mathbf{x} - \mathbf{x}'e^{-\mathcal{L}t}\mathbf{x} \right) .
$$

However, this is impossible since $(A\mathbf{x})_i + \alpha_t \left(e^{-\mathcal{L}t}\mathbf{x} \right)_i > 0$ and $x'A\mathbf{x} - \alpha_t \left(\mathbf{x}'\mathbf{x} - \mathbf{x}'e^{-\mathcal{L}t}\mathbf{x} \right) < 0$, thus proving the proposition.

4.1   Selecting Relevant Levels

One of the questions left open in [10] is how to select the values of the regularizer parameter that induce relevant partitions. Indeed, as the scale parameter α varies continuously from its maximum value down to 0, we expect the size of the extracted segments to vary almost as smoothly. Here we adopt an entropy shedding approach to the selection of the relevant levels of the hierarchy: the value of $\mathbf{x} \in \Delta$ that maximizes (2) can be considered as a probability distribution whose entropy is a measure of the size and cohesiveness of the cluster. As the parameter α is decreased we expect the first extracted cluster to steadily become less cohesive, eventually losing a few peripheral elements, until we reach a point where there is a substantial modification in the cluster structure as the current group gets split into multiple parts, thus producing a jump in the entropy value.

Figure 2 shows an example where a set of points generated from three bivariate Gaussian distribution are clustered using the original hierarchical formulation (2). The left image shows the point distribution, while the plot on the right

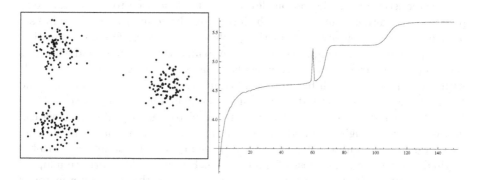

Fig. 2. Entropy value of the distribution associated with first cluster as a function of the regularizing parameter α

show the entropy for different values of α. there are three well defined plateaus corresponding to a single cluster encompassing all the data ($\alpha > 115$), a cluster without the points in the rightmost distribution which is furthest apart from the other ($75 < \alpha < 100$), and cluster encompassing only one of the three point sets ($20 < \alpha < 55$). after that even smaller subsets are extracted.

Accordingly, we start the clustering procedure at a sufficiently high time to obtain a single cluster and we reduce the time until we reach the next entropy plateau. In our implementation we use an exponential schedule for the reduction of the time parameter t, i.e., we multiply t by a constant factor $t_{\mathrm{mult}} = 0.8$. once a new plateau is reached, we partition the data and continue recursively on each cluster, until we reach the final partition at $t = 0$.

4.2   Subsampling

Despite their many advantages, pairwise clustering approaches are computationally very demanding due to their scaling behavior with the quantity of data. On a dataset containing N examples, the number of potential comparisons scales with $O(N^2)$, thereby rendering the approaches unfeasible for problems involving very large data sets as is the case of pixel based segmentation of even moderately large images. It is therefore of primary importance to develop strategies to reduce the number of comparisons required by subsampling the data and extending the grouping to out-of-sample points after the clustering process has taken place.

In [11] was proposed a subsampling approach for the dominant sets framework and a more efficient extension scheme was proposed in [17] in order to adapt the famework to spatio-temporal segmentation. The approach takes an element of a cluster S to act as a cluster centroid, namely it takes the element i which maximizes the weight $\mathrm{w}_S(i)$ with respect to the cluster S. This way, the similarity of an out-of-sample point j to a cluster S is simply the similarity to its centroid c_S. The first step of the out-of-sample segmentation is to extract the clusters from the sampled points. With the initial segmentation to hand, each pixel is then assigned to the closest cluster.

While the out of sampling approach certainly helps with the computation of the cluster structure at the various levels, it cannot be used to reduce the complexity of the kernel computation, which requires the computation of the full set of eigenvalues and eigenvectors of the Laplacian matrix $\mathcal{L}$. Further, note that subsampling techniques require a connection to all the other nodes and while this is not a problem for the matrix A which has connection at all ranges, is unusable with the Laplacian, which has only local connections and subsampling it would break the mesh connectivity structure and severely modify the eigenspaces. However, we can use the locality of the heat kernel to our advantage, as we can down-sample he original boundary information to obtain a smaller mesh from which we can compute the eigenvectors which can then just be up-scaled to the original size with minimal loss of information. Hence, in this work we adopted a mixed strategy for data reduction: we down-sampled the mesh by a factor of 8 in each direction and computed the 10 smallest eigenvectors of $\mathcal{L}$, which are then up-scaled to reconstruct a least-squares approximation of the heat kernel.

With the full up-scaled kernel to hand, we can use the subsampling procedure described in [17] both on the similarity matrix A and on the kernel $I - e^{-\mathcal{L}t}$.

5   Experimental Results

In order to assess the performance of the proposed kernel-based coarse-to-fine hierarchical segmentation approach, we tested it on the Berkeley database [9] using only color information for the similarity matrix A. Clearly, the final goal is to incorporate more descriptive pairwise similarities, but it would be hard to separate the effect of the diffusion kernel from that of other boundary-based information. In all the experiments the similarity between two pixels i and j is taken to be $a_{ij} = e^{-\frac{1}{2}d(i,j)^2/\sigma^2}$, where $d(i,j)$ is the perceptual distance between the colors of pixels i and j computed as the Euclidean distance on the CIE Luv color space.

Figures 3 and 4 show two images and the computed segment hierarchies. On the top left corner of each group we see the original images, while the other images display the segmentation hierarchy. For each segmented image we show the clusters extracted at the next level, where all pixels belonging to the same cluster are drawn using the average color of the cluster and pixels that have already being eliminated are drawn in black. the segmented image is then linked to the images showing the lower level segmentations. Note how the first separation in the image in Figure 3 is a major figure ground separation with all the details on the camel clustered together. It is only at lower levels of the hierarchy that the details form separate clusters. Further, note how in the image in Figure 4, the

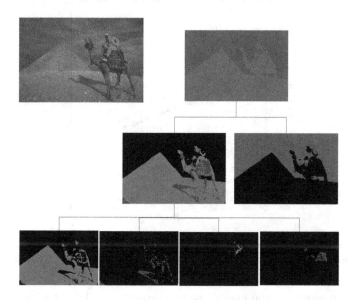

Fig. 3. Segmentation hierarchies extracted from the camel image

Fig. 4. Segmentation hierarchies extracted from the fireman image

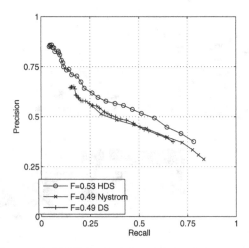

Fig. 5. Precision/Recall on the Berkeley database for our hierarchical method (HDS), the flat dominant set framework (DS) and the Nyström extension (Nystrom)

first partition separates the major image components, while at a lower scale the highlights get separated from the fireman's suit and helmet.

Next, for a more quantitative analysis, we computed the Precision and Recall for boundary detection on the full Berkeley database. The analysis was performed for our hierarchical approach, the original dominant sets framework with out-of-sample extension, and the Nyström extension [3], an out-of-sample generalization of normalized cut [15]. All pairwise segmentation approaches were based on the same color similarity matrix.

Figure 5 shows the resulting Precision/Recall curves. We can see that the original dominant sets and the Nyström extension performed almost identically, while our proposed approach shows a marginal but very clear advantage. It is worth noting that the standard precision recall curve does not evaluate the quality of the segment hierarchies, but is only concerned with a single flat segmentation. For this reason on our algorithm only the final partition, the one at $t = 0$, was used for the boundary extraction. Hence, the advantage over the standard dominant sets framework is due to the coarse-to-fine nature of the hierarchical divisive algorithm used, which extracts stronger and clearer boundaries first, thus reducing noise and detection errors around clear separations.

6   Conclusions

In this paper we have presented a coarse-to-fine hierarchical segmentation approach which uses an anisotropic diffusion kernel to generate the levels of the hierarchy. The approach is a generalization of the hierarchical dominant sets framework presented in [10] with the addition of a heat kernel-based regularizer term that enforces locality and boundary separation. We have proven that the term does indeed generate a scale-space and proposed an entropy measure to select the relevant scales. Further, we have proposed a mixed strategy for out-of-sample extension in the presence of the diffusion kernels. We compared the performance of the approach to the original flat dominant sets framework and to the Nyström extension applied to normalized cut, showing that our coarse-to-fine approach outperforms both on the standard Berkeley database.

Acknowledgment

We acknowledge the nancial support of the Future and Emerging Technology (FET) Programme within the Seventh Framework Programme for Research of the European Commission, under FET-Open project SIMBAD grant no. 213250.

References

1. Zhang, F., Hancock, E.R.: Graph Spectral Image Smoothing. In: Escolano, F., Vento, M. (eds.) GbRPR. LNCS, vol. 4538, pp. 191–203. Springer, Heidelberg (2007)
2. Fischer, B., Buhmann, J.M.: Path-based clustering for grouping of smooth curves and texture segmentation. IEEE Trans. Pattern Anal. Machine Intell. 25(4), 513–518 (2003)

3. Fowlkes, C., Belongie, S., Chun, F., Malik, J.: Spectral grouping using the Nyström method. IEEE Trans. Pattern Anal. Machine Intell. 26, 214–225 (2004)
4. Grady, L.: Random Walks for Image Segmentation. IEEE Trans. on Pattern Anal. and Machine Intell. 28(11), 1768–1783 (2006)
5. Hofmann, T., Buhmann, J.M.: Pairwise data clustering by deterministic annealing. IEEE Trans. Pattern Anal. Machine Intell. 19, 1–14 (1997)
6. Karni, Z., Gotsman, C.: Spectral compression of mesh geometry. In: SIGGRAPH 2000: Proceedings of the 27th annual conference on Computer graphics and interactive techniques, pp. 279–286. ACM Press, New York (2000)
7. Kondor, R., Lafferty, J.: Diffusion kernels on graphs and other discrete structures. In: Proceedings of the 19th Intl. Conf. on Machine Learning (ICML) (2002)
8. Malik, J., Belongie, S., Leung, T., Shi, J.: Contour and texture analysis for image segmentation. Int. J. of Computer Vision 43(1), 7–27 (2001)
9. Martin, D., Fowlkes, C., Tal, D., Malik, J.: A Database of Human Segmented Natural Images and its Application to Evaluating Segmentation Algorithms and Measuring Ecological Statistics. In: Proc. 8th Int'l Conf. Computer Vision, vol. 2, pp. 416–423 (2001)
10. Pavan, M., Pelillo, M.: Dominant sets and hierarchical clustering. In: 9th IEEE International Conference on Computer Vision – ICCV 2003, vol. I, pp. 362–369. IEEE Computer Society Press, Los Alamitos (2003)
11. Pavan, M., Pelillo, M.: Effcient out-of-sample extension of dominant-set clusters. In: Saul, L.K., Weiss, Y., Bottou, L. (eds.) Advances in Neural Information Processing Systems 17, pp. 1057–1064. MIT Press, Cambridge (2005)
12. Pavan, M., Pelillo, M.: Dominant sets and pairwise clustering. IEEE Trans. Pattern Anal. Machine Intell. 29(1), 167–172 (2007)
13. Roth, V., Laub, J., Kawanabe, M., Buhmann, J.M.: Optimal cluster preserving embedding of nonmetric proximity data. IEEE Trans. Pattern Anal. Machine Intell. 25, 1540–1551 (2003)
14. Sarkar, S., Boyer, K.: Quantitative measures of change based on feature organization: Eigenvalues and eigenvectors. Computer Vision and Image Understanding 71, 110–136 (1998)
15. Shi, J., Malik, J.: Normalized cuts and image segmentation. IEEE Trans. Pattern Anal. Machine Intell. 22, 888–905 (2000)
16. Torsello, A., Di Ges, M., Pelillo, M.: Integrating Boundary Information in Pairwise Segmentation. In: International Conference on Image Analysis and Processing - ICIAP 2007, pp. 23–28. IEEE Computer Society, Los Alamitos (2007)
17. Torsello, A., Pavan, M., Pelillo, M.: Spatio-temporal segmentation using dominant sets. In: Rangarajan, A., Vemuri, B.C., Yuille, A.L. (eds.) EMMCVPR 2005. LNCS, vol. 3757, pp. 301–315. Springer, Heidelberg (2005)

Tracking as Segmentation of Spatial-Temporal Volumes by Anisotropic Weighted TV

Markus Unger, Thomas Mauthner, Thomas Pock, and Horst Bischof

Graz University of Technology, Austria
{unger,mauthner,pock,bischof}@icg.tugraz.at
http://www.gpu4vision.org

Abstract. Tracking is usually interpreted as finding an object in single consecutive frames. Regularization is done by enforcing temporal smoothness of appearance, shape and motion. We propose a tracker, by interpreting the task of tracking as segmentation of a volume in 3D. Inherently temporal and spatial regularization is unified in a single regularization term. Segmentation is done by a variational approach using anisotropic weighted Total Variation (TV) regularization. The proposed convex energy is solved globally optimal by a fast primal-dual algorithm. Any image feature can be used in the segmentation cue of the proposed Mumford-Shah like data term. As a proof of concept we show experiments using a simple color-based appearance model. As demonstrated in the experiments, our tracking approach is able to handle large variations in shape and size, as well as partial and complete occlusions.

1 Introduction

Although frequently tackled over the last decades, robust visual object tracking is still a vital topic in computer vision. The need for handling variations of the objects appearance, changes in shape and occlusions makes it a challenging task. Additionally, robust tracking algorithms should be able to deal with cluttered and varying background and illumination variations. We formulate the tracking problem as globally optimal segmentation of an object in the spatial-temporal volume. Under the assumption that an object undergoes only small geometric and appearance changes between two consecutive frames, the object is represented as a connected volume containing similar content. Applying the segmentation on a volume instead of single frames, enhances robustness in the case of partial occlusions and similar background. Furthermore, no explicit shape model has to be learned in advance. Instead spatial and temporal consistency is enforced by a single regularization term.

1.1 Related Work

Numerous different approaches have been applied to the visual tracking problem. For a detailed review we refer to [1]. Superior results have been achieved by

D. Cremers et al. (Eds.): EMMCVPR 2009, LNCS 5681, pp. 193–206, 2009.

patch-based [2] or simple kernel-based methods such as [3]. Avidan [3] considered tracking as a binary classification problem on the pixel level. An ensemble of weak classifier is trained on-line to distinguish between object and current background, while a subsequent mean-shift procedure [4] obtains the exact object localization. Grabner et al. [2] proposed on-line AdaBoost for feature selection, where the object representation is trained on-line with respect to the current background. Although those methods have shown their robust tracking behavior in several applications, they lack an explicit representation of the objects shape, due to their representation by a simple rectangular or elliptical region. Under the assumption of affine object transformation interest point based trackers, like the work of Ozuysal et al. [5], perform excellent with fast runtimes. The drawback of such approaches is the enormous amount of needed pre-calculated training samples, and the limitation that no update is done during tracking. Shape-based [6] or contour based [7] tracking methods deliver additional information about the object state or enhance the tracking performance on cluttered background. While Donoser and Bischof [6] used MSER [8] segmentation results for tracking, Isard and Blake [7] applied the CONDENSATION algorithm on edge information. Therefore feature extraction or segmentation were independent from the tracking framework. In contrast, especially level-set methods support the unified approach of tracking and segmentation in one system [9], [10], [11], [12], [13]. [10] modeled object appearance using color and texture information while a shape prior is given by level sets. [11] incorporated Active Shape Model based on incremental PCA, which allowed the online adoption of the shape models. [9] extended the mean-shift procedure by [4], by applying fixed asymmetric kernels to estimate translation, scale and rotation. For a more detailed review on the use of level set segmentation we refer to [12]. Recently, Bibby et al. [13] proposed an approach, where they used pixel-wise posterior instead of likelihoods in a narrow band level set framework for robust visual tracking. The use of pixel-wise posterior led to sharper extrema of the cost function, while the GPU based narrow band level set implementation achieved real-time performance. All of the above approaches work on single frames. In [14], Mansouri et al. proposed a joint space-time segmentation algorithm based on level sets. The main idea of interpreting tracking as segmentation in a spatial-temporal volume is closely related to the approach presented in this paper. In contrary to our approach level set methods are used, that can easily get stuck in local minima.

A lot of work has been done on image segmentation. For contour-based image segmentation the Geodesic Active Contour (GAC) model [15] has received much attention. In the following we will shortly review some energy minimization based approaches. Graph cuts are currently widely used for computer vision applications. Boykov et al. [16], [17] used a minimum cut algorithm to solve a graph based segmentation energy. Other graph based segmentation approaches were proposed by Grady with the random walker algorithm [18], which was extended in [19]. In [20], a TV based energy was used for segmentation of moving objects. While graph cuts allow simple and fast implementations, it is well-known that the quality of the segmentation depends on the connectivity of the underlying graph, and can cause systematic metrication errors [21]. Furthermore memory consumption is usually very high. Continuous maximal flows were presented by Appleton et al. [22]. In [23], Zach et al. extended continuous maximal flows to the anisotropic setting.

Variational approaches try to obtain a segmentation based on a continuous energy formulation. Therefore the weighted Total Variation (TV) as used by Bresson et al. [24], [25], Leung and Osher [26] and Unger et al. [27], [28], has become quite popular. Continuous formulations do not suffer from metrication errors, and have become reasonable fast by implementing them on the GPU [28]. Another well known variational segmentation framework is the Mumford-Shah image segmentation model [29]. Bresson et. al. [30] showed how non-local image information can be incorporated into a variational segmentation framework. In [31], Werlberger et al. showed how shape prior information can be incorporated using a Mumford-Shah like data term.

2   Tracking as Segmentation in a Spatial-Temporal Volume

In the following we will provide some details on the concept of interpreting tracking as the segmentation of a 3D volume similar to [14]. A color image I is defined in the 2D image domain Ω as $I : \Omega \rightarrow \mathbb{R}^3$. The 2D frames of a video sequence can be viewed as a volume by interpreting the temporal domain T as the third dimension. Thus the volume is defined as $V : (\Omega \times T) \rightarrow \mathbb{R}^3$. This makes it possible to incorporate spatial and temporal regularization in an unified framework. If we assume a high enough sampling rate, adjoining frames will contain similar content. The 2D objects of a single frame I correspond to cuts of planes with the

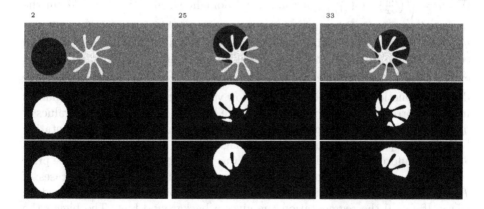

Fig. 1. Tracking of an artificial object. The first row depicts frames of the input video with frame numbers at the top. The second row shows the segmentation result using the volumetric approach. The third row shows the result of an MSER tracker implementation [6].

3D object defined in the volume V. Inherently this approach extends the forward propagation of information through time by additional backward propagation. Objects that are represented as disjoint regions in a single frame, correspond to a single volume, and are therefore tracked robustly. This concept is illustrated with an artificial example in Figure 1. Our tracking approach is compared to an MSER tracker [6], that cannot handle multiple disjoint regions. The volumetric approach does not suffer from such a shortcoming, as the regions are connected in the volume.

3   Algorithm

3.1   The Segmentation Model

We propose to use the following variational minimization problem for the task of image segmentation:

$$\min_u \left\{ E_p = \int_{\Omega \times T} \left(g_x |\nabla_x u| + g_t |\nabla_t u| \right) d\boldsymbol{x} dt + \lambda \int_{\Omega \times T} f u d\boldsymbol{x} dt \right\} . \tag{1}$$

The first term is a regularization term using anisotropic TV. The segmentation is represented by $u : (\Omega \times T) \to [0, 1]$. A binary labeling into foreground F ($u = 1$) and background B ($u = 0$) would force $u \in \{0, 1\}$. As this would make the energy non-convex, we can make use of convex relaxation [32]. For the g-weighted TV, Bresson already showed [33] that by letting u vary continuously, the regularization term becomes convex. To obtain a binary segmentation, any levelset of u can be selected using thresholding [23]. The segmentation cue $f :$ $(\Omega \times T) \to \mathbb{R}$ gives hints whether the pixel belongs to the foreground or the background. The gradient operators in the regularization term are defined as $|\nabla_x u| = \sqrt{\left(\frac{\partial u}{\partial x}\right)^2 + \left(\frac{\partial u}{\partial y}\right)^2}$ in the spatial domain Ω, and $|\nabla_t u| = \left|\frac{\partial u}{\partial z}\right|$ in the temporal domain T. Edge information is incorporated by $g_x : (\Omega \times T) \to \mathbb{R}$ and $g_t : (\Omega \times T) \to \mathbb{R}$ that subsequently represent edges in the current frame and edges from one to the next frame. The edge potential g_x is computed as $g_x = \exp\left(-a |\nabla_x V|^b\right)$. Likewise one can compute $g_t = \exp\left(-a |\nabla_t V|^b\right)$. The edge detection function maps strong edges to low values. Consequently discontinuities in u that correspond to the image region, are likely to be located at low values of g_x and g_t during the minimization process. This ensures that the segmentation boundary snaps to strong edges in the image.

The Mumford-Shah [29] like data term was already used in [31] for shape prior segmentation. For the segmentation cue f we distinguish the following cases: If $f = 0$ the data term is eliminated and segmentation is done solely based on edges. If $f > 0$ the segmentation cue gives a background hint. The bigger the value of f, the more likely it will be classified as background. In a similar manner $f < 0$ gives foreground hints. We use color features as described in Section 4.2 to compute f. Of course any other features or information can be incorporated through the segmentation cue.

Fig. 2. Comparison of anisotropic TV and standard TV regularization. The first row shows frames of the original video where the left player is tracked. In the second row the anisotropic regularization term shows a better segmentation of fast moving details than the standard weighted TV regularization in the third row.

The usage of an anisotropic weighted TV norm for regularization has the advantage that discontinuities in the spatial domain V and in the time domain T are separated. This allows a more accurate segmentation of small and fast moving objects. To illustrate this, Figure 2 shows a comparison of the regularization term as used in (1), and the standard weighted TV as used in [33] and [28]. Therefore we simply replaced the regularization term by $\int_{\Omega \times T} g|\nabla u| \, d\boldsymbol{x}dt$ with $g = \exp\left(-a\,|\nabla V|^b\right)$. It shows that the anisotropic regularization delivers finer details during fast moving parts of the video.

3.2   Solving the Minimization Problem

In the following we derive an adaption of the primal-dual algorithm of Zhu et al. [34]. To solve the energy defined in (1), we use duality by introducing the dual variable $\boldsymbol{p} : \Omega \times T \to \mathbb{R}^3$. The dual variable can be separated into a spatial and a temporal component $\boldsymbol{p} = (\boldsymbol{p_x}, p_t)^T$. Thus we get the following constrainted primal-dual formulation of the segmentation model:

$$\min_{u} \left\{ \sup_{\boldsymbol{p}} \left\{ E_{pd} = - \int_{\Omega \times T} u\nabla \cdot \boldsymbol{p}d\boldsymbol{x}dt + \lambda \int_{\Omega \times T} fu \, d\boldsymbol{x}dt \right\} \right\} \qquad (2)$$

$$s.t. \ |\boldsymbol{p_x}(\boldsymbol{x},t)| \le g_x(\boldsymbol{x},t), \ \ |p_t(\boldsymbol{x},t)| \le g_t(\boldsymbol{x},t) \ . \qquad (3)$$

The dependence on ∇u in the primal energy $E_p\left(\boldsymbol{x},t,u,\nabla u\right)$ is removed in the primal dual energy $E_{pd}\left(\boldsymbol{x},t,u,\boldsymbol{p}\right)$, but the problem is now an optimization

problem in two variables. This energy can be solved using alternating minimization with respect to u and maximization with respect to p.

When updating the primal variable u (primal update) we derive (2) according
to u and arrive at the following Euler-Lagrange equation:

$$-\nabla \cdot p + \lambda f = 0 . \tag{4}$$

Performing a gradient descent update scheme this leads to

$$u^{n+1} = \Pi_{[0,1]}\left(u^n - \tau_p\left(-\nabla \cdot p + \lambda f\right)\right) , \tag{5}$$

with τ_p denoting the timestep. The projection Π towards the binary set $[0,1]$
can be done with a simple thresholding step.

In a second step we have to update the dual variable p (dual update). Deriving
(2) according to p one gets the following Euler-Lagrange equation:

$$\nabla u = 0 \tag{6}$$

with the additional constraints on p_x and p_t as defined in (3). This results into
a gradient ascent method with a trailed re-projection to restrict the length of p:

$$p^{n+1} = \Pi_C\left(p^n + \tau_d \nabla u\right) \tag{7}$$

Here the convex set $C = \left\{q = (q_x, q_t)^T : |q_x| \leq g_x, |q_t| \leq g_t\right\}$ denotes a cylinder
centered at the origin with the radius g_x and height g_t. The re-projection onto
C can be formulated as

$$\Pi_C(q) = \left(\frac{q_x}{\max\{1, \frac{|q_x|}{g_x}\}}, \max\left\{-g_t, \min\left\{q_t, g_t\right\}\right\}\right)^T \tag{8}$$

Primal (5) and dual (7) updates are iterated until convergence. As u is a continuous variable, and the energy in (1) is not strictly convex, u may not be a
binary image. Any level set of u can be selected as a binary segmentation by
applying a threshold $\theta \in [0,1]$. We left $\theta = 0.5$ throughout this paper. An upper
boundary for the timesteps can be stated as $\tau_d \tau_p \leq \frac{1}{6}$. In conjunction with [34],
an iterative timesteps schema was chosen as:

$$\tau_d(n) = 0.3 + 0.02n , \tag{9}$$

$$\tau_p(n) = \frac{1}{\tau_d(n)}\left(\frac{1}{6} - \frac{5}{15+n}\right) , \tag{10}$$

where n is the current iteration.

As a convergence criterion the primal-dual gap is taken into account [34]. The
primal energy E_p was already defined in (1). For the dual energy E_d we have to
reformulate the primal-dual energy (2). For a fixed p, the minimization problem
of u can be determined as:

$$u(x,t) = \begin{cases} 1 \text{ for } -\nabla \cdot p(x,t) + \lambda f(x,t) < 0 \\ 0 \text{ else} \end{cases} \tag{11}$$

Thus the dual energy can be written as

$$E_d = \int_{\Omega \times T} \min\left\{-\nabla \cdot \boldsymbol{p} + \lambda f, 0\right\} d\boldsymbol{x} dt \ . \tag{12}$$

As the optimization scheme consists of a minimization and a maximization problem, E_p presents an upper boundary of the true minimizer of the energy, and E_d presents a lower boundary. The primal-dual gap is defined as

$$G\left(u, \boldsymbol{p}\right) = E_p\left(u\right) - E_d\left(\boldsymbol{p}\right) \ . \tag{13}$$

An automatic convergence criterion can be defined based on the normalized primal-dual gap, as

$$\lambda \left| \frac{G\left(u, \boldsymbol{p}\right)}{E_p\left(u\right)} \right| < \zeta \ , \tag{14}$$

with ζ the convergence threshold. It showed throughout the experiments, that $\zeta = 0.06$ is a good choice for the convergence threshold.

4 Implementation

4.1 The Segmentation Framework

Due to limitations in computer hardware such as memory, the size of volumes that can be computed at once is limited. Although modern computing hardware can handle volumes with several thousand frames, the necessity of working on the complete sequence at once restricts tracking to offline data. Multiple similar objects, or disjoint regions belonging to the same object (e.g. by occlusions) make additional information necessary. When attempting a general framework with objects of arbitrary size and shape, this becomes a difficult task.

To tackle these problems, we propose to use an incremental approach. Only n frames are segmented at once. The algorithm is initialized on the first n frames, e.g. by drawing a rectangle around the desired object. See Section 4.2 for details of the feature based segmentation approach. If multiple objects are segmented, the user can select the desired object manually. After convergence of the segmentation algorithm (Section 3.2) foreground and background models are updated. Next, the oldest $m < n$ frames are discarded, and m new frames are added to the volume V. To speed up the tracking process we compute the segmentation only on small areas around the current object. To prevent the algorithm from segmenting similar nearby objects, only regions that overlap with the segmentation mask of the last step are selected. In case of occlusions the volumetric representation of an object might be separated into several disjoint regions. Our overlap constraint causes the tracker to discard the new region. To handle occlusions in general we therefore use the following strategy: We keep track of the average region size. If the segmentation gets smaller than a certain percentage of this average region size, the object is assumed to be occluded. In case of an occlusion the region we are working on starts to grow slowly, and no updates of the foreground and background model are done. If a region is segmented that is big enough to be considered as the object, tracking is continued on this region. Any slice $k \in [1, n]$ of the volume V can be selected as the tracking result. The number of frames the tracker looks into the future is defined by $n - k$. Thus the smaller k and the bigger n, the more robust disjoint regions are tracked.

Implementation of the tracker was done mainly on the GPU using the CUDA framework [35]. The volume depth was fixed for all experiments to $n = 8$, while slice $k = 4$ was used for the segmentation result.

4.2   Color Tracking

Object appearance is represented in RGB color space using a foreground histogram $H_F : \mathbb{R}^3 \rightarrow [0, 1]$, and a background histogram $H_B : \mathbb{R}^3 \rightarrow [0, 1]$. Following the ideas presented in [13], we are using the pixel-wise posterior instead of modeling the color appearance using the likelihood like e.g. [36]. We define $M = M_F, M_B$ as the model parameter that is either foreground F or background B. From the initialization, we obtain the foreground and background likelihoods $P(H_F|M_F)$ and $P(H_B|M_B)$. Applying Bayesian rule we can estimate the posterior $P(M_F|H_F)$ of a pixel being foreground in the context of the actual background given by $P(H_B|M_B)$ and a region-prior $P(M_j)$ with $j \in F, B$ by:

$$P(M_F|H_F) = \frac{P(H_F|M_F)P(M_F)}{\sum_{j=F,B} P(H_j|M_j)P(M_j)} \tag{15}$$

We keep track of foreground and background models by updating them online using an adaption rate α with likelihoods estimated from the current frame $P_{new}(H_j|M_j)$ as:

$$P(H_j|M_j) = (1 - \alpha)P_{old}(H_j|M_j) + \alpha P_{new}(H_j|M_j) \quad with \quad j \in F, B \tag{16}$$

In contrast to [13] we do not apply marginalization. Instead we simply set the segmentation cue $f(\boldsymbol{x}, t) = 0.5 - P(M_F|H_F(V((\boldsymbol{x}, t))))$.

5   Experimental Results

The videos presented in this Section and the software binaries are available online at http://www.gpu4vision.org.

In Figure 3, a white cat is successfully tracked and segmented. The first row shows the input video with different overlays. The rectangle is indicating the current working region. The blue color of the rectangle indicates that the tracker is working normally. If the object is believed to be occluded in some slice, the rectangle becomes orange. The current object is indicated by an orange overlay. If parts of the image get segmented, but do not belong to the object, these areas are indicated in red. Note that some regions are segmented that do not

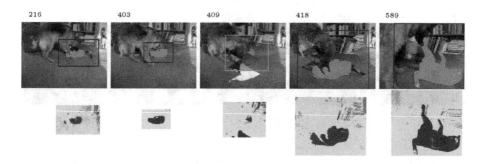

Fig. 3. Tracking example of a cat. The first row depicts the tracked object with the current segmentation and the working region as overlays to the original input image. In the second row the segmentation cue f is depicted in the range $[-0.5,\ 0.5]$.

belong to the object, but most of the incorrect regions are removed. The second row shows the segmentation cue f where the value -0.5 is mapped to black and indicates foreground, the value 0.5 is mapped to white and indicates background. Frame 409 shows a segment where a cross-fade occurs. The tracker detects the loss of the object, starts growing the search region and begins to search for the object. Frame 418 shows that the object was found correctly. Also note that the algorithm always correctly tracks the object despite large scale changes, as our tracking approach makes no restrictions on the region size.

The second example presented here shows the tracking of a fish in an aquarium. In the top row of Figure 4, the input video is shown, while the bottom row shows the extracted fish. Note that although several partially and complete occlusions occur, the tracker does not loose the object throughout the video. In case of partial occlusions the fish is still correctly segmented, as can be seen in frames 438 and 487. Also note that large shape changes do cause tracking failures, as we make no assumption on shape. In Figure 5, the video is displayed as a volume. The region corresponding to the fish is rendered using iso-surface rendering based on the segmentation mask as obtained by the tracker.

Naturally a color based tracker without any restrictions on shape and scale has its limitations. In Figure 6 a player in a volleyball game is tracked. In the beginning the tracker starts very promising by separating skin tones from the very similar sand. Around frame 337 the skin tones of other players appear in the working region, and are learned as background. As one can see in frame 377 the tracker looses the legs and arms, but still tracks the very characteristic green shirt. In frame 551 the player gets occluded by his team member, with a very similar appearance. As no additional high level information is available, both players are tracked.

In Figure 7, another video sequence is shown where the tracker fails. We tried to track the skin of the person. Due to the many occlusions the volume corresponding to skin is separated into several disjoint regions, causing problems for

Fig. 4. Tracking sequence of a fish in an aquarium, showing the ability of the tracker to handle large changes in shape, and various kinds of occlusions. The first row depicts the input video and the second row the extracted object.

the tracker. Though the tracker can recover several times, the object is permanently lost in frame 276. Other reasons for the failure in this video is the bad discrimination of foreground and background by using solely color.

Experimental results showed that a simple color tracker benefits form interpreting tracking as segmentation in 3D. The tracker successfully handled large variations in scale and shape. The examples show, that the tracker can deal with partial occlusions. Due to the incremental approach also long complete occlusions do not oppose any problem to the tracker. Figure 6 shows an improtant characteristic of the tracker to adapt foreground and background models to the most characteristic color values. This has the advantage of making the tracking of the object more robust, but also decreases segmentation performance. It also showed that multiple objects with similar appearance cannot be kept apart if occlusions occur. Here clearly high level information could help, e.g. in the volleyball example restrictions on the region size could be made, and shape information would definitely improve results.

Fig. 5. A schematic 3D rendering of the fish tracking sequence from Figure 4. The tracking result is rendered in yellow.

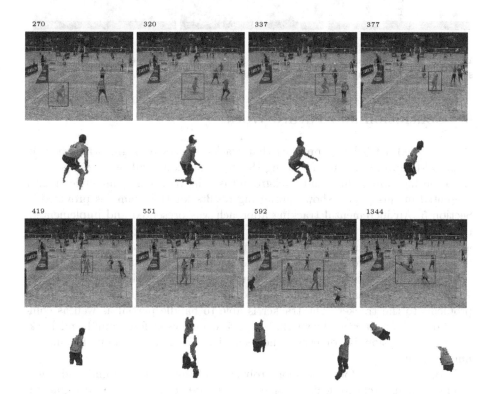

Fig. 6. Tracking of volleyball sequence, where tracker fails due to highly similar object and colors in the background. The first row shows the input video and the bottom row shows the extracted player.

Fig. 7. Video example where tracking and segmentation fail, due to too many occlusions, and bad discrimination of the color histograms

6   Conclusion and Future Work

We presented a tracking approach that tracks objects by segmenting them in a spatial-temporal volume. By using the segmentation result a pixel wise classification into foreground and background is achieved. The volumetric tracker presented in this paper, shows promising results for the examples provided in Section 5. An incremental tracking approach was presented and implemented, that works only on a small volume at a time, eliminating memory problems and allowing tracking of videos of arbitrary length. Due to the segmentation in a 3D volume, information is also propagated back through time if the regions are connected in 3D, showing improvements for tracking disjoint regions. As we make no assumptions on shape or scale even large variations cause no problems to the tracker. The tracker is able to handle partial as well as complete occlusions. It was shown that a pure color based foreground and background description is sometimes not sufficient, and leaves room for further improvement.

Future work will focus on more robust modeling of foreground and background regions. Texture features or patches would certainly improve segmentation and tracking results. Furthermore more complex appearance models with spatial modeling could improve the tracker significantly. Moreover we will focus on a more efficient implementation to achieve near realtime performance.

References

1. Yilmaz, A., Javed, O., Shah, M.: Object tracking: A survey. ACM Comput. Surv. 38(4), 13 (2006)
2. Grabner, H., Grabner, M., Bischof, H.: Real-time tracking via on-line boosting. In: British Machine Vision Conference, pp. 47–56 (2006)
3. Avidan, S.: Ensemble tracking. In: Proc. IEEE Conference on Computer Vision and Pattern Recognition, pp. 494–501 (2005)
4. Comaniciu, D., Ramesh, V., Meer, P.: Real-time tracking of non-rigid objects using mean shift. In: Proc. IEEE Conference on Computer Vision and Pattern Recognition, pp. 142–149 (2000)
5. Ozuysal, M., Fua, P., Lepetit, V.: Fast keypoint recognition in ten lines of code. In: Proc. IEEE Conference on Computer Vision and Pattern Recognition, June 2007, pp. 1–8 (2007)
6. Donoser, M., Bischof, H.: Efficient maximally stable extremal region (MSER) tracking. In: Proc. IEEE Conference on Computer Vision and Pattern Recognition, pp. 553–560 (2006)
7. Isard, M., Blake, A.: Contour tracking by stochastic propagation of conditional density. In: Proc. European Conference on Computer Vision, pp. 343–356 (1996)
8. Matas, J., Chum, O., Urban, M., Pajdla, T.: Robust wide baseline stereo from maximally stable extremal regions. In: British Machine Vision Conference, pp. 384–393 (2002)
9. Yilmaz, A.: Object tracking by asymmetric kernel mean shift with automatic scale and orientation selection. In: Proc. IEEE Conference on Computer Vision and Pattern Recognition, pp. 1–6 (2007)
10. Yilmaz, A., Li, X., Shah, M.: Contour based object tracking with occlusion handling in video acquired using mobile cameras. IEEE Trans. on Pattern Analysis and Machine Intelligence 26, 1531–1536 (2004)
11. Fussenegger, M., Roth, P., Bischof, H., Deriche, R., Pinz, A.: A level set framework using a new incremental, robust active shape model for object segmentation and tracking. Image and Vision Computing (in press)
12. Cremers, D., Rousson, M., Deriche, R.: A review of statistical approaches to level set segmentation: Integrating color, texture, motion and shape. International Journal of Computer Vision 72, 195–215 (2007)
13. Bibby, C., Reid, I.: Robust real-time visual tracking using pixel-wise posteriors. In: Proc. European Conference on Computer Vision, vol. 2, pp. 831–844 (2008)
14. Mansouri, A.R., Mitiche, A., Aron, M.: PDE-based region tracking without motion computation by joint space-time segmentation. In: Proc. International Conference on Image Processing, September 2003, vol. 2, pp. III–113–16 (2003)
15. Caselles, V., Kimmel, R., Sapiro, G.: Geodesic active contours. Intl. J. of Computer Vision 22(1), 61–79 (1997)
16. Boykov, Y., Jolly, M.P.: Interactive organ segmentation using graph cuts. In: Delp, S.L., DiGoia, A.M., Jaramaz, B. (eds.) MICCAI 2000. LNCS, vol. 1935, pp. 276–286. Springer, Heidelberg (2000)
17. Boykov, Y., Kolmogorov, V.: Computing geodesics and minimal surfaces via graph cuts. In: Proc. IEEE Conference on Computer Vision and Pattern Recognition, pp. 26–33 (2003)
18. Grady, L.: Random walks for image segmentation. IEEE Trans. on Pattern Analysis and Machine Intelligence 28(11), 1768–1783 (2006)

19. Sinop, A.K., Grady, L.: A seeded image segmentation framework unifying graph cuts and random walker which yields a new algorithm. In: Proc. International Conference on Computer Vision (October 2007)
20. Ranchin, F., Chambolle, A., Dibos, F.: Total variation minimization and graph cuts for moving objects segmentation. In: Scale Space and Variational Methods in Computer Vision, pp. 743–753 (2008)
21. Klodt, M., Schoenemann, T., Kolev, K., Schikora, M., Cremers, D.: An experimental comparison of discrete and continuous shape optimization methods. In: Forsyth, D., Torr, P., Zisserman, A. (eds.) ECCV 2008, Part I. LNCS, vol. 5302, pp. 332–345. Springer, Heidelberg (2008)
22. Appleton, B., Talbot, H.: Globally minimal surfaces by continuous maximal flows. IEEE Trans. Pattern Analysis and Machine Intelligence 28(1), 106–118 (2006)
23. Zach, C., Niethammer, M., Frahm, J.M.: Continuous maximal flows and Wulff shapes: Application to MRFs. In: Proc. IEEE Conference on Computer Vision and Pattern Recognition (2009)
24. Bresson, X., Esedoglu, S., Vandergheynst, P., Thiran, J., Osher, S.: Global minimizers of the active contour/snake model. In: Free Boundary Problems (FBP): Theory and Applications (2005)
25. Bresson, X., Esedoglu, S., Vandergheynst, P., Thiran, J., Osher, S.: Fast global minimization of the active contour/snake model. J. Math. Imaging and Vision 28(2), 151–167 (2007)
26. Leung, S., Osher, S.: Fast global minimization of the active contour model with TV-inpainting and two-phase denoising. In: 3rd IEEE Workshop on Variational, Geometric and Level Set Methods in Computer Vision, pp. 149–160 (2005)
27. Unger, M., Pock, T., Bischof, H.: Continuous globally optimal image segmentation with local constraints. In: Computer Vision Winter Workshop 2008, Moravske Toplice, Slovenija (February 2008)
28. Unger, M., Pock, T., Trobin, W., Cremers, D., Bischof, H.: TVSeg - Interactive Total Variation based image Segmentation. In: British Machine Vision Conference 2008, Leeds, UK (September 2008)
29. Mumford, D., Shah, J.: Optimal approximations by piecewise smooth functions and variational problems. Comm. on Pure and Applied Math. XLII(5), 577–685 (1988)
30. Bresson, X., Chan, T.F.: Non-local unsupervised variational image segmentation models. UCLA CAM Report 08-67 (2008)
31. Werlberger, M., Pock, T., Unger, M., Bischof, H.: A variational model for interactive shape prior segmentation and real-time tracking. In: International Conference on Scale Space and Variational Methods in Computer Vision (June 2009)
32. Nikolova, M., Esedoglu, S., Chan, T.F.: Algorithms for finding global minimizers of image segmentation and denoising models. SIAM J. on App. Math. 66 (2006)
33. Bresson, X., Esedoglu, S., Vandergheynst, P., Thiran, J.P., Osher, S.: Fast global minimization of the active contour/snake model. Journal of Mathematical Imaging and Vision 28, 151–167 (2007)
34. Zhu, M., Wright, S.J., Chan, T.F.: Duality-based algorithms for total variation image restoration. UCLA CAM Report 08-33 (2008)
35. Lindholm, E., Nickolls, J., Oberman, S., Montrym, J.: NVIDIA Tesla: A unified graphics and computing architecture. IEEE Micro 28(2), 39–55 (2008)
36. Cremers, D.: Dynamical statistical shape priors for level set-based tracking. IEEE Trans. on Pattern Analysis and Machine Intelligence 28(8), 1262–1273 (2006)

Complementary Optic Flow

Henning Zimmer[1,4], Andrés Bruhn[1], Joachim Weickert[1], Levi Valgaerts[1],
Agustín Salgado[2], Bodo Rosenhahn[3], and Hans-Peter Seidel[4]

[1] Mathematical Image Analysis Group
Faculty of Mathematics and Computer Science
Saarland University, Saarbrücken, Germany
{zimmer,bruhn,weickert,valgaerts}@mia.uni-saarland.de
[2] Departamento de Informática y Sistemas
Universidad de Las Palmas de Gran Canaria, Las Palmas de Gran Canaria, Spain
asalgado@dis.ulpgc.es
[3] Institut für Informationsverarbeitung, University of Hannover
Hannover, Germany
rosenhahn@tnt.uni-hannover.de
[4] Max-Planck Institute for Informatics, Saarbrücken, Germany
hpseidel@mpi-sb.mpg.de

Abstract. We introduce the concept of complementarity between data
and smoothness term in modern variational optic flow methods. First
we design a sophisticated data term that incorporates HSV colour representation with higher order constancy assumptions, completely separate robust penalisation, and constraint normalisation. Our anisotropic
smoothness term reduces smoothing in the data constraint direction instead of the image edge direction, while enforcing a strong filling-in effect orthogonal to it. This allows optimal complementarity between both
terms and avoids undesirable interference. The high quality of our complementary optic flow (COF) approach is demonstrated by the current
top ranking result at the Middlebury benchmark.

1 Introduction

In spite of the fact that variational optic flow methods are around for almost
three decades and that they mark the state-of-the-art in terms of accuracy, there
has been remarkably little reseach on the compatibility of their two ingredients:
the data term and the smoothness term. The data term models constancy assumptions on certain image properties, e.g. grey value constancy in the seminal Horn and Schunck model [1]. The smoothness term penalises fluctuations
in the flow field. However, these terms may contradict each other: While the
brightness constancy assumption constrains the flow only along the image gradient but not across it (aperture problem), most smoothness terms enforce their
constraints also along the image gradient. One notable exception is the Nagel-
Enkelmann model [2] where the homogeneous Horn and Schunck smoothness
term is replaced by an anisotropic one. For large image gradients the latter one
works solely orthogonal to the image gradient. Thus, both terms complement

D. Cremers et al. (Eds.): EMMCVPR 2009, LNCS 5681, pp. 207–220, 2009.

each other without undesirable interference. The fact that the Nagel-Enkelmann model outperforms the Horn and Schunck approach demonstrates the high potential of such a complementarity.

Unfortunately, this paradigm of complementary behaviour has not been explored further after 1986. Instead of this, research has focussed on improving the data or smoothness constraints independently. The goal of our paper is to propose a synergistic model for variational optic flow computation that integrates state-of-the-art data and smoothness assumptions in such a way that both terms work complementary. We will see that this can still lead to a very substantial gain in accuracy. This is demonstrated by the fact that our so-called *complementary optic flow (COF)* method ranks number one in the widely-used Middlebury benchmark.

Our paper is organised as follows: In Sec. 2 we review variational optic flow. Our data term is derived in Sec. 3 and is then used to complement the smoothness term in Sec. 4. After discussing implementation issues in Sec. 5, we show experiments proving the favourable performance of our method in Sec. 6. We then conclude with a summary and an outlook on future work in Sec. 7.

Related Work. Our model naturally incorporates many concepts that have demonstrated their usefulness over the years. Therefore let us briefly sketch the advances in data and smoothness terms that have been most influential for us.

For the data term, Black and Anandan [3] replaced the quadratic penalisation from [1] by a robust one which helps to cope with outliers caused by noise or occlusions. More recently, in order to make the data term robust under additive illumination changes, Brox et al. [4] successfully combined the classical *brightness constancy assumption (BCA)* with the *gradient constancy assumption (GCA)* [4,5,6]. Bruhn and Weickert [7] later improved this idea by introducing a separate robust penalisation of the BCA and the GCA. This gives advantages in those situations where one of the two constraints produces an outlier. Moreover, in realistic scenarios, one also has to deal with multiplicative illumination changes [8]. If colour image sequences are available, one solution to this issue can be the use of alternative colour spaces with photometric invariances, see [9] and the references therein. Besides the discussed robustification efforts, successful modifications of the data term have been reported by normalising the data term [10,11]. It prevents an undesirable overweighting of the data term at large image gradient locations.

Regarding the smoothness term, first ideas go back to Horn and Schunck [1] who used a *homogeneous* regulariser that does not respect any flow discontinuities. However, since different image objects may move in different directions, it is desirable to also permit discontinuities. This can for example be achieved by using *image-driven* regularisers that take into account image discontinuities. Alvarez et al. [12] proposed an isotropic model with a scalar-valued weight function that reduces the regularisation at image edges. An anisotropic counterpart that also exploits the directional information of image discontinuities was introduced by Nagel and Enkelmann [2]. Their method regularises the flow field

along image edges but not across them. However, as not every image edge coincides with a flow edge, image-driven methods are prone to oversegmentation
artifacts in textured image regions. To avoid this, *flow-driven* regularisers have
been proposed that respect discontinuities of the evolving flow and are thus not
misled by image textures. In the isotropic setting this comes down to the use of
robust, nonquadratic penalisers, which for discrete energy functions have been
proposed by Black and Anandan [3]. In the context of rotationally invariant
variational methods they go back to Schnörr [5], and Weickert and Schnörr [13]
later presented an anisotropic extension. Nevertheless, the problem of flow-driven
regularisers lies in less sharp and badly localised flow edges compared to image-
driven approaches. The recent discrete method of Sun et al. [14] incorporates an
anisotropic regulariser based on a Steerable Random Field [15] that uses directional flow derivatives steered by image structures. It can thus be classified as
joint image- and flow-driven (JIF), allowing to obtain sharp flow edges without
oversegmentation problems.

2   Variational Optic Flow

Let $f(\mathbf{x})$ be an image sequence with $\mathbf{x} := (x, y, t)^\top$, where $(x, y)^\top \in \Omega$ denotes
the location within a rectangular image domain $\Omega \subset \mathbb{R}^2$ and $t \geq 0$ denotes time.
We further assume that f is presmoothed by a Gaussian convolution of standard
deviation σ. The optic flow field $\mathbf{w} := (u, v, 1)^\top$ describes the displacement vector
field between two frames at time t and $t + 1$. It is found by minimising a global
energy functional of the general form

$$E(u, v) = \int_\Omega [M(u, v) + \alpha\, V(\nabla_2 u, \nabla_2 v)] \; \mathrm{d}x\, \mathrm{d}y \;, \tag{1}$$

where $\nabla_2 := (\partial_x, \partial_y)^\top$ denotes the spatial gradient operator. The term $M(u, v)$
denotes the data term, $V(\nabla_2 u, \nabla_2 v)$ the smoothness term, and $\alpha > 0$ is a
smoothness weight. According to the calculus of variations, a minimiser (u, v) of
the energy (1) necessarily has to fulfil the associated Euler-Lagrange equations

$$\partial_u M - \alpha \left(\partial_x \left(\partial_{u_x} V \right) + \partial_y \left(\partial_{u_y} V \right) \right) = 0 \;, \tag{2}$$

$$\partial_v M - \alpha \left(\partial_x \left(\partial_{v_x} V \right) + \partial_y \left(\partial_{v_y} V \right) \right) = 0 \;, \tag{3}$$

with homogeneous Neumann boundary conditions.

3   Data Term

Let us now derive our data term in a systematic way. The classical starting
point is the *brightness constancy assumption (BCA)* used by Horn and
Schunck [1]. It states that image intensities remain constant under their

displacement, i.e., $f(\mathbf{x} + \mathbf{w}) = f(\mathbf{x})$. Assuming that the displacement is sufficiently small, we can perform a first-order Taylor linearisation that yields the *optic flow constraint (OFC)*

$$0 = f_x u + f_y v + f_t = \nabla_3 f^{\top} \mathbf{w} , \tag{4}$$

where the subscripts denote partial derivatives and $\nabla_3 := (\partial_x, \partial_y, \partial_t)^{\top}$ denotes the spatio-temporal gradient operator. For a quadratic penalisation the corresponding data term is given by

$$M_1(u, v) = \left(\nabla_3 f^{\top} \mathbf{w}\right)^2 = \mathbf{w}^{\top} J_0 \mathbf{w} , \tag{5}$$

with the tensor $J_0 := \nabla_3 f \nabla_3 f^{\top}$.

The OFC is not sufficient to compute a unique solution (aperture problem), but only allows to compute the flow component orthogonal to the image edges, the so-called *normal flow*. It is defined as

$$\mathbf{w}_n := (\mathbf{u}_n^{\top}, 1)^{\top} := \left(-\frac{f_t}{|\nabla_2 f|} \frac{\nabla_2 f^{\top}}{|\nabla_2 f|}, 1\right)^{\top} . \tag{6}$$

Normalisation. Our experiments will show that normalising the data term can be beneficial. Following [10,11] and using the abbreviation $\mathbf{u} := (u, v)^{\top}$, we rewrite the data term M_1 as

$$M_1(u, v) = \left(\nabla_2 f^{\top} \mathbf{u} + f_t\right)^2 = |\nabla_2 f|^2 \left(\left(\frac{\nabla_2 f}{|\nabla_2 f|}\right)^{\top} (\mathbf{u} - \mathbf{u}_n)\right)^2 =: |\nabla_2 f|^2 d^2 . \tag{7}$$

The term d constitutes a projection of the difference between the estimated flow $\mathbf{u}$ and the normal flow $\mathbf{u}_n$ in the direction of the image gradient $\nabla_2 f$. Hence, this rewriting allows a geometric interpretation of the data constraint in terms of the distance from $\mathbf{u}$ to the line described by the OFC (4). Ideally, we would like to penalise this distance d, but in the data term M_1 it is weighted by the squared spatial image gradient. This results in a stronger enforcement of the data constraint at high gradient locations. Such an overweighting may be inappropriate as large gradients can be caused by unreliable structures, such as noise or occlusions.

As a remedy, we normalise the data term M_1 by multiplying it with a factor [10,11]

$$\theta_0 := \frac{1}{|\nabla_2 f|^2 + \zeta^2} , \tag{8}$$

where the regularisation parameter $\zeta > 0$ avoids division by zero. The normalised version of M_1 can be written as

$$M_2(u, v) = \mathbf{w}^{\top} \overline{J}_0 \mathbf{w}, \text{ with } \overline{J}_0 := \theta_0 J_0 . \tag{9}$$

Gradient Constancy Assumption. To cope with the problem of additive illumination changes, the *gradient constancy assumption (GCA)* has been proposed [4,5,6]. It states that image gradients remain constant under their displacement, i.e., $\nabla_2 f(\mathbf{x}+\mathbf{w}) = \nabla_2 f(\mathbf{x})$. A data term that combines both BCA and GCA is

$$M_3(u,v) = \mathbf{w}^\top J \mathbf{w} \ , \tag{10}$$

where we use the *motion tensor notation* [16]

$$J := \nabla_3 f \, \nabla_3 f^\top + \gamma \left(\nabla_3 f_x \, \nabla_3 f_x^\top + \nabla_3 f_y \, \nabla_3 f_y^\top \right) \ . \tag{11}$$

Here, the parameter $\gamma > 0$ steers the contribution of the GCA.

To normalise M_3, we replace the motion tensor J by

$$\bar{J} := \theta_0 \, \nabla_3 f \, \nabla_3 f^\top + \gamma \left(\theta_x \, \nabla_3 f_x \, \nabla_3 f_x^\top + \theta_y \, \nabla_3 f_y \, \nabla_3 f_y^\top \right) \ , \tag{12}$$

and obtain the data term $M_4(u,v) = \mathbf{w}^\top \bar{J} \mathbf{w}$. The two additional normalisation factors are defined as

$$\theta_x := \frac{1}{|\nabla_2 f_x|^2 + \zeta^2} \ , \quad \text{and} \quad \theta_y := \frac{1}{|\nabla_2 f_y|^2 + \zeta^2} \ . \tag{13}$$

Postponing the Linearisation. Linearisation of the data term with respect to u and v is only valid for small displacements. In order to handle large displacements correctly, Brox et al. [4] postpone any linearisation to the numerical scheme. Applying this strategy within the data term M_4 yields

$$M_5(u,v) = \left| \sqrt{\theta_0} \left(f(\mathbf{x}+\mathbf{w}) - f(\mathbf{x}) \right) \right|^2 \tag{14}$$
$$+ \gamma \left(\left| \text{diag} \left(\sqrt{\theta_x}, \sqrt{\theta_y} \right) \left(\nabla_2 f(\mathbf{x}+\mathbf{w}) - \nabla_2 f(\mathbf{x}) \right) \right|^2 \right) \ ,$$

where $\text{diag}(a,b)$ denotes the 2×2 the diagonal matrix with the entries a and b.

We wish to emphasise that the numerical solution for large displacement optic flow proceeds by computing flow increments in a multiresolution framework. The linearisation of the data term M_5 w.r.t. these small increments will give rise to the motion tensor (12) on every image scale.

Colour Image Sequences. In a next step we extend the data term M_5 to multi-channel sequences by coupling three colour channels $\left(f^1(\mathbf{x}), f^2(\mathbf{x}), f^3(\mathbf{x}) \right)$. A natural formulation for this is

$$M_6(u,v) = \sum_{i=1}^{3} \left(\left| \sqrt{\theta_0^i} \left(f^i(\mathbf{x}+\mathbf{w}) - f^i(\mathbf{x}) \right) \right|^2 \right. \tag{15}$$
$$\left. + \gamma \left| \text{diag} \left(\sqrt{\theta_x^i}, \sqrt{\theta_y^i} \right) \left(\nabla_2 f^i(\mathbf{x}+\mathbf{w}) - \nabla_2 f^i(\mathbf{x}) \right) \right|^2 \right) \ .$$

Photometric Invariant Colour Spaces. Realistic illumination models encompass a multiplicative influence [8], which cannot be captured by the GCA. This problem can be tackled by replacing the RGB colour space by the *Hue Saturation Value (HSV)* colour space [17] instead. The hue channel is invariant under multiplicative illumination changes and in particular under shadow, shading, highlights and specularities. The saturation channel is only invariant w.r.t. shadow and shading and the value channel exhibits none of these invariances. In [9], only the hue channel was used for optic flow computation. We will additionally use the saturation and value channel, because they contain information that is not encoded in the hue channel.

Robust Penalisers. To provide robustness of the data term against outliers caused by noise and occlusions, Black and Anandan [3] proposed to refrain from a quadratic penalisation. Instead they use a non-quadratic penalisation function $\Psi_M(s^2)$, where s denotes the data constraint. Good results are reported in [4] for the regularised L_1-norm, $\Psi_M(s^2) := \sqrt{s^2 + \varepsilon^2}$, with a small regularisation parameter $\varepsilon > 0$. Bruhn et al. [7] use a *separate* L_1 penalisation of the BCA and the GCA, which is advantageous if one assumption produces an outlier. In our variational framework we will go further by proposing a separate robustification of each HSV channel. It can be justified by the distinct information content of each of the three channels that drives the optic flow estimation in different ways.

Final Data Term. Incorporating our separate robustification idea into M_6 brings us to our final data term

$$M(u, v) = \sum_{i=1}^{3} \Psi_M \left(\left| \sqrt{\theta_0^i} \left(f^i(\mathbf{x} + \mathbf{w}) - f^i(\mathbf{x}) \right) \right|^2 \right) \tag{16}$$

$$+ \gamma \left(\sum_{i=1}^{3} \Psi_M \left(\left| \text{diag} \left(\sqrt{\theta_x^i}, \sqrt{\theta_y^i} \right) \left(\nabla_2 f^i(\mathbf{x} + \mathbf{w}) - \nabla_2 f^i(\mathbf{x}) \right) \right|^2 \right) \right) .$$

This data term is (i) normalised, it (ii) combines the BCA and GCA, (iii) does not linearise the constancy assumptions and (iv) uses the HSV colour space with (v) a separate robustification of all colour channels.

To derive the contributions of the data term (16) to the Euler-Lagrange equations (2) and (3), we use the abbreviations from [4]:

$$f_{**} := \partial_{**} f(\mathbf{x} + \mathbf{w}), \quad f_z := f(\mathbf{x} + \mathbf{w}) - f(\mathbf{x}), \quad f_{*z} := \partial_* f(\mathbf{x} + \mathbf{w}) - \partial_* f(\mathbf{x}), \tag{17}$$

where $** \in \{x, y, xx, xy, yy\}$ and $* \in \{x, y\}$. Then we can write the contributions $\partial_u M$ and $\partial_v M$ as

$$\partial_u M = \sum_{i=1}^{3} \Psi_M' \left(\theta_0^i \left(f_z^i \right)^2 \right) \cdot \theta_0^i f_z^i f_x^i \tag{18}$$

$$+ \gamma \left(\sum_{i=1}^{3} \Psi_M' \left(\theta_x^i \left(f_{xz}^i \right)^2 + \theta_y^i \left(f_{yz}^i \right)^2 \right) \cdot \left(\theta_x^i f_{xz}^i f_{xx}^i + \theta_y^i f_{yz}^i f_{xy}^i \right) \right) ,$$

$$\partial_v M = \sum_{i=1}^{3} \Psi_M' \left(\theta_0^i \left(f_z^i \right)^2 \right) \cdot \theta_0^i f_z^i f_y^i \tag{19}$$

$$+ \gamma \left(\sum_{i=1}^{3} \Psi_M' \left(\theta_x^i \left(f_{xz}^i \right)^2 + \theta_y^i \left(f_{yz}^i \right)^2 \right) \cdot \left(\theta_x^i f_{xz}^i f_{xy}^i + \theta_y^i f_{yz}^i f_{yy}^i \right) \right) ,$$

where $\Psi_M'(s^2)$ denotes the derivative of $\Psi_M(s^2)$ w.r.t. its argument. Here we see that the separate robustification of the HSV channels makes sense: If a specific channel violates the imposed constancy assumption at a certain location, the corresponding argument of the decreasing function Ψ_M' will be large, yielding a downweighting of this channel. The other channels that satisfy the constancy assumption will then have a dominating influence on the solution.

4   Smoothness Term

4.1   Previous Smoothness Terms

For overcoming the aperture problem and for regularising the estimated flow field, energy-based methods include a smoothness term (regulariser). It models the assumption of a smooth flow field. A quadratic smoothness term as proposed by Horn and Schunck [1] penalises the squared magnitude of the flow gradients:

$$V_1(\nabla_2 u, \nabla_2 v) = |\nabla_2 u|^2 + |\nabla_2 v|^2 \ . \tag{20}$$

In the corresponding Euler-Lagrange equations, this leads to a homogeneous diffusion term that tends to blur important flow edges. Since it is desirable that regularisers permit flow discontinuities, numerous discontinuity perserving smoothness terms have been proposed. They are classified as either *image-* or *flow-driven*, depending on whether the smoothing process is adapted to image edges or evolving flow edges. In addition, one can distinguish *isotropic* and *anisotropic* strategies. Whereas the first type makes use of a scalar valued diffusivity to reduce the smoothing at edges, the latter also takes into account the directional information by means of a diffusion tensor. For an extensive and in-depth survey on classical discontinuity preserving regularisers, we refer to [13].

Joint Image- and Flow-driven Regularisation (JIF). Recently, Sun et al. [14] presented an anisotropic smoothness term in a discrete setting. It is modelled by a Steerable Random Field [15] that uses directional flow derivatives steered by image structures. It thus combines the advantages of image- and flow-driven regularisers, a strategy that we will name *joint image- and flow-driven (JIF)* regularisation. To obtain directional information of image structures, the authors analyse the eigenvectors of the structure tensor [18] $S_\rho := K_\rho * [\nabla_2 f \, \nabla_2 f^\top]$, where K_ρ is a Gaussian of standard deviation ρ and $*$ denotes the convolution operator. The structure tensor is a symmetric, positive semidefinite 2×2 matrix that possesses two orthonormal eigenvectors $\mathbf{s}_1$ and $\mathbf{s}_2$ with corresponding eigenvalues $\mu_1 \geq \mu_2 \geq 0$. The vector $\mathbf{s}_1$ points across image structures, whereas the vector $\mathbf{s}_2$ points along them. With these notations, the regulariser from [14] can be written as

$$V_2(\nabla_2 u, \nabla_2 v) = \Psi_V\left(\left(\mathbf{s}_1^\top \nabla_2 u\right)^2\right) + \Psi_V\left(\left(\mathbf{s}_1^\top \nabla_2 v\right)^2\right) \tag{21}$$
$$+ \Psi_V\left(\left(\mathbf{s}_2^\top \nabla_2 u\right)^2\right) + \Psi_V\left(\left(\mathbf{s}_2^\top \nabla_2 v\right)^2\right) \ .$$

The corresponding Euler-Lagrange equations are

$$\partial_u M - \alpha \, \text{div} \left(D_u \left(\mathbf{s}_1, \mathbf{s}_2, \nabla_2 u \right) \nabla_2 u \right) = 0 \; , \tag{22}$$

$$\partial_v M - \alpha \, \text{div} \left(D_v \left(\mathbf{s}_1, \mathbf{s}_2, \nabla_2 v \right) \nabla_2 v \right) = 0 \; , \tag{23}$$

with the diffusion tensors

$$D_p \left(\mathbf{s}_1, \mathbf{s}_2, \nabla_2 p \right) := \left(\mathbf{s}_1 | \mathbf{s}_2 \right) \begin{pmatrix} \Psi_V' \left(\left(\mathbf{s}_1^\top \nabla_2 p \right)^2 \right) & 0 \\ 0 & \Psi_V' \left(\left(\mathbf{s}_2^\top \nabla_2 p \right)^2 \right) \end{pmatrix} \begin{pmatrix} \mathbf{s}_1^\top \\ \mathbf{s}_2^\top \end{pmatrix} \; , \tag{24}$$

for $p \in \{u, v\}$. We observe that this regulariser indeed exhibits the desired image- and flow-driven behaviour: The regularisation direction is adapted to the image structure directions $\mathbf{s}_1$ and $\mathbf{s}_2$, whereas the magnitude of the regularisation depends on the flow contrast encoded in $\nabla_2 p$. As a result, this regulariser yields the same sharp flow edges as image-driven methods but does not suffer from oversegmentation problems.

4.2   Our Novel Constraint Adaptive Regulariser (CAR)

In spite of its sophistication, the JIF model still suffers from a few shortcomings. As a remedy we will introduce three amendments that will be discussed now.

Regularisation Tensor. A first remark w.r.t. JIF regularisation is that the directional information from the structure tensor S_ρ is not consistent with the imposed constraints of our data term (16). It is more natural to take into account the directional information provided by the motion tensor (12) and to steer the anisotropic regularisation process w.r.t. "constraint edges" instead of image edges. We propose to analyse the eigenvectors $\mathbf{r}_1$ and $\mathbf{r}_2$ of the *regularisation tensor*

$$R_\rho := \sum_{i=1}^{3} K_\rho * \left[\theta_0^i \nabla_2 f^i \left(\nabla_2 f^i \right)^\top + \gamma \left(\theta_x^i \nabla_2 f_x^i \left(\nabla_2 f_x^i \right)^\top + \theta_y^i \nabla_2 f_y^i \left(\nabla_2 f_y^i \right)^\top \right) \right] \; . \tag{25}$$

The regularisation tensor integrates neighbourhood information of the motion tensor entries for every colour channel. By exploiting the invariances of the HSV colour space, it is not prone to be misled by "phantom" edges, like shadow edges.

Rotational Invariance. Unfortunately the smoothness term V_2 lacks the desirable property of rotational invariance, because the projections of $\nabla_2 u$ and $\nabla_2 v$ onto the eigenvector directions are penalised separately. As a remedy we propose to jointly penalise the projections on the eigenvector directions of the regularisation tensor, yielding

$$V_3(\nabla_2 u, \nabla_2 v) = \Psi_V \left(\left(\mathbf{r}_1^\top \nabla_2 u \right)^2 + \left(\mathbf{r}_1^\top \nabla_2 v \right)^2 \right) \tag{26}$$
$$+ \Psi_V \left(\left(\mathbf{r}_2^\top \nabla_2 u \right)^2 + \left(\mathbf{r}_2^\top \nabla_2 v \right)^2 \right) \; .$$

Single Robust Penalisation. The regulariser V_3 performs a *twofold robust penalisation* in both eigenvector directions. Because the data term mainly constraints the flow in the direction of the largest eigenvalue of the spatial motion tensor, we propose a *single robust penalisation* solely in $\mathbf{r}_1$-direction. In the orthogonal $\mathbf{r}_2$-direction, we opt for a strong quadratic penalisation. The advantages of this strong filling-in effect along constraint edges will be confirmed by our experiments in Sec. 6. Also incorporating the single robust penalisation yields the regulariser

$$V(\nabla_2 u, \nabla_2 v) = \Psi_V\left(\left(\mathbf{r}_1^\top \nabla_2 u\right)^2 + \left(\mathbf{r}_1^\top \nabla_2 v\right)^2\right) + \left(\mathbf{r}_2^\top \nabla_2 u\right)^2 + \left(\mathbf{r}_2^\top \nabla_2 v\right)^2 , \quad (27)$$

where we use the Perona-Malik regulariser (Lorentzian) [19,20] given by $\Psi_V(s^2) := \lambda^2 \log(1 + (s^2/\lambda^2))$ with a contrast parameter $\lambda > 0$. We call the regulariser V the *constraint adaptive regulariser (CAR)*. It complements the proposed robust data term M from (16) in an optimal fashion.

The corresponding Euler-Lagrange equations for are

$$\partial_u M - \alpha \operatorname{div}\left(D\left(\mathbf{r}_1, \mathbf{r}_2, \nabla_2 u, \nabla_2 v\right) \nabla_2 u\right) = 0 , \quad (28)$$

$$\partial_v M - \alpha \operatorname{div}\left(D\left(\mathbf{r}_1, \mathbf{r}_2, \nabla_2 u, \nabla_2 v\right) \nabla_2 v\right) = 0 , \quad (29)$$

where the joint diffusion tensor is given by

$$D\left(\mathbf{r}_1, \mathbf{r}_2, \nabla_2 u, \nabla_2 v\right) := \left(\mathbf{r}_1 | \mathbf{r}_2\right) \begin{pmatrix} \Psi_V'\left(\left(\mathbf{r}_1^\top \nabla_2 u\right)^2 + \left(\mathbf{r}_1^\top \nabla_2 v\right)^2\right) & 0 \\ 0 & 1 \end{pmatrix} \begin{pmatrix} \mathbf{r}_1^\top \\ \mathbf{r}_2^\top \end{pmatrix} . \quad (30)$$

Comparing our joint diffusion tensor (30) to its JIF counterparts (24), the following innovations become apparent: (i) The smoothing direction is adapted to constraint edges instead of image edges. (ii) We achieve rotational invariance by coupling the two flow components in the argument of Ψ_V'. (iii) We only reduce the smoothing *across* constraint edges. Along them, always a strong diffusion with strength 1 is performed, resembling edge-enhancing anisotropic diffusion [21].

When using $\partial_u M$ and $\partial_v M$ as given in (18) and (19) in the Euler-Lagrange equations (28) and (29), we obtain the Euler-Lagrange equations for our proposed *complementarity optic flow (COF)* method.

5   Implementation

To solve the Euler-Lagrange equations we use a coarse-to-fine multiscale warping approach [4]. On each warping level, small flow increments are computed via a linearised approach, allowing to handle large displacements correctly. The computations are speeded up by a nonlinear multigrid scheme [16] that solves the problem at each warping level based on a Gauß-Seidel type solver with alternating line relaxation [16].

Spatial image and flow derivatives are discretised via central finite differences of fourth and second order, respectively [16]. For the motion tensor, these derivatives are averaged from the two frames $f(x, y, t)$ and $f(x, y, t+1)$, whereas for the regularisation tensor, they are solely computed at the first frame.

6 Experiments

Our first experiment shows the importance of different constituents of our model. We compare our method to four modified versions of it where we have changed one distinct feature: (i) No data term normalisation. (ii) Using a regulariser with twofold instead of single robust penalisation as in V_3. (iii) Using a regulariser with single robust penalisation as in V, but based on the structure tensor instead of the regularisation tensor. (iv) Using the RGB colour space. For the latter version, we only separately robustify the BCA and the GCA, as a separate robustification of the RGB channels makes no sense. In Fig. 1, we show results for the *Urban3* sequence from the recent optic flow database [22] of the Middlebury University[1]. To visualise flow fields, we plot the magnitude of the flow vectors. Throughout our experiments we use the parameters $\zeta = 0.1, \varepsilon = 0.001, \lambda = 0.1$. Specifically for the *Urban3* sequence, we fixed the parameters $\sigma = 0.7, \gamma = 1.0, \rho = 1.5$ and only tuned the value of α, as is given in the caption of Fig. 1. There, we also state the corresponding average angular error (AAE) measures [23] in order to compare the quality of estimated flow fields to the ground truth. Note that the errors were computed for the whole image, whereas for visualisation purposes, the flow fields in Fig. 1 (c)–(h) show details. We notice a lot of artifacts for the method without data term normalisation (Fig. 1 (d)) that severely deteriorate the flow estimate. With a twofold robust penalisation (Fig. 1 (e)), artifacts at flow edges emerge due to the inhibited smoothing along edges. The results for the RGB version (Fig. 1 (f)) and our approach (Fig. 1 (h)) look rather similar due to the uncritical illumination conditions in this synthetic sequence. However, at the connection between the two buildings in the middle of the image, our approach performs better. When using the directional information from the structure tensor (Fig. 1 (g)), the results look promising, but artifacts at flow corners appear.

For a second experiment we created a real world test sequence with difficult illumination conditions caused by pronounced shadows, see Fig. 2 (a)–(b). Using this test sequence, we compare our method to the RGB version, the method of Brox et al. [4] and a version of our approach with a rotationally invariant JIF regulariser. This regulariser is similar to V_3 but uses the eigenvectors $\mathbf{s}_1$ and $\mathbf{s}_2$ instead of $\mathbf{r}_1$ and $\mathbf{r}_2$. As fixed parameters we set $\sigma = 0.5, \gamma = 20.0$ and $\rho = 2.5$. We see that the RGB method (Fig. 2 (c)) suffers from artifacts due to the shadows in the marked regions. When using the JIF regulariser (Fig. 2 (d)), the flow edges are dislocated and the shadow edges in the marked regions yield unpleasant artifacts. This demonstrates the drawbacks of the use of the structure tensor instead of the regularisation tensor, and of the twofold robust penalisation. Because of the latter, perturbing staircasing artifacts arise. The method of Brox et al. [4] (Fig. 2 (e)), that is considered to be accurate and robust, gives poor results for this sequence. Solely our method (Fig. 2 (f)) produces an agreeable flow field in spite of the difficult illumination conditions and the large displacements (up to 25 pixels) in this sequence.

---

[1] available under *http://vision.middlebury.edu/flow/data/*

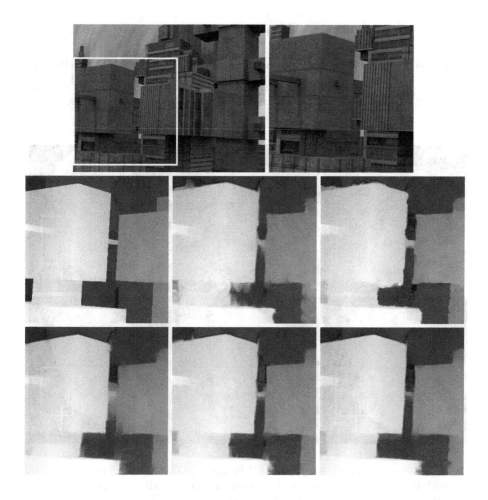

Fig. 1. The *Urban3* sequence. **First row:** (a) Frame 10. (b) Zoom in marked region. **Second row:** (c) Ground truth flow magnitude plot in marked region. (d) Corresponding flow magnitude plot without normalisation ($\alpha = 300.0$, AAE=4.57). (e) Twofold robust penalisation ($\alpha = 50.0$, AAE=3.56). **Third row:** (f) RGB colour space ($\alpha = 100.0$, AAE=3.09). (g) Structure tensor ($\alpha = 50.0$, AAE=2.99). (h) Our approach ($\alpha = 75.0$, AAE=2.95).

For a final comparison of our method to state-of-the-art approaches, we submitted our results to the Middlebury benchmark page[2]. In accordance to their guidelines, we used a fixed set of parameters for all sequences: $\alpha = 600.0, \sigma = 0.5, \gamma = 20.0$ and $\rho = 2.5$. In Tab. 1, we show the average rank of the Top 8 methods for the AAE. With the proposed method we are able to achieve the first rank. This shows that a sophisticated and transparent modelling allows to outperform other well-engineered methods that incorporate many more processing steps.

---

[2] available under *http://vision.middlebury.edu/flow/eval/results/*

Table 1. The Top 8 of the Middlebury ranking for the AAE (as of June 12, 2009)

Method	Our Method	Adaptive	Aniso. Huber-L1	Spatially variant	TV-L1- improved	Occlusion bounds	Brox et al.	Multicue MRF
Avg. rank	4.8	5.0	6.5	7.1	7.9	8.6	9.2	9.2

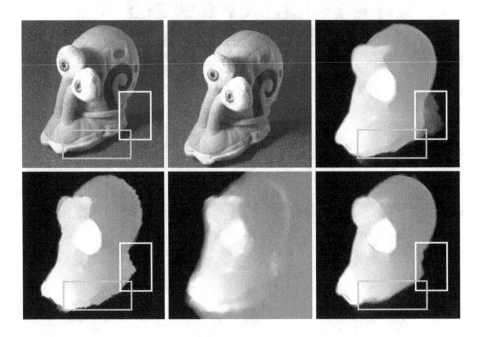

Fig. 2. The *Snail* sequence. **First row: (a)** Frame 1. **(b)** Frame 2. **(c)** Flow magnitude plot with RGB colour space ($\alpha = 800.0$). **Second row: (d)** JIF ($\alpha = 1500.0$) **(e)** Brox et al. [4] ($\alpha = 75.0$). **(f)** Our approach ($\alpha = 1500.0$).

The running time for the *Urban* sequence was 44.3 s on a standard PC (3.2 GHz Intel Pentium 4, 256 MB RAM). This proves that the used multigrid scheme [16] allows to obtain moderate runtimes for standard test sequences.

7   Conclusions and Outlook

We have presented a novel variational optic flow technique based on the concept of complementarity between data and smoothness term. By refraining from the traditional viewpoint that such terms are natural competitors within a joint energy-based framework, we succeeded to unify their advantages and achieve the currently most accurate results in the Middlebury benchmark.

Our data term integrates sophisticated components such as embedding higher order constancy assumptions in an HSV colour representation with a separate robust penalisation of each channel, renouncement of linearisations, and constraint normalisation. The directional information that results from these constraints is used in our complementary anisotropic smoothness term. This smoothness term combines the advantages of image- and flow-driven regularisation. However, compared to the approach of Sun et al. [14], it is rotationally invariant, respects constraint edges instead of image edges, and it restricts robust penalisation to the constraint direction. We have given detailed motivations showing that these model refinements arise in a natural and systematic way. Moreover, in the experiments we have proven that each of our amendments in the data and smoothness term is beneficial and contributes to the favourable accuracy of our complementary optic flow (COF) approach.

We hope that our research triggers further investigations on incorporating complementarity concepts in image processing and computer vision. This may allow to exploit similar synergies also in the context of other tasks that are currently dominated by energy-based strategies.

Acknowledgements. Our reseach was partly funded by the International Max-Planck Research School (IMPRS), the Deutsche Forschungsgemeinschaft (DFG) under the project *WE 2602/6-1*, and the German Academic Exchange Service (DAAD). This is gratefully acknowledged.

References

1. Horn, B., Schunck, B.: Determining optical flow. Artificial Intelligence 17, 185–203 (1981)
2. Nagel, H.H., Enkelmann, W.: An investigation of smoothness constraints for the estimation of displacement vector fields from image sequences. IEEE Transactions on Pattern Analysis and Machine Intelligence 8, 565–593 (1986)
3. Black, M.J., Anandan, P.: Robust dynamic motion estimation over time. In: Proc. 1991 IEEE Computer Society Conference on Computer Vision and Pattern Recognition, Maui, HI, pp. 292–302. IEEE Computer Society Press, Los Alamitos (1991)
4. Brox, T., Bruhn, A., Papenberg, N., Weickert, J.: High accuracy optical flow estimation based on a theory for warping. In: Pajdla, T., Matas, J(G.) (eds.) ECCV 2004. LNCS, vol. 3024, pp. 25–36. Springer, Heidelberg (2004)
5. Schnörr, C.: Segmentation of visual motion by minimizing convex non-quadratic functionals. In: Proc. Twelfth International Conference on Pattern Recognition, Jerusalem, Israel, vol. A, pp. 661–663. IEEE Computer Society Press, Los Alamitos (1994)
6. Tretiak, O., Pastor, L.: Velocity estimation from image sequences with second order differential operators. In: Proc. Seventh International Conference on Pattern Recognition, Montreal, Canada, July 1984, pp. 16–19 (1984)
7. Bruhn, A., Weickert, J.: Towards ultimate motion estimation: Combining highest accuracy with real-time performance. In: Proc. Tenth International Conference on Computer Vision, Beijing, China, vol. 1, pp. 749–755. IEEE Computer Society Press, Los Alamitos (2005)

8. van de Weijer, J., Gevers, T.: Robust optical flow from photometric invariants. In: Proc. 2004 IEEE International Conference on Image Processing, Singapore, vol. 3, pp. 1835–1838. IEEE Signal Processing Society (October 2004)
9. Mileva, Y., Bruhn, A., Weickert, J.: Illumination-robust variational optical flow with photometric invariants. In: Hamprecht, F.A., Schnörr, C., Jähne, B. (eds.) DAGM 2007. LNCS, vol. 4713, pp. 152–162. Springer, Heidelberg (2007)
10. Simoncelli, E.P., Adelson, E.H., Heeger, D.J.: Probability distributions of optical flow. In: Proc. 1991 IEEE Computer Society Conference on Computer Vision and Pattern Recognition, Maui, HI, pp. 310–315. IEEE Computer Society Press, Los Alamitos (1991)
11. Lai, S.H., Vemuri, B.C.: Reliable and efficient computation of optical flow. International Journal of Computer Vision 29(2), 87–105 (1998)
12. Alvarez, L., Esclarín, J., Lefébure, M., Sánchez, J.: A PDE model for computing the optical flow. In: Proc. XVI Congreso de Ecuaciones Diferenciales y Aplicaciones, Las Palmas de Gran Canaria, Spain, September 1999, pp. 1349–1356 (1999)
13. Weickert, J., Schnörr, C.: A theoretical framework for convex regularizers in PDE-based computation of image motion. International Journal of Computer Vision 45(3), 245–264 (2001)
14. Sun, D., Roth, S., Lewis, J., Black, M.: Learning optical flow. In: Forsyth, D., Torr, P., Zisserman, A. (eds.) ECCV 2008, Part III. LNCS, vol. 5304, pp. 83–97. Springer, Heidelberg (2008)
15. Roth, S., Black, M.: Steerable random fields. In: Proc. 2007 IEEE International Conference on Computer Vision, Rio de Janeiro, Brazil. IEEE Computer Society Press, Los Alamitos (2007)
16. Bruhn, A., Weickert, J., Kohlberger, T., Schnörr, C.: A multigrid platform for real-time motion computation with discontinuity-preserving variational methods. International Journal of Computer Vision 70(3), 257–277 (2006)
17. Golland, P., Bruckstein, A.M.: Motion from color. Computer Vision and Image Understanding 68(3), 346–362 (1997)
18. Förstner, W., Gülch, E.: A fast operator for detection and precise location of distinct points, corners and centres of circular features. In: Proc. ISPRS Intercommission Conference on Fast Processing of Photogrammetric Data, Interlaken, Switzerland, June 1987, pp. 281–305 (1987)
19. Black, M.J., Anandan, P.: The robust estimation of multiple motions: parametric and piecewise smooth flow fields. Computer Vision and Image Understanding 63(1), 75–104 (1996)
20. Perona, P., Malik, J.: Scale space and edge detection using anisotropic diffusion. IEEE Transactions on Pattern Analysis and Machine Intelligence 12, 629–639 (1990)
21. Weickert, J.: Theoretical foundations of anisotropic diffusion in image processing. Computing Supplement 11, 221–236 (1996)
22. Baker, S., Roth, S., Scharstein, D., Black, M., Lewis, J., Szeliski, R.: A database and evaluation methodology for optical flow. In: Proc. 2007 IEEE International Conference on Computer Vision, Rio de Janeiro, Brazil. IEEE Computer Society Press, Los Alamitos (2007)
23. Barron, J.L., Fleet, D.J., Beauchemin, S.S.: Performance of optical flow techniques. International Journal of Computer Vision 12(1), 43–77 (1994)

Parameter Estimation for Marked Point Processes. Application to Object Extraction from Remote Sensing Images*

Florent Chatelain[1], Xavier Descombes[2], and Josiane Zerubia[2]

[1] GIPSA-Lab / Grenoble Institute of Technology
961, rue de la Houille Blanche, BP 46 F - 38402 GRENOBLE Cedex
[2] Ariana, joint research group INRIA/I3S, INRIA Sophia Antipolis
2004, route des Lucioles, BP 93 06902 Sophia Antipolis Cedex, France

Abstract. This communication addresses the problem of estimating the parameters of a family of marked point processes. These processes are of interest in extraction of object networks from remote sensing images. They are defined from a combination of several energy terms. First, a data energy term controls the localization of the objects with respect to the data. Second, prior information is given by intern energy terms corresponding to geometrical constraints on the configuration to be detected. An estimation procedure of the weight associated with these energies is studied. The application to unsupervised detection of objects is finally discussed.

1 Introduction

Marked point processes have been recently studied for object extraction from remotely sensed images [1]. These processes enable one to model the configuration of an unknown number of objects in an image, within a stochastic framework. They can manage to deal with some geometrical prior information about both the objects to be extracted and the interactions such as alignments or overlappings between them. Typical applications that have been investigated include road network [2], building [3], tree crown [4] or flamingo [5] extraction.

In the considered applications, a marked point process is defined by a density function with respect to the Poisson measure. Within the framework of Gibbs point process, this density can be expressed from the combination of several energy terms: first, a data energy term, which controls the localization of the objects with respect to the data, and second, prior information about the objects is given by internal energy terms corresponding to geometrical constraints on the objects. The parameters that govern these energies are the so-called "hyperparameters" of the model. These "hyperparameters" used to be calibrated by hand. As their values depend on both the model and the image to be processed, the

---

* This work was partially funded by CNES and by associated team ODESSA, and was conducted during a postdoc fellowship of the first author at INRIA Sophia Antipolis.

D. Cremers et al. (Eds.): EMMCVPR 2009, LNCS 5681, pp. 221–234, 2009.

calibration step is often long. In order to develop fully unsupervised detection procedures, an estimation of these hyperparameters has to be performed.

The problem of estimating the parameters of a marked point process have received a great attention in the literature in the case of "complete data", for which the configuration, i.e. the set of marked points corresponding to the objects, is known. The main difficulty lies in the fact that the normalizing constant of the process density is not tractable. Thus, maximum likelihood estimators cannot be directly derived. To tackle this problem, several estimators have been proposed. A first class of methods consists of approximating the likelihood thanks to Markov Chain Monte Carlo (MCMC) methods. Corresponding estimators are then numerically obtained by maximizing the resulting estimated likelihood (see, for instance, [6,7,8]). These methods can be extended to the "missing data" case, where the process is known on a partial observation window, but is defined on a larger region [7]. Another class of estimation methods is based on the pseudolikelihood concept. This pseudolikelihood is an inference function derived from a combination of valid likelihoods associated with conditional events. Pseudolikelihood estimators are then obtained by maximizing the corresponding pseudolikelihood (see, for instance, [9,10,11,12,13,14]). The interest of these methods is that one avoids the simulation step, since the density normalizing constant does not appear in the pseudolikelihood.

Our applications rely on the framework of "incomplete data", where the configuration of the objects to be extracted is unknown. It is important to note that this framework is more general than the "missing data" case introduced above, since only the radiometry of the remote sensing image is available. Therefore, appropriate estimation methods must be derived. The main contribution of this work consists of studying an estimation method based on a Expectation-Maximization (EM) procedure. Finally, the resulting estimates are used in order to extract the objects by minimizing the energy of the process.

This paper is outlined as follows. Section 2 recalls some definitions and basic properties about spatial marked point processes. The families of marked point processes studied in this work are presented in Section 3. In Section 4, the proposed estimation method is introduced. Some numerical experiments are presented and discussed in Section 5. Section 6 looks finally at the conclusions that can be drawn from this work.

2   Marked Point Processes

In this section, the statistical background of marked point processes is briefly presented in the context of object network modeling. For a more detailed presentation of marked point processes, the interested reader is invited to consult [15].

2.1   Definitions

Let W be a compact set of $\mathbb{R}^2$ that corresponds to the definition domain of the observed image. The data, i.e. the radiometry associated with each pixel of the observed image, is denoted as y.

Point process. A configuration of points p that belong to W is an unordered set of points $p = \{p_1, \ldots, p_{n(p)}\}$ where $n(p)$ denotes the number of points in the configuration p and where $p_i \in W$ for all $1 \leq i \leq n(p)$. A point process P of points in W is a measurable mapping from a given probability space to configurations of points in W. Thus, a point process is a random variable P whose realizations are random configurations of points.

The Poisson point process is a famous example of point process. In a Poisson process realization, the points are independently distributed. These processes are interesting since they play an analog role on the configuration of points space to Lebesgue measure on $\mathbb{R}^d$. As it is explained later in Sec. 2.2, point processes can be defined by their density with respect to the probability measure of a reference Poisson point process.

Marked point process for object modeling. A marked point process living in $S = W \times M$ is a point process where some marks belonging to the set M are added to the position of the points in W. The marks represent a collection of parameters that fully describe the objects of interest. For instance, the marks reduce to the radius parameter in the case of circular objects. A configuration of objects in S is then an unordered set of objects $x = \{(p_1, m_1), \ldots, (p_{n(x)}, m_{n(x)})\}$. Therefore, a marked point process of objects in S is a random variable X whose realizations are random configurations of objects, i.e. marked points. These configurations belong to:

$$\Omega = \bigcup_{n \geq 0} \Omega_n, \tag{1}$$

where Ω_n is the set of all the configurations x with a finite number $n(x) = n$ of objects, and Ω is the set of all the configurations with a finite number of objects. Finally, it is interesting to note that, if M is a bounded set, a marked point process can be viewed as a point process defined on the space $S = W \times M$.

2.2   Density of a Marked Point Process

Let $\mu(\cdot)$ be the probability measure of a Poisson point process defined by its intensity measure $\nu(\cdot)$ on the space of configurations Ω and let $h_\theta(\cdot)$ be a mapping from Ω to $[0, +\infty)$ parameterized by $\theta \in \Theta$. Consider now the following normalizing function:

$$c(\theta) = \int_\Omega h_\theta(x) \mu(dx). \tag{2}$$

When $c(\theta) < +\infty$ for all $\theta \in \Theta$, then the following functions:

$$f_\theta(x) = \frac{h_\theta(x)}{c(\theta)}, \tag{3}$$

define a family $(f_\theta)_{\theta \in \Theta}$ of probability density functions of a point process on Ω with respect to the reference Poisson process. It is important to note that the computation of the normalizing constant $c(\cdot)$ expressed in (2) requires an integration over the set Ω of all the configurations with a finite number of objects, which is a huge space. As a consequence, except for the Poisson case, it is not possible to obtain a tractable expression of this constant from an analytical or numerical point of view.

Within the Gibbs framework, this density can finally be expressed in an energetic form:

$$f_{\boldsymbol{\theta}}(\boldsymbol{x}) = \frac{1}{c(\boldsymbol{\theta})} \exp\left(-w\, U(\boldsymbol{x})\right), \tag{4}$$

where $U(\boldsymbol{x})$ is the energy of the process and $w > 0$ is the weight parameter associated with this energy. Note that the parameter w is homogeneous to the inverse of a temperature.

3   Proposed Models for Object Extraction

This section introduces the family of marked point processes which is studied in order to extract object networks. This family is defined from its Gibbs density (4). For sake of simplicity, the dependences with respect to the parameter vector $\boldsymbol{\theta}$ and to the data $\boldsymbol{y}$ are omitted in the energy notations. The model energy is divided into two parts $U(\boldsymbol{x}) = U_d(\boldsymbol{x}) + U_p(\boldsymbol{x})$ corresponding to:

- an external energy, $U_d(\boldsymbol{x})$, which quantifies the adequacy of the configuration $\boldsymbol{x}$ with respect to the data $\boldsymbol{y}$,
- an intern energy, $U_p(\boldsymbol{x})$, which favors or penalizes some specific geometrical patterns in the configuration $\boldsymbol{x}$.

3.1   Intern Energy: Prior Term

In this work, the intern energy corresponds to the penalization of object overlappings. This is done by introducing a "hard core" constraint. Let $R(u_1, u_2)$ be the following ratio:

$$R(u_1, u_2) = \frac{|u_1 \cap u_2|}{\min\left(|u_1|, |u_2|\right)}, \tag{5}$$

where $|u|$ denotes the area of a given geometric shape u. Then, the intern energy is defined as a pairwise interaction process:

$$U_p(\boldsymbol{x}) = \sum_{1 \leq i < j \leq n(\boldsymbol{x})} U_p(x_i, x_j), \tag{6}$$

where

$$U_p(x_i, x_j) = \begin{cases} +\infty & \text{if } x_i \neq x_j \text{ and } R(x_i, x_j) > s, \\ 0 & \text{otherwise.} \end{cases} \tag{7}$$

Here $s \in [0, 1]$ is a given threshold parameter (included in $\boldsymbol{\theta}$) that characterizes the maximal overlapping ratio between the objects of the configuration. As a consequence, this intern energy allows one to avoid multiple detections of the same object.

3.2   External Energy: Data Term

Two main approaches can be adopted in order to define the data energy term. The first one consists of considering the likelihood function describing the data y for a fixed configuration x of objects. The second one is based on the definition of local energies associated with each object. These local energies can be derived from any contrast measure between the distribution of the pixels belonging to an object and the distribution of the pixel belonging to a neighborhood of this object.

Bayesian model. When a tractable expression of the likelihood $\mathcal{L}(y|x)$ of the data given an object configuration x is available, the external energy can be derived as:

$$U_d(x) = -\log \mathcal{L}(y|x). \tag{8}$$

For a given configuration x, the data energy associated with one object u is thus expressed as:

$$U_d(u|x) = \begin{cases} U_d(x) - U_d(x\setminus\{u\}) & \text{if } u \in x, \\ U_d(x \cup \{u\}) - U_d(x) & \text{otherwise.} \end{cases} \tag{9}$$

To define the likelihood, a Gaussian mixture is considered:

$$\mathcal{L}(y|x) = \prod_{y \in \mathcal{C}_o} f(y; m_o, \sigma_o^2) \prod_{y \in \mathcal{C}_b} f(y; m_b, \sigma_b^2), \tag{10}$$

where $\mathcal{C}_o$ and $\mathcal{C}_b$ are the subsets of pixels belonging respectively to the objects of the configuration x and to the background, $f(\cdot; m, \sigma^2)$ being the probability density function of the Gaussian distribution with mean m and variance σ^2. Note that the parameters of the Gaussian distributions are directly estimated from a k-means classification of the image.

Using the Bayes rule, one can see from eq. (8) and (4) that when $w = 1$, the density of the process reduces to the following posterior distribution:

$$f(x|y) = \frac{f_p(x)\mathcal{L}(y|x)}{\int_\Omega f_p(x)\mathcal{L}(y|x)dx} \propto f_p(x)\mathcal{L}(y|x), \tag{11}$$

where $f_p(x) \propto -\log(U_p(x))$ is the density of the marked point process associated with the prior term. This remark emphasizes that such a choice of the data term corresponds to a Bayesian framework when the energy weight equals one. By abuse of language, this data energy is also referred as Bayesian in the general case $w > 0$.

Detector framework. The previous approach requires to exhibit a likelihood function describing the distribution of the considered image given a configuration of objects. However, describing accurately the image distribution can be a difficult problem. Moreover, a not enough accurate likelihood may yield a data term not enough robust to provide reliable information (see [16] for a discussion on this issue).

In this "detector" model, the data energy is computed at the object level as a sum over the objects of the configuration:

$$U_d(\boldsymbol{x}) = \sum_{x \in \boldsymbol{x}} U_d(x), \tag{12}$$

where $U_d(x) \in [-1, 1]$. An object x will be attractive and therefore favored if its data energy $U_d(x)$ is negative. Conversely, a positive data energy will penalize its presence.

The data term relies on the choice of a statistical contrast measure, denoted as $J(\mathcal{O}(x), \mathcal{N}(x))$, between the distribution of set of pixels $\mathcal{O}(x)$ belonging to the object x, and the distribution of the pixels belonging to its boundary $\mathcal{N}(x)$. Finally, the data energy $U_d(x)$ is obtained as:

$$U_d(x) = Q\left(J(\mathcal{T}(x), \mathcal{N}(x))\right), \tag{13}$$

where $Q(\cdot) : \mathbb{R}^+ \mapsto [-1, 1)$ is a quality function. The following quality function, proposed in [16], is considered:

$$Q(J) = \begin{cases} 1 - \left(\frac{J}{J_0}\right)^{\frac{1}{3}} & \text{if } J < J_0, \\ \exp\left(-\frac{J - J_0}{3 J_0}\right) - 1 & \text{otherwise,} \end{cases} \tag{14}$$

where $J_0 > 0$ is a given threshold. This function is depicted in Fig. 1 for $J_0 = 5$. One can see that the role of J_0 is crucial. Indeed, it gives a positive value of the data energy to contrast measure lower than J_0, and a negative one otherwise.

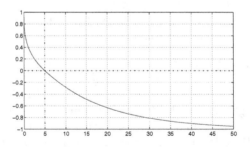

Fig. 1. Quality function Q_{J_0} ($J_0 = 5$)

The threshold J_0 controls therefore the way the objects are penalized or favored. In our applications, the radiometry of the objects to be detected is greater than the background of the image. In [16], a statistical distance $J(\cdot, \cdot)$ adapted from a Bhattacharya distance is proposed. In this work, we consider a contrast measure derived from statistics associated with the following binary hypothesis test:

$$H_0 : \quad J(\mathcal{O}(x), \mathcal{N}(x)) \leq J_0 \quad \leftarrow \text{"absence of object" hypothesis,}$$
$$H_1 : \quad J(\mathcal{O}(x), \mathcal{N}(x)) > J_0 \quad \leftarrow \text{"presence of the object } x \text{" hypothesis.}$$

Under the assumption that the samples belonging respectively to $\mathcal{O}(x)$ and $\mathcal{N}(x)$ are independent and Gaussian distributed, the so-called *Welch's t-test*, is based on the following statistics [17]:

$$t = \frac{m_\mathcal{O} - m_\mathcal{N}}{\sqrt{\frac{s_\mathcal{O}^2}{n_\mathcal{O}} + \frac{s_\mathcal{N}^2}{n_\mathcal{N}}}} \tag{15}$$

where $m_\mathcal{O}$ and $m_\mathcal{N}$ are the empirical means of $\mathcal{O}(x)$ and $\mathcal{N}(x)$ respectively, $s_\mathcal{O}^2$ and $s_\mathcal{N}^2$ their empirical variances, $n_\mathcal{O}$ and $n_\mathcal{O}$ denoting their sample size. Under the null hypothesis H_0, the distribution of the statistics t can be approximated by a Student's T distribution whose degree of freedom d is:

$$d = \frac{\left(\frac{s_\mathcal{O}^2}{n_\mathcal{O}} + \frac{s_\mathcal{N}^2}{n_\mathcal{N}}\right)^2}{\frac{s_\mathcal{O}^4}{n_\mathcal{O}^2(n_\mathcal{O}-1)} + \frac{s_\mathcal{N}^4}{n_\mathcal{N}^2(n_\mathcal{N}-1)}}.$$

Finally, the contrast measure used to construct the energy data term is:

$$J(\mathcal{O}(x), \mathcal{N}(x)) = t. \tag{16}$$

Moreover, the threshold J_0 involved in the quality function corresponds to the p-value associated with a given probability of false alarm α:

$$J_0 = F_t^{-1}(1 - \alpha), \tag{17}$$

where F_t^{-1} is the inverse cumulative distribution function of the Student's T distribution with d degrees of freedom. The advantages of this modeling are multiple:

- this contrast measure is not sensitive to a location-scale factor. Consequently, any linear transform of the image will not change the energy of the process,
- for each object, the adaptive threshold J_0 is deduced from the probability of false alarm α,
- the probability of over-detection (i.e. of false alarm) α can be controlled.

An aerial image of a flamingo colony in Camargue, France, is depicted in Fig. 2(a). The data map corresponding to the Bayesian external energy is depicted in Fig. 2(b) in case of circular objects of fixed radius. The detector data map, computed for a false alarm rate $\alpha = 10^{-6}$ is displayed in Fig. 2(c). The comparison of these two maps shows large differences of energy levels. This remark emphasizes the importance of the energy weight w in the model.

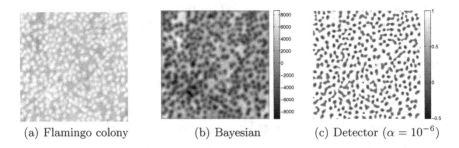

(a) Flamingo colony   (b) Bayesian   (c) Detector ($\alpha = 10^{-6}$)

Fig. 2. Remote sensing image of a flamingo colony in Camargue, France (©La Tour du Valat), and associated data maps for the two models of external energy

3.3   Extraction Algorithm

The investigated estimator, denoted as $\widehat{x} \in \Omega$, of the object configuration is the maximum likelihood estimator, which is obtained by maximizing the density of the marked point process for an appropriate value of the parameters $\theta_0 \in \Theta$. From (4), one can see that it is equivalent to the Gibbs energy minimization problem:

$$\widehat{x} = \arg\max_{x \in \Omega} f_{\theta_0}(x) = \arg\min_{x \in \Omega} U(x). \tag{18}$$

When the data term corresponds to the likelihood of the observation given an object configuration, then $\widehat{x}$ turns out to be a classical MAP estimator within the Bayesian framework.

Finding the minimum of the energy on Ω is not straightforward. In the general case, it is not possible to derive an analytical expression of $\widehat{x}$. As stated in the previous section, the density of the marked point process is defined up to an unknown normalizing constant. However, it is possible to sample some realizations of the process thanks to a Reversible Jump Monte Carlo Markov Chain (RJMCMC) algorithm [18], even if the density normalizing constant is unknown. Density optimization is then achieved by a RJMCMC sampler embedded in a Simulated Annealing (SA) scheme (see [19] for more details).

4   Parameter Estimation

In the proposed models, this unknown parameter vector θ is composed of the weight parameter $w > 0$ introduced in (4), and of the maximal overlapping ratio $s \in [0, 1]$ introduced in (6). This section addresses the problem of estimating θ from the radiometry y of a given image. In the simulations presented in the next section, the parameter s is fixed to a deterministic value. Therefore, the estimation problem reduces to the estimation of the energy weight w. In [19], it is shown that the best extraction performances are obtained with a cooling schedule concentrated around a critical temperature T_c. Since the parameter w is homogeneous to the inverse of a temperature, its estimation aims at providing an estimate of the inverse of the critical temperature. Therefore, estimating this parameter w allows one to used a cooling schedule based on normalized temperatures, which does not depend anymore on the processed image and the external energy model.

The main difficulty arising in this estimation problem is that the object configuration x, and the marginal density of the observation $f_\theta(y)$ are unknown. In such a situation, Expectation-Maximization (EM) algorithms offer an appropriate framework for parameter estimation. The EM algorithm, introduced in [20], is an iterative method that is used to determine the Maximum Likelihood Estimators (MLE) in the case of incomplete data. Each iteration can be divided into two steps. First, an expectation step, which consists of computing the expectation of the process density denoted as $f_\theta(x) \equiv f_\theta(x|y)$. Second, a maximization step with respect to parameter vector θ is performed. However, it is not possible to obtain a tractable expression of the expectation since the density normalizing constant is intractable. To tackle this problem, an approximated version of the Stochastic EM (SEM) procedure described in [21] is used.

Pseudo-Likelihood approximation. The concept of Pseudo-Likelihood (PL) has been widely studied in the literature since the seminal paper of Besag [22]. Moreover, this concept has been generalized to spatial marked point process [23,9]. For a given object configuration x, the PL is defined as:

$$\mathrm{PL}_S(\theta; x, y) = \left[\prod_{x_i \in x} \lambda_\theta(x_i; x, y)\right] \exp\left\{-\int_S \lambda_\theta(u; x, y)\nu(du)\right\}, \quad (19)$$

where $u \in S$ is an object, $\nu(\cdot)$ is the intensity of the reference Poisson process and $\lambda_\theta(\cdot; x, y) : S \to [0, +\infty)$ denotes the conditional Papangelou intensity of the process. In our models, this intensity can be expressed as:

$$\lambda_\theta(u; x, y) = \exp\left[-w\left(U_d(u|x) + \sum_{x_i \in x| \ x_i \neq u} U_p(x_i, u)\right)\right], \quad (20)$$

where $U_p(\cdot, \cdot)$ is the pairwise intern energy defined in (7). From eq. (19), one can see that the PL depends on the following normalizing constant:

$$z(\theta; x, y) = \exp\left\{-\int_S \lambda_\theta(u; x, y)\nu(du)\right\}. \quad (21)$$

Since this normalizing constant expresses as a summation over the state space S, efficient numerical quadrature scheme can be used to calculate it [13]. Thus, numerical values of the PL can be easily computed.

In order to estimate the parameters, it is finally proposed to approximate the process density, i.e. its likelihood, by its PL. Such an approximation is motivated by the following two points:

1) For a given configuration x, the PL is an inference function which is close to the likelihood (for instance, the PL of a Poisson process reduces to its likelihood).
2) In the context of complete data, the MLEs and the Maximum PL Estimators (MPLEs) exhibit quite similar properties and performances for some classical examples of processes [24].

Proposed SEM algorithm. The approximation of the process likelihood by its PL leads to the following approximate version of the SEM algorithm:

(S) Stochastic step: Simulation of $x^k \sim f_{\theta^k}(x|y)$,
(E) Expectation step: Computation of $Q(\theta, \theta^k; y) = \log \text{PL}(\theta; x^k, y)$,
(M) Maximization step: $\theta^{k+1} = \arg \max_{\theta} Q(\theta, \theta^k; y)$,

$$(22)$$

where θ^k denotes the current estimate of the parameter vector θ in the k-th iteration.

5 Simulation Results

Many simulations have been conducted in order to validate the estimation method of the energy weight parameter w. In these simulations, the modeled shapes correspond to circular object. The intensity of the reference Poisson process is normalized: $\nu(S) = 1$. In addition, the maximal overlapping ratio parameter s is set to $s = 1/2$. This value is justified by a simple geometrical reasoning: two objects of the configuration fit the same object on the data if the intersection area is greater than the half of their respective areas. Finally, the probability of false alarm is set to $\alpha = 1e - 6$ in the detector data term.

5.1 Initialization of the SEM Algorithm

To avoid converging to a local maximum of the likelihood, it is crucial to initialize the algorithm with a value w_0 close enough to the MLE of w. This can be done by considering the following function:

$$I(w) = \int_S \lambda_{\theta}(u; \emptyset, y)\nu(du),$$

where $\lambda_{\theta}(u; \emptyset, y)$ is the Papangelou intensity associated with the null configuration. Then, the function $I(w)$ represents the mean number of objects in absence of interaction. Let $\beta > 1$ be the exact number of objects to be extracted. The idea consists of considering the initial value w_0 defined such that $I_{\beta}(w_0) = \beta$

Theorem 1. *Let β be a real strictly greater than 1. The function*

$$w \mapsto I(w) - \beta \tag{23}$$

has an unique root in $(0, +\infty)$.

Proof. By construction $I(\cdot)$ is a strictly convex function (as the limit of the sum of strictly convex functions). To prove the theorem, it remains to study the function on the boundary of its definition domain. It is straightforward to obtain that $I(0) = \int_S \nu(du) = \nu(S) = 1$. Moreover, when w goes to infinity, the term $e^{-wU_d(u)}$ converges toward 0 if $U_d(u) > 0$ or toward the Dirac distribution $\delta_u(\cdot)$ if $U_d(u) < 0$. As a consequence, $I(w) \to \sum_{u \in \boldsymbol{x}_\infty} \nu'(u)$, where $\nu'(\cdot) : S \to [0, +\infty)$ is the reference Poisson process intensity function, and $\boldsymbol{x}_\infty$ denotes the configuration of all the objects possessing a strictly negative data energy. Since the data energy is a regular and continuous function, $\boldsymbol{x}_\infty$ is composed of an infinity of objects. Thus, $I(w) \to +\infty$ when $w \to +\infty$. This concludes the proof.

A realization of the logarithm of the function $I(\cdot)$ is depicted in Fig. 3. It shows that the function $I(\cdot)$ exponentially increases for large values of w. Therefore, even for larger values $\beta' >> \beta$ than the real number of objects, the unique root of the function $I(\cdot) - \beta'$ is quite close to the root of $I(\cdot) - \beta$. It leads to the following scheme in order to choose an initial value for the parameter w:

1) Coarse over-estimation of the number of objects $\beta_0 \geq 1$.
2) Computation of the root w_0 of the function $w \mapsto I(w) - \beta_0$.

The second step is performed thanks to a Newton-Raphson procedure.

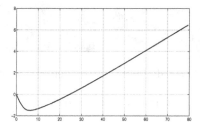

Fig. 3. $\log_{10}(I(w))$ versus w for the detector energy term applied to the flamingo colony image depicted in Fig. 2(a)

5.2   Numerical Results

Fig. 4 displays some estimation results obtained from a 640×480 synthetic image. This synthetic image has been generated according to a mixture of Gaussian distributions. The parameters of the object and background classes are respectively $(\mu_0, \sigma_o^2) = (130, 400)$ and $(\mu_f, \sigma_f^2) = (100, 400)$. The image is composed of 60 quasi-circular objects. In order to initialize the estimation algorithm, the over-estimated number of object is set to $\beta_0 = 1000$. The estimates of the weight w are $\widehat{w} = 7.19 \times 10^{-4}$ and $\widehat{w} = 37.9345$ for the respective Bayesian and detector external energy models. Note that the number of iterations done in order to satisfy the convergence criterion is $k = 3$ in the Bayesian case, and $k = 9$ for the detector model. The configurations for which the estimation method has converged are depicted (red circles) in 5(a) and 5(b). One can see that these

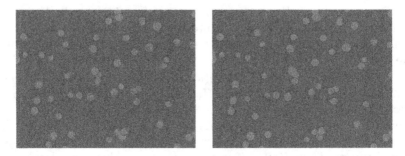

(a) Bayesian: $\widehat{w} = 7.19 \times 10^{-4}$ (42   (b) Detector: $\widehat{w} = 14.23$ (40 objects)
objects)

Fig. 4. Estimation results on an synthetic image for the Bayesian and the detector data energy models

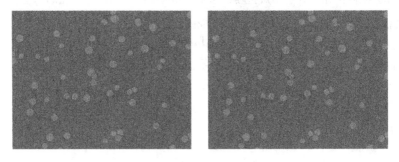

(a) Bayesian: 58 detected objects   (b) Detector: 57 detected objects

Fig. 5. Extraction results

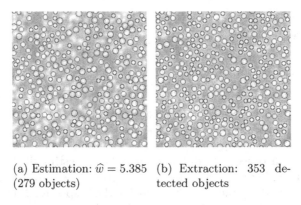

(a) Estimation: $\widehat{w} = 5.385$ (b) Extraction: 353 de-
(279 objects)   tected objects

Fig. 6. Estimation and extraction results for the detector data energy model applied to the flamingo colony image

configurations are already quite close to the real one. Based on the estimated value of the parameter w, the extraction of the object configuration is achieved by a simulated annealing scheme. The normalized initial and final temperature of the cooling schedule are set to $T_i = 2$ and $T_f = 0.05$ respectively. The extraction results are depicted in Fig. 5. These results match very well the data, the misdetections corresponding to objects that are located in the border of image. Finally, the same experimental setup is applied to the real 274×269 image of the flamingo colony presented in Fig. 2(a) in the case of the detector data energy model. The estimate of the weight, obtained after 12 iterations, is $\widehat{w} = 5.385$. The obtained configurations are depicted in Fig. 6. The extraction result, based on the estimated weight, appears to be in good agreement with the data. As a consequence, these simulations show the interest and the performance of the proposed estimation method.

6 Conclusions

This paper studied an estimation method of the energy weight parameter of a Gibbs marked point process. A method based on a SEM algorithm is proposed, the process likelihood being approximated by the pseudo-likelihood. This method has shown good performances on both synthetic and real images for different external energy models. It allows one to perform a quasi-automatic extraction of surface objects, such as tree crowns or flamingos. The application to the estimation of some interaction parameters and the extension to more complex geometrical shape models are currently under investigation.

References

1. Descombes, X., Kruggel, F., Lacoste, C., Ortner, M., Perrin, G., Zerubia, J.: Marked point process in image analysis: from context to geometry. In: Int. Conference on Spatial Point Process Modelling and its Application (SPPA), Castellon, Spain (2004)
2. Lacoste, C., Descombes, X., Zerubia, J.: Point processes for unsupervised line network extraction in remote sensing. IEEE Trans. Pattern Analysis and Machine Intelligence 27(10), 1568–1579 (2005)
3. Ortner, M., Descombes, X., Zerubia, J.: A marked point process of rectangles and segments for automatic analysis of digital elevation models. IEEE Trans. Pattern Analysis and Machine Intelligence 30(1), 105–119 (2008)
4. Perrin, G., Descombes, X., Zerubia, J.: 2d and 3d vegetation resource parameters assessment using marked point processes. In: Proc. International Conference on Pattern Recognition (ICPR), Hong-Kong (August 2006)
5. Descamps, S., Descombes, X., Béchet, A., Zerubia, J.: Automatic flamingo detection using a multiple birth and death process. In: Proc. IEEE International Conference on Acoustics, Speech and Signal Processing (ICASSP 2008), pp. 304–306 (2008)
6. Geyer, C.: On the convergence of Monte Carlo maximum likelihood calculations. Journal of the Royal Statistical Society, Series B 56, 261–274 (1994)

7. Geyer, C.: Likelihood inference for spatial point process. In: Stochastic Geometry - Likelihood and Computation. Chapman and Hall/CRC (1999)
8. Descombes, X., van Lieshout, M., Stoica, R., Zerubia, J.: Parameter estimation by a Markov chain Monte Carlo technique for the Candy-model. In: Proc. of IEEE Workshop on Statistical Signal Processing, Singapour (August 2001)
9. Jensen, J., Møller, J.: Pseudolikelihood for exponential family models of spatial point processes. The Annals of Applied Probability 1(3), 445–461 (1991)
10. Jensen, J., Kunsch, R.: On asymptotic normality of pseudo likelihood estimates for pairwise interaction processes. Ann. Inst. Statist. Math. 46, 475–486 (1994)
11. Mase, S.: Consistency of the maximum pseudo-likelihood estimator of continuous state space Gibbsian processes. The Annals of Applied Probability 5(3), 603–612 (1995)
12. Mateu, J., Montes, F.: Pseudo-likelihood inference for Gibbs processes with exponential families through generalized linear models. Statistical Inference for Stochastic Processes 4(2), 125–154 (2001)
13. Baddeley, A., Turner, R.: Practical maximum pseudolikelihood for spatial point patterns (with discussion). Australian and New Zealand Journal of Statistics 42(3), 283–322 (2000)
14. Billiot, J.M., Coeurjolly, J.F., Drouilhet, R.: Maximum pseudo-likelihood estimator for nearest-neighbours Gibbs point processes. Electronic Journal of Statistics 2, 234–264 (2008)
15. Van Lieshout, M.N.M.: Markov Point Processes and Their Applications. Imperial College Press, London (2000)
16. Perrin, G., Descombes, X., Zerubia, J.: A non-Bayesian model for tree crown extraction using marked point processes. Research Report 5846, INRIA, France (February 2006)
17. Welch, B.L.: The generalization of Student's problem when several different population variances are involved. Biometrika 34, 28–35 (1947)
18. Green, P.: Reversible jump Markov chain Monte Carlo computation and Bayesian model determination. Biometrika 82, 711–732 (1995)
19. Perrin, G., Descombes, X., Zerubia, J.: Adaptive simulated annealing for energy minimization problem in a marked point process application. In: Proc. Energy Minimization Methods in Computer Vision and Pattern Recognition (EMMCVPR), St Augustine, Florida, USA (November 2005)
20. Dempster, A.P., Laird, N.M., Rubin, D.B.: Maximum likelihood from incomplete data via the EM algorithm. Journal of the Royal Stat. Society. Series B (Methodological) 39(1), 1–38 (1977)
21. Celeux, G., Chauveau, D., Diebolt, J.: On stochastic versions of the EM algorithm. Research Report 2514, INRIA (March 1995)
22. Besag, J.: Some methods of statistical analysis for spatial data. Bulletin of the International Statistical Institute 47, 77–92 (1977)
23. Besag, J., Milne, R., Zachary, S.: Point process limits of lattice processes. Journal of Applied Probability 19, 210–216 (1982)
24. Møller, J., Waagepetersen, R.: Statistical inference and simulation for spatial point processes. Chapman and Hall/CRC, Boca Raton (2003)

Three Dimensional Monocular Human Motion Analysis in End-Effector Space

Søren Hauberg, Jerome Lapuyade, Morten Engell-Nørregård, Kenny Erleben, and Kim Steenstrup Pedersen

The eScience Center, Dept. of Computer Science, University of Copenhagen
{hauberg,lapuyade,mort,kenny,kimstp}@diku.dk

Abstract. In this paper, we present a novel approach to three dimensional human motion estimation from monocular video data. We employ a particle filter to perform the motion estimation. The novelty of the method lies in the choice of state space for the particle filter. Using a non-linear inverse kinematics solver allows us to perform the filtering in end-effector space. This effectively reduces the dimensionality of the state space while still allowing for the estimation of a large set of motions. Preliminary experiments with the strategy show good results compared to a full-pose tracker.

1 Introduction

Three dimensional human motion analysis is the process of estimating the configuration of body parts over time from sensor input [1]. One approach to this estimation is to use motion capture equipment where electromagnetic markers are attached to the body and then tracked in three dimensions. While this approach gives accurate results, it is intrusive and cannot be used outside laboratory settings.

Our long-term goal is to use human motion analysis as part of a physiotherapeutic rehabilitation system where the motion of a patient is tracked and analysed during exercise sessions performed both at the hospital and at the patient's home. The motion information will then be used to provide real-time feedback to the patient and to collect statistics on the patient's progress. The system should serve as an aid to the patient while performing a self-training programme at home, in that the patient will get instant response on whether or not the exercise is performed optimally. Furthermore, the system may act as a progress measurement tool for the physiotherapist. It is essential to develop systems that can be used by the patient at home, which rules out marker-based systems and hard-to-calibrate multi-camera solutions. Thus, the focus is on developing monocular vision-based systems.

In this paper, we present a novel approach to three dimensional monocular human motion estimation. The novelty of the system lies in the choice of state space. Monocular motion estimation in three dimensions is an inherently ill-posed problem since the observed images are two dimensional. This manifests

D. Cremers et al. (Eds.): EMMCVPR 2009, LNCS 5681, pp. 235–248, 2009.

itself in that the distribution of the human pose is multi-modal with an unknown number of modes. To reliably estimate this distribution we need methods that cope well with multi-modal distributions. Currently, the best method for such problems is the particle filter [2], which represents the distribution as a set of weighted samples. Unfortunately, the particle filter is smitten by the curse of dimensionality in that the necessary number of samples grow exponentially with the dimensionality of state space. The consequence is that the particle filter is only applicable to low dimensional state spaces. This is in direct conflict with the fact that the human body has a great number of degrees of freedom.

Our approach is inspired by recent results from the world of animation [3]. Here animators are faced with the task of posing human figures on a frame by frame basis. This time-consuming task has created the need for models that allow animators to create realistic looking figures using few parameters. *Inverse kinematic* models allow the animator to pose only a few selected parts of the human figure, called *end-effectors*, while the remaining parts can be positioned by solving a non-linear least-squares optimisation problem. The end-effectors are most often the head, the hands and the feet of the figure. Although the underlying optimisation problem is hard to solve, recent work [3] has shown that realistic looking animations can be created in real-time using inverse kinematics. In this paper, we investigate the usefulness of estimating human motion in end-effector space rather than full-pose space. This is done using inverse kinematics to infer the full pose from end-effector positions.

To simplify the measurement model of the particle filter, our measurement model is based on a simple Markov random field texture model for each limb of the current pose model. We do not attempt to handle self-occlusions in this paper as focus is on the choice of state space. When necessary, this can be introduced later by appropriately changing the measurement model.

Our main point in this paper is to show the feasibility of tracking in end-effector space compared to tracking in full-pose space and argue that the former approach allows for real-time implementations with reasonable accuracy.

1.1   Related Work

Much work has gone into human motion analysis. The bulk of the work is in locating the position of moving humans in image sequences and classifying their actions. It is, however, beyond the scope of this paper to give a review of this work. The interested reader can consult review papers such as [4].

In recent years, much work has gone into more detailed visual human motion analysis in three dimensions [1]. The main difficulty in this area is the high number of degrees of freedom in the human body, which gives rise to high dimensional state spaces. To overcome this problem, many researchers reduce the dimensionality of state space using manifold learning [5,6,7]. The basic idea is to learn a manifold from motion capture data and then perform the motion analysis on this manifold. It seems that most researchers taking this route focus on simple low-dimensional motions, such as *walking* [7,8,9,10], *golf swings* [9,10], *tennis playing* [11] etc. In this context, end-effector tracking using inverse kinematics can be interpreted as a dimensionality reduction technique that is based on domain specific knowledge about general human motion.

While kinematic skeleton models have been used a lot in motion analysis it seems that inverse kinematics have been given little attention. The approach closest to ours is that of Sminchisescu and Triggs [12] who use inverse kinematics to enumerate possible interpretations of the input, which results in a more efficient sampling scheme for their particle filter. They, however, still perform the motion analysis in full-pose space, whereas we work in end-effector space. This provides a very efficient way of reducing the dimensionality of the state space that still allows us to work with a large class of motions.

1.2   Organisation of the Paper

This paper is organised as follows. In the next two sections the theoretical background is presented. First an introduction to inverse kinematics is given in Sec. 2. A brief introduction to particle filtering is then given in Sec. 3. In Sec. 4 we discuss two choices of the state space for the motion analysis system and in Sec. 5 we describe the measurement system used in our implementation. Preliminary results and a discussion of the pros and cons of the method are presented in Sec. 6 and the paper is concluded in Sec. 7.

2   Posing with Inverse Kinematics

Inverse kinematics is the problem of manipulating the pose of a skeleton in order to achieve a desired pose disregarding inertia and forces. The problem can be posed as a non-linear optimisation problem.

In the context of human modeling a skeleton is often modeled as a collection of rigid bodies connected by rotational joints of 1–3 degrees of freedom such as the one shown in Fig. 1. All joints are constrained in their rotation, as exemplified by joint i in Fig. 1 with l_i and u_i showing the limits of the angle θ_i. To compute the position and orientation of a joint in space we perform a transformation of the bone relative to its parent joint. The transformation consist of a rotation and a translation corresponding to the shape and orientation of the joint relative to its parent. These transformations are then nested to create chains of joints. Each chain ends in an end-effector, which can be regarded as the handle for controlling the chain. Thus, the full transformation of a joint from local space to global space can be performed.

The problem can be formally stated as follows. Given the set of joint parameters $\boldsymbol{\theta}$ we can change the values of $\boldsymbol{\theta}$ and gain explicit control over all joint angles. This in turn controls the position and orientation of the end-effector $\boldsymbol{a} = \boldsymbol{F}(\boldsymbol{\theta})$. This is commonly known as forward kinematics. Given a desired end-effector goal position, $\boldsymbol{g}$, one seeks the value of $\boldsymbol{\theta}$ such that

$$\boldsymbol{\theta} = \boldsymbol{F}^{-1}(\boldsymbol{g}) \ . \tag{1}$$

This is known as inverse kinematics. Closed form solutions exist for models with less than 7 degrees of freedom. However, the human body has a lot more than

Fig. 1. An illustration of the kinematic model. End-effector positions are shown as green dots, while the desired positions (goals) are shown as red dots.

7 degrees of freedom and has a large degree of interdependency between joints. Thus, it makes sense to solve the problem globally for the entire skeleton. High performance implies the need for iterative numerical methods. By posing the problem as a constrained least-squares fitting problem we are given a number of possible methods to solve the problem. Given a skeleton containing K kinematic chains, each with exactly one end-effector, we agglomerate the K end-effector functions into one function

$$a = \left[a_1^T \ldots a_K^T\right]^T = \left[F_1(\theta)^T \ldots F_K(\theta)^T\right]^T = F(\theta) \ , \tag{2}$$

where a_j is the world coordinate position of the jth end-effector and $F_j(\theta)$ is the end-effector function corresponding to the jth kinematic chain. Using the agglomerated end-effector function, we create the objective function

$$f(\theta) = (g - F(\theta))^T (g - F(\theta)) \ , \tag{3}$$

where $g = \left[g_1^T \ldots g_K^T\right]^T$ is the agglomerated vector of end-effector goals. The optimisation problem is then

$$\theta^* = \arg\min_{\theta} f(\theta) \quad \text{s.t.} \quad l \le \theta \le u \ . \tag{4}$$

Here l is a vector containing the minimum joint limits and u is a vector of the maximum joints limits. For a more thorough description of the inverse kinematics problem and the constraint model see [13,3]. Any constrained non-linear optimisation method may be used to solve the problem. In this paper we use a simple, yet effective, gradient projection method with line search [14]. To compute the gradient we need the Jacobian matrix of $f(\theta)$ denoted J. For rotational joints this can be easily computed. For each chain, the Jacobian matrix contains a 3×1 entry for each rotational degree of freedom. This entry is computed as the cross product of the rotational axis and the vector from the joint to the end-effector as shown in Fig. 2. Given this Jacobian matrix the gradient can be computed as $(g - a)J^T$. A more thorough description of the calculation of the Jacobian can be found in [15].

Fig. 2. Finding the rotational derivative of a joint. The derivative is found as the cross product of the rotational axis (red) and the vector from joint to end-effector (green) the resulting tangent vector is shown in blue.

3   Bayesian Filtering

Bayesian filtering is concerned with estimating the unobserved state of a system from observations. In terms of human motion analysis, it is concerned with estimating a sequence of human poses given a sequence of images. This section provides a brief overview of the topic. For more details, the interested reader should consult papers such as [2] and the references therein.

In Bayesian filtering it is assumed that the observation I_t at time t is solely governed by a hidden variable s_t. The distribution of this variable is in turn assumed to form a Markov chain, such that

$$p(s_{1:T}, I_{1:T}) = p(s_1)p(I_1|s_1) \prod_{t=1}^{T-1} p(s_{t+1}|s_t)p(I_{t+1}|s_{t+1}) \ . \tag{5}$$

Here $s_{1:T} = \{s_1, \ldots, s_T\}$ denotes a sequence of state variables and likewise for $I_{1:T}$.

The objective of estimating the current state given all observations is expressed as estimating the *filtering distribution* $p(s_t|I_{1:t})$. This can be estimated recursively as [2]

$$p(s_t|I_{1:t}) \propto \int p(s_{t-1}|I_{1:t-1})p(s_t|s_{t-1})p(I_t|s_t)\mathrm{d}s_{t-1} \ . \tag{6}$$

Unfortunately, this integral can only be computed exactly in simple cases. For instance, if the state is finite valued the corresponding algorithm is called Baum-Welsh filtering [16] and if the model is linear and Gaussian, it can be computed by the Kalman filter [17]. In more general cases approximations are necessary. In recent years the particle filter [2] has seen growing popularity. It represents the filtering distribution as a set of weighted samples, which allows for multi-modal distributions and non-linear processes.

In general the filtering objective is to estimate moments of the filtering distribution instead of estimating the distribution itself. In particle filtering the basic idea is to draw samples, also called *particles*, from the distribution and estimate the moments using these, i.e.

$$\int h(s_t)p(s_t|I_{1:t})\mathrm{d}s_t \approx \frac{1}{N}\sum_{j=1}^{N} h(s_t^{(j)}) \ , \tag{7}$$

where $s_t^{(j)}$ are the samples, and $h(\cdot)$ is any function of interest. In practice, we cannot draw samples from the filtering distribution as it is unknown. We therefore draw samples from an instrumental distribution $q(s_t|I_t, s_{t-1})$. These samples are then weighted such that the weighted sum of the samples provides an unbiased estimate of the moments. It can be proved [2] that these weights can be recursively updated as

$$\omega_t^{(j)} \propto \omega_{t-1}^{(j)} \cdot \frac{p(I_t|s_t^{(j)})p(s_t^{(j)}|s_{t-1}^{(j)})}{q(s_t^{(j)}|I_t, s_{t-1}^{(j)})} \quad . \tag{8}$$

Most often the instrumental distribution is chosen to be the predictive distribution $p(s_t|s_{t-1})$ as this simplifies the weight update. The resulting algorithm is known as the Bootstrap filter, which we are employing in this paper. This algorithm performs the following iterative steps for all particles

- Draw new samples $s_t^{(j)}$ from $p(s_t|s_{t-1}^{(j)})$;
- Compute normalised weights $\omega_t^{(j)} \propto \omega_{t-1}^{(j)} p(I_t|s_t^{(j)})$.

From a practical point of view this approach is not numerically stable as all but one of the weights tends towards zero. To overcome this issue it is common to only keep samples with large weights, which is done by resampling the samples. In the Bootstrap filter a sample is kept in the next iteration with a probability equal to its weight. This can result in the same sample appearing several times after resampling.

To model the dynamics of the state process it is necessary to include either second order information or velocity. This can be done by either extending the state with a velocity vector or by changing the state distribution from a first to a second order Markov chain. The latter approach keeps the dimensionality of the state space at a minimum allowing for a computationally efficient filter, while the former approach allow us to estimate velocities. Since this is not needed in the current application we choose the latter approach. This corresponds to drawing new samples from $p(s_t|s_{t-1}^{(j)}, s_{t-2}^{(j)})$ instead of $p(s_t|s_{t-1}^{(j)})$, while the weight update remains unchanged.

Although the particle filter in general is able to cope with multi-modal distributions and non-linear state changes it is not without flaws. Unless very good predictive models are available the filter generally needs $\mathcal{O}(N^D)$ samples to reliably estimate the moments [2], where D is the dimensionality of the state space.

4   Human Motion Analysis

As mentioned in the introduction, we wish to infer the full pose of the human in the scene from the image sequence. To simplify matters, we will restrict ourselves to working on the upper body, i.e. torso, head and hands. We will also assume known limb-sizes. This effectively reduces the full-pose space to being the space of joint angles. The most straight-forward choice of state space is then the space of angles, which will be discussed in Sec. 4.1. An alternative low dimensional state space will be discussed in Sec. 4.2.

4.1   Full-Pose Motion Analysis

We have seen in Sec. 3 that Bayesian motion analysis can be performed using a particle filter. The obvious way of realising such a filter is to perform the filtering in the space of all poses $\boldsymbol{\theta}$. This only requires that we provide a predictive distribution $p(\boldsymbol{\theta}_t|\boldsymbol{\theta}_{t-1}, \boldsymbol{\theta}_{t-2})$ and a likelihood function $p(I_t|\boldsymbol{\theta}_t)$.

The likelihood function will be described in detail in Sec. 5, so here we focus on prediction. This can simply be performed by extrapolating the two previous states and adding noise. In more detail we define

$$p(\boldsymbol{\theta}_t|\boldsymbol{\theta}_{t-1}, \boldsymbol{\theta}_{t-2}) = \mathcal{N}\left(\boldsymbol{\theta}_t|\boldsymbol{\theta}_{t-1} + \boldsymbol{\Delta}_{t-1}, \boldsymbol{\Sigma}\right) \ , \tag{9}$$

where $\boldsymbol{\Delta}_{t-1} = \boldsymbol{\theta}_{t-1} - \boldsymbol{\theta}_{t-2}$ represents the current displacement and $\boldsymbol{\Sigma}$ is the covariance matrix of the prediction noise. In absence of additional knowledge about the motion, we choose the least committed model and define $\boldsymbol{\Sigma} = \sigma^2\boldsymbol{I}$, with σ being a parameter and $\boldsymbol{I}$ being the identity matrix.

This simple setup provides a full system for Bayesian motion analysis. The problem with this approach is the high dimensionality of the state space. In this paper we restrict ourselves to studying an upper body model that has 47 degrees of freedom. Since the necessary number of particles grows exponentially with the dimensionality of the state space, we cannot expect to reliably use a particle filter in this state space. We therefore seek a more low dimensional state space with a similar amount of expressive power.

4.2   End-Effector Motion Analysis

As an alternative to the full-pose state space, we propose to use inverse kinematics to reduce the dimensionality of the state space. This is essentially done by performing the tracking in end-effector space. Here we define the end-effectors as the head and the hands. The end-effector space is thus the three dimensional positions of the three end-effectors, i.e. $\mathbb{R}^9$. The strategy is then to use inverse kinematics to infer the full pose. Once the full pose has been infered it can be measured just like the system working in full-pose space (see Sec. 5). This allows us to compare results from both systems.

In more details we define $\boldsymbol{x}_{head}$, $\boldsymbol{x}_{hand_0}$ and $\boldsymbol{x}_{hand_1}$ as the three dimensional positions of the end-effectors. To simplify matters, we will assume that the hands are conditionally independent given the position of the head. That is,

$$p(\boldsymbol{x}_{head}, \boldsymbol{x}_{hand_0}, \boldsymbol{x}_{hand_1}) = p(\boldsymbol{x}_{head})p(\boldsymbol{x}_{hand_0}|\boldsymbol{x}_{head})p(\boldsymbol{x}_{hand_1}|\boldsymbol{x}_{head}) \ . \tag{10}$$

This simply means we represent the hands relative to the position of the head. Thus we define the state as $\boldsymbol{s} = (\boldsymbol{x}_{head}, \boldsymbol{x}_{hand_0} - \boldsymbol{x}_{head}, \boldsymbol{x}_{hand_1} - \boldsymbol{x}_{head})$. This factorisation has the consequence that we can treat the end-effectors separately in the filtering.

To be able to make measurements we need to be able to compute $p(I_t|\boldsymbol{s}_t)$. As described in Sec. 2 we can compute the full pose $\boldsymbol{\theta}_t$ from the state $\boldsymbol{s}_t$. This allows us to perform the measurement in full-pose space rather than simply measuring the position of the head and the hands. So we define $p(I_t|s_t) \equiv p(I_t|\theta_t)$.

The prediction can be performed much like the full-pose system (9). This boils down to linear extrapolation of the end-effectors followed by addition of Gaussian noise. In more detail

$$p(s_t|s_{t-1}, s_{t-2}) = \mathcal{N}\left(s_t|s_{t-1} + \Delta_{t-1}, \sigma^2 I\right) , \tag{11}$$

where $\Delta_{t-1} = s_{t-1} - s_{t-2}$ represents the current displacement.

5   Visual Measurements

In this section we present a method for computing the likelihood of a full pose, i.e. $p(I_t|\theta_t)$. To avoid notational clutter we will drop the t subscript in the following. The basic idea is to assume that the individual limbs are independent, i.e.

$$p(\theta) = \prod_{n=1}^{N} p(\theta^{(n)}) \quad \text{and} \quad p(\theta|I) = \prod_{n=1}^{N} p(\theta^{(n)}|I) , \tag{12}$$

where $\theta^{(n)}$ are the parameters of the nth limb. From this assumption we see that

$$p(I|\theta) = \frac{p(I)p(\theta|I)}{p(\theta)} \tag{13}$$

$$= p(I)^{-(N-1)} \prod_{n=1}^{N} \frac{p(I)p(\theta^{(n)}|I)}{p(\theta^{(n)})} \tag{14}$$

$$\propto \prod_{n=1}^{N} p(I|\theta^{(n)}) . \tag{15}$$

That is, we only need to be able to evaluate the likelihood $p(I|\theta^{(n)})$ of individual limbs.

The basic assumption in the measurement model is that limbs can be treated independently. However, if one limb occludes another this assumption no longer holds. In this paper we do not attempt to model this situation as we are concerned with the choice of state space rather than the visual measurements.

5.1   Modeling the Likelihood of a Limb

We use a simple Markov random field (MRF) model to describe the appearance of a limb, in which the limb appearance statistics is described by histograms of a set of descriptive features capturing texture and colour information. The likelihood function $p(I|\theta^{(n)})$ thus takes the form of a Gibbs distribution with an energy functional consisting of terms for texture, colour and background. Each of these are considered independent and we thus define the likelihood as

$$-\log p(I|\theta^{(n)}) = \alpha_T d_T^2(H^{Tm}, H^T|\theta^{(n)}) + \alpha_C d_C^2(H^{Cm}, H^C|\theta^{(n)})$$
$$+ \alpha_B d_B^2(H^{Bm}, H^B|\theta^{(n)}) + \text{constant} , \tag{16}$$

where $d_T^2(H^{T_m}, H^T | \theta^{(n)})$ is the distance between a texture model H^{T_m} and the observed texture H^T. $d_C^2(\cdot)$ and $d_B^2(\cdot)$ are the similar counterparts of the colour and background models. The α parameters control the relative importance of the individual terms.

The texture and colour models are based on the same principle, which boils down to computing a normalised histogram of a descriptive feature within the limb and comparing that with a normalised model histogram. When constructing these histograms we use a Gaussian aperture to weigh the individual pixel contributions. This aperture will be described in the next section. When comparing the histograms we are using the earth mover's distance [18], which takes into account small perturbations and shifts of the histograms.

The background model is based on a simple thresholding of the absolute difference between a background image and the current image. This binary image B is then compared to a simple rendering R of the pose. The rendered pose is created by thresholding the Gaussian apertures used in the histogram creation. We thus define $d_B^2(H^{B_m}, H^B | \theta^{(n)})$ as the number of pixels where the two binary images do not agree.

The final objective of tracking is to provide an estimate of the current pose. When using Bayesian filtering, one such estimate is the maximum a posterior solution (MAP), which corresponds to finding the maximum of the filtering distribution. With our specific choice of MRF measurement model, the MAP solution will correspond to minimising the energy functional (16) with respect to the pose. Hence, the objective is to find the pose for which the earth mover's distance between the model histograms and the actual observation histograms is minimised.

5.2 Gaussian Limb Aperture

In the kinematic model each limb correspond to a line segment as was illustrated in Fig. 1. Hence, in order to create texture and colour histograms we need to define a spatial extend of each limb. This is done by forming a Gaussian aperture around each limb line segment. This aperture is used to weight individual pixel contributions in the histograms.

To compute the aperture, the line segment is projected onto the image plane and its mean point $\boldsymbol{\mu}_n$ is computed. From the orientation vector $\boldsymbol{v}_1$ along the projected line segment and its perpendicular counter-part $\boldsymbol{v}_2$ a covariance matrix

$$\boldsymbol{\Sigma}_n = \lambda_1 \boldsymbol{v}_1 \boldsymbol{v}_1^T + \lambda_2 \boldsymbol{v}_2 \boldsymbol{v}_2^T \tag{17}$$

is formed. The Gaussian aperture can then be defined as

$$L_n(\boldsymbol{x} | \theta^{(n)}) = \exp\left(-\frac{1}{2}(\boldsymbol{x} - \boldsymbol{\mu}_n)^T \boldsymbol{\Sigma}_n^{-1}(\boldsymbol{x} - \boldsymbol{\mu}_n)\right) . \tag{18}$$

The first eigenvalue λ_1 is chosen such that $2.5\sqrt{\lambda_1} = d/2$, where d is the length of the projected line segment. The second eigenvalue is computed as $\lambda_2 = w_n \lambda_1$, where w_n controls the width of the nth limb. The entire process is illustrated in Fig. 3.

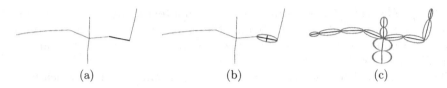

$$(a) \qquad\qquad (b) \qquad\qquad (c)$$

Fig. 3. An illustration of the Gaussian apertures. (a) The kinematic skeleton with one high-lighted limb. (b) The direction of the limb is computed and a Gaussian aperture stretching in this direction is formed. (c) All Gaussian apertures of the same pose. Notice that the apertures of different limbs can have different widths.

5.3   Texture and Colour Features

In each pixel we compute both colour and texture features. Specifically, we compute colour saturations and gradient orientations. The colour saturation is computed as the S-channel of the HSV representation of the image. Each pixel entry in the histograms are weighted with the value of the Gaussian limb aperture in the pixel. In more detail, we collect the colour saturation histogram as

$$H_k^C(C\{I\}|\theta^{(n)}) = \sum_{x=1}^{M} L_n(x|\theta^{(n)})\mathbf{1}_{\Delta_k}[C\{I\}(x)], \quad k = 1,\ldots,K \ , \qquad (19)$$

where k is the index of the histogram bin, $\mathbf{1}_{\Delta_k}$ is the indicator function of the bin interval Δ_k and $C\{I\}$ is the colour saturation of the image I.

The texture model is computed in much the same manner, except the gradient orientations are also weighted with the gradient lengths. Specifically, the histogram is computed as

$$H_k^T(\psi\{I\}|\theta^{(n)}) = \sum_{x=1}^{M} \beta(x)L_n(x|\theta^{(n)})\mathbf{1}_{\Delta_k}[\psi\{I\}(x)], \quad k = 1,\ldots,K \ , \qquad (20)$$

where β is the gradient length and ψ is its orientation. To make the features independent of the orientation of the limb, we compute the gradient in the principal coordinate system of the Gaussian aperture.

To control the relative importance of the features we compute the parameters α_T and α_C in an *ad-hoc* manner as

$$\alpha_T = \frac{\mathcal{H}(H^{C_m})}{\mathcal{H}(H^{T_m}) + \mathcal{H}(H^{C_m})} \quad \text{and} \quad \alpha_C = \frac{\mathcal{H}(H^{T_m})}{\mathcal{H}(H^{T_m}) + \mathcal{H}(H^{C_m})} \ , \qquad (21)$$

where $\mathcal{H}(\cdot)$ is the entropy of a histogram. The intuition behind this choice is that more peaked model histograms should have a greater influence on the likelihood. The importance α_B of the background model has been selected manually.

The training of the model requires collecting texture and colour histograms for each limb. This is done by letting the human in front of the camera take a known pose, which allows us to collect the histograms from the first frame. The known starting pose also provides us with an initial state for the motion analysis.

6   Discussion

In the previous sections two different motion analysis systems have been described: one working on full-pose space and one working in end-effector space. Both use the same measurement system and their prediction systems are as similar as possible. This enables a comparison of the two trackers. It should be noted that the visual measurements are by far the computationally most expensive part of the tracking. Since each particle requires one visual measurement, the number of particles is proportional to the final computational time.

Fig. 4 shows selected frames from an image sequence with tracking results superimposed.[1] The result is computed as the mean of all particles, as this seems to stabilise the estimate when only few particles are used. First, we tracked the motion in full-pose space using 100 particles. Here the system quickly looses track of one arm and produces a large amount of "jitter" in the motion estimation. The computations took approximately five minutes on standard PC hardware. We then increased the number of particles to 5000, which resulted in a successful tracking with only little jitter. Unfortunately, this required more than 10 hours of computation time. As a final experiment we ran the tracking in end-effector space using 25 particles. The result of this experiment is a successful tracking that is comparable in quality with the previous experiment, but with somewhat more jitter. This jitter is a direct consequence of the small amount of particles. If this is increased the jittering decreases. The computations took approximately five minutes.

The results show that tracking in end-effector space is possible and that large speed-ups can be achieved using this approach. When creating a tracker for use in a physiotherapeutic rehabilitation programme it is essential to have real-time performance in order to provide feedback to the patient. Tracking in end-effector space makes this requirement more plausible compared to tracking in full-pose space.

Tracking in end-effector space does provide speed-ups while it still allows for a large set of motions. The resulting tracker is thus more versatile than low dimensional trackers that are tuned towards very specific motion types, such as walking, golf or tennis playing and so forth. However, the tracking will most often be less precise when it is performed in end-effector space. When infering the full pose from the end-effectors, the inverse kinematics solver finds one out of several minima. Hence, a particle can get a low weight even if it is in the correct part of end-effector space due to limitations of the inverse kinematics solver. This problem will result in a loss of accuracy.

The choice of state space basically boils down to a choice between accuracy, versatility and speed. Tracking in full-pose space allows for a high accuracy and versatility, but sacrifices speed. By learning manifolds in full-pose space it is possible to track on these. This allows for high accuracy and speed, but sacrifices versatility as the manifolds can only describe single types of motion. Tracking in end-effector space allows for great versatility and speed, but comes with a loss of accuracy.

---

[1] The sequences are available on-line at http://humim.org/emmcvpr2009/

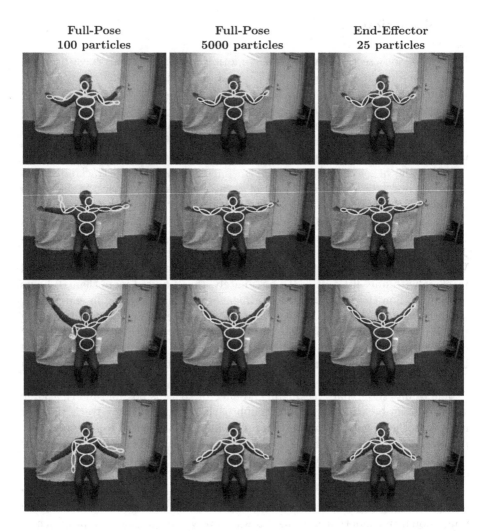

| Full-Pose
100 particles | Full-Pose
5000 particles | End-Effector
25 particles |

Fig. 4. Images from tracking sessions. The first column corresponds to tracking using 100 particles in full-pose space and the second column corresponds to 5000 particles in the same space. The third column corresponds to 25 particles in end-effector space. The rows correspond to the 32nd, 94th, 126th and the 196th frame of the sequence.

7   Conclusion and Future Work

In this paper we presented a low dimensional state space — the end-effector space — suitable for fast and versatile human motion estimation. Experiments with tracking in this space shows that good results can be achieved using only few particles. Due to the use of an inverse kinematics solver the approach can be expected to have less accuracy than when working in full-pose space, but makes real-time tracking of humans feasible.

In the immediate future we plan on improving the measurement model such that it can handle self-occlusions. The work of Sidenbladh and Black [19] and Roth et. al. [20] seems like good sources of inspiration. Also, a more detailed experimental validation will be performed. Specifically, we will validate our method on the ground truth data set of Knossow et.al. [21] and compare with their method, which should allow us to quantitatively evaluate the accuracy of our approach.

References

1. Poppe, R.: Vision-based human motion analysis: An overview. Computer Vision Image Understing 108(1-2), 4–18 (2007)
2. Cappé, O., Godsill, S.J., Moulines, E.: An overview of existing methods and recent advances in sequential Monte Carlo. Proceedings of the IEEE 95(5), 899–924 (2007)
3. Engell-Nørregård, M., Erleben, K.: A Projected Non-linear Conjugate Gradient Method for Interactive Inverse Kinematics. In: MATHMOD 2009 - 6th Vienna International Conference on Mathematical Modelling (2009)
4. Moeslund, T.B., Hilton, A., Krüger, V.: A survey of advances in vision-based human motion capture and analysis. Comput. Vis. Image Underst. 104(2), 90–126 (2006)
5. Wang, J.M., Fleet, D.J., Hertzmann, A.: Gaussian Process Dynamical Models for Human Motion. IEEE Transactions on Pattern Analysis and Machine Intelligence 30(2), 283–298 (2008)
6. Urtasun, R., Fleet, D.J., Fua, P.: 3D People Tracking with Gaussian Process Dynamical Models. In: CVPR 2006: Proceedings of the 2006 IEEE Computer Society, Conference on Computer Vision and Pattern Recognition, Washington, DC, USA, pp. 238–245. IEEE Computer Society, Los Alamitos (2006)
7. Lu, Z., Carreira-Perpinan, M., Sminchisescu, C.: People Tracking with the Laplacian Eigenmaps Latent Variable Model. In: Platt, J., Koller, D., Singer, Y., Roweis, S. (eds.) Advances in Neural Information Processing Systems 20, pp. 1705–1712. MIT Press, Cambridge (2008)
8. Sidenbladh, H., Black, M.J., Fleet, D.J.: Stochastic tracking of 3d human figures using 2d image motion. In: Vernon, D. (ed.) ECCV 2000. LNCS, vol. 1843, pp. 702–718. Springer, Heidelberg (2000)
9. Elgammal, A.M., Lee, C.S.: Tracking People on a Torus. IEEE Transaction on Pattern Analysis and Machine Intelligence 31(3), 520–538 (2009)
10. Urtasun, R., Fleet, D.J., Hertzmann, A., Fua, P.: Priors for people tracking from small training sets. In: Tenth IEEE International Conference on Computer Vision, ICCV 2005, vol. 1, pp. 403–410 (2005)
11. Loy, G., Eriksson, M., Sullivan, J., Carlsson, S.: Monocular 3D Reconstrunction of Human Motion in Long Action Sequences. In: Pajdla, T., Matas, J(G.) (eds.) ECCV 2004. LNCS, vol. 3024, pp. 442–455. Springer, Heidelberg (2004)
12. Sminchisescu, C., Triggs, B.: Kinematic Jump Processes for Monocular 3D Human Tracking. In: IEEE International Conference on Computer Vision and Pattern Recognition, pp. 69–76 (2003)
13. Engell-Nørregård, M., Erleben, K.: Estimation of Joint types and Joint Limits from Motion capture data. In: WSCG 2009: 17-th International Conference in Central Europe on Computer Graphics, Visualization and Computer Vision (2009)

14. Nocedal, J., Wright, S.J.: Numerical optimization. Springer Series in Operations Research. Springer, New York (1999)
15. Zhao, J., Badler, N.I.: Inverse kinematics positioning using nonlinear programming for highly articulated figures. ACM Trans. Graph. 13(4), 313–336 (1994)
16. Baum, L.E.: An inequality and associated maximization technique in statistical estimation of probabilistic functions of markov processes. Inequalities 3, 1–8 (1972)
17. Kalman, R.: A new approach to linear filtering and prediction problems. Transactions of the ASME-Journal of Basic Engineering 82(D), 35–45 (1960)
18. Rubner, Y., Tomasi, C., Guibas, L.: A metric for distributions with applications to image databases. In: Sixth International Conference on Computer Vision, pp. 59–66 (1998)
19. Sidenbladh, H., Black, M.J.: Learning image statistics for bayesian tracking. In: Proc. of International Conference on Computer Vision, Vancouver, Canada, vol. II, pp. 709–716 (2001)
20. Roth, S., Sigal, L., Black, M.J.: Gibbs likelihoods for bayesian tracking. In: CVPR 2004 (2004)
21. Knossow, D., Ronfard, R., Horaud, R.P.: Human motion tracking with a kinematic parameterization of extremal contours. International Journal of Computer Vision 79(2), 247–269 (2008)

Robust Segmentation by Cutting across a Stack of Gamma Transformed Images

Elena Bernardis[1] and Stella X. Yu[2]

[1]Department of Computer and Information Science
University of Pennsylvania, Philadelphia, PA 19104, USA
[2]Computer Science Department
Boston College, Chestnut Hill, MA 02467, USA

Abstract. Medical image segmentation appears to be governed by the global intensity level and should be robust to local intensity fluctuation. We develop an efficient spectral graph method which seeks the best segmentation on a stack of gamma transformed versions of the original image. Each gamma image produces two types of grouping cues operating at different ranges: Short-range attraction pulls pixels towards region centers, while long-range repulsion pushes pixels away from region boundaries. With rough pixel correspondence between gamma images, we obtain an aligned cue stack for the original image. Our experimental results demonstrate that cutting across the entire gamma stack delivers more accurate segmentations than commonly used watershed algorithms.

1 Introduction

Hair cells of the inner ear transduce mechanical signals into electrical signals [1]. Each hair bundle is composed of tens of stereocilia organized in an organ-pipe-like formation of increasing height (Fig. 1). Automatic segmentation of these stereocilia in their fluorescent images is vital for medical research on hearing.

Segmentation of such medical images often appears to be governed by global intensity levels, yet imaging noise and local intensity fluctuation presents considerable challenges. Two scenarios are illustrated in Fig. 2.

Morphological methods and energy-driven methods are widely used in medical image segmentation. While the former prescribes a local computational procedure, e.g. watershed algorithms [2,3], the latter involves the minimization of a global energy function formulated based on either regions [4,5] or contours, e.g. active contours [6] and level set methods [7].

While morphological methods are computationally efficient but prone to local noise, energy-driven approaches are computationally costly and critically dependent on initial seed solutions. Various techniques have been proposed to combine their benefits, e.g. watersnakes [8] and level sets for watershed [3].

Graph cuts methods have also been employed to overcome the limitations of watershed algorithms, which are essentially local segmentation methods. These include segmenting a single connected component with isoperimetric graph partitioning [9] and thin structures with augmented banded graph cuts [10].

D. Cremers et al. (Eds.): EMMCVPR 2009, LNCS 5681, pp. 249–260, 2009.

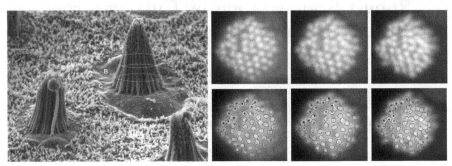

a: 3D view of hair cells   b: 2D fluorescent slices & their segmentations

Fig. 1. Stereocilia segmentation. **a)** Hair cells are composed of tens of stereocilia organized in an organ-pipe-like formation of increasing height.**b)** Fluorescent images (Row 1) and their segmentations (our results, Row 2) at multiple heights show the cross sections (e.g. A,B,C in **a**) of individual stereocilia (marked by colored dots).

We present a graph cuts approach that is robust to local intensity fluctuation and can extract several regions of interest without any user initialization. We encode the impact of high and low intensities, which we will refer to as peaks and valleys, in pairwise grouping cues that encourage peak regions to stay together and valley regions to divide apart. It is the job of global integration to decide where region boundaries should be.

Our key idea is that regions of an image appear stable with respect to the gamma transformation of the image, while cues in each gamma transformed version reflect an ever changing balance between peaks and valleys, as peaks shrink and valleys expand with an increasing gamma. The desired segmentation must be the global consensus of local cues from a stack of these gamma images.

Illustrated in Fig. 3, given an image I, we first create several gamma transformed versions: $I_n = I^{\gamma_n}$. For each I_n, we define two complementary local grouping cues: a short-range attraction between nearby pixels with similar intensities and a long-range repulsion between distant pixels with similar intensities but separated by valleys. The former occurs most likely for pixels belonging to the same stereocilium and the latter for pixels belonging to adjacent stereocilia. Large repulsion demands single boundaries to occur somewhere between two distant pixels, whereas large attraction discourages the formation of boundaries between two nearby pixels, preventing the oversegmentation problems in Fig. 2. We establish rough local alignment between gamma images and project cues derived from each I_n to the original image I through pixel correspondences. We seek the optimal graph cuts across the cue stack of attraction and repulsion, producing segmentation X_k for the original image I at a granularity determined by the number of eigenvectors k.

We will address the integration of multiple cues in Section 2, formulate our pairwise grouping cues for stereocillia images in Section 3, present experimental results in Section 4, and conclude the paper in Section 5.

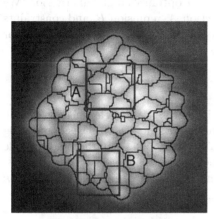

watershed segmentation

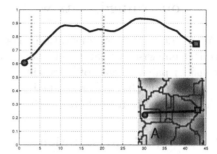

A: intensity fluctuation at boundaries

B: intensity fluctuation inside regions

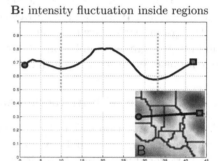

Fig. 2. Local intensity fluctuation presents considerable challenges in medical image segmentation. **A)** Fluctuation at boundaries weakens the separation between two intensity peaks. **B)** Fluctuation inside regions tends to break up an otherwise well defined intensity peak. Both cases cause oversegmentations in watershed approaches. The solid black line plots the 1D intensity profile along the line connecting the two pixels in the inset, which shows the image in a labeled window on the left. The dotted green lines mark the desired boundaries between intensity peaks.

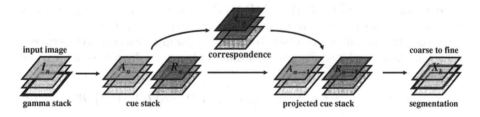

Fig. 3. Method Overview. Given an image, we build a stack of its gamma transformed versions, i.e., $I_n = I^{\gamma_n}$. For each gamma image I_n, we derive pairwise attraction A_n and repulsion R_n between pixels. We compute pixel correspondences C_n between adjacent gamma layers, and project cues at each layer to the reference layer I_1: $A_{n \to 1}$ and $R_{n \to 1}$. Cutting across the aligned cue stack produces segmentation X_k that is invariant to gamma transformations, k indicating the granularity of segmentation.

2   Constrained Cuts with Attraction and Repulsion

We formulate the segmentation in a spectral graph-theoretic framework. We collect pairwise cues and seek the solution that optimizes a global criterion. We consider pairwise cues of three kinds: attraction A, repulsion R, and constraints U. These cues have been studied separately in [11,12,13]. We combine them for the first time in a single framework.

2.1   Graph Representation

In spectral graph methods, an image I is represented by a weighted graph $G(V, E, W)$, where V denotes the set of nodes, E the set of edges connecting the nodes, and W the weights attached to edges. A pixel then becomes a node in the graph, each pairwise grouping cue becomes a weight between two nodes, and image segmentation becomes a graph node partitioning problem: We seek k partitions of node set V such that $V = \cup_{l=1}^{k} V_l$ and $V_i \cap V_j = \emptyset$, $\forall i \neq j$.

2.2   Criterion with Attraction and Repulsion

A good segmentation should have strong within-group attraction and between-group repulsion, and weak between-group attraction and within-group repulsion.

Characterizing this intuition with *linkratio* allows us to achieve both objectives simultaneously [14]. *linkratio* L of two node sets (P, Q) measures the fraction of connections from P to Q among all the connections that P has:

$$\text{linkratio} \qquad L(P, Q; W) = \frac{C(P, Q; W)}{C(P, V; W)} \qquad (1)$$

$$\text{connections} \qquad C(P, Q; W) = \sum_{i \in P, j \in Q} W(i, j) \qquad (2)$$

In particular, we have $L(P, P; W) + L(P, V \setminus P; W) = 1$, i.e. maximizing a within-group *linkratio* is equivalent to minimizing its between-group *linkratio*.

We seek to maximize *linkratios* from within-group attraction and between-group repulsion, combined linearly according to their total degree of connections:

$$\text{max} \qquad \varepsilon = \sum_{l=1}^{k} \alpha\, L(V_l, V_l; A) + (1 - \alpha)\, L(V_l, V \setminus V_l; R) \qquad (3)$$

$$\text{where} \qquad \alpha = \frac{C(V_l, V; A)}{C(V_l, V; A) + C(V_l, V; R)} \qquad (4)$$

α is a number between 0 and 1, indicating the total degree of attraction. $1 - \alpha$ indicates the total degree of repulsion.

2.3   Partial Grouping Constraints

We represent the partitioning by partition indicator $X = [X_1, \ldots, X_k]$, where X_l is an $N \times 1$ binary indicator for partition V_l, $X_l(i) = 1$ if pixel $i \in V_l$, and 0 otherwise, $l = 1, \ldots, k$. N is the number of pixels in the image.

We consider partial grouping constraints which require pixels a and b to belong in the same region, i.e. $X(a) = X(b)$. The collection of c such constraints can be written as $U^T X = 0$, where U is an $N \times c$ matrix, and each column of U has only two non-zero numbers, $+1$ and -1.

2.4   Optimal Solution

Our criterion ε with pairwise attraction A and pairwise repulsion R, subject to grouping constraints U can be written in a compact matrix form:

$$\text{maximize} \qquad \varepsilon(X) = \sum_{l=1}^{k} \frac{X_l^T W X_l}{X^T D X_l} \tag{5}$$

$$\text{subject to} \qquad X \in \{0,1\}^{N \times k}, \; X 1_k = 1_N \tag{6}$$

$$U^T X = 0 \tag{7}$$

$$\text{where} \qquad W = A - R + D_R \tag{8}$$

$$D = D_A + D_R \tag{9}$$

1_n denotes the $n \times 1$ vector of all 1's. $D_W = \text{Diag}(W 1_N)$ is an $N \times N$ diagonal matrix, and its diagonal contain the total degree of W connections for each node.

Relaxing the binary constraints, we can solve this optimization problem [11] with the eigenvectors of $H D^{-1} W H$, where $H = I - D^{-1} U (U^T D^{-1} U)^{-1} U^T$. We then discretize the eigenvectors to obtain the final segmentation [14].

3   Pairwise Grouping Cues from Image Intensities

The success of global integration depends on the local cues that feed into it. We define a short-range attraction that pulls pixels towards region centers, a long-range repulsion that pushes pixels away from region boundaries, and partial grouping constraints that force peripheral background pixels to belong together. With pixel correspondence between gamma images, we obtain a cue stack for the original image.

3.1   Short-Range Attraction within Individual Peaks

Attraction $A(i,j)$ between pixels i and j encodes local intensity similarity. The straightforward definition

$$A(i,j) = e^{-\frac{|I_i - I_j|^2}{2\sigma_a^2}} \tag{10}$$

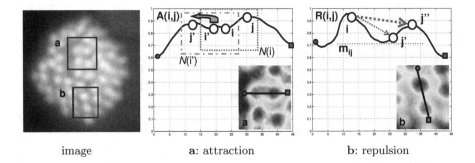

image   a: attraction   b: repulsion

Fig. 4. Pairwise attraction and repulsion. **a)** Our attraction is adaptive to the local intensity range within each neighborhood $\mathcal{N}(i)$, so that $A(i',j') \approx A(i,j)$, enhancing the discrimination of two nearby similar peaks. **b)** Our repulsion is strongest for nearby peaks and gets reduced as two pixels approach the inbetween valley: $R(i,j'') > R(i,j')$. m_{ij} is the minimal intensity level between pixels i and j.

requires fine parameter tuning and tends to merge nearby peaks of similar intensities. We introduce a new definition that is asymmetrical between two pixels and acts to pull pixels towards intensity peaks.

For pixels i and j, $A(i,j)$ is inversely proportional to the the maximal intensity difference between i and any pixel on the line ij, with sensitivity regulated by local intensity range δ_i in i's neighborhood $\mathcal{N}(i)$:

$$A(i,j) = e^{-\frac{\max_{t\in\text{line}(i,j)} |I_i - I_t|^2}{2\delta_i^2 \cdot \sigma_a^2}} \tag{11}$$

$$\delta_i = \max_{t\in\mathcal{N}(i)} I_t - \min_{t\in\mathcal{N}(i)} I_t \tag{12}$$

We choose $\mathcal{N}(i)$ to be slightly larger than a stereocilium so that δ_i is estimated between the peak and surrounding valleys (Fig. 4a). With adaptive scaling by local intensity range δ_i, $A(i,j)$ effectively enhances attraction within weak peaks and allows a single parameter setting for σ_a to work on a variety of images.

3.2   Long-Range Repulsion between Peaks

Adjacent peaks provide a strong cue as to where the boundaries should lie. This cue is encoded by long-range repulsion. Intuitively, two pixels of similar intensity should belong to different peaks if they are separated by a valley. We define repulsion $R(i,j)$ between pixels i and j to be proportional to the difference with the minimal intensity m_{ij} encountered on the line ij:

$$R(i,j) = 1 - e^{-\frac{\min(|I_i - m_{ij}|, |I_j - m_{ij}|)}{\sigma_r}} \tag{13}$$

$$m_{ij} = \min_{t\in\text{line}(i,j)} I_t. \tag{14}$$

The farther away the pixels are from the valley, the larger the intensity difference with the minimum, and the larger the repulsion (Fig. 4b).

3.3 Pixel Correspondence and Cue Projection

With each gamma transformation, while peaks remain peaks and valleys remain valleys, their regions of influence change: Peaks shrink and valleys expand; pixels belonging to one peak region could become part of the background. Local grouping cues derived from gamma images consequently do not completely agree with each other. We establish rough pixel correspondence and project cues on individual gamma image back to the original image.

Let $A_n(i,j)$ be the affinity (i.e. attraction) between pixels i and j at gamma image I_n. We follow the approach in [15] by computing the corresponding pixel location $C_n(i)$ as the center of mass of i's affinity field and composing them recursively to obtained the cue stack for the original image $I = I_1$:

$$A_{n\to1}(i,j) = A_n(C_n(i), C_n(j)) \tag{15}$$

$$R_{n\to1}(i,j) = R_n(C_n(i), C_n(j)) \tag{16}$$

$$C_n(i) = \sum_{j\in N(i)} A_n(i,j)C_{n-1}(i) \tag{17}$$

where $C_1(i)$ is pixel i's location in the original image I.

Cutting across the cue stack is equivalent to cutting a single graph with the following total attraction A and total repulsion R:

$$A = \sum_n D_{A,n}^{-1}A_{n\to1} + A_{n\to1}D_{A,n}^{-1} \tag{18}$$

$$R = \sum_n D_{R,n}^{-1}R_{n\to1} + R_{n\to1}D_{R,n}^{-1}. \tag{19}$$

where $D_{A,n}$ and $D_{R,n}$ are the degree matrices for A_n and R_n respectively.

3.4 Partial Grouping Constraints

We obtain a crude background mask by intensity thresholding on the original image. This mask is translated into our graph cuts framework as partial grouping constraints where two pixels in the peripheral background must belong together in the final segmentation. We form the constraint matrix U from the collections of these pairwise grouping constraints.

3.5 Algorithm

Given image I, we compute a segmentation using the following procedure:

1. Build a gamma image stack where $I_n = I^{\gamma_n}$ and $I_1 = I$;
2. For each gamma image I_n,
   - (2.1) compute attraction A_n and repulsion R_n,
   - (2.2) compute pixel correspondence C_n,
   - (2.3) compute $A_{n\to1}, R_{n\to1}$ by projecting A_n, R_n to the original image I;

3. Compute total attraction A and repulsion R by collapsing the stack;
4. Form partial grouping constraints U from a background mask;
5. Solve the eigenvectors of weights $W = A - R + D_R$ with constraints U;
6. Obtain a discrete segmentation from the eigenvectors.

4   Experiments

We implement our algorithm in MATLAB. The same set of parameters are used for all our images ($\sim 300 \times 300$): $\gamma = \{1, 2, 4\}$, $\sigma_a = 0.3$, $\sigma_r = 2\sigma_a$, neighbourhood radius 8 and 16 for attraction and repulsion respectively. We choose the number of eigenvectors k according to the expected number of stereocilia in the image.

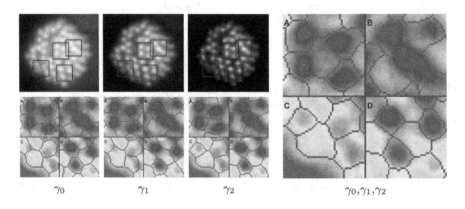

γ_0 $\qquad\qquad$ γ_1 $\qquad\qquad$ γ_2 $\qquad\qquad\qquad$ $\gamma_0, \gamma_1, \gamma_2$

Fig. 5. Better segmentation is obtained by cutting across the gamma stack instead of a single gamma image. Left shows 3 individual gamma images and their segmentations in 4 labeled windows. Right shows the segmentations based on all 3 γ images.

Fig. 5 shows that better segmentation is achieved by integrating cues over the entire gamma stack instead of an individual gamma image. Single peaks originally faint or without clear boundaries are enhanced in gamma transformed images. However, with an increasing gamma, valleys are widened and boundaries become less precise. Cutting across the gamma image stack allows segmenting out weak peaks while retaining precise boundaries throughout the image.

Fig. 6 shows our coarse to fine segmentations. When the number of eigenvectors k is small, our segmentations resemble the watershed results. However, our segmentations are not disrupted by local intensity fluctuation and do not break up salient peaks. When k increases, our segmentions locate each peak with tighter delineation. Our method successfully segments out weak peaks without utilizing the near regularity of the spatial layout of stereocilia.

Fig. 7 shows additional results on images of poor imaging quality.

We measure the goodness of segmentation by scoring it with respect to the ground-truth center locations of stereocilia. Let `disk`(i) denote a disk of some

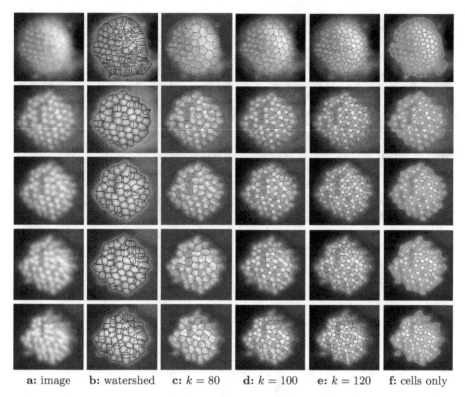

a: image   **b:** watershed   **c:** $k = 80$ **d:** $k = 100$ **e:** $k = 120$ **f:** cells only

Fig. 6. Coarse-to-fine stereocilia segmentations. For each image (Column **a**), we show watershed segmentations (Column **b**) and our results (Columns **c-e**) as the number of eigenvectors k increases. Extracted stereocilia (Column **f**) show that our method is robust to local intensity fluctuation, can discover weak peaks and precise boundaries.

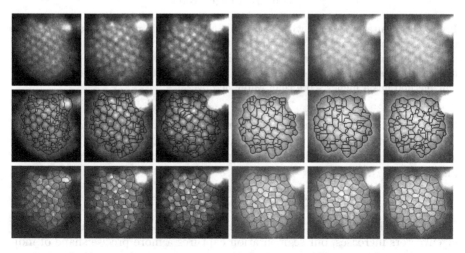

Fig. 7. Our method works equally well on noisy and low-contrast images. $k = 40$. Rows 1-3 show images, watershed results and our results respectively.

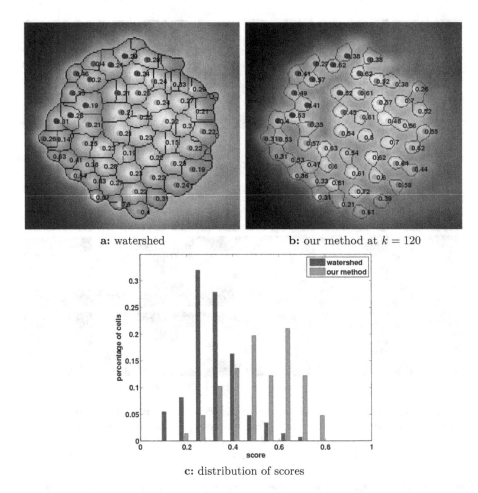

a: watershed   **b:** our method at $k = 120$

c: distribution of scores

Fig. 8. Segmentation scores with respect to ground-truth stereocilia centers. These center locations are marked by colored dots. Each number indicates the score of a particular segment that contains a stereocilium center. **a** and **b** show a score example for watershed and our method. **c** shows the distribution of scores from all the images. Our method has a higher score than watershed overall.

fixed radius throughout the haircell bundles, located at stereocilium center i. Let $\mathtt{segment}(i)$ denote the segment of maximal overlap with $\mathtt{disk}(i)$. Our score is a number between 0 and 1, measuring the extent of overlap between $\mathtt{disk}(i)$ and $\mathtt{segment}(i)$:

$$\mathtt{score}(i) = \frac{\mathtt{disk}(i) \cap \mathtt{segment}(i)}{\mathtt{disk}(i) \cup \mathtt{segment}(i)}. \tag{20}$$

The higher the score, the more precise the segmentation. As the number of eigenvectors increases, our segmentation captures a more precise shape of individual stereocilia. Fig. 8 shows with both image examples and statistics that our method overall scores higher than watershed.

Our method segments the background into multiple valley regions, which are of little interest to medical researchers. By requiring the mean intensity in the region center to be higher than the periphery, we get rid of valleys and automatically extract the stereocilia, as shown in Fig. 6f.

5   Conclusions

The segmentation of medical images appears to be governed by the global intensity level, yet local intensity fluctuation poses considerable challenges to both local methods such as watershed and global methods such as level sets.

We develop a spectral graph-theoretic method which finds the best segmentation on a stack of gamma transformed versions of the original image. Each gamma image produces two kinds of local grouping cues: short-range intensity similarity cues that pull pixels towards stereocilia centers, and long-range intensity difference cues that push pixels away from stereocilia boundaries. We obtain a cue stack for the original image using pixel correpondences between gamma images. We then seek the optimal graph cuts across the aligned cue stack which maximize within-group attraction and between-group repulsion. The near-global optimal solution can be found efficiently using eigendecomposition.

Our method has only a few parameters and requires little tuning. We obtain accurate and robust results on a variety of images with the same set of parameters, demonstrating the advantage of cutting across the entire gamma stack instead of the original image or any gamma image alone, and achieving better performance than watershed algorithms.

The segmentation issues we investigate in this paper are not restricted to stereocilia images. Our approach of making a global decision based on two types of local cues operating at different spatial ranges and from multiple gamma images provides a robust and efficient alternative to watershed or level set methods in many medical image applications.

References

1. Vollrath, M.A., Kwan, K.Y., Corey, D.P.: The micromachinery of mechanotransduction in hair cells. Annual Review of Neuroscience 30, 339–365 (2007)
2. Meyer, F.: Topographic distance and watershed lines. Signal Process. 38(1), 113–125 (1994)
3. Tai, X.-C., Hodneland, E., Weickert, J., Bukoreshtliev, N.V., Lundervold, A., Gerdes, H.-H.: Level set methods for watershed image segmentation. In: Sgallari, F., Murli, A., Paragios, N. (eds.) SSVM 2007. LNCS, vol. 4485, pp. 178–190. Springer, Heidelberg (2007)
4. Mumford, D., Shah, J.: Optimal approximations by piecewise smooth functions and associated variational problems. Comm. Pure Math., 577–684 (1989)
5. Geman, S., Geman, D.: Stochastic relaxation, gibbs distributions, and the bayesian restoration of images, 452–472 (1990)
6. Xu, C., Prince, J.L.: Snakes, shapes, and gradient vector flow. IEEE Transactions on Image Processing (1998)

7. Malladi, R., Sethian, J.A.: Level set methods for curvature flow, image enchancement, and shape recovery in medical images. In: Proc. of Conf. on Visualization and Mathematics, pp. 329–345. Springer, Heidelberg (1997)
8. Nguyen, H.T., Worring, M., van den Boomgaard, R., Worring, H.T.N.M.: Watersnakes: Energy-driven watershed segmentation (2003)
9. Grady, L.: Fast, quality, segmentation of large volumes – isoperimetric distance trees. In: Leonardis, A., Bischof, H., Pinz, A. (eds.) ECCV 2006. LNCS, vol. 3953, pp. 449–462. Springer, Heidelberg (2006)
10. Sinop, A.K., Grady, L.: Accurate banded graph cut segmentation of thin structures using laplacian pyramids. In: Larsen, R., Nielsen, M., Sporring, J. (eds.) MICCAI 2006. LNCS, vol. 4191, pp. 896–903. Springer, Heidelberg (2006)
11. Yu, S.X., Shi, J.: Segmentation given partial grouping constraints. PAMI 26(2), 173–183 (2004)
12. Yu, S.X., Shi, J.: Understanding popout through repulsion. In: CVPR, Kauai Marriott, Hawaii, USA, December 9-15 (2001)
13. Yu, S.X., Shi, J.: Segmentation with pairwise attraction and repulsion. In: ICCV, Vancouver, Canada, July 9-12 (2001)
14. Yu, S.X., Shi, J.: Multiclass spectral clustering. In: ICCV, Nice, France, October 11-17 (2003)
15. Yu, S.X.: Segmentation induced by scale invariance. In: CVPR, San Diego, June 20-26 (2005)

Integrating the Normal Field of a Surface in the Presence of Discontinuities

Jean-Denis Durou[1,2], Jean-François Aujol[1], and Frédéric Courteille[3]

[1] CMLA, ENS Cachan, CNRS, Universud, Cachan, France
[2] IRIT, Université Paul Sabatier, Toulouse, France
[3] LGC, Université Paul Sabatier, Toulouse, France

Abstract. We show how to integrate the normal field of a surface in the presence of discontinuities by three different ways. We obtain very satisfactory 3D-reconstructions, from the point of view of the accuracy of the reconstructions. As an important consequence, no prior segmentation of the scene into parts without discontinuity is required anymore. Finally, we test the three proposed methods of integration in the framework of photometric stereo, a technique which aims at computing the normal field of a scene surface from several images of this scene lighted under different directions.

1 Introduction

Computing the 3D-shape of a surface from a collection of normals is not so straightforward as it could appear, even in the case of a dense normal field i.e., when the normal to the surface is known at each pixel of an image. This classical problem of 3D-reconstruction, which is usually called normal field integration, has been solved using either the calculus of variations [1], direct integration [2] or frequency-domain methods [3,4]. In a previous work [5], we improved the original algorithm by Horn and Brooks [1] in two ways: we showed that the knowledge of the height on the boundary, a knowledge which is usually not available, is not required; we also showed how to take perspective into account. In the present paper, we propose a novel improvement of this algorithm which is compatible with the previous ones: we show how to deal with discontinuous surfaces, a situation which occurs in practice as soon as there are occlusions. This improvement allows us to integrate the normal field on a whole dense normal field, without need for prior segmentation into several parts without discontinuity.

In Section 2, we recall the basic equations of normal integration. In Section 3, Horn and Brooks' algorithm and our previous improvements are presented. In Section 4, three new methods of integration of a normal field are exhibited, compared and tested on a normal field computed by photometric stereo from real images. Section 5 summarizes the main contributions of the paper.

2 Basic Equations of Normal Integration

Suppose that, in each point $Q = (x, y)$ in the image of a surface $\mathcal{S}$, we know the unit outgoing normal $\mathbf{n}(x, y) = [n_X(x, y), n_Y(x, y), n_Z(x, y)]^t$. Then, the function

D. Cremers et al. (Eds.): EMMCVPR 2009, LNCS 5681, pp. 261–273, 2009.

n is a dense normal field. Integrating a normal field **n** consists in searching for the shape $\mathcal{S}$ i.e., for three functions X, Y and Z such that the normal to $\mathcal{S}$ at the point $P = [X(x,y), Y(x,y), Z(x,y)]^t$ conjugated with Q is **n**(x,y). Due to the lack of space, no rigorous state-of-the-art on the integration of a normal field is done (see e.g. [6,7,8]). Let us also cite [9], in which the problem of integrating a sparse normal field is addressed.

Let us first recall the fundamental equations of normal integration. For the sake of simplicity, we will omit the dependences in (x,y). Either for orthographic or for perspective projection, it is easy to show that X and Y can be deduced from Z. Under the assumption of orthographic projection, the depth Z can be computed thanks to the following elementary PDE:

$$\nabla Z = \frac{1}{g}\,[p, q]^t, \tag{1}$$

where $p = -n_X/n_Z$ and $q = -n_Y/n_Z$, and g denotes the image magnification. Thus, the problem of integrating a normal field is that of integrating the gradient of Z. It has been shown in [5] that a strict analogy exists between the perspective case and the orthographic case, provided that a change in the unknown is done:

$$\mathcal{Z} = \ln |Z|. \tag{2}$$

The new unknown $\mathcal{Z}$ satisfies the following PDE, which is similar to (1):

$$\nabla \mathcal{Z} = [r, s]^t, \tag{3}$$

with the following definitions of r and s:

$$r = -\frac{n_X}{x\,n_X + y\,n_Y + f\,n_Z},$$
$$s = -\frac{n_Y}{x\,n_X + y\,n_Y + f\,n_Z}, \tag{4}$$

where f denotes the focal length of the camera. Here again, the problem of integrating a normal field is that of integrating the gradient of $\mathcal{Z}$.

In order to ensure that the normal field is integrable i.e., that Eqs. (1) or (3) can be integrated whatever the integration path, it is necessary and sufficient that p and q (in the orthographic case) or r and s (in the perspective case) satisfy the Schwartz equations $\partial p/\partial y = \partial q/\partial x$ or $\partial r/\partial y = \partial s/\partial x$. In practice, a normal field is never rigorously integrable. There are two classical ways of dealing with this problem. The first one consists in using several integration paths, and then to mean the different integrals [2]. The second solution consists in considering Eqs. (1) or (3) as optimization problems [1]. Apart from their slowness, these last methods of integration have two main advantages: on the one hand, they are more robust to noise; on the other hand, in the case where the Schwartz equation is not satisfied, they provide however an acceptable shape. In the following of the paper, we will focus on this second solution. It is noteworthy that considering the orthographic case is enough, since Eqs. (1) and (3) are similar. The only difference is that Eq. (3) requires the knowledge of the focal distance f, as well as the location of the principal point, because these parameters occur in the definitions (4) of r and s: explicitly for f; implicitly for the location of the principal point, since the coordinates x and y of a pixel depend on it.

3   Integration Using Quadratic Regularization

3.1   Continuous Formulation

For the sake of simplicity, let us suppose that $g = 1$. The resolution of Eq. (1) using quadratic regularization amounts to minimizing the following functional:

$$\mathcal{Q}(Z) = \iint\limits_{(x,y)\in\Omega} \|\nabla Z(x,y) - \mathbf{v}(x,y)\|^2 \, dx \, dy, \tag{5}$$

where Ω is the "domain of reconstruction", $\nabla Z = [Z_x, Z_y]^t$ stands for the gradient of Z, and $\mathbf{v} = [p, q]^t$ is the "reduced normal field" i.e., the datum. Quadratic regularization is known to work well in the case of smooth surfaces. A straightforward computation gives:

$$\nabla \mathcal{Q}(Z) = -2 \operatorname{div}(\nabla Z - \mathbf{v}). \tag{6}$$

It follows that the Euler-Lagrange equation associated to $\mathcal{Q}(Z)$ is:

$$\operatorname{div} \nabla Z = \operatorname{div} \mathbf{v}. \tag{7}$$

This is a Poisson equation, which is easy to solve, even analytically [10]. Nevertheless, solving Eq. (7) is equivalent to searching for an extremum of $\mathcal{Q}(Z)$ only if Z is constrained on the boundary $\partial\Omega$ of Ω. Otherwise, this equation has to be complemented with the "natural boundary equation" at each point of the boundary $\partial\Omega$, which is here the Neumann boundary condition $(\nabla Z - \mathbf{v}) \cdot \mathbf{N} = 0$, where the vector $\mathbf{N}$ is normal to $\partial\Omega$ in the image plane.

3.2   Improved Horn and Brooks' Scheme

Horn and Brooks propose in [1] a resolution of Eq. (7) that comes from the following approximation of the expression (5) of $\mathcal{Q}(Z)$:

$$\mathcal{E}(\mathbf{Z}) = \sum\sum_{(i,j)\in\Omega'} \left[\frac{Z_{i+1,j} - Z_{i,j}}{\delta} - \frac{p_{i+1,j} + p_{i,j}}{2} \right]^2 + \left[\frac{Z_{i,j+1} - Z_{i,j}}{\delta} - \frac{q_{i,j+1} + q_{i,j}}{2} \right]^2. \tag{8}$$

In this expression, δ denotes the distance between neighbouring pixels, Ω' the set of pixels $(i,j) \in \Omega$ such that $(i+1,j) \in \Omega$ and $(i,j+1) \in \Omega$, and $\mathbf{Z}$ the vector $[Z_{i,j}]_{(i,j)\in\mathring{\Omega}}$, where $\mathring{\Omega} = \Omega \backslash \partial\Omega$ is the set of pixels $(i,j) \in \Omega$ whose four nearest neighbours are in Ω. The values $Z_{i,j}$ of the pixels lying on $\partial\Omega$ are not considered as unknowns, since Horn and Brooks use a Dirichlet boundary condition. For the sake of simplicity, let us suppose that $\delta = 1$. For a pixel $(i, j) \in \overset{\circ}{\Omega}$, one gets from the characterization $\nabla \mathcal{E} = 0$ of an extremum and from (8):

$$4 \, Z_{i,j} - (Z_{i+1,j} + Z_{i,j+1} + Z_{i-1,j} + Z_{i,j-1}) + \frac{p_{i+1,j} - p_{i-1,j} + q_{i,j+1} - q_{i,j-1}}{2} = 0.$$

$$(9)$$

This equation is a discrete approximation of Eq. (7). Horn and Brooks solve Eq. (9) using the following iteration [1]:

$$Z_{i,j}^{k+1} = \frac{Z_{i+1,j}^k + Z_{i,j+1}^k + Z_{i-1,j}^k + Z_{i,j-1}^k}{4} - \frac{p_{i+1,j} - p_{i-1,j} + q_{i,j+1} - q_{i,j-1}}{8}.$$

$$(10)$$

The initialization is not a cause for concern, since the functional $\mathcal{Q}(Z)$ is convex. In our experiments, we use $Z^0 = 0$.

In order to avoid the need for Z on the boundary, it suffices to consider that all the values $Z_{i,j}$, for $(i, j) \in \Omega$, are unknowns. This implies that the equations $\partial \mathcal{E} / \partial Z_{i,j} = 0$, for $(i, j) \in \partial \Omega$, are not written under the form (9). For example, if $(i, j) \in \Omega' \cap \partial \Omega$, then (10) has to be replaced with:

$$Z_{i,j}^{k+1} = \frac{Z_{i+1,j}^k + Z_{i,j+1}^k}{2} - \frac{p_{i+1,j} + p_{i,j} + q_{i,j+1} + q_{i,j}}{4}.$$

$$(11)$$

In fact, equations such as (11) are nothing else than a discrete version of the natural boundary condition. This improvement of Horn and Brooks' scheme [5] is denoted IS_{L_2}.

3.3   Limits of the Improved Horn and Brooks' Scheme

Some of the computer vision techniques for 3D-reconstruction, as shape-from-shading, photometric stereo or shape-from-texture, first compute a normal field, and then need to integrate this normal field. Among them, photometric stereo is particularly interesting, since the computation of the normals is well-posed as soon as at least three images, taken using the same camera pose but different lightings, are available. Therefore, photometric stereo, a technique which has known a renewal in the last years [8,11,12], is well indicated to evaluate the methods of integration.

In [5], IS_{L_2} was tested on three photographs of a Beethoven's bustle (see Fig. 1-left) which are available on the web[1]. Moreover, estimates of the three lightings are provided as well. The computed shape is qualitatively very good (see Fig. 1-right). Nevertheless, the goal of this paper is to propose some improvements for scheme IS_{L_2}. In fact, it is well-known that quadratic regularization is not well adapted to discontinuities. Let us now test IS_{L_2} on the reduced normal field $\mathbf{v}_b$ of the benchmark surface $\mathcal{S}_b$ shown in Fig.2-left. The 3D-reconstruction which is obtained after 100×128 iterations of IS_{L_2} is qualitatively very bad (see Fig. 2-right). Nevertheless, we will see further the usefulness of this final

---

[1] http://www.ece.ncsu.edu/imaging/Archives/ImageDataBase/Industrial/

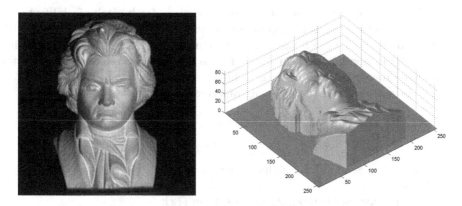

Fig. 1. Left: one of the three photographs of a Beethoven's bustle lighted under different directions. Right: 3D-reconstruction obtained from these three photographs using photometric stereo at each pixel on the bustle, then integrating the computed normal field using IS_{L_2} (the depth Z of the background is arbitrarily put to 0).

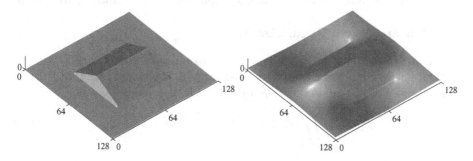

Fig. 2. Left: benchmark surface $\mathcal{S}_b$ with discontinuous depth Z_b and discontinuous reduced normal field $\mathbf{v}_b$. Right: 3D-reconstruction obtained using IS_{L_2}: $\|\Delta Z\|_2 = 2.48$. The integration is performed on the whole domain $[1, 128] \times [1, 128]$. This depth function is denoted Z_{L_2}.

configuration Z_{L_2}. More precisely, let us introduce, as a quantitative evaluation of the reconstructions, the minimal root mean square error $\|\Delta Z\|_2$ between a 3D-reconstruction and the ground truth i.e., the root mean square error which corresponds to their best matching. Between Z_{L_2} and the ground truth Z_b of $\mathcal{S}_b$, we find $\|\Delta Z\|_2 = 2.48$. In the following of the paper, the use of other regularizers will allow us to reach much lower values for $\|\Delta Z\|_2$.

4   Integration Using Non-quadratic Regularization

4.1   Introduction

By analogy with regularization methods in image processing, it is tempting to consider other regularization choices. In image restoration, quadratic regularization is indeed famous for its ease of use, but notorious for its lack of ability to recover sharp edges. It has been proposed to use regularization functions ϕ which both smooth the data in homogeneous regions but keep sharp edges by avoiding smoothing in non-homogeneous regions (see e.g. [13] and references therein for a detailed review of such methods in image restoration).

Let us now consider the following functional:

$$\mathcal{F}_\phi(Z) = \iint\limits_{(x,y)\in\Omega} \phi(\|\nabla Z(x,y) - \mathbf{v}(x,y)\|) \, dx \, dy. \tag{12}$$

Of course, this general form includes the quadratic regularization case (5) when $\phi(s) = s^2$.

A straightforward computation gives, from (12):

$$\nabla \mathcal{F}_\phi(Z) = -\mathrm{div}\left(\frac{\phi'(\|\nabla Z - \mathbf{v}\|)}{\|\nabla Z - \mathbf{v}\|}(\nabla Z - \mathbf{v})\right). \tag{13}$$

The optimality condition (Euler-Lagrange equation) $\nabla \mathcal{F}_\phi(Z) = 0$ can then be used to compute a numerical solution, as explained in the following subsections.

4.2   Linear Growth Regularization

For the sake of clarity, we detail here a first example of non-quadratic regularization. Let us consider the case of a linear growth functional [14], that is let us choose $\phi(s) = \sqrt{s^2 + \alpha^2}$ in functional (12) which then reads:

$$\mathcal{L}(Z) = \iint\limits_{(x,y)\in\Omega} \sqrt{\|\nabla Z(x,y) - \mathbf{v}(x,y)\|^2 + \alpha^2} \, dx \, dy. \tag{14}$$

Ideally, we would choose $\alpha = 0$. For image restoration, $\alpha = 1$ is a good choice when the greylevel values are in $[0, 255]$. The gradient (13) of the functional in this particular case becomes:

$$\nabla \mathcal{L}(Z) = -\mathrm{div}\left(\frac{\nabla Z - \mathbf{v}}{\sqrt{\|\nabla Z - \mathbf{v}\|^2 + \alpha^2}}\right). \tag{15}$$

Therefore, the Euler-Lagrange equation associated with the functional $\mathcal{L}(Z)$ is:

$$\mathrm{div}\left(\frac{\nabla Z}{\sqrt{\|\nabla Z - \mathbf{v}\|^2 + \alpha^2}}\right) = \mathrm{div}\left(\frac{\mathbf{v}}{\sqrt{\|\nabla Z - \mathbf{v}\|^2 + \alpha^2}}\right), \tag{16}$$

with Neumann boundary condition on $\partial\Omega$. Eq. (16) is much less tractable than Eq. (7). A way of computing its solution consists in using a semi-implicit scheme: we implicit the linear part of the equation, but its non-linear parts remain explicit. This gives:

$$\mathrm{div}\left(\frac{\nabla Z^{k+1}}{\sqrt{\|\nabla Z^k - \mathbf{v}\|^2 + \alpha^2}}\right) = \mathrm{div}\left(\frac{\mathbf{v}}{\sqrt{\|\nabla Z^k - \mathbf{v}\|^2 + \alpha^2}}\right). \tag{17}$$

Our first new scheme of integration, denoted IS_{L_1}, is as follows for $(i,j) \in \overset{\circ}{\Omega}$:

$$
Z_{i,j}^{k+1} = \frac{1}{2d_{i,j}^k + d_{i-1,j}^k + d_{i,j-1}^k}[d_{i,j}^k(Z_{i+1,j}^k + Z_{i,j+1}^k) + d_{i-1,j}^k Z_{i-1,j}^k + d_{i,j-1}^k Z_{i,j-1}^k
$$
$$
-d_{i,j}^k \frac{p_{i,j} + p_{i+1,j} + q_{i,j} + q_{i,j+1}}{2} + d_{i-1,j}^k \frac{p_{i,j} + p_{i-1,j}}{2} + d_{i,j-1}^k \frac{q_{i,j} + q_{i,j-1}}{2}],
$$

(18)

where the factors $d_{i,j}^k$ denote the following discrete approximations of the denominators of Eq. (17):

$$
d_{i,j}^k = \frac{1}{\sqrt{(Z_{i+1,j}^k - Z_{i,j}^k - \frac{p_{i,j} + p_{i+1,j}}{2})^2 + (Z_{i,j+1}^k - Z_{i,j}^k - \frac{q_{i,j} + p_{i,j+1}}{2})^2 + \alpha^2}}.
$$

(19)

For the same reasons as in the case of quadratic regularization, $Z^0 = 0$ is used as initial configuration, and Eq. (18) has to be modified for pixels (i,j) lying on the boundary $\partial\Omega$.

4.3   Non-convex Regularization

Let us now consider functional (12) in general. In image restoration [13], the regularization functions ϕ are usually called "ϕ-functions". Such functions are required to have a linear growth around zero (to preserve edges), and a sublinear growth at infinity (so that high values of the argument are not penalized too much). We will consider both following classical ϕ-functions:

$$
\phi_1(s) = \ln(s^2 + \beta^2) \quad \Rightarrow \quad \phi_1'(s) = \frac{2s}{s^2 + \beta^2},
$$
$$
\phi_2(s) = \frac{s^2}{s^2 + \gamma^2} \quad \Rightarrow \quad \phi_2'(s) = \frac{2\gamma^2 s}{(s^2 + \gamma^2)^2}.
$$

(20)

We a priori prefer this last choice, since $\phi_2(s)$ remains less than 1 when s tends towards $+\infty$. Moreover, with this last choice:

$$
\phi_2(s) = \frac{(s/\gamma)^2}{1 + (s/\gamma)^2}.
$$

(21)

This means that the parameter γ controls the large values of s. In the case of noisy data, we will use a greater value for γ than in the case of non-noisy data.

Notice that with the choices of either ϕ_1 or ϕ_2, functional (12) is no longer convex (contrary to both functionals (5) and (14)). There may then be several minimizers. In our numerical experiments, we will have to face this problem.

The Euler-Lagrange equations associated to the functionals $\mathcal{F}_{\phi_1}(Z)$ and $\mathcal{F}_{\phi_2}(Z)$ are, respectively:

$$
\mathrm{div}\left[\frac{\nabla Z}{\|\nabla Z - \mathbf{v}\|^2 + \beta^2}\right] = \mathrm{div}\left[\frac{\mathbf{v}}{\|\nabla Z - \mathbf{v}\|^2 + \beta^2}\right],
$$
$$
\mathrm{div}\left[\frac{\nabla Z}{(\|\nabla Z - \mathbf{v}\|^2 + \gamma^2)^2}\right] = \mathrm{div}\left[\frac{\mathbf{v}}{(\|\nabla Z - \mathbf{v}\|^2 + \gamma^2)^2}\right],
$$

(22)

with Neumann boundary conditions on $\partial\Omega$. There are strong similarities between both these equations and Eq. (16). The numerical schemes that we use to solve them are the same as (18), except that the factors $d_{i,j}^k$ must be replaced, respectively, with $e_{i,j}^k$ and $f_{i,j}^k$:

$$e_{i,j}^k = \frac{1}{(Z_{i+1,j}^k - Z_{i,j}^k - \frac{p_{i,j} + p_{i+1,j}}{2})^2 + (Z_{i,j+1}^k - Z_{i,j}^k - \frac{q_{i,j} + p_{i,j+1}}{2})^2 + \beta^2},$$

$$f_{i,j}^k = \frac{1}{\left[(Z_{i+1,j}^k - Z_{i,j}^k - \frac{p_{i,j} + p_{i+1,j}}{2})^2 + (Z_{i,j+1}^k - Z_{i,j}^k - \frac{q_{i,j} + p_{i,j+1}}{2})^2 + \gamma^2\right]^2}.$$

$$(23)$$

These schemes will be denoted, respectively, by IS_{ϕ_1} and IS_{ϕ_2}.

4.4   Numerical Evaluation of the New Algorithms

In this subsection, we are going to test the three new schemes IS_{L_1}, IS_{ϕ_1} and IS_{ϕ_2} on the reduced normal field $\mathbf{v}_b$ of the benchmark surface $\mathcal{S}_b$ (see Fig. 2-left). As each of these schemes depends on one parameter, respectively α, β and γ, we must first study the influence of this parameter on the accuracy of the reconstruction.

The accuracy $\|\Delta Z\|_2$ of the 3D-reconstruction obtained using the scheme IS_{L_1}, in function of α, is plotted on the left of Fig. 3. The best reconstruction, which is represented in Fig. 4, corresponds to $\|\Delta Z\|_2 = 1.66$ and is reached when $\alpha = 0.055$. It looks indeed a little more similar to $\mathcal{S}_b$ than the surface represented in Fig. 2-right, with a lower value of $\|\Delta Z\|_2$. However, this result is not fully satisfactory. The evolution of $\|\Delta Z\|_2$ in function of the number of iterations, which is represented on the right of Fig. 3, shows the convergent behaviour of IS_{L_1} (we proved that IS_{L_1} is a convergent scheme, but due to lack of space, the proof is not included in the paper; see [15] for further details). This curve also tells us that a fixed number of 100×128 iterations gives a good approximation

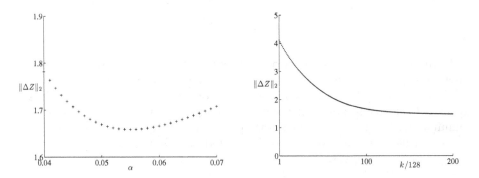

Fig. 3. Left: $\|\Delta Z\|_2$ in function of α, after 100×128 iterations of IS_{L_1}. Right: $\|\Delta Z\|_2$ in function of the number of iterations of IS_{L_1}, for the optimal value $\alpha^* = 0.055$. In these tests, the initialization $Z^0 = 0$ is used.

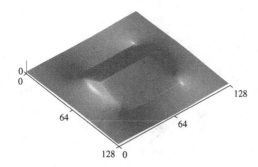

Fig. 4. 3D-reconstruction obtained using IS_{L_1}, for the optimal value $\alpha^* = 0.055$ and $Z^0 = 0$: $\|\Delta Z\|_2 = 1.66$

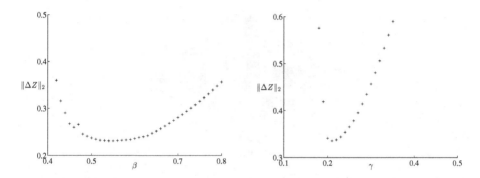

Fig. 5. Left: $\|\Delta Z\|_2$ in function of β, after 100×128 iterations of IS_{ϕ_1}. Right: $\|\Delta Z\|_2$ in function of γ, after 100×128 iterations of IS_{ϕ_2}. In these tests, the initialization $Z^0 = Z_{L_2}$ is used.

of the limit. Since similar curves are obtained for both other schemes, we decide to fix the number of iterations at 100×128 for all the tests. Of course, better stopping criteria could have been used, but this was not our main concern. Moreover, this first scheme is quite slow, due to the complexity of the formula (18) and (19): the CPU time of 100×128 iterations is equal to $17s$ on a P4 2.4 GHz, for a domain of reconstruction Ω which contains 128×128 points (the CPU times of both other new schemes is approximately the same). The resort to classical acceleration techniques, as for instance multi-grid methods, would probably be welcome.

The accuracy $\|\Delta Z\|_2$ of the 3D-reconstruction which is obtained after 100×128 iterations of the schemes IS_{ϕ_1}, in function of β, is plotted on the left of Fig. 5. The best reconstruction, which is represented in Fig. 6, corresponds to $\|\Delta Z\|_2 = 0.23$ and is reached when $\beta = 0.55$ and $Z^0 = Z_{L_2}$ (see Fig. 2-right). It is qualitatively and quantitatively much better than the previous ones.

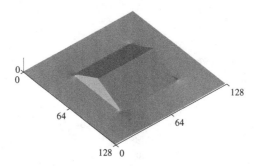

Fig. 6. 3D-reconstruction obtained using IS_{ϕ_1}, for the optimal value $\beta^* = 0.55$ and $Z^0 = Z_{L_2}$: $\|\Delta Z\|_2 = 0.23$

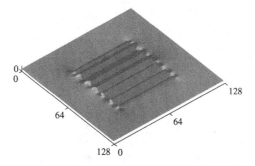

Fig. 7. 3D-reconstruction obtained using IS_{ϕ_1}, for the optimal value $\beta^* = 0.55$ and $Z^0 = 0$: $\|\Delta Z\|_2 = 4.02$

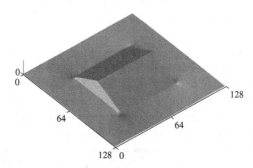

Fig. 8. 3D-reconstruction obtained using IS_{ϕ_2}, for the optimal value $\gamma^* = 0.21$ and $Z^0 = Z_{L_2}$: $\|\Delta Z\|_2 = 0.36$

An important feature of the scheme IS_{ϕ_1} is its high sensitivity to the initial configuration Z^0. A second reconstruction, which is obtained with the same value of β but $Z^0 = 0$, is represented in Fig. 7. It illustrates the existence of local minima for $\mathcal{F}_{\phi_1}(Z)$, as claimed in Section 4.3.

Fig. 9. 3D-reconstruction obtained using IS_{ϕ_2}, for the optimal value $\gamma^* = 0.21$ and $Z^0 = 0$: $\|\Delta Z\|_2 = 3.39$

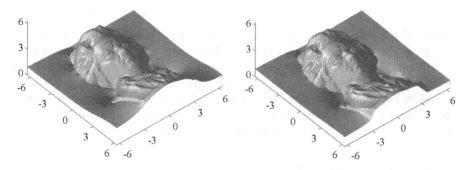

Fig. 10. 3D-reconstructions from the three photographs of the Beethoven's bustle, integrating the normal field using IS_{L_2} (left) and IS_{L_1} with $\alpha = 0.1$ (right) on $[1, 256] \times [1, 256]$

Finally, the accuracy $\|\Delta Z\|_2$ of the 3D-reconstruction which is obtained after 100×128 iterations of the scheme IS_{ϕ_2}, in function of γ, is plotted on the right of Fig. 5. The best reconstruction, which is represented in Fig. 8, corresponds to $\|\Delta Z\|_2 = 0.36$ and is reached when $\gamma = 0.21$ and $Z^0 = Z_{L_2}$. It is approximately as satisfactory as the previous one. Nevertheless, a qualitative comparison between both curves in Figs. 5-left and 5-right tells us that IS_{ϕ_2} is more sensitive to γ than IS_{ϕ_1} to β. A second reconstruction, which is a local minimum of $\mathcal{F}_{\phi_2}(Z)$, is represented in Fig. 9.

4.5 Application to Photometric Stereo

The part of the Beethoven's bustle which is visible in the photograph of Fig. 1-left does not contain self-occlusion. On the other hand, there is a discontinuity between the silhouette of the bustle and the background. Unfortunately, the background looks black in all these images, so that photometric stereo cannot be used at such pixels. In order to test the schemes IS_{L_1}, IS_{ϕ_1} and IS_{ϕ_2} on real data,

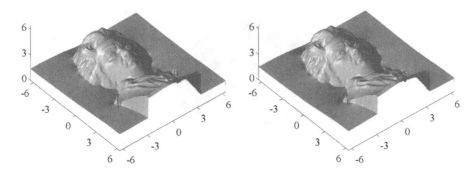

Fig. 11. 3D-reconstructions from the three photographs of the Beethoven's bustle, integrating the normal field using IS_{ϕ_1} with $\beta = 2$ (left) and IS_{ϕ_2} with $\gamma = 1$ (right) on $[1, 256] \times [1, 256]$

we consider the background as a plane with uniform normal $\mathbf{n} = [0, 0, 1]^t$. The reconstruction on the left of Fig. 10 is obtained using IS_{L_2} on the whole domain $[1, 256] \times [1, 256]$, without prior segmentation. Obviously, the discontinuity is not well handled. On the other hand, the three reconstructions on the right of Fig. 10 and in Fig. 11 are obtained using IS_{L_1}, IS_{ϕ_1} and IS_{ϕ_2}. The big gaps along the silhouette are rather well reconstructed, without any prior segmentation, as this was the case for the reconstruction of Fig. 1-right.

5   Conclusion and Perspectives

In this paper, after a theoretical study, we improve an existing method of normal integration. We show how to avoid prior segmentation, and consequently how to deal with possible discontinuities along silhouettes. More specifically, we prove the efficiency of non-convex regularization using ϕ-functions. As an application, we successfully use two new methods of integration in the framework of photometric stereo. A first perspective of this work is to avoid the empirical estimation of the optimal value of the parameters, since all of the proposed methods of integration are parametric. Another perspective is to test other ϕ-functions. As third perspective, we must question the way how the proposed methods face noisy normal fields (as this is done in [16]).

A last, but not least, perspective is to deal with multiview photometric stereo techniques, in order to produce complete 3D-reconstructions, and not only 2.5D-reconstructions (as, for instance, the surfaces shown in Figs. 1-right, 10 or 11). In [17], an interesting method of integration of a multiview normal field using a level set method is proposed, but the use of quadratic regularization makes it clearly impossible to retrieve the fine details of the 3D-shapes. In that work, as in [18], the silhouettes of the objects have to be segmented. Our new methods of integration could help avoiding this prior processing.

References

1. Horn, B.K.P., Brooks, M.J.: The Variational Approach to Shape From Shading. Comp. Vis., Grap., and Im. Proc. 33(2), 174–208 (1986)
2. Wu, Z., Li, L.: A Line-Integration Based Method for Depth Recovery from Surface Normals. Comp. Vis., Grap., and Im. Proc. 43(1), 53–66 (1988)
3. Frankot, R.T., Chellappa, R.: A Method for Enforcing Integrability in Shape from Shading Algorithms. IEEE Trans. Patt. Anal. Mach. Intell. 10(4), 439–451 (1988)
4. Wei, T., Klette, R.: Depth Recovery from Noisy Gradient Vector Fields Using Regularization. In: Petkov, N., Westenberg, M.A. (eds.) CAIP 2003. LNCS, vol. 2756, pp. 116–123. Springer, Heidelberg (2003)
5. Durou, J.D., Courteille, F.: Integration of a Normal Field without Boundary Condition. In: Proc. 11th IEEE Int. Conf. Comp. Vis., 1st Workshop on Photometric Analysis for Computer Vision, Riode Janeiro, Brazil (October 2007)
6. Klette, R., Schlüns, K.: Height data from gradient fields. In: Proc. Machine Vision Applications, Architectures, and Systems Integration, Proceedings of the International Society for Optical Engineering, Boston, Massachusetts, USA, November 1996, vol. 2908, pp. 204–215 (1996)
7. Schlüns, K., Klette, R.: Local and global integration of discrete vector fields. In: Solina, F., Kropatsch, W.G., Klette, R., Bajcsy, R. (eds.) Advances in Computer Vision. Advances in Computing Science, pp. 149–158. Springer, Heidelberg (1997)
8. Horowitz, I., Kiryati, N.: Depth from Gradient Fields and Control Points: Bias Correction in Photometric Stereo. Im. and Vis. Comp. 22(9), 681–694 (2004)
9. Lempitsky, V., Boykov, Y.: Global Optimization for Shape Fitting. In: Proc. IEEE Conf. Comp. Vis. and Patt. Recog., Workshop on Beyond Multiview Geometry: Robust Estimation and Organization of Shapes from Multiple Cues, Minneapolis, Minnesota, USA (June 2007)
10. Simchony, T., Chellappa, R., Shao, M.: Direct Analytical Methods for Solving Poisson Equations in Computer Vision Problems. IEEE Trans. Patt. Anal. Mach. Intell. 12(5), 435–446 (1990)
11. Tankus, A., Kiryati, N.: Photometric Stereo under Perspective Projection. In: Proc. 10th IEEE Int. Conf. Comp. Vis., Beijing, China, October 2005, vol. I, pp. 611–616 (2005)
12. Basri, R., Jacobs, D.W., Kemelmacher, I.: Photometric Stereo with General, Unknown Lighting. Int. J. Comp. Vis. 72(3), 239–257 (2007)
13. Aubert, G., Kornprobst, P.: Mathematical Problems in Image Processing. Applied Mathematical Sciences, vol. 147. Springer, Heidelberg (2002)
14. Charbonnier, P., Blanc-Féraud, L., Aubert, G., Barlaud, M.: Deterministic edge-preserving regularization in computed imaging. IEEE Trans. Im. Proc. 6(2), 298–311 (1997)
15. Aujol, J.F., Durou, J.D.: Realistic integration of a dense normal field (submitted, 2009)
16. Fraile, R., Hancock, E.R.: Combinatorial Surface Integration. In: Proc. 18th Int. Conf. Patt. Recog., Hong Kong, August 2006, vol. I, pp. 59–62 (2006)
17. Chang, J.Y., Lee, K.M., Lee, S.U.: Multiview Normal Field Integration Using Level Set Methods. In: Proc. IEEE Conf. Comp. Vis. and Patt. Recog., Workshop on Beyond Multiview Geometry: Robust Estimation and Organization of Shapes from Multiple Cues, Minneapolis, Minnesota, USA (June 2007)
18. Hernández, C., Vogiatzis, G., Brostow, G.J., Stenger, B., Cipolla, R.: Multiview Photometric Stereo. IEEE Trans. Patt. Anal. Mach. Intell. 30(3), 548–554 (2008)

Intrinsic Second-Order Geometric Optimization for Robust Point Set Registration without Correspondence

Dirk Breitenreicher and Christoph Schnörr

University of Heidelberg
Image & Pattern Analysis Group (IPA)
Heidelberg Collaboratory for Image Processing (HCI)
Speyerer Strasse 6, 69115 Heidelberg, Germany

Abstract. Determining Euclidean transformations for the robust registration of noisy unstructured point sets is a key problem of model-based computer vision and numerous industrial applications. Key issues include accuracy of the registration, robustness with respect to outliers and initialization, and computational speed.

In this paper, we consider objective functions for robust point registration without correspondence. We devise a numerical algorithm that fully exploits the intrinsic manifold geometry of the underlying Special Euclidean Group SE (3) in order to efficiently determine a local optimum. This leads to a quadratic convergence rate that compensates the moderately increased computational costs per iteration. Exhaustive numerical experiments demonstrate that our approach exhibits significantly enlarged domains of attraction to the correct registration. Accordingly, our approach outperforms a range of state-of-the-art methods in terms of robustness against initialization while being comparable with respect to registration accuracy and speed.

1 Introduction

1.1 Overview and Motivation

Registration of point sets is an important task in many 3D vision applications including quality inspection [1,2], reverse engineering [3], object recognition and detection [4,5,6,7]. In each case, robustness against noise, imprecise initialization and accuracy of registration are important as well as sufficiently short runtimes. Additionally, large pose variations between model and shape together with outliers and unstructured point measurements often render this problem quite challenging.

Each approach amounts to the design of an objective function and a numerical algorithm for computing an optimal registration. We review related work in Sec. 1.2. Generally speaking, since the correspondence between model and measurements is assumed to be unknown, the overall optimization problem is inherently nonconvex. Hence, robustness against poor initializations is a crucial issue.

In this context, we focus on an objective criterion that does not require to determine point correspondences explicitly. The domain of definition is therefore just the entire set of Euclidean transformations of 3D space, i.e. the Special Euclidean Group SE (3).

D. Cremers et al. (Eds.): EMMCVPR 2009, LNCS 5681, pp. 274–287, 2009.
© Springer-Verlag Berlin Heidelberg 2009

Fig. 1. Left: State-of-the-art methods that do not sufficiently take into account the manifold structure of Euclidean transformations are susceptible to imprecise initializations of the model (blue) and the scene (red) to be registered, and may reach a poor local optimum. Right: Taking the intrinsic geometry of the underlying manifold into account significantly increases robustness with respect to poor initialization.

Regarded as a matrix group, this set is a smooth manifold embedded in the corresponding ambient matrix space. We devise a Newton-like optimization algorithm that fully exploits the intrinsic manifold geometry (up to second order) to efficiently determine a locally optimal transform representing the registration.

Comparison with a range of state-of-the-art methods (see next section) reveals a significantly enlarged domain of attraction to the correct registration, thus alleviating the problem of poor initializations. Figure 1 illustrates this point, confirmed by exhaustive numerical experiments reported in section 5.

1.2 Related Work

Registration with Point Correspondence. The problem to register two point sets amounts to the chicken-and-egg problem of determining simultaneously both point correspondences and a rigid transformation. Having solved either problem, the other one becomes trivial. Consequently, most approaches proceed in an alternating fashion: given an estimate of the transformation, correspondence can be determined followed by improving the estimated transformation, and so forth. The prototypical representant is the Iterative Closest Point (ICP) algorithm [8] that due to its simplicity still is a state-of-the-art algorithm [1,2].

It is well known that this two-step iteration is susceptible to noise and poor initialization. While numerous robust variants including [9,10] have been suggested, a major drawback concerning the *representation* of the problem remains, in particular when dealing with unstructured point sets: explicit correspondences increase both the nonconvexity and nonsmoothness of the objective function, and gaining insight into the optimization problem is hampered by complicated structure of the domain of optimization comprising *both* Euclidean transformations and correspondence variables.

Registration without Explicit Correspondence. In order to obtain an optimization criteria that avoids computing corresponding points in each iterate, Mitra et al. [11] as well as Pottmann et al. [6] approximate the objective distance by local quadratic functions that represent the distance of certain points to the scene.

Another way to avoid the explicit determination of correspondence has been suggested by Tsin and Kanade [12], Jian and Vemuri [13] and Wang et al. [14]. By representing point clouds of both the scene and the model by mixture distributions, registration can be achieved by minimizing the squared L_2 distance [12,13] or the Jensen-Shannon divergence [14] between two distributions. Compared to [11,6] this avoids exhaustive pre-computation of the local distance approximation at the cost of more expensive function evaluations.

Because we consider this class of approaches as advantageous in connection with *unstructured* noisy point sets, we adopt mixture models of model and scene points in this paper.

Geometric Optimization. However, in this case, the optimal rigid body transformation cannot be determined in closed form. In order to minimize distance measures between mixture distributions representing unstructured point sets, methods of continuous optimization like gradient descent or Newton-like schemes have to be applied. This task differs from standard applications because the underlying domain where an optimum has to be computed is a curved space (manifold).

Concerning manifolds related to the orthogonal group (Grassmann and Stiefel manifolds) continuous optimization methods are considered in [15]. Adler et al. [4], for instance, proposed a corresponding Newton-like algorithm for human spine alignment. Concerning Euclidean transformations, Li and Hartley [16] presented a Branch and Bound algorithm that determines the optimal registration of two 3D point sets together with the correspondence in terms of a permutation matrix.

A decisive advantage of this approach is its independence from initialization because the global optimum is always found. On the other hand, the runtime scales badly, e.g. nearly 20 min for 200 points, which excludes industrial applications with hundreds of points. Furthermore, point sets of *equal* cardinality are required as input which is not the case in the standard scenario of matching a model (small point set) with a scene (large point set).

Pottmann et al. [6] as well as Taylor and Kriegman [17] suggested an iterative registration algorithm based on successive local approximations of the manifold Euclidean transformations in terms of the tangent space at the current iterate.

In a similar way, Krishnan et al. [3] proposed an algorithm for multiple point set alignment. We consider this approximation in more detail below (Section 4) and work out differences to our approach (Sections 4, 5).

1.3   Contribution

In this paper we devise and study a second-order optimization method that fully exploits the geometry of the manifold SE (3) of Euclidean transformations in order to minimize a distance measure between two mixture distributions representing two unstructured point sets. Additionally, we show that our novel algorithm

- outperforms state of the art algorithms including ICP and Softassign [8,9,1,2] in terms of speed of convergence,

– and has a significantly larger basin of quadratic convergence to the correct registration than previous work based on local approximations of $SE(3)$ [6].

1.4 Organization

In Sect. 2 we recall objective criteria used for point set registration with and without explicit representation of point correspondences. Section 3 collects elements of differential geometry needed to detail our optimization approach in Sect. 4, and to point out differences to related work based on approximate Newton methods.

In Sect. 5, we compare our approach to state-of-the-art point set registration algorithms with respect to runtime and robustness against poor initializations, i.e. the size of the region of convergence to the correct registration. We conclude and point out further work in Sect. 6.

2 Objective Functions

Let $\{u_i, i = 1, \ldots, N\} \subset \mathbb{R}^3$ denote a set of scene points obtained by a scanning device and, let $\{v_j, j = 1, \ldots, M\} \subset \mathbb{R}^3$ be a point set specified by a given model description, i.e. a CAD file or a sample scan.

The objective of registration is to find a rigid body transformation $Y \in SE(3)$ such that model and scene points are aligned best. Here, $SE(3)$ denotes the special Euclidean group parametrized by a proper rotation matrix $R \in SO(3)$ and a translation vector $t \in \mathbb{R}^3$. There are multiple ways to parametrize rotations R like Euler angles, quaternions etc. For optimization and numerical algorithm design, however, working with the matrix representation of the group $SE(3)$ Euclidean transforms is most appropriate.

2.1 Explicit Point Correspondences

The most common criterion for point registration is the sum of squared Euclidean distances of corresponding points given by

$$\min_{Y=\{R,t\}\in SE(3)} \sum_{i=1}^{N} \|u_i - Rv_{\eta(i)} - t\|_2^2 , \tag{1}$$

where $\eta(i) : \{1, \ldots, N\} \to \{1, \ldots M\}$ denotes a correspondence function that assigns a scene point to its model counterparts. As this function is assumed to be unknown, apart from the optimization with respect to Y, we have to determine the optimal η as well. To do so, the correspondence function is replaced by weights w_{ij} assigning closest point pairs to each other, that is

$$\min_{Y\in SE(3)} \sum_{i=1}^{N} \sum_{j=1}^{M} w_{ij} \|u_i - Rv_j - t\|_2^2 , \tag{2}$$

where $w_{ij} = 1$ if $j = \arg\min_k \|u_i - Rv_k - t\|_2^2$ and 0 otherwise. Note, that $w_{ij} = w_{ij}(Y)$ depends on the current estimate of the transformation which complicates the optimization of (2) considerably.

The common approach is to solve alternately for the transformation parameters R, t and correspondences w_{ij}. Drawbacks of related work in connection with unstructured point sets are discussed in Sect. 1.2.

2.2 Implicit Point Correspondences

An alternative class of approaches [13,12] for registration utilizes kernel estimates of functions in terms of given points sets,

$$s(x) := \sum_{i=1}^{N} \mu_i K\left(\frac{1}{2\sigma_s^2}\|x - u_i\|_2^2\right), \tag{3a}$$

$$m(x) := \sum_{j=1}^{M} \nu_j K\left(\frac{1}{2\sigma_m^2}\|x - Rv_j - t\|_2^2\right), \tag{3b}$$

where $K(\cdot)$ denotes a smoothing kernel integrating to 1, and σ_s, σ_m are parameters related to the noise levels. The values $\mu_i, \nu_j \geq 0, \sum_i \mu_i = \sum_j \nu_j = 1$ signal the importance of related samples if such information is available and otherwise are set to be constant, $1/N, 1/M$, as in this paper. Thus, $s(x), m(x)$ can be regarded as probability density estimates with respect to the assignment of points x to the scene and the model, respectively. We henceforth assume that all user parameters have been fixed beforehand.

Following [12], registrations of model and scene can now be evaluated by probabilistic distance measures of the respective distributions (3) including the Kullback-Leibler divergence

$$D(s\|m) = \int_x s(x) \log \frac{s(x)}{m(x)}$$

$$= \int_x s(x) \log s(x) - \int_x s(x) \log m(x). \tag{4}$$

We ignore the first term of (4) in the following because it does not depend on the transformation to be determined.

A further and reasonable simplification results from taking into account the noise level only in terms of a *single* smoothing parameter σ_m in (3). Correspondingly, choosing the Gaussian kernel for K in (3) and considering $\sigma_s \rightarrow 0$, function $s(x)$ becomes a sum of Dirac distributions. Insertion into the second term of (4) yields

$$\sum_{i=1}^{N} \log\left(\frac{1}{M}\sum_{j=1}^{M} \exp\left(-\frac{1}{2\sigma_m^2}\|u_i - Rv_j - t\|_2^2\right)\right), \tag{5}$$

where we dropped the constant $1/N$ and the factor normalizing the Gaussian because it does not depend on the transformation to be determined.

We point out that in contrast to the objective function (2), (5) only depends on the rigid body transformation and not on further variables representing point correspondences.

Furthermore, (5) parallels smoothed objective functions for prototypical clustering [18] in terms of the log of a sum of Gaussians. A corresponding optimization scheme, therefore, is given by the fixed point iteration

$$\underset{Y \in \mathsf{SE}(3)}{\arg\min} \sum_{i=1}^{N} \sum_{j=1}^{M} \rho_{ij} \left(Y^{(k)} \right) \| u_i - Rv_j - t \|_2^2 , \tag{6}$$

where

$$\rho_{ij} (Y) = \frac{\exp \left(- \frac{1}{2\sigma_m^2} \| u_i - Rv_j - t \|_2^2 \right)}{\sum_{l=1}^{M} \exp \left(- \frac{1}{2\sigma_m^2} \| u_i - Rv_l - t \|_2^2 \right)} . \tag{7}$$

This procedure is a variant of the Softassign procedure [9] (without annealing) that is significantly more robust than procedures based on hard assignments as in (2).

On the other hand, due to the structure of (6) only a linear convergence rate is achieved as will be confirmed in Sect. 5. This motivates the study of Newton algorithms that exhibit quadratic convergence rates in general. Furthermore, by fully exploiting the geometry of the underlying manifold, we increase robustness against poor initializations.

3   The Manifold of Euclidean Transformations

We collect in this section few basic concepts needed to specify and discuss our optimization approach in Sect. 4. For the mathematical background, we refer to e.g. [19,20].

3.1   The Lie Group SE (3)

Euclidean transformations $Y = \{R, t\} \in \mathsf{SE}(3)$ map a point x to $Yx = Rx + t$ and form a group via concatenation: $Y_1 Y_2 = \{R_1, t_1\}\{R_2, t_2\} = \{R_1 R_2, t_1 + R_1 t_2\}$. The inverse element is $Y^{-1} = \{R^{-1}, -R^{-1}t\}$.

For the purpose of optimization and numerical analysis, it is convenient to identify $\mathsf{SE}(3) \subset \mathsf{GL}(4)$ with a subgroup of all 4×4 regular matrices with respect to matrix multiplication. Keeping symbols for simplicity, this representation reads

$$Y = \begin{pmatrix} R & t \\ 0^\top & 1 \end{pmatrix}, \quad Y^{-1} = \begin{pmatrix} R^\top & -R^\top t \\ 0^\top & 1 \end{pmatrix} . \tag{8}$$

In this way $\mathsf{SE}(3)$ becomes a differentiable manifold embedded into $\mathsf{GL}(4)$, hence a Lie group.

3.2   Tangents

With each Lie group is associated its Lie algebra, the vector space tangent to the manifold at I. In case of $\mathsf{SE}(3)$ it reads

$$\mathfrak{se}(3) = \left\{ \begin{pmatrix} \Phi_R & \Phi_t \\ 0^\top & 0 \end{pmatrix} \middle| \Phi_R{}^\top = -\Phi_R, \ \Phi_t \in \mathbb{R}^3 \right\} , \tag{9}$$

which is easily deduced from the fact that $\mathfrak{se}\,(3)$ contains all matrices $\boldsymbol{\Phi}$ such that for all $t \in \mathbb{R}$, the matrix exponential $\exp(t\boldsymbol{\Phi}) \in \mathsf{SE}\,(3)$ is a Euclidean transformation, and $\boldsymbol{R} = \exp(\boldsymbol{\Phi_R})$ for some skew-symmetric $\boldsymbol{\Phi_R}$. The latter is just Rodrigues' formula for rotations in 3D.

In the following, we denote the vector space (9) equipped with the canonical inner product $\langle \boldsymbol{\Phi}_1, \boldsymbol{\Phi}_2 \rangle = \mathrm{tr}(\boldsymbol{\Phi}_1{}^{\top} \boldsymbol{\Phi}_2)$ with $\mathcal{T} := \mathfrak{se}\,(3)$. Furthermore, functions F and its derivatives defined on $\mathsf{SE}\,(3)$ are evaluated at $\boldsymbol{Y} = \boldsymbol{I}$ without loss of generality, because during iterative optimization the current iterate $\boldsymbol{Y}$ can be regarded as offset redefining the model's original pose.

3.3   Gradients

The gradient ∇F of a function $F \colon \mathsf{SE}\,(3) \to \mathbb{R}$ is defined by the relation [19]

$$\langle \nabla F, \boldsymbol{\Phi} \rangle = \langle \partial F, \boldsymbol{\Phi} \rangle \,, \ \forall \boldsymbol{\Phi} \in \mathcal{T} \,, \tag{10}$$

where ∂F is the usual matrix derivative of F given by $(\partial F)_{ij} = \frac{\partial}{\partial \boldsymbol{Y}_{ij}} F$. Because $\mathsf{SE}\,(3)$ is embedded into $\mathsf{GL}(4)$, eqn. (10) shows that $\nabla F - \partial F$ is orthogonal to all $\boldsymbol{\Phi} \in \mathcal{T}$. Thus, $\nabla F \in \mathcal{T}$ is the orthogonal projection π mapping ∂F to $\mathcal{T}$. Using the same block-factorization as in (9),

$$\partial F = \begin{pmatrix} \partial F_{11} & \partial F_{12} \\ \partial F_{21} & \partial F_{22} \end{pmatrix} \,, \tag{11}$$

this projection can be computed in closed form:

$$\nabla F = \pi(\partial F) = \begin{pmatrix} \frac{1}{2} \left(\partial F_{11} - \partial F_{11}{}^{\top} \right) & \partial F_{1,2} \\ \boldsymbol{0}^{\top} & 0 \end{pmatrix} \,. \tag{12}$$

3.4   Hessians

In addition to the gradient, optimization with the Newton method requires to compute the Hessian of a given objective function $F(\boldsymbol{Y})$ defined on $\mathsf{SE}\,(3)$. Similar to determining the gradient in the previous section, the usual definition valid for Euclidean spaces has to be adapted to the manifold $\mathsf{SE}\,(3)$.

The Hessian of a function $F \colon \mathsf{SE}\,(3) \to \mathbb{R}$, evaluated at $\boldsymbol{Y} = \boldsymbol{I}$, is a linear mapping from $\mathcal{T}$ onto itself [20] given by $\overline{\nabla}_{\boldsymbol{\Phi}}(\nabla F)$, $\forall \boldsymbol{\Phi} \in \mathcal{T}$, where the gradient ∇F is given by (12) and $\overline{\nabla}$ is the Levi-Civita connection defining the covariant derivative $\overline{\nabla}_{\boldsymbol{\Phi}}$ of the vector field ∇F.

To obtain a more explicit expression in terms of the ordinary first- and second-order derivatives, we denote by $\{\mathcal{L}_k\}_{k=1,\ldots,6}$ the canonical basis spanning the translational and skew-symmetric components of tangents $\boldsymbol{\Phi} = \sum_k \phi_k \mathcal{L}_k \in \mathcal{T}$ defined by eqn. (9). Then the quadratic form of the Hessian with respect to any $\boldsymbol{\Phi}$ is given by [15]

$$\langle \overline{\nabla}_{\boldsymbol{\Phi}}(\nabla F), \boldsymbol{\Phi} \rangle = \partial^2 F\,(\boldsymbol{\Phi}, \boldsymbol{\Phi}) - \langle \partial F, \Gamma(\boldsymbol{\Phi}, \boldsymbol{\Phi}) \rangle \tag{13}$$

with $\partial^2 F\left(\boldsymbol{\Phi}, \boldsymbol{\Phi}\right)$ denoting the bilinear form $\sum_{ij,kl} \frac{\partial^2 F}{\partial Y_{ij} \partial Y_{kl}} \boldsymbol{\Phi}_{ij} \boldsymbol{\Psi}_{kl}$ and

$$\Gamma(\boldsymbol{\Psi}, \boldsymbol{\Phi}) = \sum_{i,j,k} \psi_i \phi_j \Gamma_{ij}^k \boldsymbol{\mathcal{L}}_k . \tag{14}$$

We list the so-called Christoffel symbols Γ_{ij}^k defining the connection $\overline{\nabla}$ in the Appendix.

4   Second Order Optimization on SE (3)

In the usual Euclidean space $\mathbb{R}^n$, second-order optimization of some objective function $F : \mathbb{R}^n \to \mathbb{R}$ using the Newton method is based on the local quadratic model

$$F(\boldsymbol{x}^k + \boldsymbol{x}) \approx F(\boldsymbol{x}^k) + \boldsymbol{x}^\top \partial F + \frac{1}{2} \boldsymbol{x}^\top \boldsymbol{H} \boldsymbol{x}, \tag{15}$$

where $\partial F, \boldsymbol{H}$ denote the (ordinary) gradient and Hessian of F evaluated at $\boldsymbol{x}^k$, respectively. The gradient of (15) vanishes if $\boldsymbol{x}$ solves the linear system

$$\boldsymbol{H}\boldsymbol{x} = -\partial F , \tag{16}$$

leading to the update $\boldsymbol{x}^{k+1} = \boldsymbol{x}^k + \boldsymbol{x}$.

In this section, we discuss two ways to generalize this iteration to the case of objective functions $F(\boldsymbol{Y}) : \mathsf{SE}\,(3) \to \mathbb{R}$.

4.1   Truncating the Exponential Map

It is well known that a geodesic path $\boldsymbol{Y}(t) \in \mathsf{SE}\,(3)$ with $\boldsymbol{Y}(0) = \boldsymbol{I}$ and tangent $\dot{\boldsymbol{Y}}(0) = \boldsymbol{\Phi}$ is locally given by the exponential mapping $\exp \colon \mathcal{T} \to \mathsf{SE}\,(3)$,

$$\exp(t\boldsymbol{\Phi}) = \sum_{k=0}^{\infty} \frac{(t\boldsymbol{\Phi})^k}{k!} . \tag{17}$$

Accordingly, it makes sense to consider local approximations

$$\boldsymbol{Y}_{lin}(t) \approx \boldsymbol{I} + t\,\boldsymbol{\Phi} \tag{18a}$$

$$\boldsymbol{Y}_{quad}(t) \approx \boldsymbol{I} + t\,\boldsymbol{\Phi} + \frac{t^2}{2}\boldsymbol{\Phi}^2 , \tag{18b}$$

respectively, as suggested by Pottmann et al. [6], and to determine the optimal tangent vector $t\,\boldsymbol{\Phi}$. By inserting the approximations (18a) and (18b) into $F(\boldsymbol{Y})$, and by expanding $\boldsymbol{\Phi}$ with respect to the basis $\{\boldsymbol{\mathcal{L}}_k\}_{k=1,\ldots,6}$ introduced above, the objective function $F(\boldsymbol{Y})$ is restricted to the 6-dimensional vector space $\mathcal{T}$ in terms of the coefficients $t\phi = t(\phi_1, \ldots, \phi_6)^\top$ as variables.

As a result, the linear system (16) defining the Newton iteration is replaced by (we keep the symbols $\boldsymbol{H}$ and ∂F for simplicity)

$$\boldsymbol{H}(t\,\phi) = -\partial F , \tag{19}$$

where $(\partial F)_i = \frac{\partial}{\partial \phi_i} F$ and $H_{ij} = \frac{\partial^2}{\partial \phi_i \partial \phi_j} F$ evaluated at $\phi = 0$.

However, because (18a) and (18b) are only local approximations of the Euclidean group, inserting the solution $t\, \boldsymbol{\Phi} = \sum_k (t\, \phi_k) \mathcal{L}_k$ of the linear system (19) does not give an element of $\mathsf{SE}\,(3)$ in general. Rather, the Newton update $\boldsymbol{Y} \in \mathsf{SE}\,(3)$ is determined by inserting $t\, \boldsymbol{\Phi}$ into the exponential map (17).

4.2 Intrinsic Newton Updates

Instead of restricting the objective function F to the tangent space $\mathcal{T}$ through the local manifold approximations (18) first, and then computing Newton updates by solving (19), we may base the Newton iteration directly on the intrinsic gradient and Hessian of the manifold $\mathsf{SE}\,(3)$.

This means that the linear system (16) in the Euclidean case is replaced by the linear system defined by the variational equation

$$\langle \overline{\nabla}_{\boldsymbol{\Phi}}(\nabla F), \boldsymbol{\Psi} \rangle = -\langle \nabla F, \boldsymbol{\Psi} \rangle \,, \quad \forall \boldsymbol{\Psi} \in \mathcal{T} \,, \tag{20}$$

with the gradient ∇F given by (12) and the Hessian defined in (13).

As in the previous section, the tangent vector $\boldsymbol{\Phi}$ solving (20) does not directly result in a Euclidean transformation $\boldsymbol{Y}$ as Newton update. Rather, here we use the exponential mapping

$$\boldsymbol{Y} = \exp(\boldsymbol{\Phi}) \tag{21}$$

defined by (17).

4.3 Local vs. Intrinsic Approximation

While both schemes require to solve linear systems (19) and (20) in each iteration, respectively, there are major differences in terms of convergence properties. We address this issue in this section and take it up again in connection with discussing experimental results in Sect. 5.

Recall that the objective function to be studied in this paper reads

$$F(\boldsymbol{Y}) = -\sum_{i=1}^{N} \log \left(\frac{1}{M} \sum_{j=1}^{M} \exp\left(-h_{ij}(\boldsymbol{Y}) \right) \right) , \tag{22}$$

where $h_{ij}(\boldsymbol{Y}) = \frac{1}{\sigma^2} \| \boldsymbol{u}_i - \boldsymbol{R}\boldsymbol{v}_j - \boldsymbol{t} \|_2^2$ and $\boldsymbol{Y} \in \mathsf{SE}\,(3)$.

Approximating the rigid body transformation by truncating (17) after the linear term (18a) yields a redefinition of h_{ij} such that optimization of F is restricted to the tangent space $\mathcal{T}$. Because this approach provides an accurate approximation only within a small neighborhood around the current iterate, however, convergence to the correct local optimum is unlikely if it lies outside this neighborhood [6].

In contrast, second order truncation (18b) provides a more accurate approximation of the manifold $\mathsf{SE}\,(3)$ locally. On the other hand, inserting the quadratic approximation into h_{ij} maps $\boldsymbol{R}\boldsymbol{v}_j + \boldsymbol{t}$ to

$$\left(\boldsymbol{v}_j + \boldsymbol{\Phi}_t + \boldsymbol{\Phi}_R \boldsymbol{v}_j + \frac{1}{2} \boldsymbol{\Phi}_R (\boldsymbol{\Phi}_t + \boldsymbol{\Phi}_R \boldsymbol{v}_j) \right) . \tag{23}$$

Using the fact that $\boldsymbol{\Phi}_R$ is skew symmetric, the latter part rewrites as

$$\frac{1}{2}\left(\boldsymbol{\Phi}_R\boldsymbol{\Phi}_t + (\boldsymbol{\phi}^\top v_j)\boldsymbol{\phi} - (\boldsymbol{\phi}^\top\boldsymbol{\phi})v_j\right), \qquad (24)$$

where $\boldsymbol{\phi}$ are the coefficients of the expansion $\boldsymbol{\Phi}_R = \sum_k \phi_k \mathcal{L}_k$.

As a consequence, when the rotation components of Newton updates happen to become large in magnitude, the nonconvexity of the objective function due to the quadratic terms in (24) may cause Newton updates to step into wrong directions. This will be confirmed by numerical experiments in the following section.

Finally, the intrinsic second-order approximation (20) computes update directions within the tangent space, as do the approaches discussed above based on (18). A notable difference, however, is that in this case the Hessian and the gradient utilize information of the *embedding of* SE (3) *into the ambient space* in terms of the connection and covariant derivatives, respectively, moving nearby tangents along the manifold.

We will show in the next section that this difference is relevant in practice, too.

5   Numerical Evaluation and Comparison

In 3D vision computer vision applications both robustness against poor initializations and sufficiently short processing times are of utmost importance. In this section, we apply the proposed Newton algorithm to rigid point set registration and experimentally demonstrate the major benefits and drawbacks of our scheme: *fast* convergence in a *large* region of quadratic attraction at the cost of slightly more expensive function evaluations.

Moreover, we compare the proposed scheme to a range of state-of-the-art algorithms including Iterative Closest Point [8], the fix-point iteration (6) as a special case of Softassign [9], and the Newton procedure based on local approximation of the Euclidean group [6].

In our experiments we only considered the case of perfect 3D point measurements in this paper, i.e. no noise or occlusion, in order to clearly separate for each scheme the effect of poor initializations from noise sensitivity. We point out, however, that by adjusting the kernel parameters of (5) or introducing background kernels to handle occlusion, extensions to noisy scenarios are straightforward.

5.1   Speed of Convergence

Algorithms like ICP [8] or Softassign [9] return less accurate registrations in cases where the underlying point set has no salient regions. This often occurs in industrial applications where smooth surfaces have to be registered accurately. To compare the ability of the approaches to cope with such scenarios, we generated 2500 data points by randomly sampling points from the smooth function $3(x-1)^2 + 3\sin(2y)$ on the unit interval.

Next, we transformed a copy of the model slightly according to a rigid body transformation (about 4 degree in each rotation and by a total of 0.12 in translation), such

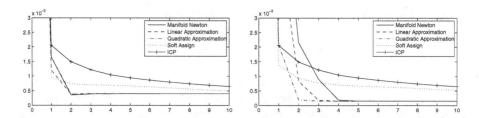

Fig. 2. Evaluation of the performance of Newton algorithms based on linear and quadratic motion approximation [6], and based on manifold structure (this paper) as well as ICP [8] and Softassign [9] for different values of σ (left: 0.3, right: 0.15). Each plot visualizes the value of the cost function (2) in the corresponding iterate. ICP and Softassign converge linearly while the remaining approaches converge quadratically.

that all approaches ICP [8], Softassign [9], the Newton schemes based on local approximation [6] and the approach proposed in this paper always converge. Figure 2 reveals that the convergence rates differ significantly.

While for varying σ, the Newton procedures based on approximation of the Euclidean group converge slightly faster than the approach presented in this paper, all of them exhibit quadratic convergence properties. In contrast, ICP and Softassign only converge linearly to the optimal configuration. As a result, they return less accurate registrations under tight runtime constraints (fixed number of iterations).

This superior performance of the Newton schemes require more expensive computations of the Hessians in each iteration. While ICP requires $O(M \log N)$ computations in each iteration using K-D trees, the evaluation of the gradient and the Hessian of (5) causes costs of $O(MN)$. Putting this into numbers, one round of ICP requires about 1 second. In contrast, the computation of the derivatives, using MatLab research code, needs between 8 (linear and quadratic approximation [6]) and 12 seconds (our approach). This difference is primarily due to the higher dimension of the space in which the gradient and the Hessian are computed. We expect that using a more careful C-tuned implementation will return more accurate registrations if the maximum runtime is fixed beforehand, as is required in industrial applications.

5.2 Basin of Convergence

Additionally to fast convergence, robustness to poor initializations is important in many applications. The region of attraction for ICP [8] has already been analyzed in [11]. Thus we only consider Newton procedures here.

For comparison, we used the same initial setup as [11], i.e. a model of the Stanford Bunny, visualized in Fig. 1, rotated around the z-axis and shifted in the x-y plane by the size of the model. As scene we used a copy of the model placed in the origin. Moreover, since we are primarily interested in quadratic and fast convergence and the resulting accuracy after a fixed runtime, we terminated the algorithms after 25 iterations.

We observe that especially for transformations with rotational initialization error, the Newton approach proposed in this work has a significantly larger domain of attraction to

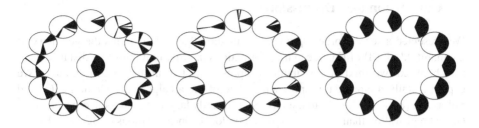

Fig. 3. Evaluation of the region of quadratic convergence for Newton algorithms based on linear (left) and quadratic (middle) local approximation [6], and based on the intrinsic local approximation (right) (this paper) for fixed $\sigma = 0.1$. Each circle center corresponds to the initial translation offset in the x-y plane, where the middle circle center is the origin. The slices in each circle refer to the initial rotation around the z-axis. These slices are colored black if the model converges to the scene within the first few iterations and otherwise remains white. The results illustrate that the approach proposed in this paper is significantly more robust against poor initializations.

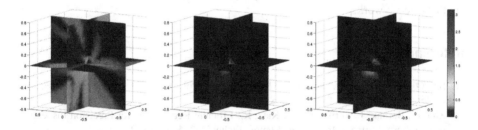

Fig. 4. Visualization of the angular error of the translational update computed with the linear (left) and quadratic (middle) local approximation approach [6], and with the intrinsic local approximation (right) (this paper), as a function of the translational offset (ground truth) model ↔ scene in 3D-space. No rotation was applied. Each graphics depicts slices through the three-dimensional "error fields". While the linear local approximation fails again in this simpler scenario, both quadratic approximations are more robust against this type of initialization error. Figure 3 shows however that only the intrinsic approximation (this paper) remains stable if rotational initialization errors additionally occur.

the correct solution than the procedures based on local approximations of the Euclidean group, as visualized in Fig. 3 and explained in more detail in the corresponding caption.

In a related experiment we examined the update direction of a single iterate of each scheme, cf. Fig. 4. By only translating the Stanford Bunny in $\mathbb{R}^3$ we found that quadratic local motion approximation as well as our approach exhibit a lower angular error then the scheme based on linear local approximation. We point out that the angular error of all approaches near the origin is primarily due the nature of the objective function (5), that is a slight detrimental effect of the smoothed objective function discussed in Sect. 2.2. Decreasing the value of σ after few iteration steps would fix this minor issue.

6   Conclusion and Discussion

We presented a second-order optimization method that fully exploits the geometry of the manifold SE (3) of Euclidean transformations in order to minimize a distance measure between two mixture distributions representing two unstructured point sets. We experimentally compared this approach to state-of-the-art algorithms including ICP and Softassign [8,9] and showed that it has a significantly larger basin of convergence to the correct registration than recent work based on local approximations of SE (3) [6].

This better performance comes at the cost of slightly more expensive computations required in each iteration. Thus, in further work we want to address this issue by considering approximating schemes of the objective that allow faster evaluation of the objective function.

Additionally we want to analyze the region of quadratic attraction more carefully in order to derive bounds that guarantee convergence to the desired local optimum.

Acknowledgements. The authors would like to thank the VMT Vision Machine Technic Bildverarbeitungssysteme GmbH, a company of the Pepperl+Fuchs Group, for supporting this reseach work.

References

1. Shi, Q., Xi, N., Chen, Y., Sheng, W.: Registration of Point Clouds for 3D Shape Inspection. In: Int. Conf. Intelligent Robots and Systems (2006)
2. Zhu, L., Barhak, J., Shrivatsan, V., Katz, R.: Efficient Registration for Precision Inspection of Free-Form Surfaces. Int. J. Adv. Manuf. Technol. 32, 505–515 (2007)
3. Krishnan, S., Lee, P.Y., Moore, J.B., Venkatasubramanian, S.: Optimisation-on-a-Manifold for Global Registration of Multiple 3D Point Sets. Int. J. Intell. Syst. Technol. Appl. 3(3/4), 319–340 (2007)
4. Adler, R.L., Dedieu, J.P., Margulies, J.Y., Martens, M., Shub, M.: Newton's Method on Riemannian Manifolds and a Geometric Model for the Human Spine. IMA J. Numer. Anal. 22(3), 359–390 (2002)
5. Frome, A., Huber, D., Kolluri, R., Bülow, T.: Recognizing Objects in Range Data using Regional Point Descriptors. In: Proc. Europ. Conf. Comp. Vision. (2004)
6. Pottmann, H., Huang, Q.X., Yang, Y.L., Hu, S.M.: Geometry and Convergence Analysis of Algorithms for Registration of 3D Shapes. Int. J. Computer Vision 67(3), 277–296 (2006)
7. Rodgers, J., Anguelov, D., Pang, H.C., Koller, D.: Object Pose Detection in Range Scan Data. In: Proc. Conf. Comp. Vision Pattern Recogn. (2006)
8. Besl, P.J., McKay, N.D.: A Method for Registration of 3-D Shapes. IEEE Trans. Pattern Anal. Mach. Intell. 14(2), 239–256 (1992)
9. Rangarajan, A., Chui, H., Bookstein, F.L.: The Softassign Procrustes Matching Algorithm. In: Proc. Int. Conf. Inf. Process. Med. Imaging (1997)
10. Rusinkiewicz, S., Levoy, M.: Efficient Variants of the ICP Algorithm. In: Proc. Int. Conf. 3D Digital Imaging and Modeling (2001)
11. Mitra, N.J., Gelfand, N., Pottmann, H., Guibas, L.: Registration of Point Cloud Data from a Geometric Optimization Perspective. In: Proc. Sym. Geom. Process. (2004)
12. Tsin, Y., Kanade, T.: A Correlation-Based Approach to Robust Point Set Registration. In: Proc. Europ. Conf. Comp.Vision (2004)

13. Jian, B., Vemuri, B.C.: A Robust Algorithm for Point Set Registration Using Mixture of Gaussians. In: Proc. Int. Conf. Comp. Vision (2005)
14. Wang, F., Vemuri, B.C., Rangarajan, A., Schmalfuss, I.M., Eisenschenk, S.J.: Simultaneous Nonrigid Registration of Multiple Point Sets and Atlas Construction. In: Proc. Europ. Conf. Comp. Vision (2006)
15. Edelman, A., Arias, T.A., Smith, S.T.: The Geometry of Algorithms with Orthogonality Constraints. SIAM J. Matrix Anal. Appl. 20, 303–353 (1999)
16. Li, H., Hartley, R.: The 3D-3D Registration Problem Revisited. In: Proc. Int. Conf. Comp. Vision (2007)
17. Taylor, C.J., Kriegman, D.J.: Minimization on the Lie Group SO(3) and Related Manifolds. Technical Report 9405, Center for Systems Sciene, Dept. of Electrical Engineering, Yale University (1994)
18. Teboulle, M.: A Unified Continuous Optimization Framework for Center-Based Clustering Methods. J. Mach. Learn. Res. 8, 65–102 (2007)
19. Matsushima, Y.: Differentiable Manifolds. Marcel Dekker, Inc., New York (1972)
20. do Carmo, M.P.: Riemannian Geometry. Birkhäuser, Boston (1992)

A   Christoffel Symbols Defining the Connection $\overline{\nabla}$

The non-zero Christoffel symbols of (14) are

$$\Gamma_{12}^3 = \Gamma_{23}^1 = \Gamma_{31}^2 = \frac{1}{2}, \tag{25a}$$

$$\Gamma_{13}^2 = \Gamma_{21}^3 = \Gamma_{32}^1 = -\frac{1}{2}, \tag{25b}$$

$$\Gamma_{15}^6 = \Gamma_{26}^4 = \Gamma_{34}^5 = 1, \tag{25c}$$

$$\Gamma_{16}^5 = \Gamma_{24}^6 = \Gamma_{35}^4 = -1. \tag{25d}$$

Geodesics in Shape Space via Variational Time Discretization

Benedikt Wirth[1], Leah Bar[2], Martin Rumpf[1], and Guillermo Sapiro[2]

[1] Institute for Numerical Simulation, University of Bonn, Germany
[2] Department of Electrical and Computer Engineering, University of Minnesota, Minneapolis, U.S.A.

Abstract. A variational approach to defining geodesics in the space of implicitly described shapes is introduced in this paper. The proposed framework is based on the time discretization of a geodesic path as a sequence of pairwise matching problems, which is strictly invariant with respect to rigid body motions and ensures a 1-1 property of the induced flow in shape space. For decreasing time step size, the proposed model leads to the minimization of the actual geodesic length, where the Hessian of the pairwise matching energy reflects the chosen Riemannian metric on the shape space. Considering shapes as boundary contours, the proposed shape metric is identical to a physical dissipation in a viscous fluid model of optimal transportation. If the pairwise shape correspondence is replaced by the volume of the shape mismatch as a penalty functional, for decreasing time step size one obtains an additional optical flow term controlling the transport of the shape by the underlying motion field. The implementation of the proposed approach is based on a level set representation of shapes, which allows topological transitions along the geodesic path. For the spatial discretization a finite element approximation is employed both for the pairwise deformations and for the level set representation. The numerical relaxation of the energy is performed via an efficient multi–scale procedure in space and time. Examples for 2D and 3D shapes underline the effectiveness and robustness of the proposed approach.

1 Introduction

This paper deals with the computation of geodesic paths and distances between (possibly non-rigid) shapes represented via *level sets* in 2D and 3D. Such computations are fundamental for problems ranging from computational anatomy to object recognition, warping, and matching. The aim is to reliably and effectively evaluate distances between non-parametrized geometric shapes of possibly different topology. We investigate the close link between abstract geometry on the infinite-dimensional space of shapes and the continuum mechanical viewpoint of shapes as being boundary contours of physical objects, e. g. identifying the Riemannian metric on shape space with the physical dissipation — the loss of energy due to friction. Thereby, we simultaneously address the following major challenges:

D. Cremers et al. (Eds.): EMMCVPR 2009, LNCS 5681, pp. 288–302, 2009.

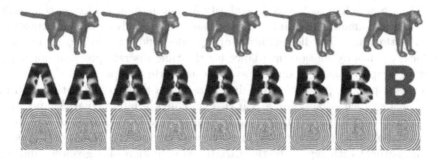

Fig. 1. Time-discrete geodesics between a cat and a lion and the letters A and B. Geodesic distance is measured on the basis of viscous dissipation inside the objects (color-coded in the middle row from blue, low dissipation, to red, high dissipation), which is induced by a pairwise 1-1 deformation map between consecutive shapes along the discrete geodesic path. Shapes are represented via level set functions, whose level lines are texture-coded in the bottom row for the 2D example.

- a physically sound modeling of the geodesic flow of shapes given as boundary contours of objects on a void background,
- the need for a coarse time discretization which is nevertheless invariant with respect to rigid body motions, ensures a 1-1 object correspondence, and relates to the corresponding continuous geodesic path,
- a numerically effective multi–scale treatment of the resulting time and space discrete energy.

Our approach is closely linked to the concept of optimal transportation [1]. The motion field v governing the flow in shape space vanishes outside the object bounded by the corresponding shape contour. The field is optimal in the sense that it minimizes an accumulated physical dissipation — a quadratic functional depending on the first order local variation of a flow field, representing the rate at which mechanical energy is converted into heat in a viscous fluid per unit volume. Thus, the Riemannian metric on the shape space is defined to coincide with this rate of dissipation. If we assume frame indifference as first principle (rigid body motion invariance), then the dissipation depends only on the symmetric part $\epsilon[v] = \frac{1}{2}(\mathcal{D}v^{\mathrm{T}} + \mathcal{D}v)$ of the Jacobian $\mathcal{D}v$ of the underlying motion field v. Under the additional assumption of isotropy, a typical model for the local rate of dissipation is given by $\mathbf{Diss}[v] = \int_0^1 \int_{\mathcal{O}} g(v, v)\,\mathrm{d}x\,\mathrm{d}t$ with

$$g(v, v) = \frac{\lambda}{2}\left(\mathrm{tr}\,\epsilon[v]\right)^2 + \mu\,\mathrm{tr}(\epsilon[v]^2) \tag{1}$$

(cf. Fuchs et al. [2]), where $\mathcal{O}$ describes the deformed object. Here $\mathrm{tr}(\epsilon[v]^2)$ measures the averaged local change of length and $(\mathrm{tr}\,\epsilon[v])^2$ the local change of volume (obviously $\mathrm{div}\,v = \mathrm{tr}(\epsilon[v]) = 0$ represents an incompressible flow), induced by the transport by v. In their pioneering paper Miller et al. [3] exploited the fact that in case of sufficient Sobolev regularity for the motion field v on the whole surrounding domain, the induced flow consists of a family of diffeomorphisms. A straightforward time discretization of a geodesic flow would neither guarantee local rigid body motion invariance for the time discrete problem nor a 1-1 mapping property between objects at consecutive time steps.

In this paper, we present a time discretization of the squared path length in shape space which is based on a pairwise matching of intermediate shapes corresponding to subsequent time steps. In fact, such a discretization of a path as concatenation of short connecting lines between consecutive points on the path is most natural with regard to the variational definition of a geodesic and for instance underlies the algorithm by Schmidt et al. [4]. Our approach is inspired both by work in mechanics [5] and in geometry [6]. Here, a suitable deformation energy will measure the deformation between subsequent shapes. This can be regarded as the infinite-dimensional counterpart of the following time discretization for a geodesic between two points s_A and s_B on a finite-dimensional Riemannian manifold: Consider a sequence of points $s_A = s_0, s_1, \ldots, s_K = s_B$ connecting two fixed points s_A and s_B and minimize $\sum_{k=1}^{K} \mathrm{dist}^2(s_{k-1}, s_k)$, where $\mathrm{dist}(\cdot, \cdot)$ is a suitable approximation of the Riemannian distance. In our case, the squared approximate distance is replaced by the deformation energy, for which we will employ a particular class of so-called polyconvex energies [7] to ensure both exact frame indifference (observer independence and thus rigid body motion invariance) and a global 1-1 property. We will also discuss the corresponding continuous problem when the time discretization step vanishes.

Even though the functionals are borrowed from nonlinear elasticity, the underlying physics is only related to elasticity in the sense that a viscous deformation can be regarded as the limit of infinitely small elastic deformations with subsequent stress relaxation. Indeed, different from elasticity, none of the shapes is in a stressed configuration since local stresses are immediately absorbed via dissipation, which in a physical context reflects a local heating.

Both the built-in exact frame indifference and the 1-1 mapping property ensure that fairly coarse time discretizations already lead to an accurate approximation of geodesic paths (cf. Fig. 2). The actual convergence is dealt with later in this paper. Careful consideration is required with respect to the effective minimization of the time discrete path length. Already in the case of low dimensional Riemannian manifolds the need for efficient minimization strategies is apparent. To give a conceptual sketch of the proposed algorithm on the actual shape space, Fig. 3 depicts the proposed procedure in the case of $\mathbb{R}^2$ considered as the stereographic projection of the two-dimensional sphere and outlines the advantage of our proposed optimization framework.

1.1   Related Work

Conceptually, in the last decade, the distance between shapes has been been studied on the basis of a general framework of a space of shapes and its intrinsic structure. The notion of shape space has been introduced already in 1984 by Kendall [8].

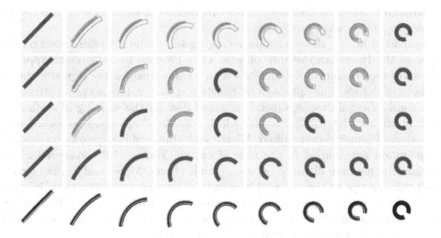

Fig. 2. Discrete geodesics between a straight and a rolled up bar, from first row to fourth row based on 1, 2, 4, and 8 time steps. The light gray shapes in the first row show the linear interpolation of the deformations connecting the dark gray shapes. The shapes from the finest time discretization are overlayed over the others as thin black lines. In the last row the rate of viscous dissipation is rendered on the shape domains $\mathcal{O}_1, \ldots, \mathcal{O}_{K-1}$ from the previous row, color-coded as ███ ██.

Fig. 3. Different refinement levels of discrete geodesics ($K = 1, 2, 4, \ldots, 256$) from Johannisburg to Kyoto in the steoreographic projection (right) and backprojected on the globe (left). A single-level nonlinear Gauss-Seidel on the finest resolution with successive relaxation of the different vertices requires 917235 elementary relaxation steps, whereas in a cascadic relaxation from coarse to fine resolution in time, only 2593 of these elementary minimization steps are needed.

An isometrically invariant distance measure between two objects $\mathcal{S}_A$ and $\mathcal{S}_B$ in (different) metric spaces is the Gromov–Hausdorff distance, which is (in a simplified form) defined as the minimizer of $\frac{1}{2} \sup_{y_i=\phi(x_i),\psi(y_i)=x_i} |d(x_1, x_2) - d(y_1, y_2)|$ over all maps $\phi : \mathcal{S}_A \to \mathcal{S}_B$ and $\psi : \mathcal{S}_B \to \mathcal{S}_A$, matching point pairs (x_1, x_2) in $\mathcal{S}_A$ with pairs (y_1, y_2) in $\mathcal{S}_B$. It evaluates — globally and based on an L^∞ type functional — the lack of isometry between two different shapes. Mémoli and Sapiro [9] introduced this concept into the shape analysis community and discussed efficient numerical algorithms based on a robust notion of intrinsic distances $d(\cdot, \cdot)$ on the shapes given by point clouds. Bronstein et al. incorporate the Gromov–Hausdorff distance concept in various classification and modeling approaches in geometry processing [10].

Charpiat et al. [11] discuss shape averaging and shape statistics based on the notion of the Hausdorff distance of sets. They propose to use smooth approximations of the Hausdorff distance based on a comparison of the signed distance functions of shapes. The approach by Eckstein et al. [12] is conceptually related. They consider regularized geometric gradient flow for the warping of surfaces.

There are a variety of approaches which consider shape space as an infinite-dimensional Riemannian manifold. Michor and Mumford [13] gave a corresponding definition exemplified in the case of curves. Younes [14] considered a left invariant Riemannian distance between planar curves. Miller and Younes generalized this concept to the space of images [15]. Klassen and Srivastava [16] proposed a framework for geodesics in the space of arclength parametrized curves and suggested a shooting type algorithm for the computation, whereas Schmidt et al. [4] presented an alternative variational approach.

Dupuis et al. [17] and Miller et al. [18] defined the distance between shapes based on a flow formulation in the embedding space. The underlying motion fields v are globally defined, and as Riemannian metric they considered $\int_\Omega Lv \cdot v \, dx$, where L was chosen as a higher order elliptic operator [19,14]. This operator ensures sufficient regularity along paths of finite length and thus implies a diffeomorphic property for the flow map ϕ generated via integration of the motion fields v.

Fuchs et al. [2] proposed a Riemannian metric on shape space motivated by linearized elasticity, leading to the same quadratic form (1), which is in their approach evaluated on a displacement field. They used a B-spline parametrization of the shape contour together with a finite element approximation for the displacements on an accompanying triangulation of one of the two objects. Due to the linearization this approach is not rigid body motion invariant, and they do not consider a hierarchical treatment. The explicitly parametrized shapes on a geodesic path share the same topology. A Riemannian metric in the space of surface triangulation in 3D of fixed mesh topology has been investigated by Kilian et al. [20], where an inner product of deformations fields as the underlying metric measures the local distance from a rigid body motion.

1.2   Key Contributions

Key contributions of our approach are: • The presented time discretization strictly ensures rigid body motion invariance and a 1-1 mapping property. • The implicit treatment of shapes via level sets allows for topological transitions and enables to compute geodesics in the context of partial occlusion. • Robustness and effectiveness of the algorithm is ensured via a cascadic multi–scale relaxation strategy. • The approach mathematically rigorously links consecutive pairwise shape matching and a flow perspective on a Riemannian shape space. • A formal connection between physics-motivated and geometry-motivated shape spaces is provided, with an intuitive physical interpretation of the framework.

2   Variational Time Discretization: Preamble and the Discrete Geodesic Model

In this section, we present the time discretization of a geodesic path in shape space, whereas the induced Riemannian distance will be investigated in Sec. 3. We do not consider a purely geometric notion of shapes as curves in 2D or surfaces in 3D. In fact, motivated by physics, we consider shapes $\mathcal{S}$ as boundaries $\partial \mathcal{O}$ of sufficiently regular, open object domains $\mathcal{O} \subset \mathbb{R}^d$ for $d = 2, 3$.

Discrete path in shape space. Given two shapes $\mathcal{S}_A$, $\mathcal{S}_B$, we define a discrete path of shapes as a sequence of shapes $\mathcal{S}_0, \mathcal{S}_1, \ldots, \mathcal{S}_K$ with $\mathcal{S}_0 = \mathcal{S}_A$ and $\mathcal{S}_K = \mathcal{S}_B$. For the time step $\tau = \frac{1}{K}$ the shape $\mathcal{S}_k$ is supposed to be an approximation of $\mathcal{S}(t_k)$ $(t_k = k\tau)$, where $\mathcal{S}(t), t \in [0, 1]$, is a continuous path connecting $\mathcal{S}_A = \mathcal{S}(0)$ and $\mathcal{S}_B = \mathcal{S}(1)$, e.g. a geodesic between these two shapes (continuous results will be presented in the next section).

Pairwise deformations between consecutive shapes. Now, we introduce a matching deformation ϕ_k for each pair of consecutive shapes $\mathcal{S}_{k-1}$ and $\mathcal{S}_k$ such that $\phi_k(\mathcal{S}_{k-1}) = \mathcal{S}_k$, and a corresponding deformation energy

$$\mathcal{E}_{\text{deform}}[\phi_k, \mathcal{S}_{k-1}] = \int_{\mathcal{O}_{k-1}} W(\mathcal{D}\phi_k) \, dx, \tag{2}$$

where W is an energy density (cf. [21]). As in the axiom of elasticity, the energy is assumed to depend only on the local deformation, reflected by the Jacobian $\mathcal{D}\phi$. But different from elasticity, we suppose the material to relax immediately so that the object at the next time step is again in a stress-free configuration. Let us emphasize that the stored energy does not depend on the deformation history as in most plasticity models in engineering. If we postulate as fundamental assumption on the time discretization the invariance of the deformation energy with respect to rigid body motions,[1] i.e.

$$\mathcal{E}_{\text{deform}}[Q \circ \phi_k + b, \mathcal{S}_{k-1}] = \mathcal{E}_{\text{deform}}[\phi_k, \mathcal{S}_{k-1}] \tag{3}$$

for $Q \in SO(d)$ and $b \in \mathbb{R}^d$ (the axiom of frame indifference in continuum mechanics), one deduces that the energy density only depends on the first Cauchy–Green deformation tensor $\mathcal{D}\phi^T \mathcal{D}\phi$ (which geometrically represents the metric measuring the deformed length in the reference configuration), i.e. $W(A) = \bar{W}(A^T A)$ for some $\bar{W}$. Now, we assume that the deformation is chosen such that the Hessian at the identity coincides with the desired local dissipation rate or metric tensor (1) (cf. Sec. 3). For an isotropic material the energy can be rewritten as a function solely depending on the principal invariants of the Cauchy–Green tensor, namely $I_1 = \text{tr}(\mathcal{D}\phi^T \mathcal{D}\phi)$, controlling the local average change of length, $I_2 = \text{tr}(\text{cof}(\mathcal{D}\phi^T \mathcal{D}\phi))$ $(\text{cof}\,A := \det A \, A^{-T})$, reflecting the local average change of area, and $I_3 = \det(\mathcal{D}\phi^T \mathcal{D}\phi)$, which controls the

---

[1] Our general framework can be extended to other invariances as well.

Fig. 4. Discrete geodesic for two different examples from [2] and [23] where the local rate of dissipation is color-coded as ▰▰▰ ▰▰. In the right example the local preservation of isometries is clearly visible, whereas in the left example stretching is the major effect.

local change of volume. Furthermore let us assume that the energy is a convex function of $\mathcal{D}\phi$, cof$\mathcal{D}\phi$, and det $\mathcal{D}\phi$ and that isometries, i. e. deformations with $\mathcal{D}\phi^{\mathrm{T}}(x)\mathcal{D}\phi(x) = \mathbb{1}$, are local minimizers [7] (Fig. 4 provides an example of good local isometry preservation). A template in this class of energy densities is $\bar{W} = \alpha_1 I_1^{\frac{p}{2}} + \alpha_2 I_2^{\frac{q}{2}} + \Gamma(I_3)$ with $p > 0$, $q \geq 0$, α_1, $\alpha_2 > 0$, and Γ convex. Indeed, by straightforward computation one verifies that for any dissipation rate (1), there is a nonlinear energy density of the above type such that the dissipation rate is the corresponding Hessian. In our computations this energy was chosen so that $\frac{\lambda}{\mu} = 3$. A built-in penalization of volume shrinkage, i.e. $\Gamma \xrightarrow{I_3 \to 0} \infty$, ensures local injectivity [22], and thus the sequence of deformations ϕ_k linking objects $\mathcal{O}_{k-1}$ and $\mathcal{O}_k$ actually represents homeomorphisms (rigorously proved for deformations with finite energy under mild assumptions for sufficiently large p, q, certain growth conditions on Γ, and very soft instead of void material on $\Omega\backslash\mathcal{O}$ with Dirichlet boundary conditions on $\partial\Omega$). Let us remark that self-contact at the boundary is still possible, so that the mapping from $\mathcal{S}_{k-1} = \partial\mathcal{O}_{k-1}$ to $\mathcal{S}_k = \partial\mathcal{O}_k$ does not have to be homeomorphic. By interpreting such self-contact as a closing of the gap between two object edges in the sense that the viscous material flows together, this will indeed allow for certain topological transitions along a discrete path in shape space [7] (cf. Fig. 1 for an example). Based on these mechanical preliminaries we can now define a time discrete geodesic path.

Definition (Discrete Geodesic). A discrete path $\mathcal{S}_0, \mathcal{S}_1, \ldots, \mathcal{S}_K$ connecting two shapes $\mathcal{S}_A$ and $\mathcal{S}_B$ is a discrete geodesic, if there exists an associated family of deformations ϕ_k with $\phi_k(\mathcal{S}_{k-1}) = \mathcal{S}_k$ which minimizes the total energy $\sum_{k=1}^{K} \mathcal{E}_{\mathrm{deform}}[\phi_k, \mathcal{S}_{k-1}]$.

Relaxed formulation of the consecutive matching. Computationally, the constraint $\phi_k(\mathcal{S}_{k-1}) = \mathcal{S}_k$ for a 1-1 matching of consecutive shapes is difficult to treat and non-robust (e. g. , not allowing for the handling of noise). Hence, we utilize a relaxed formulation adding a mismatch penalty

$$\mathcal{E}_{\mathrm{match}}[\phi_k, \mathcal{S}_{k-1}, \mathcal{S}_k] = \mathrm{vol}(\mathcal{O}_{k-1}\triangle\phi_k^{-1}(\mathcal{O}_k)),\tag{4}$$

where $A\triangle B = A\backslash B \cup B\backslash A$ defines the symmetric difference between two sets. One might want to further restrict the set of possible shapes $\mathcal{S}_k$ along a discrete geodesic adding an additional surface energy term $\mathcal{E}_{\mathrm{area}}[\mathcal{S}] = \int_{\mathcal{S}} da$. Finally, we end up with the total discrete energy

$$\mathcal{E}_\tau[(\phi_k, \mathcal{S}_{k-1}, \mathcal{S}_k)_{k=1,\ldots,K}]$$
$$= \sum_{i=1}^{K} \left(\frac{1}{\tau} \mathcal{E}_{\mathrm{deform}}[\phi_k, \mathcal{S}_{k-1}] + \eta \mathcal{E}_{\mathrm{match}}[\phi_k, \mathcal{S}_{k-1}, \mathcal{S}_k] + \nu\tau \mathcal{E}_{\mathrm{area}}[\mathcal{S}_k] \right), \quad (5)$$

where η, ν are parameters, and a minimizer of this energy describes a *relaxed discrete geodesic path* between two shapes $\mathcal{S}_A$ and $\mathcal{S}_B$.

3   Viscous Fluid Model in the Limit for $\tau \to 0$

We now investigate the relation of the above-introduced relaxed discrete geodesic paths and a time continuous model for geodesics in shape space.

At first, let us derive from a time discrete sequence of deformations $(\phi_k)_{k=1,\ldots,K}$ and shapes $(\mathcal{S}_k)_{k=0,\ldots,K}$ a time continuous deformation field ϕ_τ, a corresponding motion field v_τ, and a continuous path $\mathcal{S}_\tau$ in shape space:

$$v_\tau(t) := \frac{1}{\tau}(\phi_k - \mathbb{1}), \quad (6)$$
$$\phi_\tau(t) := (\mathbb{1} + (t - t_{k-1})v_\tau) \circ \phi_{k-1} \circ \ldots \phi_1, \quad (7)$$
$$\mathcal{S}_\tau(t) := (\mathbb{1} + (t - t_{k-1})v_\tau)(\mathcal{S}_{k-1}), \quad (8)$$

for $t \in [t_{k-1}, t_k)$. If we now let $\tau \to 0$ and assume that $\mathcal{S}_\tau(t) \to \mathcal{S}(t)$ for a regular family of shapes $(\mathcal{S}(t))_{0 \le t \le 1}$ and that $v_\tau(t) \to v(t)$ with an induced sufficiently regular flow $(\phi(t))_{0 \le t \le 1}$ with $\dot{\phi} = v$, the following limit behavior can be observed: The first term in the global discrete energy representing the deformation energy (2) turns into a time continuous dissipation functional,

$$\mathbf{Diss}[v] = \int_0^1 \int_{\mathcal{O}(t)} \mathbf{C}\epsilon[v] : \epsilon[v] \, dx \, dt, \quad (9)$$

where $\mathbf{C} = 2\,\mathrm{Hess}\,\bar{W}$, $\epsilon[v] = \frac{1}{2}(\mathcal{D}v^{\mathrm{T}} + \mathcal{D}v)$, and $A{:}B = \mathrm{tr}(A^{\mathrm{T}}B)$ for $A, B \in \mathbb{R}^{d \times d}$. To see this, we observe that $\mathcal{D}\phi_\tau(t)^{\mathrm{T}}\mathcal{D}\phi_\tau(t) = \mathbb{1} + 2(t - t_k)\,\epsilon[v_\tau(t)] + O(\tau^2)$ for $t \in [t_k, t_{k+1})$, and by second order Taylor expansion of $\bar{W}$ at the identity, $\bar{W}(\mathcal{D}\phi^{\mathrm{T}}\mathcal{D}\phi) = \tau^2 \mathbf{C}\epsilon[v] : \epsilon[v] + O(\tau^3)$. Here, we have used the fact that $\bar{W}$ attains its minimum 0 at the identity. Thus, the resulting Riemannian structure given by the rate of dissipation is indeed associated with the Hessian of our in general nonlinear deformation energy at the identity. For the well-known exemplary metric (1), the length control based on the first invariant I_1 of $\mathcal{D}\phi_\tau$ turns into the infinitesimal length control via $\mathrm{tr}(\epsilon[v]^2)$, and the volume control based on the third invariant I_3 of $\mathcal{D}\phi_\tau$ turns into the control of compression via $\mathrm{tr}(\epsilon[v])^2$ (cf. Fig. 5 for the impact of these two terms on the shapes along a geodesic path).

Fig. 5. Two geodesic paths between dumb bell shapes varying in the size of the ends. In the left example the ratio λ/μ between the parameters of the dissipation is 0.01 (leading to rather independent compression and expansion of the ends since the associated change of volume implies relatively low dissipation), and 100 in the right example (now mass is actually transported from one end to the other). The underlying texture on the shape domains $\mathcal{O}_0, \ldots, \mathcal{O}_{K-1}$ is aligned to the transport direction, and the absolute value of the velocity v is color-coded as ▇▇▇▇.

In the limit the term for the mismatch energy (4) converges to an optical flow type energy

$$\mathcal{E}_{\mathrm{OF}} = \int_0^1 \int_{\mathcal{S}(t)} \left| (1, v(t))^{\mathrm{T}} \cdot n_{\mathcal{S}(t)} \right| \, \mathrm{d}a \, \mathrm{d}t, \tag{10}$$

where $n_{\mathcal{S}(t)}$ denotes the normal on the shape tube $\cup_{t \in [0,1]} \mathcal{S}(t)$ in space time and $(1, v(t))$ is the underlying space time motion field (cf. L^1 type optimal flow functionals like in [24,25]). To see this, we have to consider $\mathrm{vol}(\mathcal{O}_{k-1} \triangle \phi_k^{-1}(\mathcal{O}_k))$ as the time discrete mismatch induced by a motion field v_τ which is not consistent with the actual time discrete flow of the shape $\mathcal{S}_\tau$. Indeed, $(1, v_\tau)^{\mathrm{T}} \cdot n_{\mathcal{S}_\tau}$ is the local rate with which $(\mathbb{1} + (t - t_{k-1}) v_\tau)(\mathcal{S}_{k-1})$ and the tube of shapes $\cup_{t \in [0,1]} \mathcal{S}(t)$ diverge on the time interval $[t_{k-1}, t_k)$.

The third term of the global energy measuring the shape perimeter turns into the time integral over the perimeter. Finally, as $\mathcal{S}_\tau(t) \to \mathcal{S}(t)$ and $v_\tau(t) \to v(t)$, the energy converges against

$$\mathcal{E}[v, \mathcal{S}] = \mathbf{Diss}[v] + \eta \, \mathcal{E}_{\mathrm{OF}}[v, \mathcal{S}] + \nu \int_0^1 \mathcal{E}_{\mathrm{area}}[\mathcal{S}(t)] \, \mathrm{d}t. \tag{11}$$

The convergence of the time-discrete energy functional (5) for $\tau \to 0$ in the sense of Γ-convergence involves further considerations and is not treated here. In the limit $\eta \to \infty$, the optical flow term will act as a mere penalty which ensures that the family of shapes $\mathcal{S}(t)$ is exactly generated by the flow associated with $v(t)$. In this case and if we set $\nu = 0$, $\mathbf{Diss}[v]$ indeed represents the first fundamental form for the desired Riemannian metric. Thus, the notion of our time discrete geodesics is consistent with this both geometrically and physically sound time continuous geodesic path model in a Riemannian shape space.

4   Regularized Level Set Approximation

To numerically solve the minimization problem for the energy (5), we assume the object domains $\mathcal{O}$ to be represented by zero super level sets $\{x \in \Omega : u(x) > 0\}$ of a scalar function u. Similar representations of shapes have been used for shape matching and warping in [26,11]. We follow the approximation proposed by Chan and Vese [27] and encode the partition of the domain into object and background in the different energy terms via a regularized Heaviside function $H_\epsilon(u_k)$. As in [27] we consider the function $H_\epsilon(x) := \frac{1}{2} + \frac{1}{\pi} \arctan\left(\frac{x}{\epsilon}\right)$, where ϵ is a scale parameter representing the width of the smeared-out shape contour. Hence, the mismatch energy is replaced by the approximation

$$\mathcal{E}^\epsilon_{\text{match}}[\phi_k, u_{k-1}, u_k] = \int_\Omega (H_\epsilon(u_k(\phi_k)) - H_\epsilon(u_{k-1}))^2 dx, \tag{12}$$

and the area of the kth shape $\mathcal{S}_k$ is replaced by the total variation $\mathcal{E}^\epsilon_{\text{area}}[u_k] = \int_\Omega |\nabla H_\epsilon(u_k)|\, dx$ of $H_\epsilon \circ u_k$. With respect to the deformation energy we assume that the whole computational domain is deformed, but with a material which is several orders of magnitude softer on the complement set $\Omega \setminus \mathcal{O}_k$ than inside $\mathcal{O}_k$. Hence, the elastic energy (2) is replaced by the energy

$$\mathcal{E}^{\epsilon,\delta}_{\text{deform}}[\phi_k, u_{k-1}] = \int_\Omega ((1-\delta)H_\epsilon(u_{k-1}) + \delta)\, W(\mathcal{D}\phi_k) dx, \tag{13}$$

where $\delta = 10^{-4}$ in our implementation. Let us emphasize that in the energy minimization algorithm, the guidance of the initial zero level lines towards the final shapes relies on the nonlocal support of the regularized Heaviside function (cf. [28]). Finally, we end up with the approximation of the total energy,

$$\mathcal{E}^{\epsilon,\delta}_\tau[(\phi_k, u_k)_k] = \sum_{k=1}^K \left(\frac{1}{\tau} \mathcal{E}^{\epsilon,\delta}_{\text{deform}}[\phi_k, u_{k-1}] + \eta \mathcal{E}^\epsilon_{\text{match}}[\phi_k, u_{k-1}, u_k] + \nu\tau \mathcal{E}^\epsilon_{\text{area}}[u_k] \right). \tag{14}$$

In our applications we have chosen $\eta = 200$ and $\nu = 0$ except for Fig. 7, where $\nu = 0.005$. The essential formulas for the variation of the energy can be found in the appendix.

5   Finite Element Discretization in Space

For the spatial discretization of the energy $\mathcal{E}^{\epsilon,\delta}_\tau$ in (14) the finite element method has been applied. The level set functions u_k and the different components of the deformations ϕ_k are represented by continuous, piecewise multilinear (trilinear in 3D and bilinear in 2D) finite element functions U_k and Φ_k on a regular grid superimposed on the domain $\Omega = [0,1]^d$. For the ease of implementation we consider dyadic grid resolutions with $2^L + 1$ vertices in each direction and a grid size $h = 2^{-L}$. In 2D we considered $L = 7, \ldots, 10$ and in 3D $L = 7$.

Single level minimization algorithm. For fixed time step τ and fixed spatial grid size h, let us denote by $\mathcal{E}^{\epsilon,\delta}_{\tau,h}[(\Phi_k, U_k)_k]$ the discrete total energy depending on the set of K discrete deformations $\Phi_1, \ldots, \Phi_K$ and $K + 1$ discrete level set functions $U_0, \ldots, U_K$, where U_0 and U_K describe the shapes $\mathcal{S}_A$ and $\mathcal{S}_B$ and are fixed. This is a nonlinear functional both in the discrete deformations Φ_k

(due to the concatenation $U_k \circ \Phi_k$ with the discrete level set function U_k and the nonlinear integrand $W(\cdot)$ of the deformation energy $\mathcal{E}_{\text{deform}}^{\epsilon,\delta}$) and in the discrete level set functions U_k (due to the concatenation with the regularized Heaviside function $H_\epsilon(\cdot)$). In our energy relaxation algorithm for fixed time step and grid size, we consider a gradient descent approach. We constantly alternate between performing a single gradient descent step for all deformations and one for all level set functions. The step sizes are chosen according to Armijo's rule. This simultaneous relaxation with respect to the whole set of discrete deformations and discrete level set functions, respectively, already outperforms a simple non-linear Gauss-Seidel type relaxation (cf. Fig. 3). Nevertheless, the capability to identify a globally optimal shortest path between complicated shapes depends on an effective multi–scale relaxation strategy (see below).

Numerical quadrature. Integral evaluations in the energy descent algorithm are performed by Gaussian quadrature of third order on each grid cell. For various terms we have to evaluate pushforwards $U \circ \Phi$ of a discretized level set function U or a test function under a discretized deformation Φ. In our algorithm, this evaluation is performed exactly at the quadrature points.

Cascadic multi–scale algorithm. The variational problem considered here is highly nonlinear, and for fixed time step size the proposed scheme is expected to have very slow convergence; also it might end up in some nearby local minimum. Here, a multi-level approach (initial optimization on a coarse scale and successive refinement) turns out to be indispensable in order to accelerate convergence and not to be trapped in local minima far from the global minimum. Due to our assumption of a dyadic resolution $2^L + 1$ in each grid direction, we are able to build a hierarchy of grids with $2^l + 1$ nodes in each direction for $l = L, \ldots, 0$. Via a simple restriction operation we restrict every finite element function to any of these coarse grid spaces. Starting the optimization on a coarse grid, the results from coarse scales are successively prolongated onto the next grid level for a refinement of the solution [29]. Hence, the construction of a multigrid hierarchy allows to solve coarse scale problems in our multi-scale approach on coarse grids. Since the width ϵ of the diffusive shape representation $H_\epsilon \circ u_k$ should naturally scale with the grid width h, we choose $\epsilon = h$.

On a 3 GHz Pentium 4, still without runtime optimization, 2D computations for $L = 8$ and $K = 8$ require ~ 1 h. Based on a parallelized implementation we observed almost linear scaling.

6   Further Results and Generalizations

We have computed discrete geodesic paths for 2D and 3D shape contours. The method is both robust and flexible due to the underlying implicit shape description via level sets (cf. Fig. 1 and Fig. 6). Indeed, neither topologically equivalent meshes on the initial shapes are required, nor need the shapes themselves be topologically equivalent. In addition, we can easily restrict the approach

Fig. 6. Geodesic path between the cat and the lion, with the local rate of dissipation on the shapes $\mathcal{S}_0, \ldots, \mathcal{S}_{K-1}$ color-coded as ▬▬▬ (top) and a transparent slicing plane with texture-coded level lines (bottom)

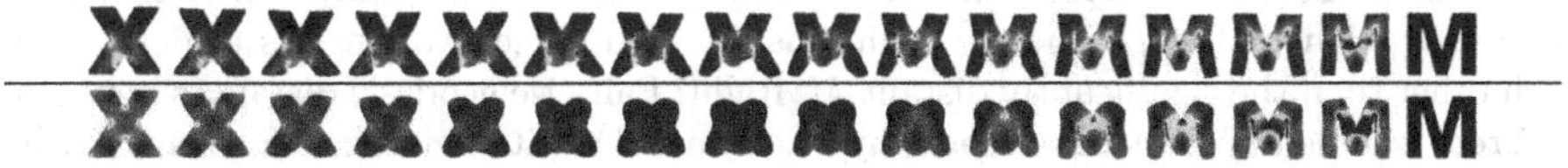

Fig. 7. Geodesic paths between an X and an M, without a contour length term ($\nu = 0$), allowing for crack formation, (top rows) and with this term damping down cracks and rounding corners (bottom rows). In the bottom rows we additionally enforced area preservation along the geodesic.

Fig. 8. A discrete geodesic connecting different poses of a matchstick man can be computed (from left to right starting with the second), even though part of one arm and one leg of $\mathcal{S}_0$ (left) are occluded.

to the submanifold of 2D area or 3D volume preserving objects based on a predictor corrector scheme. Fig. 7 shows an example of two different geodesics between the letters X and M, demonstrating the impact of the term $\mathcal{E}_{\text{area}}$ controlling the $d - 1$ dimensional area of the shapes.

In many shape classification applications, one would like to evaluate the distance of a partially occluded shape from a given template shape. As a proof of concept, Fig. 8 depicts a corresponding discrete geodesic path. This requires a minor modification of our model, i. e. solely for $k = 0$ in $\mathcal{E}^{\epsilon}_{\text{match}}$ we insert a smooth function as a mask for $\mathcal{S}_0$.

Furthermore, we evaluated distances between different 2D letters based on the discrete geodesic path length. The resulting clustering is shown in Fig. 9 left. Finally, in Fig. 9 right we studied distances between four different foot level sets converted from 3D scans. Surprisingly, the observed clustering is different from the criterion based on the enclosed volume.

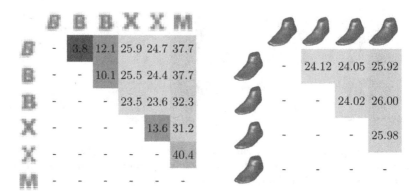

Fig. 9. Left: Pairwise geodesic distances between (also topologically) different letter shapes. Obviously, the *B*s and *X*s form clusters, and these two clusters are closer to each other than the significantly distant *M*. Right: Pairwise geodesic distances between different scanned 3D feet (data courtesy of adidas). Despite being geometrically fairly close, the computed geodesic distance allows to single out the fourth foot as being significantly farther away from the other three, which are almost at equal distance, even though feet 1 and 4 are of equal volume and feet 2 and 3 have 13 % less volume.

7   Conclusions and Future Work

We have proposed a novel variational time discretization of geodesics in shape space. The key ingredients are the 1-1 mapping property between consecutive time steps and the rigid body motion invariance. The approach is physically motivated and based on measuring flow-induced dissipation in the interior of shape contours. The proposed formulation allows to weight the effect of the local change of length and volume separately, leading to significantly different geodesic paths. Both physically and with respect to the shape description, geodesic paths can undergo certain topological transitions. A cascadic multi–scale relaxation strategy renders the computation robust and effective. Future generalization of the model might deal with the incorporation of prior statistical knowledge and the space of general image morphologies. Furthermore, we would like to rigorously investigate the time discrete to time continuous limit via the concept of Γ convergence.

Acknowledgement. This work has been partially supported by NSF, ARO, ONR, NGA, and DARPA. Benedikt Wirth was supported by BIGS Mathematics and the Hausdorff Center for Mathematics at Bonn University. Furthermore, the authors thank Sergio Conti for valuable hints on the physical model.

References

1. Zhu, L., Yang, Y., Haker, S., Allen, T.: An image morphing technique based on optimal mass preserving mapping. IEEE T. Image Process 16(6), 1481–1495 (2007)
2. Fuchs, M., Jüttler, B., Scherzer, O., Yang, H.: Shape metrics based on elastic deformations. Journal of Mathematical Imaging and Vision (to appear, 2009)

3. Miller, M.I., Younes, L.: Group actions, homeomorphisms and matching: a general framework. International Journal of Computer Vision 41(1-2), 61–84 (2001)

4. Schmidt, F.R., Clausen, M., Cremers, D.: Shape matching by variational computation of geodesics on a manifold. In: Franke, K., Müller, K.-R., Nickolay, B., Schäfer, R. (eds.) DAGM 2006. LNCS, vol. 4174, pp. 142–151. Springer, Heidelberg (2006)

5. Zhao, H.K., Chan, T., Merriman, B., Osher, S.: A variational level set approach to multiphase motion. J. Comp. Phys. 127, 179–195 (1996)

6. Luckhaus, S., Sturzenhecker, T.: Implicit time discretization for the mean curvature flow equation. Calc. Var. 3, 253–271 (1995)

7. Ciarlet, P.G.: Three-dimensional elasticity. Elsevier Science Publisers B. V., Amsterdam (1988)

8. Kendall, D.G.: Shape manifolds, procrustean metrics, and complex projective spaces. Bull. London Math. Soc. 16, 81–121 (1984)

9. Mémoli, F., Sapiro, G.: A theoretical and computational framework for isometry invariant recognition of point cloud data. Found. Comput. Math. 5, 313–347 (2005)

10. Bronstein, A., Bronstein, M., Kimmel, R.: Numerical Geometry of Non-Rigid Shapes. Monographs in Computer Science. Springer, Heidelberg (2008)

11. Charpiat, G., Faugeras, O., Keriven, R.: Approximations of shape metrics and application to shape warping and empirical shape statistics. Foundations of Computational Mathematics 5(1), 1–58 (2005)

12. Eckstein, I., Pons, J., Tong, Y., Kuo, C., Desbrun, M.: Generalized surface flows for mesh processing. In: Eurographics Symposium on Geometry Processing (2007)

13. Michor, P.W., Mumford, D.: Riemannian geometries on spaces of plane curves. J. Eur. Math. Soc. 8, 1–48 (2006)

14. Younes, L.: Computable elastic distances between shapes. SIAM J. Appl. Math. 58, 565–586 (1998)

15. Miller, M.I., Younes, L.: Group actions, homeomorphisms and matching: a general framework. Technical report, John Hopkins University, Maryland (1999)

16. Klassen, E., Srivastava, A., Mio, W., Joshi, S.: Analysis of planar shapes using geodesic paths on shape spaces. IEEE T. Pattern Anal. 26(3), 372–383 (2004)

17. Dupuis, D., Grenander, U., Miller, M.: Variational problems on flows of diffeomorphisms for image matching. Quarterly of Applied Mathematics 56, 587–600 (1998)

18. Miller, M., Trouvé, A., Younes, L.: On the metrics and Euler-Lagrange equations of computational anatomy. Ann. Rev. Biomed. Eng. 4, 375–405 (2002)

19. Sundaramoorthi, G., Yezzi, A., Mennucci, A.: Sobolev active contours. International Journal of Computer Vision 73(3), 345–366 (2007)

20. Kilian, M., Mitra, N.J., Pottmann, H.: Geometric modeling in shape space. ACM Transactions on Graphics 26(64), 1–8 (2007)

21. Droske, M., Rumpf, M.: Multi scale joint segmentation and registration of image morphology. IEEE Trans. Pattern Anal. 29(12), 2181–2194 (2007)

22. Ball, J.: Global invertibility of Sobolev functions and the interpenetration of matter. Proc. Roy. Soc. Edinburgh 88A, 315–328 (1981)

23. Charpiat, G., Maurel, P., Pons, J.P., Keriven, R., Faugeras, O.: Generalized gradients: Priors on minimization flows. Int. J. Comput. Vision 73(3), 325–344 (2007)

24. Kornprobst, P., Deriche, R., Aubert, G.: Image sequence analysis via partial differential equations. Journal of Mathematical Imaging and Vision 11, 5–26 (1999)

25. Black, M.J., Anandan, P.: A framework for the robust estimation of optical flow. In: Fourth International Conference on Computer Vision, ICCV 1993, pp. 231–236 (1993)

26. Kapur, T., Yezzi, L., Zöllei, L.: A variational framework for joint segmentation and registration. In: IEEE CVPR - MMBIA, pp. 44–51 (2001)
27. Chan, T.F., Vese, L.A.: Active contours without edges. IEEE Transactions on Image Processing 10(2), 266–277 (2001)
28. Caselles, V., Kimmel, R., Sapiro, G.: Geodesic active contours. International Journal of Computer Vision 22(1), 61–79 (1997)
29. Bornemann, F., Deuflhard, P.: The cascadic multigrid method for elliptic problems. Num. Math. 75(2), 135–152 (1996)

Appendix. Here, we give explicit formulas for the variation of the different energy contributions in directions of the unknown functions u_k $(k = 1, \ldots, K-1)$ and ϕ_k $(k = 1, \ldots, K)$, required in the numerical implementation. Let us denote by $\langle \delta_w \mathcal{E}, \vartheta \rangle$ a variation of an energy $\mathcal{E}$ with respect to a parameter function w in a direction ϑ. Using straightforward differentiation, for sufficiently smooth u_k and ϕ_k we obtain

$$\langle \delta_{\phi_k} \mathcal{E}^\epsilon_{\mathrm{match}}[\phi_k, u_{k-1}, u_k], \psi \rangle = 2 \int_\Omega \left(H_\epsilon(u_k \circ \phi_k) - H_\epsilon(u_{k-1}) \right) \delta_\epsilon(u_k \circ \phi_k) \nabla u_k \circ \phi_k \cdot \psi \, dx \, ,$$

$$\langle \delta_{u_{k-1}} \mathcal{E}^\epsilon_{\mathrm{match}}[\phi_k, u_{k-1}, u_k], \vartheta \rangle = -2 \int_\Omega \left(H_\epsilon(u_k \circ \phi_k) - H_\epsilon(u_{k-1}) \right) \delta_\epsilon(u_{k-1}) \vartheta \, dx \, ,$$

$$\langle \delta_{u_k} \mathcal{E}^\epsilon_{\mathrm{match}}[\phi_k, u_{k-1}, u_k], \vartheta \rangle = 2 \int_\Omega \left(H_\epsilon(u_k \circ \phi_k) - H_\epsilon(u_{k-1}) \right) \delta_\epsilon(u_k \circ \phi_k) \vartheta \circ \phi_k \, dx \, ,$$

$$\langle \delta_{\phi_k} \mathcal{E}^{\epsilon,\delta}_{\mathrm{deform}}[\phi_k, u_{k-1}], \psi \rangle = \int_\Omega \left((1-\delta) H_\epsilon(u_{k-1}) + \delta \right) W_{,A}(\mathcal{D}\phi_k) : \mathcal{D}\psi \, dx$$

for test functions ϑ and test displacements ψ, where $W_{,A}$ denotes the derivative of W with respect to its matrix argument. For the variation of $\mathcal{E}^\epsilon_{\mathrm{area}}[u_k]$ we refer to [27].

Image Registration under Varying Illumination: Hyper-Demons Algorithm

Mehran Ebrahimi and Anne L. Martel

Department of Medical Biophysics, University of Toronto
Imaging Research, Sunnybrook Health Sciences Centre
Toronto, Ontario, Canada
`mehran.ebrahimi@sri.utoronto.ca, anne.martel@sri.utoronto.ca`

Abstract. The goal of this paper is to present a novel recipe for deformable image registration under varying illumination, as a natural extension of the demons algorithm. This generalization is derived directly from the optical-flow constraints in a variational formulation. Furthermore, our approach provides a new mathematical interpretation of the demons algorithm via fixed-point iterations in a consistent framework.

1 Introduction

Since the appearance of Thirion's work [13,14], the so-called *demons method* has become a popular deformable registration technique in medical imaging. A reason for this success may be attributed to the simplicity, speed, and performance of Thirion's proposed technique. Over the years, the demons method has been extended by several authors [6,2,15,16] who have proposed different variations.

Although Thirion's seminal work [13,14] intuitively borrows physical ideas of thermodynamics, deeper mathematical insight of these ideas has been introduced in [10,11] which focuses on understanding the algorithm.

In this paper, we shall provide a novel mathematical interpretation of the demons algorithm, originating directly from the optical flow constraints [7,5,17]. The optical flow equations [7] inspired Thirion to introduce his so-called *forces* in his proposed scheme. Furthermore, to the best of our knowledge, none of these existing extensions of the demons algorithm focuses directly on the varying illumination problems introduced by changes in image intensity such as those due to contrast enhancement or to changes in coil sensitivity between images. Our formulation shall address this varying illumination problem.

To proceed, we need to rigorously introduce the required background material which we will be employed in our formulation.

In Section 2, we will introduce an intensity-based deformation model. For consistency, we adapt our notation from [10] and shall mention our distinctive differences when necessary. In Section 3, we briefly cover the background on optical flow constraints and associated methods, including the demons algorithm. Section 4 is dedicated to developing our extension. Finally, we will present various computational experiments and concluding remarks in Section 5.

D. Cremers et al. (Eds.): EMMCVPR 2009, LNCS 5681, pp. 303–316, 2009.

2   An Intensity-Based Deformation Model

An *intensity-based deformation process* is modeled via a real-valued multivariate intensity function $E(X(x,t),t)$ in which $t \in [0,1]$ is an artificial time-step, $x = (x_1, x_2, \ldots, x_d)$ represents the initial coordinate of a particle in the image domain $\Omega \subset \mathbb{R}^d$, and the function

$$X : \Omega \times [0,1] \to \mathbb{R}^d, \quad X = (X_1, X_2, \ldots, X_d)$$

defines the path of the particle [10]. Based on the fact that the coordinate of the particle at time $t = 0$ is x, we write $X(x,0) = x$. We also assume that the coordinate of the particle at time $t = 1$ is defined by the transformation $\phi(x)$ meaning that $X(x,1) = \phi(x)$. The known measurements of such deformation process are the target image $f(x)$ and the source image $g(x)$ defined in $L^2(\Omega)$, under the two main assumptions

$$E \mid_{t=0} = f(X(x,0)) = f(x),$$

$$E \mid_{t=1} = g(X(x,1)) = g \circ \phi(x).$$

[We have slightly modified the model presented in [10] (pp. 159) that assumes $E \mid_{t=1} = g(x)$.] We are now able to define deformations both at a time instance and over a time interval. More formally, we define a time dependent instantaneous deformation vector

$$\frac{dX}{dt} = (u_1, u_2, \ldots, u_d) = u$$

and the total displacement vector by

$$U(x) = X(x,1) - X(x,0) = \phi(x) - x. \tag{1}$$

Equation (1) yields $\phi(x) = U(x) + x$. Furthermore, by the fundamental theorem of calculus we can write

$$\int_0^1 [\frac{dE}{dt}]dt = E \mid_{t=1} - E \mid_{t=0} = g \circ \phi - f, \tag{2}$$

$$\int_0^1 u \, dt = \int_0^1 [\frac{dX}{dt}]dt = X(x,1) - X(x,0) = U(x). \tag{3}$$

These equations will be used in the following sections.

3   Optical Flow Constraints and Associated Methods

3.1   Conserved Intensity

In the simplest case, the optical flow equation (see for example Horn-Schunck [7]) relies on the conservation (or constancy) of intensity at a time instance t

and a spatial location x, i.e., $\frac{dE}{dt} = 0$. Hence,

$$\frac{dE}{dt} = \frac{dE(X(x,t),t)}{dt} = E_t + \sum_{i=1}^{d} E_{X_i} \frac{dX_i}{dt}$$

$$= E_t + \sum_{i=1}^{d} u_i E_{X_i} = 0.$$

This equation contains d unknowns $\{u_i\}_{1 \leq i \leq d}$ and does not possess a unique solution. A possibility to overcome this ill-posed problem is to regularize the solution by exploiting spatial correspondence in the image domain Ω at any time instance t. Horn and Schunck [7] use a uniform smoothness constraint in the regularization expression and form

$$\arg\min_{\{u_i\}} \int_{x \in \Omega} \left[\lambda^2 \sum_{i=1}^{d} \sum_{j=1}^{d} (\frac{\partial u_i}{\partial x_j})^2 \right] + \left[E_t + \sum_{i=1}^{d} u_i E_{X_i} \right]^2 dx.$$

To compute the d unknown instantaneous deformation vectors $\{u_i\}_{1 \leq i \leq d}$, the corresponding Euler-Lagrange equations are computed, a discretization of time with $\Delta t = 1$ is used and the solution is estimated iteratively. It is worth mentioning that the original model of [7] does not distinguish between x and X.

3.2 Non-conserved Intensity

Gennert-Negahdaripour [5] propose a non-conserved intensity model in which the rate of change of intensity is modeled as a linear (polynomial of degree $m = 1$) expression of intensity with spatially-varying coefficients c_k as

$$\frac{dE}{dt} = \sum_{k=0}^{m} c_k E^k.$$

The corresponding minimization is

$$\min_{\{u_i\}\{c_k\}} \int_{\Omega} \left[\lambda^2 \sum_{i=1}^{d} \sum_{j=1}^{d} (\frac{\partial u_i}{\partial x_j})^2 + \sum_{k=0}^{m} \lambda_k^2 \sum_{j=1}^{d} (\frac{\partial c_k}{\partial x_j})^2 \right] +$$

$$\left[E_t + \sum_{i=1}^{d} u_i E_{X_i} - \sum_{k=0}^{m} c_k E^k \right]^2 dx, \tag{4}$$

which contains $d + m + 1$ unknowns that are the instantaneous deformation vectors $\{u_i\}_{1 \leq i \leq d}$ and the coefficients of the intensity shift $\{c_k\}_{0 \leq k \leq m}$. The approach in [5] is very similar to Horn-Schunck's. However, the original model in [5] only addresses a linear intensity shift, i.e., $m = 1$, and does not consider the case where $d > 2$ which involves larger matrix inversions.

A series of methods introduced in [1,4,9] propose constructing a very large system of equations directly based on the expression (4). Iterative solvers (e.g., conjugate gradient method) are used to minimize the objective functional and estimate the deformation vectors. However, these methods require enormous memory and their hierarchical extensions are not obvious in general.

3.3   Demons Method

The demons method proposed by Thirion [13,14] introduces a notation adapted from Maxwell's demons in thermodynamics. As a special type of demon-type Thirion introduces "a complete grid of demons" in which the "pushing forces" are computed for demons placed at every pixel of the image and computed based on the minimum-norm solution of $\frac{dE}{dt} = E_t + \sum_{i=1}^{d} u_i E_{X_i} = 0$. Thirion approximates such force expressions as $u_i = \frac{-f_{x_i}(g-f)}{\epsilon^2 + \sum_{i=1}^{d} f_{x_i}^2}$ that leads to Alg. 1 for a fixed number of iterations. It is worth mentioning that no distinction between instantaneous and total deformations (u and U) is made in the original approach [13,14]. Also, ϵ is replaced by $f - g$ which may cause instabilities. Some of these issues have been clarified in [10].

Alg. 1.   Demons algorithm, complete grid of demons

> **read** target image $f(x)$ and source image $g(x)$.
> **choose** a linear smoothing operator G, a number ϵ, and some large N.
> **set** $U_i^{(1)}(x) = 0$,   for $1 \le i \le d$ and $x \in \Omega$.
> **for n $= 1 : N$ do**   // Iterations
> $$\phi(x) = U^{(n)}(x) + x$$
> $$U_i^{(n+1)} = G\left[U_i^{(n)} - \frac{f_{x_i}\left[g \circ \phi - f\right]}{\epsilon^2 + \|\nabla f\|^2}\right], \qquad 1 \le i \le d$$
> **end**
> **return** the displacement $U^{(n+1)}$, and the registered source image $g \circ \phi$.

---

4   Building a General Model

4.1   The Problem Set-Up

We plan to derive Alg. 1 and include the intensity correction terms in the demons algorithm. To do so, we revisit the Gennert-Negahdaripour model [5]

$$\min_{\{u_i\}\{c_k\}} \int_\Omega \left[\lambda^2 \sum_{i=1}^{d}\sum_{j=1}^{d}\left(\frac{\partial u_i}{\partial x_j}\right)^2 + \sum_{k=0}^{m}\lambda_k^2\sum_{j=1}^{d}\left(\frac{\partial c_k}{\partial x_j}\right)^2\right] +$$

$$\left[E_t + \sum_{i=1}^{d} u_i E_{X_i} - \sum_{k=0}^{m} c_k E^k\right]^2 dx.$$

Due to an anticipated improved representation, we change the scale of the coefficients defining $\alpha_k = \lambda/\lambda_k$, $w_k = -c_k/\alpha_k$ and obtain the new functional

$$L = \int_\Omega \lambda^2 \Big[\sum_{i=1}^{d} \sum_{j=1}^{d} (\frac{\partial u_i}{\partial x_j})^2 + \sum_{k=0}^{m} \sum_{j=1}^{d} (\frac{\partial w_k}{\partial x_j})^2 \Big] +$$
$$\Big[E_t + \sum_{i=1}^{d} u_i E_{X_i} + \sum_{k=0}^{m} \alpha_k w_k E^k \Big]^2 dx.$$

This change of variable, will enable us to use a simplified version of Sherman-Morrison-Woodbury matrix inversion lemma [18,12] in deriving the solution in the next section. The new objective is to minimize L with respect to $\{u_i\}_{1 \le i \le d}$ and $\{w_k\}_{0 \le k \le m}$. The corresponding Euler-Lagrange equations of this minimization yields

$\forall\ 1 \le i \le d$,

$$\lambda^2 (\sum_{j=1}^{d} \frac{\partial^2 u_i}{\partial x_j{}^2}) = E_{X_i} \Big[E_t + \sum_{i=1}^{d} u_i E_{X_i} + \sum_{k=0}^{m} \alpha_k w_k E^k \Big], \qquad (5)$$

$\forall\ 0 \le k \le m$,

$$\lambda^2 (\sum_{j=1}^{d} \frac{\partial^2 w_k}{\partial x_j{}^2}) = \alpha_k E^k \Big[E_t + \sum_{i=1}^{d} u_i E_{X_i} + \sum_{k=0}^{m} \alpha_k w_k E^k \Big]. \qquad (6)$$

Similar to [7], assume that for any v we approximate the Laplacian of v with

$$\nabla^2 v = \sum_{j=1}^{d} \frac{\partial^2 v}{\partial x_j{}^2} \approx \kappa^2 (Gv - v)$$

in which G is some linear smoothing operator and κ is a constant real number. One can verify that assuming G to be a Gaussian with zero mean and standard deviation $\frac{1}{\sqrt{\ln 4}}$ and $\kappa = 2$ leads to the corresponding approximations presented in [7]. We also define the set of vectors

$$\mathbf{M} = (E_{X_1}, E_{X_2}, \ldots, E_{X_d}, \alpha_0, \alpha_1 E^1, \ldots, \alpha_m E^m)$$

$$\mathbf{V} = (u_1, u_2, \ldots, u_d, w_0, w_1, \ldots, w_m)^T$$

$$\bar{\mathbf{V}} = G[\mathbf{V}] = (Gu_1, Gu_2, \ldots, Gu_d, Gw_0, Gw_1, \ldots, Gw_m)^T$$

which will be used in finding the solution.

4.2   Deriving the Solution

The set of Euler-Lagrange Equations (5) and (6) can be summarized as

$$(\lambda\kappa)^2(\bar{\mathbf{V}} - \mathbf{V}) = \mathbf{M}^T(E_t + \mathbf{MV}).$$

Arranging the terms to form a linear equation of $\mathbf{V}$ yields

$$\left[\mathbf{M}^T\mathbf{M} + (\lambda\kappa)^2\mathbf{I}\right]\mathbf{V} = (\lambda\kappa)^2\bar{\mathbf{V}} - E_t\mathbf{M}^T. \tag{7}$$

Solving Equation (7) for $\mathbf{V}$ using Lemma 1 (simplified Sherman-Morrison-Woodbury matrix inversion formula [18,12]), for $\epsilon = \lambda\kappa$ yields [See the proofs in the Appendix]

$$\mathbf{V} = \bar{\mathbf{V}} - \frac{\mathbf{M}^T(E_t + \mathbf{M}\bar{\mathbf{V}})}{(\lambda\kappa)^2 + \mathbf{MM}^T}. \tag{8}$$

Lemma 1. If $\mathbf{M}$ is a row-vector, then for any nonzero ϵ

$$[\mathbf{M}^T\mathbf{M} + \epsilon^2\mathbf{I}]^{-1} = \frac{1}{\epsilon^2}[\mathbf{I} - \frac{\mathbf{M}^T\mathbf{M}}{\epsilon^2 + \mathbf{MM}^T}].$$

Finally, taking G from both sides of Equation (8) yields

$$\bar{\mathbf{V}} = G\left[\mathbf{V}\right] = G\left[\bar{\mathbf{V}} - \frac{\mathbf{M}^T(E_t + \mathbf{M}\bar{\mathbf{V}})}{(\lambda\kappa)^2 + \mathbf{MM}^T}\right].$$

Hence, we wish to approximate $\bar{\mathbf{V}}$ that satisfies

$$\bar{\mathbf{V}} = G\left[\bar{\mathbf{V}} - \frac{\mathbf{M}^T(E_t + \mathbf{M}\bar{\mathbf{V}})}{(\lambda\kappa)^2 + \mathbf{MM}^T}\right].$$

Momentarily, renaming $\bar{\mathbf{V}}$ as $\mathbf{V}$ we find an estimate of $\mathbf{V}$ that instead satisfies

$$\mathbf{V} = G\left[\mathbf{V} - \frac{\mathbf{M}^T(E_t + \mathbf{MV})}{(\lambda\kappa)^2 + \mathbf{MM}^T}\right]. \tag{9}$$

From now on, we remember that any approximation of $\mathbf{V}$ requires a deconvolution step with respect to G (if G is a convolution operator) due to this change of variable. Such deconvolution is not performed assuming $\mathbf{V}$ is smooth enough. To estimate $\mathbf{V}$ that satisfies Equation (9) converting to the original notation we would like to find $\{u_i\}_{1\leq i\leq d}$, $\{w_k\}_{0\leq k\leq m}$ that

$\forall \ 1 \leq i \leq d,$

$$u_i = G\left[u_i - \frac{E_{X_i}\left[E_t + \sum_{i=1}^d u_i E_{X_i} + \sum_{k=0}^m \alpha_k w_k E^k\right]}{(\lambda\kappa)^2 + \sum_{i=1}^d (E_{X_i})^2 + \sum_{k=0}^m (\alpha_k E^k)^2}\right], \tag{10}$$

$\forall \ 0 \leq k \leq m,$

$$w_k = G\left[w_k - \frac{\alpha_k E^k\left[E_t + \sum_{i=1}^d u_i E_{X_i} + \sum_{k=0}^m \alpha_k w_k E^k\right]}{(\lambda\kappa)^2 + \sum_{i=1}^d (E_{X_i})^2 + \sum_{k=0}^m (\alpha_k E^k)^2}\right]. \tag{11}$$

Now we are ready to move from instantaneous deformation vectors to total displacements. Integrating Equation (10) with respect to time t over $[0, 1]$ yields

$$\int_0^1 u_i \, dt = G\left[\int_0^1 u_i \, dt - \right.$$

$$\left.\int_0^1 \frac{E_{X_i}\left[E_t + \sum_{i=1}^d u_i E_{X_i} + \sum_{k=0}^m \alpha_k w_k E^k\right]}{(\lambda\kappa)^2 + \sum_{i=1}^d (E_{X_i})^2 + \sum_{k=0}^m (\alpha_k E^k)^2} dt\right],$$

for any $1 \leq i \leq d$. Hence, from Equation (3), $\frac{dE}{dt} = E_t + \sum_{i=1}^d u_i E_{X_i}$, and we obtain

$$U_i = G\left[U_i - \int_0^1 \frac{E_{X_i}\left[\frac{dE}{dt} + \sum_{k=0}^m \alpha_k w_k E^k\right]}{(\lambda\kappa)^2 + \sum_{i=1}^d (E_{X_i})^2 + \sum_{k=0}^m (\alpha_k E^k)^2} dt\right]. \tag{12}$$

Choosing $\Delta t = 1$ and approximating the integrals using left Riemann sum evaluating E and its partial derivatives yields for any $1 \leq i \leq d$

$$U_i = G\left[U_i - \frac{f_{x_i}\left[\int_0^1 \frac{dE}{dt} dt + \sum_{k=0}^m \alpha_k f^k \int_0^1 w_k dt\right]}{(\lambda\kappa)^2 + \sum_{i=1}^d (f_{x_i})^2 + \sum_{k=0}^m (\alpha_k f^k)^2}\right].$$

[Note that the choice of $\Delta t = 1$ is the extreme case of Riemann sum where the interval is partitioned to only one element. The authors do not claim that this provides a superior approximation. This approximation is employed for consistency with [7].] Hence, using Equation (2) yields for any $1 \leq i \leq d$

$$U_i = G\left[U_i - \frac{f_{x_i}\left[g \circ \phi - f + \sum_{k=0}^m \alpha_k f^k W_k\right]}{(\lambda\kappa)^2 + \sum_{i=1}^d (f_{x_i})^2 + \sum_{k=0}^m (\alpha_k f^k)^2}\right],$$

in which $W_k = \int_0^1 w_k dt$. Similarly, if we proceed the same approach for w_k, $0 \leq k \leq m$, starting from Equation (11) we obtain the following pair of equations.

$\forall \ 1 \leq i \leq d,$

$$U_i = G\left[U_i - \frac{f_{x_i}\left[g \circ \phi - f + \sum_{k=0}^m \alpha_k f^k W_k\right]}{(\lambda\kappa)^2 + \|\nabla f\|^2 + \sum_{k=0}^m (\alpha_k f^k)^2}\right], \tag{13}$$

$\forall \ 0 \leq k \leq m,$

$$W_k = G\left[W_k - \frac{\alpha_k f^k\left[g \circ \phi - f + \sum_{k=0}^m \alpha_k f^k W_k\right]}{(\lambda\kappa)^2 + \|\nabla f\|^2 + \sum_{k=0}^m (\alpha_k f^k)^2}\right], \tag{14}$$

in which $\phi(x) = U(x) + x$. The solution of these two equations will be approximated numerically using the *fixed-point iterations*.

4.3   Hyper-Demons Algorithm

Approximating the solution of Equations (13) and (14) using fixed-point iterations, we obtain the Hyper-Demons Alg. 2. Assuming the normalization factors $[\alpha_0, \ldots, \alpha_m]$ to be all zero makes Alg. 2 equivalent to the demons Alg. 1. Hence, the demons algorithm is a special case of our proposed hyper-demons algorithm. A simple case of the algorithm with 0-degree polynomial is given in Alg. 3.

Alg. 2.   Hyper-Demons Algorithm

---

read target image $f(x)$ and source image $g(x)$.

choose a linear smoothing operator G, a number ϵ, polynomial degree m, normalizing factors $[\alpha_0, \ldots, \alpha_m]$, and some large N.

set $U_i^{(1)}(x) = W_k^{(1)}(x) = 0$, for any $1 \le i \le d$, $0 \le k \le m$, $x \in \Omega$.

for n = 1 : N do   // *Fixed point iterations*

$\phi(x) = U^{(n)}(x) + x$

$$U_i^{(n+1)} = G\left[U_i^{(n)} - \frac{f_{x_i}\left[g\circ\phi - f + \sum_{k=0}^{m} \alpha_k f^k W_k^{(n)}\right]}{\epsilon^2 + \|\nabla f\|^2 + \sum_{k=0}^{m}(\alpha_k f^k)^2}\right], \qquad 1 \le i \le d$$

$$W_k^{(n+1)} = G\left[W_k^{(n)} - \frac{\alpha_k f^k\left[g\circ\phi - f + \sum_{k=0}^{m} \alpha_k f^k W_k^{(n)}\right]}{\epsilon^2 + \|\nabla f\|^2 + \sum_{k=0}^{m}(\alpha_k f^k)^2}\right], \quad 0 \le k \le m.$$

end

return the displacement $U^{(n+1)}$, and the registered source image $g \circ \phi$.

Alg. 3.   Hyper-Demons Algorithm: 0-Degree Contrast Polynomial

---

read target image $f(x)$ and source image $g(x)$.

choose a linear smoothing operator G, a number ϵ, a normalizing factor α, and some large number N.

set $U_i^{(1)}(x) = W^{(1)}(x) = 0$, for any $1 \le i \le d$, $x \in \Omega$.

for n = 1 : N do   // *Fixed point iterations*

$\phi(x) = U^{(n)}(x) + x$

$$U_i^{(n+1)} = G\left[U_i^{(n)} - \frac{f_{x_i}\left[g\circ\phi - f + \alpha W^{(n)}\right]}{\epsilon^2 + \|\nabla f\|^2 + \alpha^2}\right], \qquad 1 \le i \le d$$

$$W^{(n+1)} = G\left[W^{(n)} - \frac{\alpha\left[g\circ\phi - f + \alpha W^{(n)}\right]}{\epsilon^2 + \|\nabla f\|^2 + \alpha^2}\right].$$

end

return the displacement $U^{(n+1)}$, and the registered source image $g \circ \phi$.

Hyper-Demons algorithm with right or central approximation: The terms f_{x_i}, ∇f, and f^k in the hyper-demons algorithm is a consequence of a left approximation of E_{X_i}, ∇E, and E^k in Equation (12) for t over $[0, 1]$. If we wish to use the right approximation, we need to replace the terms f_{x_i}, ∇f, and f^k in the Hyper-Demons algorithm respectively by $g_{x_i} \circ \phi$, $(\nabla g) \circ \phi$, and $(g \circ \phi)^k$. Similarly, the terms f_{x_i}, ∇f, and f^k in the Hyper-Demons algorithm are respectively replaced by $\frac{f_{x_i} + g_{x_i} \circ \phi}{2}$, $\frac{\nabla f + (\nabla g) \circ \phi}{2}$, and $(\frac{f + g \circ \phi}{2})^k$ starting from a central approximation.

5   Results and Conclusions

5.1   Theoretical Results

Our approach presents a novel and distinct interpretation of the demons algorithm compared to previous works [11,10] in this direction. Furthermore, our approach presents a generalization of the Gennert-Negahdaripour formulation [5] in a natural fashion, because we based our approach on their model yet included arbitrary image dimensions and arbitrary degrees of the polynomial. Our approach generalizes the demons algorithm to include the intensity shift, because assuming $\alpha_k \to 0$, $0 \leq k \leq m$, corresponds to $\lambda_k = \infty$, $0 \leq k \leq m$ in the Gennert-Negahdaripour formulation [5] and yields the original demons algorithm [13,14]. Furthermore, the algorithm resulting from the "right" and "central" approximations corresponds to similar ones in the literature [11,2,15,16] obtained from different point of views.

5.2   Computational Results

To better understand the introduced algorithm, we start from two 8-bit, 256×256 images shown in Fig. 1(a1-a2). The target image is generated by adding contrast enhancement to the source image and the source image is then deformed using a finite element model (FEM) [9]. The magnitude of this deformation is shown is Fig. 1(b1) and the intensity change is shown in Fig. 1(b3). If we use the central-scheme demons algorithm over various integer values of the parameter ϵ and compute the euclidean distance between the computed and the real deformation we yield the dashed curve shown in Fig. 1(e). It turns out that the minimum of such deformation distance occurs at $\epsilon = 13$ marked by "d" on Fig. 1(e). The magnitude of the corresponding deformation, the registered source, the difference between the registered source and the target, and the distance between the computed deformation and the real deformation are shown in Fig. 1(d1-d4) respectively. In all of our experiments in this section, $N = 100$ iterations were performed, G is chosen as a Gaussian of size 15×15 and of standard deviation $\sigma = 5$, and cubic interpolation is used in the algorithms when necessary.

Now if we use the central-scheme hyper-demons algorithm with zero-degree intensity shift, i.e. $m = 0$, and choose $\epsilon = 0$ varying α with the same parameters as the previous experiment we yield the solid curve plotted in Fig. 1(e).

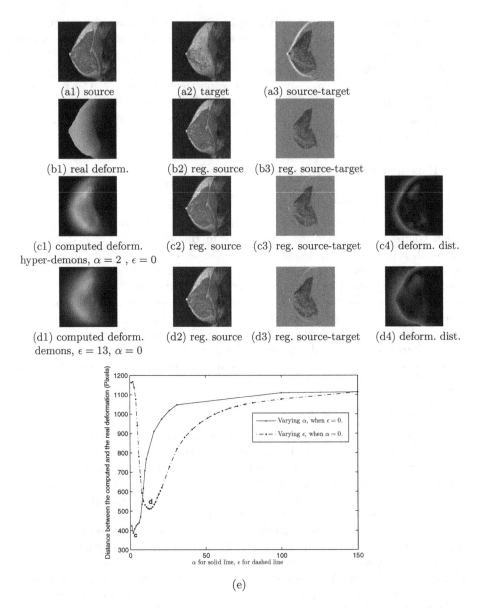

Fig. 1. Comparison of the demons and hyper-demons algorithms for simulated data

The minimum deformation distance occurs at $\alpha = 2$ marked as "c" on the plot. The corresponding deformation related images in this case are shown Fig. 1(c1-c4). It can be observed that the "best" deformation computed via changing α (i.e., a special case of hyper-demons) is slightly more precise that the "best" deformation vector computed varying ϵ (i.e., the demons algorithm).

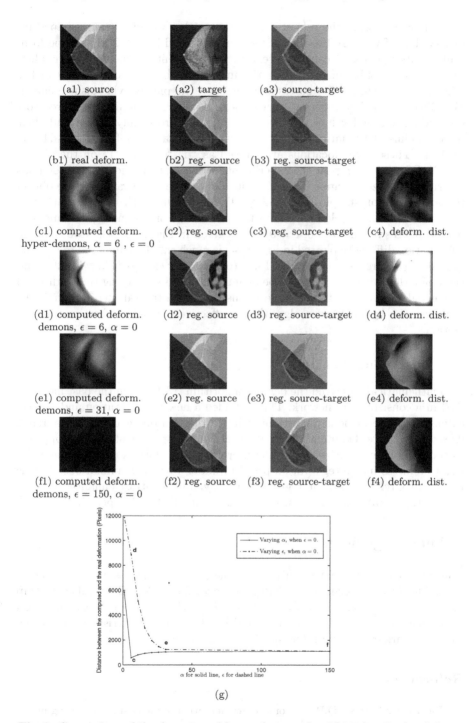

Fig. 2. Comparison of the demons and hyper-demons algorithms for simulated data

To further observe the role of α in the algorithm we add a ramp-shaped intensity shift of value 160 to the source image denoted in Fig. 2(a1) and perform similar experiments. It can be observed that the central-scheme demons algorithm is not capable of estimating the motion in this case due to this rather large intensity shift. The corresponding curve in plotted via dashed line in Fig. 2(g). It can be seen that curve does not attain a minimum. Three corresponding nominal values of $\epsilon = 6$, 31, and 150 are simply picked and their corresponding deformation-related images are displayed in Fig. 2 (d1-d4, e1-e4,f1-f4). These three locations are labeled as "d","e", and "f" on the plot of Fig. 2(g). Now if we repeat the experiment for the central-scheme hyper-demons algorithm with zero-degree intensity shift and choose $\epsilon = 0$ varying α we obtain the solid curve on the plot of Fig. 2(g). The minimum of this curve is attained at $\alpha = 6$ which is labeled as "c" on the plot. The corresponding deformation-related figures of this $\alpha = 6$, $\epsilon = 0$ case is given in Fig. 2(c1-c4). Note that the deformation difference plotted in Fig. 2(c4) is much smaller compared to the corresponding differences displayed in Fig. 2(d4,e4,f4). It can be deduced that for no value of ϵ the correct motion can be estimated in this case using the traditional demons algorithm due to the large intensity change (or varying illumination). However, the hyper-demons algorithm is effectively capable of estimating the motion.

5.3   Concluding Remarks

We mathematically derived the demons algorithm [13,14] from the past methods [5,7] in a consistent framework. This provided a new interpretation and a novel extension of the demons algorithm. It is worth mentioning that our approach does not separate the intensity correction stage from the registration as opposed to some of the existing works in the literature (e.g., [6]). A focus of our work was to address image registration problem under varying illuminations. This is particularly relevant for the registration of dynamic contrast enhanced images. A recent, general, and related work to this manuscript will appear in [3].

Acknowledgements

This research was supported in part by the Natural Sciences and Engineering Research Council of Canada (NSERC) in the form of a Post-Doctoral Fellowship for Mehran Ebrahimi. This work was also supported by the Terry Fox Foundation for Cancer Research. The authors would like to thank Dr. Kristy Brock of the Princess Margaret Hospital for the FEM simulations.

References

1. Barber, D.C., Hose, D.R.: Automatic segmentation of medical images using image registration: diagnostic and simulation applications. Journal of Medical Engineering & Technology 29(2), 53–63 (2005)

2. Cachier, P., Bardinet, E., Dormont, D., Pennec, X., Ayache, N.: Iconic feature based nonrigid registration: the pasha algorithm. Computer Vision and Image Understanding 89(2-3), 272–298 (2003)
3. Ebrahimi, M., Martel, A.L.: A general pde-framework for registration of contrast enhanced images. In: MICCAI (2009)
4. Froh, M.S., Barber, D.C., Brock, K.K., Plewes, D.B., Martel, A.L.: Piecewise-quadrilateral registration by optical flow - applications in contrast-enhanced mr imaging of the breast. In: MICCAI, vol. 2, pp. 686–693 (2006)
5. Gennert, M.A., Negahdaripour, S.: Relaxing the brightness constancy assumption in computing optical flow. Technical Report 975, MIT AI Lab. Memo (1987)
6. Guimond, A., Roche, A., Ayache, N., Meunier, J.: Three-dimensional multimodal brain warping using the demons algorithm and adaptive intensity corrections. IEEE Transactions on Medical Imaging 20(1), 58–69 (2001)
7. Horn, B.K.P., Schunck, B.G.: Determining optical flow. Artificial Intelligence 17, 185–203 (1981)
8. Lai, S.-H.: Computation of optical flow under non-uniform brightness variations. Pattern Recognition Letters 25(8), 885–892 (2004)
9. Martel, A.L., Froh, M.S., Brock, K.K., Plewes, D.B., Barber, D.C.: Evaluating an optical-flow-based registration algorithm for contrast-enhanced magnetic resonance imaging of the breast. Phys. Med. Biol. 52, 3803–3816 (2007)
10. Modersitzki, J.: Numerical Methods for Image Registration. Oxford University Press, Oxford (2003)
11. Pennec, X., Cachier, P., Ayache, N.: Understanding the "demon's algorithm": 3d non-rigid registration by gradient descent. In: Taylor, C., Colchester, A. (eds.) MICCAI 1999. LNCS, vol. 1679, pp. 597–605. Springer, Heidelberg (1999)
12. Sherman, J., Morrison, W.J.: Adjustment of an inverse matrix corresponding to a change in one element of a given matrix. Annals of Mathematical Statistics 21(1), 124–127 (1950)
13. Thirion, J.P.: Fast non-rigid matching of 3d medical images. In: Medical Robotics and Computer Aided Surgery (MRCAS), Baltimore, pp. 47–54 (1995)
14. Thirion, J.P.: Image matching as a diffusion process: an analogy with maxwell's demons. Medical Image Analysis 2(3), 243–260 (1998)
15. Vercauteren, T., Pennec, X., Malis, E., Perchant, A., Ayache, N.: Insight into efficient image registration techniques and the demons algorithm. In: Karssemeijer, N., Lelieveldt, B. (eds.) IPMI 2007. LNCS, vol. 4584, pp. 495–506. Springer, Heidelberg (2007)
16. Vercauteren, T., Pennec, X., Perchant, A., Ayache, N.: Non-parametric diffeomorphic image registration with the demons algorithm. In: Ayache, N., Ourselin, S., Maeder, A. (eds.) MICCAI 2007, Part II. LNCS, vol. 4792, pp. 319–326. Springer, Heidelberg (2007)
17. Weickert, J., Bruhn, A., Brox, T., Papenberg, N.: A survey of variational optic flow methods for small displacement, in Mathematical Models for Registration and Applications to Medial Imaging. Springer, Heidelberg (2005)
18. Woodbury, M.A.: Inverting modified matrices. Memorandum rept. 42, Statistical Research Group, Princeton University, Princeton, NJ (1950)

Appendix

Proof of Lemma 1

Proof. Note that $\mathbf{MM}^T$ is a scalar if $\mathbf{M}$ is a row-vector. Multiply $[\mathbf{M}^T\mathbf{M} + \epsilon^2\mathbf{I}]$ with its "potential" inverse $\frac{1}{\epsilon^2}[\mathbf{I} - \frac{\mathbf{M}^T\mathbf{M}}{\epsilon^2+\mathbf{MM}^T}]$ from left (and then right) and obtain the identity matrix $\mathbf{I}$. The left-hand-side calculations are given below. The right-hand-side case is shown similarly.

$$\frac{1}{\epsilon^2}[\mathbf{I} - \frac{\mathbf{M}^T\mathbf{M}}{\epsilon^2 + \mathbf{MM}^T}] \times [\mathbf{M}^T\mathbf{M} + \epsilon^2\mathbf{I}]$$

$$= \frac{1}{\epsilon^2}\left[\mathbf{M}^T\mathbf{M} - \frac{\mathbf{M}^T\mathbf{MM}^T\mathbf{M}}{\epsilon^2 + \mathbf{MM}^T} + \epsilon^2\mathbf{I} - \frac{\epsilon^2\mathbf{M}^T\mathbf{M}}{\epsilon^2 + \mathbf{MM}^T}\right]$$

$$= \frac{1}{\epsilon^2}\left[\frac{(\epsilon^2 + \mathbf{MM}^T)\mathbf{M}^T\mathbf{M} - \mathbf{MM}^T\mathbf{M}^T\mathbf{M} - \epsilon^2\mathbf{M}^T\mathbf{M}}{\epsilon^2 + \mathbf{MM}^T} + \epsilon^2\mathbf{I}\right]$$

$$= \frac{1}{\epsilon^2}\left[0 + \epsilon^2\mathbf{I}\right] = \mathbf{I}.$$

Proof of Equation (8)

Choosing $\epsilon = \lambda\kappa$, Equation (7) gives

$$[\mathbf{M}^T\mathbf{M} + \epsilon^2\mathbf{I}]\mathbf{V} = \epsilon^2\bar{\mathbf{V}} - E_t\mathbf{M}^T.$$

Hence, using Lemma 1

$$\mathbf{V} = [\mathbf{M}^T\mathbf{M} + \epsilon^2\mathbf{I}]^{-1}[\epsilon^2\bar{\mathbf{V}} - E_t\mathbf{M}^T]$$

$$= \frac{1}{\epsilon^2}[\mathbf{I} - \frac{\mathbf{M}^T\mathbf{M}}{\epsilon^2 + \mathbf{MM}^T}][\epsilon^2\bar{\mathbf{V}} - E_t\mathbf{M}^T]$$

$$= \frac{1}{\epsilon^2}[\epsilon^2\bar{\mathbf{V}} - \frac{\epsilon^2\mathbf{M}^T\mathbf{M}\bar{\mathbf{V}}}{\epsilon^2 + \mathbf{MM}^T} - E_t\mathbf{M}^T + \frac{E_t\mathbf{M}^T\mathbf{MM}^T}{\epsilon^2 + \mathbf{MM}^T}]$$

$$= \frac{1}{\epsilon^2}[\epsilon^2\bar{\mathbf{V}} - \frac{\epsilon^2\mathbf{M}^T\mathbf{M}\bar{\mathbf{V}}}{\epsilon^2 + \mathbf{MM}^T} - \frac{E_t(\epsilon^2 + \mathbf{MM}^T)\mathbf{M}^T}{\epsilon^2 + \mathbf{MM}^T} + \frac{E_t\mathbf{MM}^T\mathbf{M}^T}{\epsilon^2 + \mathbf{MM}^T}]$$

$$= \frac{1}{\epsilon^2}[\epsilon^2\bar{\mathbf{V}} - \frac{\epsilon^2\mathbf{M}^T\mathbf{M}\bar{\mathbf{V}}}{\epsilon^2 + \mathbf{MM}^T} - \frac{\epsilon^2\mathbf{M}^T E_t}{\epsilon^2 + \mathbf{MM}^T}]$$

$$= \bar{\mathbf{V}} - \frac{\mathbf{M}^T(E_t + \mathbf{M}\bar{\mathbf{V}})}{\epsilon^2 + \mathbf{MM}^T}.$$

[Note that $\mathbf{MM}^T$ is a scalar and has commuted its order in the above computation.]

Hierarchical Vibrations: A Structural Decomposition Approach for Image Analysis

Karin Engel and Klaus D. Toennies

Dept. of Simulation and Graphics
Otto–von–Guericke University Magdeburg, Germany

Abstract. We present results demonstrating that using a hierarchy of finite element vibration modes in an evolutionary deformable shape search provides a new interesting approach for the localization and segmentation of specific objects in 2D images. The design and coupling of the different levels of the shape hierarchy results in a multi–resolution shape space, which can be exploited in top–down part–based shape matching. The proposed strategy allows for segmenting complex objects from images, classification, as well as localization of the desired object under occlusions. It avoids misregistration by resolving several drawbacks inherent to standard shape–based approaches, which either cannot adequately represent non–linear variations, or rely on exhaustive prior training.

1 Introduction

Modeling global and local aspects of shape is useful for many image processing tasks including object detection, recognition and segmentation, pose estimation and motion tracking, as indicated by [1–19], among others. Studies on the human visual perception also provide evidence that a representation suitable for object detection and recognition should include a structural decomposition of the object into parts and a description of parts and relations between them [1, 20]. Such representation should cover variations and irregularities in shape and structure due to image noise, object deformation and possibly change of view point, and should allow the representation of objects under occlusion.

The main contribution of this work is a representation of complex variable objects that contain multiple parts for localizing and segmenting specific objects in an image. It is inspired by [1], but uses the structural decomposition into specific shapes in a top–down manner to overcome both, the instability inherent to structural approaches and the difficulty to extract generic parts (e.g. geons) from images in a robust way. Our model is conceptually similar to hierarchical probabilistic models, e.g. [3, 9, 10], with the main difference that it employs the finite element structural decomposition of specific shapes. As a result it combines noise robustness from energy–minimizing deformable shape models and validation of structure from structural models. This reduces the complexity of the distribution function–which would be needed to model non–linear dependencies between the shape parameters statistically–while modeling valid variation under the following assumptions. First, the desired object is (at least partially) visible

D. Cremers et al. (Eds.): EMMCVPR 2009, LNCS 5681, pp. 317–330, 2009.

in the image. Second, variation due to change of view point is negligible if the set of poses is limited. Finally, shape classes can be differentiated based on their structural configuration and/or morphology of the shape parts.

2   Structural Image Analysis: Related Work

Some methods construct complex shapes in a bottom–up fashion from segmented images. For example, in [18] the shape of the human body is parsed in a bottom–up process that employs a rigid shape–based comparison with exemplars of increasingly more complete body parts for evaluating the proposed segmentations. Other approaches to object detection start with the detection of salient features, such as edges, which are grouped, e.g. to obtain the silhouettes of regions, and evaluated using a prior model (e.g., a compositional model[21–23]) of the object. A majority of top–down structural approaches uses pre–defined combinatorial constraints between simplified object parts to encode compound shapes, e.g. in a structural description graph[4–6, 10, 13, 22, 23], using coupled/split shapes [8, 13, 14, 16], or an expert model [24]. These approaches either cannot describe structurally variable shapes, or they capture only relatively weak structural properties of shape in their *tree–structured models*. This may not always be appropriate, since structural variations might influence the morphology of the shape parts. Several approaches use a trained model of shape locations [19, 25], or probabilistic constraints between prototypical sub–shapes for generating expectation maps in a sequential recognition process [12]. Other popular methods employ Bayesian inference to compute the most probable interpretation of the image over a hypothesis space defined using joint/mixtures of probabilistic distributions [7, 9, 13, 14, 17]. Representing shape and structure by different models can be a drawback of some of these methods, as it does not allow structural deformations to directly influence morphological variation, and vice versa. This, however, may be required because often the structural aspect of shape is not independent from local shape variations. *Probabilistic models*, on the other hand, require training with representative example data for separating valid from invalid variation.

We address these issues through specific properties of the proposed method for representing and segmenting complex objects of specific classes from images. A major issue here is to abstract objects into a simplified representation which alleviates comparison between shapes based on qualitative and discriminative features. We therefore adopted the hierarchical ASM approach [7, 9], and propose a hierarchical Finite Element Model (FEM) that provides a natural framework for the multi–scale decomposition of non–linear deformation into variation of specific parts and sub–parts, etc. Shape information is represented using a combination of a set of basis functions, where the basis is defined in an a-priori manner[26, 27] (cf. section 3.1). Such prototypical parametric models are specifically suitable because they can represent variation of the desired objects in terms of physically plausible deformations at multiple scales of resolution (similar to, e.g. [7, 21, 23]). Moreover, adding statistical information is straightforward [28]. Finally, the quality of matched shapes can be evaluated and provides information for eliminating false interpretations of the image, as described in sections 3.2 and 3.3. For demonstrating its utility in structural image analysis we present in section 4 two case studies for applying our method to object localization, detection and classification tasks.

3   Method

Our method builds on the assumption that valid instances of the desired compound object can be reconstructed from a set of a-priori constrained model parameters. We therefore employ the hierarchical finite element decomposition of shape, which supports an efficient simulation of deformation. To account for variability in the relationships between shape parts and local shape deformation, the decomposition is applied in a hierarchical manner, i.e. a class–specific prototype is represented as a hierarchy of FEM. The quality of such a model instance projected into the image domain and deformed according to external (image–based) model forces can be evaluated and provides contextual shape information for eliminating false interpretations of the data in a top–down manner.

3.1   Hierarchical Decomposition–Based Shape Modelling

A parametric deformable template $\mathcal{T}(\mathbf{p})$ represents the objects undeformed shape and a set of parameters $\mathbf{p} = (\theta, \mathbf{q})$ that define how it deforms under applied forces. Model matching can be viewed as a local search for the optimum values $\mathbf{p}^t$ of the deformed model, which is commonly implemented as an optimization problem based on internal and external energies of the model (cf. sect. 3.2).

In our case, the rest shape a n–dimensional object is modeled as a continuous domain $\Omega \subset \mathbb{R}^n$, and its deformation is described by a boundary value partial differential equation that is solved for the unknown displacement field $u(\mathbf{x})$, $\mathbf{x} \in \Omega$ using the Finite Element Method. The dynamic equilibrium equation has the form

$$\frac{\partial^2 \mathbf{u}}{\partial t^2}\big|_{t>0} = \mathbf{M}^{-1}(-\mathbf{C}\frac{\partial \mathbf{u}}{\partial t}\big|_{t>0} - \mathbf{K}\mathbf{u}(t) + \mathbf{f}(t)), \tag{1}$$

where $\mathbf{K}(E, \nu)$ encapsulates the stiffness properties as well as the type of mesh and discretization used, $\mathbf{C}$ approximates a velocity–dependent damping force, and $\mathbf{M}$ may represent a constant function of material density [26]. The deformed positions $\mathbf{x}(t) = \mathbf{x}^0 + \mathbf{u}(t)$ at time $t \geq 0$ are expressed in terms of a linear mixture of $m = m_2 - m_1$ displacement fields,

$$\mathbf{x}(t) = \mathbf{x}^0 + \sum_{k=m_1}^{m_2} \phi_k \mathbf{q}_k(t), \tag{2}$$

where $m_1 \geq 1, m_2 \leq nN$, for nN degrees of freedom (DOF) of the system, and $\mathbf{x}^0$ denotes the rest positions of the N nodes. The modal vectors ϕ_k are solutions

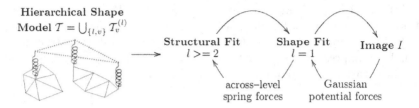

Fig. 1. Algorithm overview. In the bottom–up flow of information specific features are extracted from the underlying image. These are combined to a more complex object using a hierarchy of deformable shape models. An example of such model is depicted on the left. Here, the shapes $T_1^{(1)}, T_2^{(1)}$ of level $l = 1$ (solid lines) and the top–level model $T_3^{(2)}$ (dotted lines) contribute to a shape–structure hierarchy. The shapes are coupled across different levels l using virtual springs between specific link nodes allowing for the top–down propagation of deformations.

to the eigenproblem $(\mathbf{K} - \omega_k^2 \mathbf{M})\phi_k = 0$ and $\mathbf{q}_k$ contains the nodal coordinates in embedded space (cf. [27]).

In contrast to, e.g. Ullman et al.[3] and Yuille et al. [24], we employ a *hierarchical mixing* process to model *elastic* co–variations in shape. In our case, a hierarchical shape model $T(\mathbf{p})$ represents the decomposition of a complex shape into V discrete, linear shapes,

$$T(\mathbf{p}) = \bigcup_v T_v^{(l)}(\mathbf{p}_v), \tag{3}$$

which contribute to different hierarchy levels l. The shape parts (which also refer to morphological components or sub–shapes) at each level $l - 1$ are coupled to form a higher level l of the hierarchical shape model (figure 1), while any sub–shape may represent a compound shape on its own (see sect. 4).

As the spatial configuration of such system is described by its DOF, the desired structural constraints on the displacement fields are introduced hierarchically by across–level spring forces subject to pairs of specific link nodes at the consecutive levels. These forces allow propagating the displacement of top–level link nodes to the lower level link nodes, such that local displacements will cause deformation of the top–level shape, and vice versa. Thus it is possible to separately analyze the deformation behavior of the sub–shapes and their structural relations. In contrast to [8], this yields a hierarchy of FEM, whose nodes are subject to external model forces, which are derived from the image in a bottom–up fashion (sect. 3.2). Thereby, the model is capable of representing local and structural variability of shapes within a uniform framework.

As a result, instances of the hierarchical shape $T(\mathbf{p})$ can be compared based on their coordinates in the shape space $\mathcal{S}(T(\mathbf{p}))$[29]. The valid shape region $\mathcal{S}_V \subset \mathcal{S}(T(\mathbf{p}))$ is characterized by a mixture of parameter vectors $\mathbf{p}_v$ that span linear sub–spaces, which hierarchically depend from each other.

3.2   Hierarchical Shape Matching

Such a hierarchical model deforms into an object instance supported by image features. It should be able to localize instances of an object class based on the amount of structural (at level l) and morphological (at levels $l-1, \ldots, 1$) deformation necessary to fit the features. External model forces $\mathbf{f}(t)$ shall attract the finite element nodes to characteristic object features in the image. Such dynamic loads are created by a sensor–based sparse sampling of a scalar potential field $\mathcal{P}$, whose local minima coincide with features of interest, i.e. $\mathbf{f}(t) = -\nabla\mathcal{P}(\mathbf{x})$.

In our case parametrization of the shape parts is constrained by the top level model, which defines the (initial) placement of the FEM on the lower levels, whose deformation will cause the sub–shapes to fit local image features. In the bottom–up flow of information, the input for the finite element nodes of level $l > 1$ does not stem directly from the underlying image, but from the output of the lower hierarchy levels, i.e. only sub–shapes of level $l = 1$ have direct access to the image I. Here, external model forces subject to the nodes may be defined based on normalized[1] feature maps $I^{\mathcal{N}} \in [0,1]$, computed by linear filtering, such that the Gaussian potential forces are

$$\mathbf{f}^{(1)}(t) = \kappa \nabla I^{\mathcal{N}}(\mathbf{x}^{(1)}(t)), \tag{4}$$

where $\kappa > 0$ is a constant weight. For image segmentation the sensors usually sample either a Gaussian low pass filtered version $I^{\mathcal{N}} = G_\sigma * I$ of the image, or a gradient magnitude map $I^{\mathcal{N}} = |\nabla(G_\sigma * I)|^2$, where σ denotes the standard deviation of the low pass filter.

The input for the higher level sensors $\mathbf{x}_{w,j}^{(l)}, l > 1$, of the structural FEM with index $w \in \{1, \ldots, \mathcal{V}\}$ (cf. Eq. 3) depends on the behavior of the underlying morphological FEM. More specifically, their deformation is used to define across–level spring forces,

$$\mathbf{f}_{w,j}^{(l)}(t) = \kappa_s(\mathbf{x}_{v,i}^{(l-1)}(t) - \mathbf{x}_{w,j}^{(l)}(t-1)), \tag{5}$$

where $\mathbf{x}_{v,i}^{(l-1)}(t)$ denotes the position of the link node of the associated sub–shape model $\mathcal{T}_v^{(l-1)}$, $v \in \{1, \ldots, \mathcal{V}\}$, and $\kappa_s = \kappa f(\Delta \mathcal{Q}(\mathbf{p}_v^t, l-1)), \kappa > 0$. f maps the high–level feature gradient $\Delta \mathcal{Q}(\mathbf{p}_v^t, l-1) = \mathcal{Q}(\mathbf{p}_v^t, l-1) - \mathcal{Q}(\mathbf{p}_v^{t-1}, l-1)$ to values from the interval $[-1, 1]$, e.g. using the identity.

The high–level features are computed based on a combined objective function for estimating the energy of a model instance,

$$\mathcal{Q}(\mathbf{p}_v^t, l) = \zeta Q^d(\mathbf{p}_v^t, l) + (1-\zeta)Q^s(\mathbf{p}_v^t, l), \quad \zeta \in [0, 1], \tag{6}$$

based on a deformation criterion Q^d and a measure Q^s of correlation with the expected image data. Using the mean value of the sensor input at the node positions $\mathbf{x}_v^{(l)}(t)$ to estimate Q^s (as in [12, 30]) may not always be appropriate

---

[1] By normalization into the interval $[0, 1]$, e.g. using Gaussian normalization, the influence of different features are rendered independent from the specific filters used.

(for example if the sensor inputs do not stem from a deterministic potential field). We therefore directly employ the force formulation, and let

$$Q^s(\mathbf{p}_v^t, l) = \mathcal{F}\big(\mu(|\mathbf{f}_v^{(l)}(t)|^2)\big) \tag{7}$$

indicate the correspondence of the v–th model with the data. The function μ computes the mean value, and $\mathcal{F}(x) = \exp(-\alpha x^2), 0 < \alpha \leq 1$, normalizes the resulting values to the interval $[0, 1]$, where values close to 1 indicate high quality.

The model deformation criterion Q^d measures the degree of discrepancy between $\mathcal{T}_v^{(l)}(\mathbf{p}_v^0)$ and $\mathcal{T}_v^{(l)}(\mathbf{p}_v^t)$, i.e. the non–rigid deformation of the shape model instance in its un–rotated reference frame. In our case the strain energy is adapted from [27], i.e.

$$Q^d(\mathbf{p}_v^t, l) = \mathcal{F}\big(\mu((\mathbf{q}_{v,k}^{(l)}(t))^2(\omega_{v,k}^{(l)})^{-2})\big). \tag{8}$$

Since the low–order modes represent global variations, including rigid body modes, while the high–frequency modes are sensitive to noise, we only consider the modal amplitudes corresponding to the m intermediate vibration modes, which explain a proportion, e.g. $\beta = 0.25$, of the total variation (cf. sect. 3.1).

The hierarchical constraints facilitate initialization and deformable shape fit such that contextual shape information can be used for eliminating false interpretations of the data as follows: Each shape fit is achieved by deformations which are determined by a set of constraints corresponding to finite element vibration modes (equations 1 and 2). This concept for local optimization is extended such that our matching algorithm fits the structural model instances to the data in a hierarchical manner.

Each global shape model $\mathcal{T}_w^{(l)}, l > 1$, restricts the parametrization of the associated morphological FEM $\mathcal{T}_v^{(l-1)}$ according to the displacements of the $j = 1, ..., N_w$ top–level nodes. After initializing an instance of the global model, the instances of the local models are aligned to it by propagating the displacements of the link nodes $\mathbf{x}_{w,j}^{(l)}$ in the global model to the linked low–level nodes $\mathbf{x}_{v,i}^{(l-1)}$ (figure 2a). In this case, the displacement of the j–th top–level link node directly affects the DOF associated with a specific low–level link node, and is imposed as displacement boundary condition (BC) on the particular finite element equations of motion (static case). More specifically, this results in a transform $\theta_{vw}(\mathbf{x})$ that maps a point $\mathbf{x}$ defined in the v–th local coordinate frame to a point defined in the global coordinate frame by the position of link node $\mathbf{x}_{w,j}^{(l)}(t)$ at time $t = 0$, such that

$$\mathbf{x}_{v,i}^{(l-1)}(t) = \theta_{vw}^{-1}(\mathbf{x}_{w,j}^{(l)}(t)). \tag{9}$$

The first steps of the iterative hierarchical shape matching algorithm then account for the bottom–up flow of information between the l levels of the model. It is implemented using the hierarchy of forces derived from the particular feature maps (equations 4 and 5, figs. 2b-2c). Deformation of a FEM at level $l = 1$ uses external model forces computed by spatial filtering the underlying image, and for $l > 1$ it uses across–level spring forces between pairs of link nodes of levels $l - 1$ and l. The final step defines the top–down flow of information, which is–similar to the initialization step–realized through displacement BC (figure 2d).

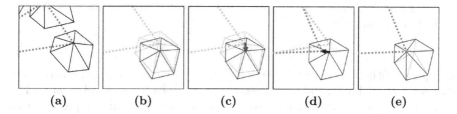

Fig. 2. Shape interactions during the hierarchical shape fit according to sect. 3.2 are exemplarily depicted for a detail of a shape (solid lines)–structure (dotted lines) hierarchy of two levels (a). (b) Result of the morphological shape fit of $\mathcal{T}_v^{(1)}$ (initial configurations in light gray). (c) The resulting across–level spring force subject to top–level node $\mathbf{x}_{w,j}^{(2)}$ (equation 5) is depicted by the red arrow. The respective top–level nodal displacement determines a displacement boundary condition (black arrow) subject to the particular first–level link node $\mathbf{x}_{v,i}^{(1)}$ (d), which causes a deformation of the sub–shape (e).

Thereby, characteristic features are hierarchically derived from the image. Their significance and semantics increase from the bottom to the top level. A computationally efficient coarse–to–fine implementation of this matching algorithm uses feature maps of different scales of resolution. On the lowest level, Gaussian potential fields are computed with dynamically decreasing values for the standard deviation σ. The high–level features are extracted using a dynamically increasing number of $m \leq n$ vibration modes, e.g. starting with the rigid-body modes. Values for σ and m are updated after each step of the matching.

3.3 Hierarchical Shape Search

As suggested in the work of Yuille et al.[24] and in [20], our method aims at computing the most plausible explanation of the image content given a (set of) prior model(s) in terms of segmentation with maximum quality. This process uses an evolutionary deformable shape search for the desired object that is inspired by [12, 30]. It initializes and optimizes several model instances (with different parametrisation) in parallel by employing the criterion function defined in equation 6 for evaluating the quality of matched model instances.

Part–based Localization and Segmentation. Model instances are initialized by transformation θ of the prototype $\mathcal{T}$ from the model coordinate frame to the image coordinate frame, i.e. $\mathcal{T}(\mathbf{p}^t) = \mathbf{x}(t) = \theta(\mathbf{x}^0 + \sum_{k=m_1}^{m_2} \phi_k \mathbf{q}_k(t)), t = 0$. For a hierarchical model (as defined in equation 3), θ includes dependent transformations according to equation 9, i.e. $\theta = \{\theta_v, \theta_{vw}\}, v, w \in \{1, \ldots, \mathcal{V}\}$.

Although it is straightforward to use other sets of transformations θ_v, we only consider translation, rotation and scaling, characterized by the set of geometric parameters position $\mathbf{c}_v$, orientation ψ_v w.r.t. a predefined axis, and scaling $\mathbf{s}_v$ of the model instance $\mathcal{T}_v^{(l)}$ in the image. The parameters $\xi_v = \{\mathbf{c}_v, \mathbf{s}_v, \psi_v\}$ might be considered as variates with a presumed Gaussian[2] distribution $\xi_v \sim N(\mu, \varsigma)$, such that random samples

$$x_v = \hat{\mu}(\xi_v) + z\sqrt{\hat{\varsigma}(\xi_v)}, \quad z \sim N(0,1). \tag{10}$$

Compared with the sequential search in [12, 31], each model instance is randomized w.r.t. all its levels, and then fitted to the data in a hierarchical manner. This will reduce the risk that false negatives in the feature detection step prevent parts from being properly localized. As the proposed model naturally considers the relationships between the different object features in the image, we only need a prior of the parameter values for the top–level model as well as for all sub–shapes, whose parametrization is only partially constrained by it (according to equation 9). Since samples for estimating the parameters $\hat{\mu}$ and $\hat{\varsigma}$ of the (conditional) probability density functions (PDF) are in our case not available, we specify an initial region of parameter values we are interested in. More specifically, we use pre–set tolerances $\hat{\varsigma}$ from the parameter values x'_v of the set of model instances $\mathcal{T}_v^{(l)}, v \in \{1, ..., \mathcal{V}\}$, generated from a representative manual segmentation, which serve as estimates for $\hat{\mu}$ (see sect. 4 for settings we used in our experiments). Dependent relations between pairs of shapes v and w can be characterized by means of the parametric transforms, $\mathbf{s}_{w \to v} = \mathbf{s}_w^{-1}\mathbf{s}_v, \psi_{w \to v} = \psi_w - \psi_v$, which are–for sake of simplicity–likewise considered to have a Gaussian PDF.

Each of the multiple model instances initiates an optimization process in order to adapt to the local conditions in the data (sect. 3.2). We organize the search by employing a priority queue of regions within the search space, where we use the quality–of–fit–values $Q(\mathbf{p}^t, l)$ for the top–level model according to equation 6 of the current model instances as the priority. Solutions with high QOF–values are selected by applying a threshold τ_Q, and further evolved until the overall quality of the current model instances, q, converges. New shape generations are generated based on the parametrization of the regionally best fitting shapes. More specifically, each selected shape is randomized in a top–down manner, such that we use in equation 10 for its v–th shape part, $\hat{\mu}(\xi_v) = x'_v$ and $\hat{\varsigma}(\xi_v) \in [0, 1]$. Here x'_v is estimated based on the parameters $\mathbf{p}^t_v$.

Shape model instances $\mathcal{T}_v^{(l)}$ with low quality are replaced by new instances based on the initial settings accordingly. Misleading shape searches due to an insufficient parametrization as well as an exponential increase in the number of shape instances can thus be avoided, while the additional "new" trials keep the search independent of known solutions. For determining $q = \mu_{\mathcal{T} \in b}(\max Q(\mathbf{p}^t, l))$ and $\tau_Q = q - \tau$, where e.g. $\tau = 0.1$, clusters of model instances with high energy are built using a regular grid of bins b over the image. The multi–resolution shape search continues until q converges, such that the desired shape is finally represented as the best rated structural configuration of shapes in the image. Our algorithm can determine the $M > 1$ best matches. If, however, the QOF–values of the best matches are below a pre–defined threshold τ'_Q, it is highly possible that no instance of the desired object could be detected in the image.

---

[2] Alternatively, a uniform distribution may be initially assumed. Later this information can be refined by employing a match list of known solutions (importance sampling).

Classification. Model–based approaches that use prior knowledge about specific shapes offer a complete characterization of the fitted shapes and imply classification. Each object is identified under a given model $\mathcal{T}_X = \{\mathcal{T}_A, ..., \mathcal{T}_Z\}$ with a probability depending on a discriminant measure $\mathcal{D}$ associated with each $\mathcal{T}_X$. If matching and classification share the same criterion, i.e. $\mathcal{D} = Q$, then pattern matching and classification may simply be implemented as one single integrated step. Hence, the competitive use of different class–specific shape models allows for classification of objects within the image by comparing the $Q(\mathbf{p}_X^t, l)$–values for the best fitting instances of each prototype.

4   Experimental Results

We selected two example applications in order to explore the ability of the proposed a-priori constrained hierarchical models of shape variation to localize complex objects in images. In the first example application, a shape–structure hierarchy of two levels is applied to the segmentation and classification of ants from specific species in 2D color images from a database. This particular application is well–suited for analyzing insensitivity w.r.t. hidden object parts, and allows us to compare our results with the results presented in [12], who used a statistical model to recognize specimen from the same database. In the second case study we demonstrate the versatility of our approach by using a 3–level FEM for the part–based detection and localization of facial features in 2D images.

4.1   Segmentation and Classification of Ants in 2D–Color Images

Our particular database of 260 images was obtained from MCZ database of the Museum of Comparative Zoology at Harvard University[3] and AntWeb by the Californian Academy of Sciences[4]. We used the lateral views that allow for segmentation using a (single-view) 2D–shape model. Furthermore, exactly one ant was displayed in each image, although parts of it may be missing.

For each of the different classes $\mathcal{A}$nochetus, $\mathcal{C}$erapachys and $\mathcal{P}$heidole, a prototypical template $\mathcal{T}_X$, $X = \mathcal{A}, \mathcal{C}, \mathcal{P}$, was generated based on a manually segmented example image of the database. Each ant shape was therefore subdivided into multiple sub–shapes of level $l = 1$, such as head, thorax, back, et cetera. The top–level models $\mathcal{T}_X^{(2)}$ constrain structural variation for anatomical reasons. Besides from the class–specific kind and number of ant body parts, the standardized positioning of the ants, e.g. on wooden sticks, causes a curved organization of the parts, which determines the specific subdivision of the shape domain, as depicted in figure 3. In contrast to [12], a statistical model of the ant color distribution was not available. We assumed that each ant can be extracted from the background by exploiting the fact that due to the standardized acquisition most background in the images will be homogeneously gray. The internal low–level

---

[3] http://mcz-28168.oeb.harvard.edu/mcztypedb.htm
[4] http://www.antweb.org/

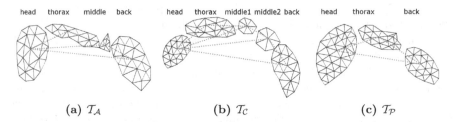

(a) $\mathcal{T}_{\mathcal{A}}$ (b) $\mathcal{T}_{\mathcal{C}}$ (c) $\mathcal{T}_{\mathcal{P}}$

Fig. 3. Class–specific 2–level FEM for three different ant genera. The number and kind of morphological components (solid lines) differ between the three prototypical structural FEM (dotted lines).

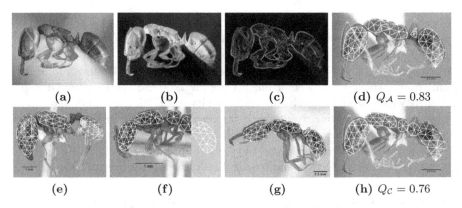

(a)   (b)   (c)   (d) $Q_{\mathcal{A}} = 0.83$

(e)   (f)   (g)   (h) $Q_{\mathcal{C}} = 0.76$

Fig. 4. The "ant color" image (b) estimates the difference in color from homogeneously gray background. Typical segmentation results are shown in (d)–(h). Our algorithm utilizes shape information in a hierarchical manner. This increases its robustness to partial occlusion, such as depicted in (f). $\mathcal{A}$. madagascarensis was classified correctly with a probability of 83% in fig. 4d, which is higher than the QOF–value for the best match with the $\mathcal{C}$erapachys–model (h).

sensors were therefore mapped to "ant color" intensity images, which were computed by suppressing homogeneous regions. The contour sensors were mapped to gradient magnitude maps (figs. 4a–4c). We used as initial settings $m = 4, \sigma = 10$ (sect. 3.2), $\hat{\varsigma}(\xi_w) = \hat{\varsigma}(\xi_{w\rightarrow v}) = 10\%$, and $10°, \forall v, w$, based on the parameterized example segmentation (sect. 3.3). According to our working assumptions, we used the image center to set up $\hat{\mu}(\mathbf{c})$ for $\mathcal{T}_X^{(2)}$, while $\hat{\varsigma}(\mathbf{c})$ is set to 20% of the image width. We further set the elastic moduli to $E = 2, \nu = 0.4$, let $\lambda = 100$, $\alpha = 0.1$ and $\zeta = 0.5$ (sect. 3.2).

Under the appropriate (manually selected) prototype $\mathcal{T}_X$ the ant was successfully segmented in all test images (figure 4). Localization and segmentation was successful under occlusions of either part of the ant and up to 30% of the ant shape (i.e. in some cases more than one part was missing, cf. fig. 4f). This makes our method superior to the approach of Bergner et al. [12], as their sequential shape search required the "head" sub–shapes to be found, and failed otherwise.

For classification we first performed the multi–resolution shape search with all three class-specific models $\mathcal{T}_X, X = \mathcal{A}, \mathcal{C}, \mathcal{P}$, using a subset of 75 images that were selected for clear appearance of exemplars from one specific genus ($\mathcal{P}$heidole), and merged the ordered lists of solutions resulting from the parallel shape searches using the different models. We found a positive classification rate of 95%, i.e. only 4 images were misinterpreted. Bergner et al. [12] reported 84% correct classifications on the same set of test images. This shows that in direct comparison restricting shape variation to local vibration modes was superior to their statistical approach to structural image interpretation.

Next, we randomly selected a subset of 20 images per class from our data base. In 93.3% of all cases the correct model exhibited a higher QOF–value (figure 4d). The difference in the QOF–values w.r.t. the correct class was significant ($p <$ 0.01, one–sided t–test), indicating that ants can be classified using our model-based segmentation given that kind, number and spatial configuration of the morphological components differ between classes. Classification might improve by training a color–classifier. However, our assumption of gray background was sufficient for reducing irrelevant input for the low–level sensors, while avoiding misclassification due to poor response to feature detectors, as reported in [12].

In another classification experiment we used all test images and applied only the $\mathcal{P}$heidole–model $\mathcal{T}_\mathcal{P}$ in order to select all images of ants from this genus. We computed 16.7% false positive solutions and 92.2% true positives.

4.2 Localization and Segmentation of Facial Features

We used a prototypical hierarchy of sub–shapes to represent the iris, eye contour, nostrils, bridge of the nose, upper and lower lips at level $l = 1$, the eyes, nose and mouth at level $l = 2$, and their structural arrangement in a face at level $l = 3$ (figure 5). To consider a large number of factors including pose, illumination, facial expression and background variation, 100 example images were collected, among others, from The Yale Face Database[5], BioID Face Database[6], Calltech Face Database[7] and MIT–CBCL Face Recognition Database[8]. In our experiment we used the same initial settings as in sect. 4.1, with the exception that the scaling of $\mathcal{T}_\mathcal{F}^{(3)}$ was varied between $10 - 80\%$ of the image width.

The facial features were in all cases properly localized and in 89% also accurately segmented. In the remaining cases, segmentation was inaccurate due to variations, e.g. in size and pose, which were not covered with the single view deformable model. In 10 additional images that contained the faces of a group of people (38 faces in total), 74% of the faces were correctly detected based on the facial features (fig. 6h). We therefore used a user–specified value for M for selecting the desired number of segmentations from the priority queue (sect. 3.3).

---

[5] http://cvc.yale.edu/projects/yalefacesB/yalefacesB.html

[6] http://www.bioid.com/

[7] http://www.vision.caltech.edu/Image-Datasets/faces/

[8] http://cbcl.mit.edu/software-datasets/heisele/

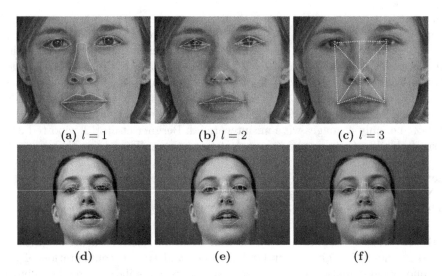

Fig. 5. The top row provides a visual display of the 3–level FEM $\mathcal{T}_{\mathcal{F}}$ used for face localization. To demonstrate its ability to locate and segment facial features, the best fitting model instance is depicted in the bottom row at different stages of the deformable shape search described in sect. 3.3, i.e. after initialization (d), first iteration (e) and at its equilibrium configuration (f).

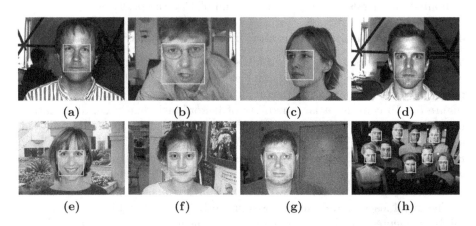

Fig. 6. Typical results of our algorithm for face localization. The boxes are used for reasons of simplicity and indicate $M \geq 1$ proposed solutions with high quality. Note that the faces in the background of figure (f) were too small to be detected.

Our shape search cannot guarantee to find the globally optimal parameter set. This would require a recursive subdivision of the parameter space and analysis of all possible matchings for transformations with parameters contained in a sufficiently large region. At present, however, there is a real possibility for an increase in the number false positive solutions with high priority.

5   Conclusion

We proposed a template–based approach to the localization and segmentation of complex objects from 2D images. Our method extends the ability of finite element models of shape to capture structural variability based on the superposition principle. Using specific energy–minimizing deformable shapes that span several levels of a shape hierarchy helps to avoid reconstruction of invalid shapes, which may result when the model parameters are chosen independently. In contrast to many existing approaches no pre–segmentation, e.g. in terms of edge detection [22, 23] or color classification [12], is employed by our algorithm. The proposed shape matching instead extracts image features with increasing semantics by introducing hierarchical constraints to the segmentation result. The constraints, which control the preference of the template to deform into similar compound shapes, are in our case derived from a single example segmentation per class. Experimental results indicate that prior estimates on the variation parameters can sufficiently separate information about variation within an object class and between classes. We can conclude from our results that the former is mainly determined by sub–shape deformation, while the latter is given structural variation parameters.

Currently, our method needs more detailed evaluation, e.g. in comparison with existing face detection methods (such as [32]). In order to improve its utility for part–based image analysis, future work will focus on the analysis of the influence of the parameters on the resulting segmentations and their evaluation.

References

1. Biedermann, I.: Human image understanding: Recent research and a theory. Comp. Vis. Graph Imag Proc. 32, 29–73 (1985)
2. Pentland, A.: Recognition by parts. In: Proc. IEEE ICCV, pp. 612–620 (1987)
3. Ullman, S., Basri, R.: Recognition by linear combinations of models. IEEE Trans. Patt. Anal. Mach. Intell. 13(10), 992–1005 (1991)
4. Christmas, W., et al.: Structural matching in computer vision using probabilistic relaxation. IEEE Trans. Patt. Anal. Mach. Intell. 17(8), 749–764 (1995)
5. Rivlin, E., et al.: Recognition by functional parts. Comp. Vis. Imag. Understand 62(2), 164–176 (1995)
6. Zhu, S., Yuille, A.: Forms, A flexible object recognition and modeling system. Int. J. Comp. Vis. 20(3), 187–212 (1996)
7. Heap, A., Hogg, D.: Improving Specificity in PDMs using a Hierarchical Approach. In: Proc. BMVC, pp. 80–89 (1997)
8. Tao, H., Huang, T.: Connected vibrations: a modal analysis approach to non-rigid motion tracking. In: Proc. IEEE CVPR, pp. 1424–1426 (1998)
9. Al-Zubi, S., Tönnies, K.D.: Extending active shape models to incorporate a-priori knowledge about structural variability. In: Van Gool, L. (ed.) DAGM 2002, vol. 2449, p. 338. Springer, Heidelberg (2002)

10. Felzenszwalb, P., Huttenlocher, D.: Pictorial structures for object recognition. Int. J. Comp. Vis. 61(1), 53–79 (2003)
11. Weyrauch, B., et al.: Component–based Face Recognition with 3D Morphable Models. In: Proc. IEEE Face Proc. Video, pp. 1–5 (2003)
12. Bergner, S., et al.: Deformable structural models. In: Proc. IEEE ICIP, pp. 1875–1878 (2004)
13. Sigal, L., et al.: Tracking Loose–Limbed People. In: Proc. IEEE CVPR, pp. 421–428 (2004)
14. Zhang, J., et al.: Bayesian body localization using mixture of nonlinear shape models. In: Proc. IEEE ICCV, pp. 725–732 (2005)
15. Ren, X., et al.: Recovering human body configurations using pairwise constraints between parts. In: Proc. IEEE ICCV, pp. 824–831 (2005)
16. Zambal, S., et al.: Improving segmentation of the left ventricle using a two-component statistical model. In: Larsen, R., Nielsen, M., Sporring, J. (eds.) MICCAI 2006. LNCS, vol. 4190, pp. 151–158. Springer, Heidelberg (2006)
17. Zhe, L., et al.: Hierarchical Part–Template Matching for Human Detection and Segmentation. In: Proc. IEEE ICCV, pp. 1–8 (2007)
18. Srinivasan, P., Shi, J.: Bottom-up Recognition and Parsing of the Human Body. In: Proc. IEEE CVPR, pp. 1–8 (2007)
19. Wang, Y., Mori, G.: Multiple tree models for occlusion and spatial constraints in human pose estimation. In: Forsyth, D., Torr, P., Zisserman, A. (eds.) ECCV 2008, Part III. LNCS, vol. 5304, pp. 710–724. Springer, Heidelberg (2008)
20. Edelman, S.: Computational theories of object recognition. Trends Cogn. Sci. 1, 296–304 (1997)
21. Bienenstock, E., et al.: Compositionality, MDL Priors, and Object Recognition. In: Proc. NIPS, pp. 838–844 (1996)
22. Zhu, L., Yuille, A.: A Hierarchical Compositional System for Rapid Object Detection. In: Proc. NIPS (2005)
23. Felzenszwalb, P., Schwartz, J.: Hierarchical Matching of Deformable Shapes. In: Proc. IEEE CVPR, pp. 1–8 (2007)
24. Yuille, A., et al.: Feature extraction from faces using deformable templates. Int. J. Comp. Vis. 8(2), 99–111 (1992)
25. Felzenszwalb, P., et al.: A discriminatively trained, multiscale, deformable part model. In: Proc. IEEE CVPR, pp. 1–8 (2008)
26. Pentland, A., Sclaroff, S.: Closed-form solutions to physically based shape modeling and recognition. IEEE Trans. Patt. Anal. Mach. Intell. 13(7), 715–729 (1991)
27. Sclaroff, S., Pentland, A.: Modal matching for correspondence and recognition. IEEE Trans. Patt. Anal. Mach. Intell. 17(6), 545–561 (1995)
28. Cootes, T., Taylor, C.: Combining point distribution models with shape models based on finite-element analysis. Imag. Vis. Comp. 13(5), 403–409 (1995)
29. Kendall, D.: A survey of the statistical theory of shape. Stat. Sci. 4, 87–120 (1989)
30. Hill, A., Taylor, C.: Model-based image interpretation using genetic algorithms. Imag. Vis. Comp. 10(5), 295–300 (1992)
31. Felzenszwalb, P.: Representation and detection of deformable shapes. IEEE Trans. Patt. Anal. Mach. Intell. 27(2), 208–220 (2005)
32. Zheng, Z., et al.: Facial feature localization based on an improved active shape model. Inform. Sci. 178(9), 2215–2223 (2008)

Exemplar-Based Interpolation of Sparsely Sampled Images

Gabriele Facciolo[1], Pablo Arias[1], Vicent Caselles[1], and Guillermo Sapiro[2]

[1] Universitat Pompeu Fabra, DTIC, 08018 Barcelona, Spain
[2] University of Minnesota, ECE, Minneapolis, MN 55455, USA
{gabriele.facciolo,pablo.arias,vicent.caselles}@upf.edu,guille@umn.edu

Abstract. A nonlocal variational formulation for interpolating a sparsely sampled image is introduced in this paper. The proposed variational formulation, originally motivated by image inpainting problems, encourages the transfer of information between similar image patches, following the paradigm of exemplar-based methods. Contrary to the classical inpainting problem, no complete patches are available from the sparse image samples, and the patch similarity criterion has to be redefined as here proposed. Initial experimental results with the proposed framework, at very low sampling densities, are very encouraging. We also explore some departures from the variational setting, showing a remarkable ability to recover textures at low sampling densities.

1 Introduction

The terms image *inpainting* and *interpolation* refer to the problem of recovering missing information in an image, in a visually plausible manner exploiting available image information. This is an ill-posed inverse problem, and as such, some sort of prior knowledge is needed for its solution. The literature on this topic is vast, since it lies in the heart of many relevant applications, such as zooming, demosaicing, super-resolution and image editing, among others.

For the purpose of this paper we distinguish two interpolation cases: when the available data consists of a set of isolated samples (be regular or irregular) and when it is given on a (not necessarily connected) region of the image. For the former we will use the term *interpolation*, reserving *inpainting* to denote the dense case.

In the case of inpaiting the available information usually allows to determine the image derivatives on the region with known data. First approaches to inpainting took advantage of this, completing the image by means of PDEs [1,2] or variational methods [3] that continued the image gradients or the level lines inside the inpainting domain. These schemes involving only interactions between *local* pixels, fail with textured images or large inpainting domains. Advances in the field of texture synthesis [4] served as inspiration for new inpainting strategies, based on the hypothesis that natural images are redundant, and self similar: The value of a pixel is synthesized from known pixels with similar neighborhoods (*patches*). These methods are often refereed to as *non-local* or *exemplar-based* (see for instance [5,6,7] and references therein). A current trend in research is

D. Cremers et al. (Eds.): EMMCVPR 2009, LNCS 5681, pp. 331–344, 2009.
© Springer-Verlag Berlin Heidelberg 2009

the combination of both, local and non-local strategies *e.g.* [8,9]. We refer to [9] for an account of this active area of research.

If the only available data consists of a nonuniform and sparse (as opposed to dense) set of samples then: 1. The gradients as well as the directions of the level lines are unknown, 2. There are no complete patches available on the image. In this setting PDE based methods cannot be directly applied and exemplar-based inpainting methods need to be adapted. This scenario appears in image super resolution, since after registering the low resolution images the overlapped grids may be seen as a non regular one.

Existing interpolation approaches consider priors based on smoothness or regularity assumptions, which can be imposed by restricting the solution to be, for instance, band limited [10], of bounded variation [11], expanded over a base of functions (*e.g.* splines [12], radial basis functions [13]), among others.

A recent front of activity is given by the techniques based on the *sparseland* model [14,15], in which the image is restricted to have a sparse representation over an overcomplete basis or *dictionary* [16,15,17]. The main difference between dictionary-based and exemplar-based methods lies in where the missing information is obtained from. Dictionary based methods look for the missing data in the dictionary (as a linear combination of a few atoms), whereas exemplar-based methods assume that the information needed lies elsewhere in the image itself (or in a database of images [18]).

A non-local prior is used in [19]. In this work the set of image patches with their similarity relations is modeled as a weighted graph and the interpolation is done by imposing regularity in this graph [20,21]. This corresponds to a non-local regularization on the image. A successful PDE approach using an anisotropic diffusion process was proposed in [22].

Our contribution. We address the problem of image interpolation from non-uniformly sparsely sampled data via a non-local exemplar-based variational approach that exploits the self-similarity of the image. In this approach, and just to prove the applicability of the self-similarity principle, we consider the simple case where the samples are arranged on a discrete (but non regular) grid, and leave the sub-pixel case for future development. The proposed variational formulation is a generalization of the inpainting framework presented in [23], which exhibits a good performance, but only for dense inpainting domains. As in [23], we set up a functional to model the nonlocal means iterations both for the image and the weights. Thus, besides the data attachment term, we include a regularization term for the weights given in terms of its entropy. The functional is then minimized with respect to both variables, the unknown image and the weights. The data attachment term is tailored to compare only the known pixel positions in one or both patches under comparison. Finally, both terms are balanced by a temperature parameter h and letting $h \to 0+$ (as in [24]) permits to iteratively improve the results. Let us mention that we have also explored a non variational model suggested by our approach that exhibits a faster convergence. The preliminary experiments suggest that exemplar-based methods can be successfully applied to sparse data interpolation.

Related work. Our work is related to the nonlocal techniques applied to demosaicing in [24,25] and super-resolution in [26], problems that can be cast as image interpolation from a regular sampling set. These methods work by averaging known pixels according to the similarity of their neighborhoods, and are closely related with our approach. More detailed comments on them will be given in subsequent Sections. Similar ideas can be also found in the field of 3D tomographic imaging [27], where incomplete 3D volumes are reconstructed via grouping them by similarity and averaging the exemplars in each cluster.

Let us mention that the problem of interpolation from a set of sparsely sampled images could be approached with the techniques of compressed sensing [14,28]. Even if the standard approach uses a set of random measurements (e.g. projections on a random basis, or noiselets) one could apply the corresponding reconstruction schemes with a random sampling of the image, as in our case. As far as we know, there is no detailed comparison between exemplar-based methods and compressed sensing in the context of image interpolation. On the other hand, as shown in this paper, exemplar-based methods can address the problem of interpolating non uniformly sampled images with large unsampled regions.

Finally, the work [25] combines sparsity and non-local techniques. There, the image self-similarity is used to obtain more robust sparse representations over a given dictionary, by assigning a common representation to similar patches.

Notation. Images are denoted as functions $u : \Omega \rightarrow \mathbb{R}$, where Ω denotes the image domain, usually a rectangle in $\mathbb{R}^2$. Pixel positions are denoted by x, x', z, z' or y, the latter for positions inside the patch. A patch of u centered at x, is denoted by $p_u(x) = p_u(x, \cdot) : \Omega_p \rightarrow \mathbb{R}$, where Ω_p is a disk (or a square) centered at $(0, 0)$. The patch is defined by $p_u(x, y) = u(x + y)$, with $y \in \Omega_p$. $O \subset \Omega$ is the set of unknown image pixels or the domain to be interpolated, and $O^c = \Omega \setminus O$ is the known portion of the domain. For simplicity we will assume that the image is defined on an extended domain $\widetilde{\Omega} = \Omega + \Omega_p$ (*i.e. widetildeΩ* is a dilation of Ω) and we work in Ω, hence a patch can be centered at any pixel in Ω without escaping the image domain. Additional notation will be introduced in the text.

2 From Inpainting to Interpolation

The framework we present here is an adaptation of the non-local inpainting functional recently introduced in [23]. In this section we briefly review this work and discuss the modifications that have to be done to allow its application to the problem of image interpolation from sparse samples addressed in this paper.

2.1 Review: Non-local Functional for Image Inpainting

In [23] we proposed the functional

$$\widetilde{E}(u, w) = \frac{1}{h}\widetilde{F}_w(u) - \sum_{x \in \widetilde{O}} \widetilde{H}_w(x) \tag{1}$$

whose minimization yields a non-local exemplar-based inpainting method. The first term is given by

$$\widetilde{F}_w(u) = \sum_{x\in\widetilde{O}}\sum_{x'\in\widetilde{O}^c} w(x,x')\|p_u(x) - p_u(x')\|_{\Omega_p}, \tag{2}$$

and it is inspired by a functional presented in [21] in the context of non-local image denoising/regularization. $\widetilde{F}_w$ measures the coherence between the patches in $\widetilde{O}$ and those in $\widetilde{O}^c$, given the similarity weight function w and a patch norm-like function $\|\cdot\|_{\Omega_p}$. $\widetilde{O}$ is an extension of O containing the centers of all patches intersecting O. In doing so, patches $p_u(x')$ centered in $x' \in \widetilde{O}^c$ consist entirely of known pixels. The term (2) promotes the similarity between the image patches centered at $x \in \widetilde{O}$ and $x' \in \widetilde{O}^c$. Indeed, minimizing $\widetilde{F}_w$ w.r.t. the image u, for a given fixed weight function w, forces pairs of patches for which $w(x,x')$ is high to be similar. Since $p_u(x')$ lies outside the inpainting domain, it is fixed and the similarity can only be enforced by modifying $p_u(x)$. Thus, incomplete patches *receive* information from outside the inpainting domain.

The weight function $w : \widetilde{O} \times \widetilde{O}^c \to \mathbb{R}^+$ measures the similarity between patches centered in the inpainting domain and in its complement. *Gaussian* weights are commonly used, i.e. $w(x,x') = \exp\left(-\frac{1}{h}\|p_u(x) - p_u(x')\|^2\right)$, where $\|\cdot\|$ is a weighted L_2 norm in the space of patches and h is the scale. In the frameworks described in [21] the weights are known and remain fixed through all the iterations. While this might be appropriate in case of denoising applications, where the weights can be estimated from the noisy image, in the image inpainting/interpolation scenario, weights are not available and have to be inferred together with the image. This idea has been applied before for super-resolution [26], denoising [29] and in a more general regularization framework [19]. None of these works present a variational justification for the weight update.

This issue was addressed in [30,23]. In [23] we consider the weight function w as an additional unknown. Instead of prescribing explicitly the Gaussian functional dependence of w w.r.t. u we do it implicitly, as a component of the optimization process. This results in a simpler functional, avoiding to deal with the complex, non-linear dependence between w and u. To this end, $w(x,\cdot)$ is constrained to be a probability density function, $\sum_{x'\in O^c} w(x,x') = 1$, and a second term given by $\sum_{x\in\widetilde{O}} \widetilde{H}_w(x)$ is added (the second term in (1)), where

$$\widetilde{H}_w(x) = - \sum_{x'\in\widetilde{O}^c} w(x,x') \log w(x,x'), \tag{3}$$

is the entropy of the probability $w(x,\cdot)$ for $x \in \widetilde{O}$. Summarizing, the first term of (1) permits the estimation of the image u from the weights w, whereas the second one allows us to compute the weights given the image.

2.2   Generalization to Interpolation

We will discuss in this section the modifications needed to adapt the inpainting formalism to the problem of image interpolation. The mechanism for adapting the similarity weight function remains unchanged, thus we will focus our attention on the image energy term. Let us assume for the moment that we know a weight function w which measures the similarity of the pairs of incomplete patches. We will detail later the issues related with the computation of these weights.

The main difference between inpainting and interpolation is the available data and its geometric organization in the image. In a typical inpainting problem, large regions of the image are known, and transfer occurs between the available information and the patches inside the interpolation domain. In the interpolation application here addressed the image is known only at some isolated positions distributed through all the image. We can still have entire continuous regions of missing information (in contrast with typical approaches addressed in compressed sensing), but we do not have at all entire patches of *available* information. This does not allow the direct application of the inpainting energy (1) to the interpolation problem, since every image patch contains unknown pixels, and thus needs information from other patches. At the same time any patch may have information to transfer to potentially all other patches. This suggests that the summation domains in Eq. (2), as well as the patch comparison metric, have to be modified. We address this next.

For the sake of generality we will use generic summation domains and denote them by D_1 and D_2. For instance, the corresponding definitions for the inpainting functional (2) are $D_1 = \widetilde{O}$ and $D_2 = \widetilde{O}^c$, while for all methods implemented below we used $D_1 = \Omega$ and $D_2 = O^c$, *i.e.* D_2 the set of known pixels. The weight function is thus defined over $D_1 \times D_2$ such that for each $x \in D_1$, $w(x, \cdot)$ is a probability over D_2.

A general description of the image term in the interpolation functional is the following:

$$F(u, w) = \sum_{x \in D_1} \sum_{x' \in D_2} w(x, x') V_\varphi(p_u(x), p_u(x')). \qquad (4)$$

We have introduced a general *pair-wise patch similarity potential* V_φ, substituting the patch norm-like function $\| \cdot \|_{\Omega_p}$. Since we deal with sparsely sampled patches, the pair-wise patch potential V_φ is based only on the known pixels around x and x':

$$V_\varphi(p_u(x), p_u(x')) = \sum_{y \in \Omega_p} \frac{g_\sigma(y)}{\rho(x, x')} (\alpha \mathcal{X}_{O^c}(x+y) + \beta \mathcal{X}_{O^c}(x'+y)) \varphi(u(x+y) - u(x'+y))$$

$$(5)$$

where g_σ is a Gaussian centered at the origin with standard deviation σ, $\mathcal{X}_S$ denotes the characteristic function of the set S and $\varphi(r) = |r|^p, r \in \mathbb{R}, 1 \leq p < \infty$ (a more general function could be considered). For instance, taking $p = 1$ leads to an algorithm based on medians (see [23]), here due to space limitations we will restrict us to the case $p = 2$. The constant parameters $\alpha, \beta \in \{0, 1\}$ are set by the user. They control whether known positions around x or x' are used in the computation of the similarity potentials (at least one of them has to be 1).

If $\alpha = 1$ the positions with known data around x are used for the computation of the similarity potential (5). This happens whether the corresponding locations

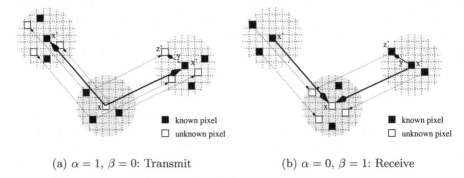

(a) $\alpha = 1$, $\beta = 0$: Transmit   (b) $\alpha = 0$, $\beta = 1$: Receive

Fig. 1. Visualization of transmission and reception processes due respectively to terms V_φ^α and V_φ^β in the patch similarity potential (see Eq. (6))

around x' belong to the data set or not. If $\beta = 1$ the similarity potential accounts for the known pixels around x'. If both of them are 1, in which case V_φ is computed from the locations known in both patches. This last case coincides with the patch comparison criterion defined in [24] in the context of demosaicing.

The normalization factor $\rho(x, x')$ is such that $\sum_{y \in \Omega_p} \frac{g_\sigma(y)}{\rho(x,x')}(\alpha \mathcal{X}_{O^c}(x + y) + \beta \mathcal{X}_{O^c}(x' + y)) = 1$ for all $x \in D_1$, $x' \in D_2$. Considering the overlap between known positions in both patches (see for instance [27]) would also make sense for the comparing patches with missing data. However, this cannot be applied to the current formulation since this eliminates the dependency of the energy on the unknown image (recall that the energy depends on the image though the similarity potential V_φ).

The proposed functional can be easily understood by splitting the pairwise patch potential into two terms $V_\varphi = V_\varphi^\alpha + V_\varphi^\beta$, with

$$V_\varphi^\alpha(p_u(x), p_u(x')) = \alpha \sum_{y \in \Omega_p} \frac{g_\sigma(y)}{\rho(x, x')} \mathcal{X}_{O^c}(x + y)\varphi(u(x + y) - u(x' + y)), \quad (6)$$

and analogously for V_φ^β. The energy F can be split accordingly in two terms. The first potential measures differences between known pixels in $p_u(x)$, with $x \in D_1$, and the corresponding pixels in $p_u(x')$, with $x' \in D_2$. Since known pixels are fixed, its minimization implies the modification of unknown pixels around x', thus transferring information from $p_u(x)$ to $p_u(x')$. On the other hand, V_φ^β considers differences between known pixels in $p_u(x')$ and the corresponding locations in $p_u(x)$. In this case known information flows from $p_u(x')$ centered at D_2 to $p_u(x)$ centered at D_1.

Since the weights $w(x, \cdot)$ are a probability over D_2 for each $x \in D_1$, we will adopt subsequently the point of view of the patch $p_u(x)$ centered at $x \in D_1$. We refer to these patches as *central* patches, and to patches centered in D_2 as *peripheral* patches. From this perspective, the minimization of the term with V_φ^α implies the *transmission* of the information (the pixel values) of known positions in the central patch $p_u(x)$ towards the unknown positions in peripheral patches

$p_u(x') \in D_2$ (see Figure 1(b)). Whereas the minimization of the term with V_φ^β implies *receiving* known pixel values from peripheral patches at D_2 (see Figure 1(b)). We refer to these processes as *transmission* and *reception*.

The complete functional for the interpolation problem becomes:

$$E(u, w) = \frac{1}{h} F(u, w) - \sum_{x \in D_1} H_w(x), \tag{7}$$

where as before $H_w(x) = -\sum_{x' \in D_2} w(x, x') \log w(x, x')$ is again the entropy of the probability $w(x, \cdot)$ for $x \in D_2$. As in the case of inpainting, this term allows to model the estimation of the weights together with the image.

A similar functional for image super-resolution was considered in [26] without explicitly modeling the weight updating step. The functional in [26] is related to the case where $\alpha = 0$ and $\beta = 1$ (or to the case of a full patch comparison).

2.3   Reinterpretation of the Image Term F

Let us rewrite the energy term (4) in a different way, in which the image values appear directly, and not as part of patches. This formulation will be useful for posterior analysis. After the change of variables $z = x + y$, $z' = x + y'$, the energy can be rewritten by adding up the pair-wise pixels differences as

$$F(u, w) = \sum_{z \in \tilde{\Omega}} \sum_{z' \in \tilde{\Omega}} m(z, z')(\alpha \mathcal{X}_{O^c}(z) + \beta \mathcal{X}_{O^c}(z')) \varphi(u(z) - u(z')), \tag{8}$$

where $\tilde{\Omega} = \Omega + \Omega_p$ (since $D_1, D_2 \subseteq \Omega$, we have that $D_1 + y, D_2 + y \subseteq \tilde{\Omega}$ for all $y \in \Omega_p$), and we have defined the *pixel-wise* influence weights $m(z, z')$ as

$$m(z, z') = \sum_{y \in \Omega_p} \mathcal{X}_{D_1}(z - y) \mathcal{X}_{D_2}(z' - y) w(z - y, z' - y) \frac{g_\sigma(y)}{\rho(z - y, z' - y)}. \tag{9}$$

These weights integrate the similarities of patches centered at $z - y \in D_1$ containing z and those centered at $z' - y \in D_2$ containing z' for $y \in \Omega_p$.

The formulation given by Eq. (4) accumulates the pair-wise potentials for each pair of patches centered in D_1 and D_2. The potentials are given by the addition of pixel value differences. In (8), the energy is rewritten by explicitly computing the contribution of each pixel value difference. The characteristic functions $\mathcal{X}_{D_1}(z - y)$ and $\mathcal{X}_{D_2}(z' - y)$ in (8) are zero if neither z nor z' are known. Only those differences involving at least one known pixel are taken into account. It becomes clear that pixel differences for which we have a large value of $m(z, z')$ are penalized. This implies the modification of $u(z)$ or $u(z')$, depending on which of them is given and which is unknown. This shows again the difference with more frequently used patch distances, where only pixels available in both patches are considered for the computation. Certainly if such approaches are iterated, as sometimes done [27,6], pixels with originally only "one side" available start to influence the computation as well after the first iteration or the first time yet are "filled".

3   Minimization of E

We have formulated the interpolation problem as the constrained optimization

$$(u^*, w^*) = \arg\min_{u,w} E(u, w) \quad \text{subject to} \tag{10}$$

$$\sum_{x' \in D_2} w(x, x') = 1 \quad \text{for all} \quad x \in D_1. \tag{11}$$

To minimize the energy E, we use an alternate coordinate descent algorithm. At each iteration, two optimization steps are solved: The constrained minimization of E with respect to w while keeping u fixed; and the minimization of E with respect to u with w fixed. This procedure yields the following iteration

1. [*Initial Condition*]   Given $u_0(x)$ with $x \in O$.
2. [*Weights Update Step*] $w_k = \arg\min_w E(u_k, w)$, subject to (11).
3. [*Image Update Step*] $u_{k+1} = \arg\min_u E(u, w_k)$.
4. [*Stopping Criterion*]   If $\|u_{k+1} - u_k\| > \tau$, go back to step 2.

In the weights updating step, the minimization of E w.r.t. w yields $w_k(x, x') = \frac{1}{q(x)} \exp\left[-\frac{1}{h} V_\varphi(p_{u_k}(x), p_u(x'))\right]$, where $q(x)$ is a normalization factor such that $\sum_{x' \in D_2} w(x, x') = 1$ for each patch $p_{u_k}(x)$. The parameter h determines the selectivity of the similarity. If h is large, maximizing the entropy becomes more relevant, yielding weights which are less selective. In the limit, when $h \to \infty$, $w_k(x, \cdot)$ becomes a uniform distribution over D_2. On the other hand, a small h yields weights more concentrated on the patches that are similar to $p_u(x)$. In fact, when $h \to 0$ the weights are given by $\lim_{h \to 0} w(x, x') = \frac{1}{\#n(x)} \mathcal{X}_{n(x)}(x')$, where $n(x) \subseteq O^c$ is the set of minimizers of $V_\varphi(p_u(x), \cdot)$, i.e. $n(x) = \{x' \in O^c : V_\varphi(p_u(x), p_u(x')) = V_{\min}(x)\}$, where $V_{\min}(x)$ is the minimum potential w.r.t. $p_u(x)$. In other words, when $h \to 0+$ the weights encode a multivalued assignment of patches with centers in D_2 for each $x \in D_1$.

The image updating step deserves more attention and is described next.

3.1   Image Updating Step

The equilibrium equation for E results in

$$\sum_{z' \in O^c} (\alpha m(z', z) + \beta m(z, z')) \varphi'(u(z) - u(z')) = 0 \quad \text{for all } z \in O. \tag{12}$$

This equation specifies the information transferred from the datum $u(z')$ to the unknown $u(z)$. This information can be transferred in any of the two modes discussed previously, *i.e.* reception, by a patch in D_1 covering z, of data coming from a patch in D_2 covering z', and/or transmission, of data from a patch in D_1 covering z', to a patch in D_2 covering z. The term $m(z, z')$ gathers all contributions by reception, whereas the term $m(z', z)$ considers all transmissions.

When $\varphi(t) = t^2$ we call the resulting method *patch-wise non-local means*. In this case Eq. (12) can be written as

$$u(z) = \frac{1}{C(z)} \sum_{z' \in O^c} (\alpha m(z', z) + \beta m(z, z'))u(z'), \qquad (13)$$

for each $z \in O$, where the normalization constant $C(z)$ is given by $C(z) = \sum_{z' \in O^c}(\alpha m(z', z) + \beta m(z, z'))$. Let us say in passing that due to our variational formulation, the image updating step is different from [24], since only the central pixel of the patch is updated in [24]. Taking $\varphi(t) = |t|$, we get the *patch-wise non-local medians*. In this case, the Euler equation (12) for u, given w, becomes $\sum_{z' \in O^c}(\alpha m(z', z) + \beta m(z, z'))\text{sign}(u(z) - u(z')) = 0$, and its solution $u(z)$ is obtained as a weighted median of the known values $u(z')$.

4   A Departure from Variational Model

We have seen that three different schemes can be derived from the proposed variational model, by changing the values of α and β. We have interpreted them, by observing the effect over the unknown pixels of u, as transmission ($\alpha = 1$, $\beta = 0$), reception ($\alpha = 0$, $\beta = 1$) and combined ($\alpha = 1$, $\beta = 1$). But each scheme also forces the manner to compute w. Now, if we abandon the variational framework, we can combine different update schemes of w and u.

We now propose a new scheme by updating the weights w according to the transmission scheme ($\alpha = 1$, $\beta = 0$), and the image u using the combined scheme ($\alpha = 1$, $\beta = 1$). The resulting algorithm was experimentally found to be numerically stable, and for relatively high sampling densities to behave like the combined scheme ($\alpha = 1$, $\beta = 1$). However for low sampling densities it exhibits a remarkable ability to speed up the convergence. An intuitive reason that may explain this scheme relies on the fact that using the transmission potential ($\alpha = 1$, $\beta = 0$), the weights $w(x, \cdot)$ are always computed using coordinates around x, with known values. Adding known positions around x' may provide a poorer estimate of the weights, specially if the current interpolation around x is bad.

5   Experimental Results

We now present experimental results with both synthetic and natural images randomly sampled with densities from 20% to 5% of the image points. The four schemes derived from the potential (5) in Section 2, are referred here as A (for $\alpha = 1$, $\beta = 0$), B ($\alpha = 0$, $\beta = 1$), AB ($\alpha = 1$, $\beta = 1$), and O for the *departure from the variational model* (which is a variant of AB). All of them have a computational cost proportional to $\mathcal{A}(D_1) \times \mathcal{A}(D_2)$ (where $\mathcal{A}(D_i)$ is the number of pixels of D_i). Since D_2 is a fraction of D_1 (the density of the sampling) then the algorithm is $\mathcal{O}(T \times \mathcal{A}(D_1)^2)$, where T is the number of iterations (usually $T < 200$). A single iteration for a 256×256 pixels image takes about 3 min on a 3GHz processor. However, with the *coarse to fine* scheme described below, the convergence is generally attained with less than 40 iterations. This amounts to a computational time of 120 minutes.

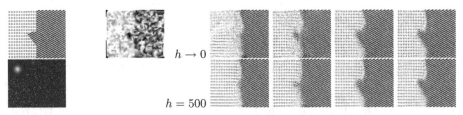

Fig. 2. Synthetic. The first column shows the original image, and the random samples (5% of the image) with the window g_σ (see (5)) depicted in the upper left corner. The second column from left shows a linear interpolation over the Delaunay triangulation of the samples. Remaining columns (from left to right) correspond to results of the schemes A, B, AB and O; The rows correspond to two different values of h.

Role of locality in the non-local algorithms. A common strategy to improve the computational performance of nonlocal methods is to reduce the size of the *search window* (subset of D_2 around the central pixel x), thereby reducing the number of comparisons performed for each pixel. As a desirable side effect, this enforces the ergodicity assumption over the data. In other words, the patches needed to estimate the current point are assumed to be found in the vicinity of it, not far away. As a consequence, the size of the search window is a very important parameter, and it may be itself subject of optimization as in [31]. In our experiments we choose the search windows to have a reasonable size (containing 100 to 500 samples) with respect to the density of the image.

Choice of V_φ. The experiments shown in Figure 2 are aimed to compare the performance of the different schemes. The best results for this data are obtained with the scheme AB. Therefore, since the experiments are also consistent with these results, from now on we will mainly show AB and O. Also notice that the textures are recovered in great detail, while the interface between them, is very imprecise. This evidences the exemplar-based nature of the algorithms, since there are plenty of examples of textures, but only few of the interface.

Initial condition and h. If the initial condition has artifacts, then for small h these methods tend to reinforce them. To reduce the dependence on the initial condition we adopt the *coarse to fine* scheme proposed in [24], where a decreasing sequence of h is used to recover first large scale structures and later refine them (as h decreases). Figures 3 and 4(b) show the results of applying the algorithms O and AB to natural images with sampling densities from 20% to 5%. For high densities the performances of both schemes is similar. For lower densities (5% for instance) O exhibits less dependence on the initial condition than AB. In particular, we can obtain with O results similar to those obtained with AB even without the coarse to fine scheme. Using $h > 0$ produces smooth results with blurred details, while using $h \to 0$ introduces a staircase effect; we expect to improve these results by using $h > 0$ in the median case (which corresponds to $p = 1$ in φ (5)). In the first two columns of Figure 3 we display: a set of random samples, and an optimal dithered set of the same image (optimal for the Laplacian-based interpolation as described in [22]). Both sets contain 10% of the image points. The Laplacian interpolation from dithered samples takes

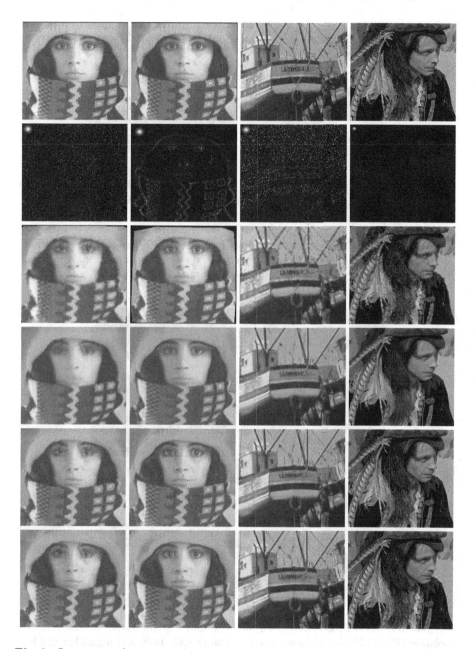

Fig. 3. *Sparse sampling interpolation.* 1st row: original images. 2nd row: input data with sample densities of 10%, 10% dithered [22], 20% and 9%. 3rd row: linear interpolation over the Delaunay triangulation of the samples; PSNRs: 25.8, 30.6 (not considering the black frame), 25.0 and 22.74. 4rt row: results of method AB with $h = 100$; PSNRs: 22.5 ,22.7, 25.5 and 22.79. 5th row: results of AB with $h \rightarrow 0$; PSNRs: 22.6, 23.0, 25.5 and 22.56. 6th row: results of the method O with $h \rightarrow 0$; PSNRs: 22.6,21.7,25.1 and 22.69. (Details can be better appreciated by zooming on a computer screen)

(a) *Cylinders*, 200×113. Inpainting a hole in a subsampled image with algorithm *B*. The algorithm makes no distinction between the hole and the sampled regions. The sampling density is 20% yielding a global sampling density of 14%.

(b) *Barbara*, 512×512 *with* 5% *of the samples*. 2nd line are results of: linear interpolation (PSNR 20.1), Laplacian interpolation (PSNR 19.9) and algorithm *O* with $h = 100$ (PSNR 22.8).

Fig. 4. Experiments with lower sampling densities. Each figure shows the original image (top left), the available samples (top right), the result of linear interpolation over the Delaunay triangulation (bottom left) and the output the algorithm specified in each figure (bottom right).

advantage of the distribution of the samples along the edges and permits to recover a visually pleasant smooth image with clear edges (see [22]), while our method is less fitted for this task (second column of Figure 3). However, for random samplings the results of the Laplacian interpolation are less convincing, while our method recovers most edges and textures of the image (see Figure 4).

Interpolation of large holes. In Figure 4(a) we show a preliminary result using method B (only reception process) applied to the interpolation of a hole in a sampled image, this choice of the potential leads to a functional similar to the inpainting one shown in [23]. Let us remark that the method was applied "as it is" to the problem, without making any distinction between the hole and the sampled areas. Other methods that involve the transmission process (AB or O) fail to fill the large holes, although all manage to recover the sparsely sampled area. We attribute the non regularity of the solution to the low frequency of the texture, which implies less exemplars to copy from, showing the main limitation of exemplar-based methods. A local regularization term can be used to impose smoothness on the result [26]. The results shown here are also available at: http://gpi.upf.edu/static/vnli

6   Conclusions and Future Work

A variational formulation for non-local example-based image interpolation was introduced in this paper. The obtained results show a promising performance. In subsequent work we will extend the present model to cover the case of samples located at non-entire positions and we will explore some variants of it.

Acknowledgements. VC acknowledges partial support by PNPGC project, ref. MTM2006-14836 and by "ICREA Académia" prize for excellence in research funded by the Generalitat de Catalunya. PA acknowledges support by the FPI grant BES-2007-14451. GS is partially supported by NSF, ONR, DARPA, NGA, and ARO.

References

1. Bertalmío, M., Sapiro, G., Caselles, V., Ballester, C.: Image inpainting. In: Proc. of SIGGRAPH (2000)
2. Bertalmío, M., Bertozzi, A., Sapiro, G.: Navier-stokes, fluid-dynamics and image and video inpainting. In: Proc. of the IEEE Conf. on CVPR (2001)
3. Masnou, S., Morel, J.M.: Level lines based disocclusion. In: Proc. of IEEE ICIP (1998)
4. Efros, A., Leung, T.: Texture synthesis by non-parametric sampling. In: Proc. of the IEEE Conf. on CVPR, pp. 1033–1038 (1999)
5. Levin, A., Zomet, A., Weiss, Y.: Learning how to inpaint from global image statistics. In: Proc. of ICCV (2003)
6. Criminisi, A., Pérez, P., Toyama, K.: Region filling and object removal by exemplar-based inpainting. IEEE Trans. on IP 13(9), 1200–1212 (2004)
7. Sun, J., Yuan, L., Jia, J., Shum, H.Y.: Image completion with structure propagation. In: Proc. of SIGGRAPH (2005)
8. Bertalmío, M., Vese, L., Sapiro, G., Osher, S.: Simultaneous structure and texture inpainting. IEEE Trans. on Image Processing 12(8), 882–889 (2003)

9. Cao, F., Gousseau, Y., Masnou, S., Pérez, P.: Geometrically-guided exemplar-based inpainting (submitted, 2008)
10. Gröchenig, K., Strohmer, T.: Numerical and theoretical aspects of non-uniform sampling of band-limited images. In: Marvasti, F. (ed.) Theory and Practice of Nonuniform Sampling. Kluwer/Plenum (2000)
11. Chan, T., Shen, J.H.: Mathematical models for local nontexture inpaintings. SIAM J. App. Math. 62(3), 1019–1043 (2001)
12. Arigovindan, M., Suhling, M., Hunziker, P., Unser, M.: Variational image reconstruction from arbitrarily spaced samples: a fast multiresolution spline solution. IEEE Trans. on IP 14(4), 450–460 (2005)
13. Shepard, D.: A two-dimensional interpolation function for irregularly-spaced data. In: Proc. of 23rd ACM national conf., pp. 517–524. ACM Press, New York (1968)
14. Candes, E.J., Wakin, M.B.: An introduction to compressive sampling. IEEE Signal Processing Magazine 25(2), 21–30 (2008)
15. Aharon, M., Elad, M., Bruckstein, A.: The K-SVD: An algorithm for designing of overcomplete dictionaries for sparse representatio. IEEE Trans. on Signal Processing 54(11), 4311–4322 (2006)
16. Mairal, J., Sapiro, G., Elad, M.: Learning multiscale sparse representations for image and video restoration. SIAM MMS 7(1), 214–241 (2008)
17. Elad, M., Starck, J., Querre, P., Donoho, D.: Simultaneous cartoon and texture image inpainting using morphological component analysis (MCA). Applied and Computational Harmonic Analysis 19(3), 340–358 (2005)
18. Hays, J., Efros, A.: Scene completion using millions of photographs (2008)
19. Peyré, G., Bougleux, S., Cohen, L.: Non-local regularization of inverse problems. In: Forsyth, D., Torr, P., Zisserman, A. (eds.) ECCV 2008, Part III. LNCS, vol. 5304, pp. 57–68. Springer, Heidelberg (2008)
20. Lezoray, O., Elmoataz, A., Bougleux, S.: Graph regularization for color image processing. Comput. Vis. Image Underst. 107(1-2), 38–55 (2007)
21. Gilboa, G., Osher, S.: Nonlocal linear image regularization and supervised segmentation. SIAM Mult. Mod. and Sim. 6(2), 595–630 (2007)
22. Belhachmi, Z., Bucur, D., Burgeth, B., Weickert, J.: How to choose interpolation data in images. Technical Report No. 205, Department of Mathematics, Saarland University, Saarbrücken, Germany (2008)
23. Arias, P., Caselles, V., Sappiro, G.: A variational framework for non-local image inpainting. In: Proc. of EMMCVPR. Springer, Heidelberg (2009)
24. Buades, A., Coll, B., Morel, J., Sbert, C.: Self similarity driven color demosaicing. IEEE TIP 18(6), 1192–1202 (2009)
25. Mairal, J., Bach, F., Ponce, J., Sapiro, G., Zisserman, A.: Non-local sparse models for image restoration. In: Proc. of ICCV (2009)
26. Protter, M., Elad, M., Takeda, H., Milanfar, P.: Generalizing the non-local-means to super-resolution reconstruction. IEEE Trans. on IP 18(1), 36–51 (2009)
27. Bartesaghi, A., Sprechmann, P., Liu, J., Randall, G., Sapiro, G., Subramaniam, S.: Classification and 3d averaging with missing wedge correction in biological electron tomography? Journal of Structural Biology 162(3), 436–450 (2008)
28. Candès, E.J., Recht, B.: Exact matrix completion via convex optimization. CoRR abs/0805.4471 (2008)
29. Awate, S., Whitaker, R.: Unsupervised, information-theoretic, adaptive image filtering for image restoration. IEEE Trans. on PAMI 28(3), 364–376 (2006)
30. Brox, T., Kleinschmidt, O., Cremers, D.: Efficient nonlocal means for denoising of textural patterns. IEEE Trans. on IP 17(7), 1057–1092 (2008)
31. Kervrann, C., Boulanger, J.: Optimal spatial adaptation for patch-based image denoising. IEEE Trans. on IP 15(10), 2866–2878 (2006)

A Variational Framework for Non-local Image Inpainting

Pablo Arias[1], Vicent Caselles[1], and Guillermo Sapiro[2]

[1] Universitat Pompeu Fabra, DTIC, 08018 Barcelona, Spain
[2] University of Minnesota, ECE, Minneapolis, MN 55455, USA
{pablo.arias,vicent.caselles}@upf.edu,guille@umn.edu

Abstract. Non-local methods for image denoising and inpainting have gained considerable attention in recent years. This is in part due to their superior performance in textured images, a known weakness of purely local methods. Local methods on the other hand have demonstrated to be very appropriate for the recovering of geometric structure such as image edges. The synthesis of both types of methods is a trend in current research. Variational analysis in particular is an appropriate tool for a unified treatment of local and non-local methods. In this work we propose a general variational framework for the problem of non-local image inpainting, from which several previous inpainting schemes can be derived, in addition to leading to novel ones. We explicitly study some of these, relating them to previous work and showing results on synthetic and real images.

1 Introduction

Image inpainting, also known as *image completion* or *disocclusion*, is an active research area in image processing. The purpose of inpainting is to obtain a visually plausible image interpolation in a region in which data are missing due to damage or occlusion. Usually, to solve this problem, the only available data is the image outside of the region to be inpainted. In addition to its theoretical importance, image inpainting is a very important problem due to its applications to image and video editing and restoration.

Inpainting methods found in the literature can be classified into two groups: *geometry-* and *texture-oriented* methods.

Geometry-oriented methods. Images are modeled as functions with some degree of smoothness (expressed for instance in terms of the curvature of the level lines or the total variation of the image), and the interpolation is performed by continuing and imposing this model inside the inpainting domain. This has been performed either using variational techniques, as for instance in [3,11,12,19,28,29], or with PDEs [4,7,36]. These methods show a good performance in propagating smooth level lines or gradients. However they fail in the presence of texture. This is often referred to as *structure* or *cartoon* inpainting.

Texture-oriented methods. Texture-oriented inpainting was born as an application of texture synthesis, e.g., [18,21]. Its recent development was triggered in part by [18,37] using non-parametric sampling techniques. In these works texture is modeled as a two dimensional probabilistic *graphical model*, in which the

D. Cremers et al. (Eds.): EMMCVPR 2009, LNCS 5681, pp. 345–358, 2009.

value of each pixel is conditioned by its neighborhood. These approaches rely directly on a sample of the desired texture to perform the synthesis.

In practice these methods work progressively by expanding a region of synthesized texture. The value for a target pixel x is copied from the center of a square patch in the sample image, chosen among those that best match the available portion of the patch centered at x. Levina and Bickel [26] recently provided a probabilistic theoretical justification for this strategy.

This method (as explained above or with some modifications) has been extensively used for inpainting [5,6,14,17,18,31]. As opposed to geometry-oriented inpainting, these so-called *exemplar-based* approaches, are *non-local*: To determine the value at x, the whole image may be scanned in the search of a matching patch.

Since these texture approaches are *greedy* procedures (each hole pixel is visited only once), the results are very sensitive to the order in which pixels are processed [14]. This issue was addressed in [24,38] where the inpainting problem is stated as the optimization of an energy derived from probabilistic graphical models (see also [25]).

A variational justification for texture-based methods was presented in [16], where the inpainting problem is reformulated as that of finding a *correspondence map* $\Gamma : O \to O^c$, O being the inpainting domain and O^c its complement w.r.t. the image domain. Denoting the image by u, the inpainted value at position $x \in O$ is then given by $u(x) = u(\Gamma(x))$, $\Gamma(\cdot)$ being the correspondence map. The authors proposed a continuous energy functional in which the unknown is the correspondence map itself:

$$E(\Gamma) = \int_O \int_{\Omega_p} (u(\Gamma(x-y)) - u(\Gamma(x) - y))^2 \mathrm{d}y \mathrm{d}x,$$

where Ω_p is the patch domain (centered at $(0,0)$). Thus Γ should map a pixel x and its neighbors in such a way that the resulting patch is close to the one centered at $\Gamma(x)$. This model has been the subject of further (theoretical) analysis by Aujol *et al.*[1].

A different variational model was presented in [32]. Images are modeled as ensembles of patches on a given patch manifold. For inpainting, the patch manifold can be learned from the set of patches in the hole's complement. The method is iterative, with each iteration having two steps. First, the patches in the hole are projected onto the manifold. Since this is done for each patch independently, the projected patches are not necessarily coherent with each other, *i.e.*overlapping patches may have different values in the overlap region. Therefore, in the second step, an image is computed by averaging the patches in the ensemble.

Exemplar-based methods provide impressive results in recovering textures and repetitive structures. However, their ability to recreate the geometry without any example is limited and not well understood. Therefore, different strategies have been proposed which combine geometry and texture inpainting [5,10,17,23]. These methods usually decompose the image in some sort of structure and texture components. Structure is reconstructed using some geometry-oriented scheme, and this is used to guide the texture inpainting.

Contributions of this work. Despite these combined methods, geometry and texture inpainting are still quite separate fields, each one with its own analysis and implementation tools. Variational models as the one introduced in this paper can provide common tools allowing a unified treatment of both trends. We therefore propose a variational framework for non-local image inpainting as a contribution to the modeling and analysis of texture-oriented methods. Our formulation is rather general and different inpainting schemes can be derived naturally from it, via the selection of the appropriate patch metric.

In the present work we study three of them, *patch NL-means*, *-medians*, and *-Poisson*. The former is related to the method of [38] and can be interpreted in terms of the *mean shift* [13] and the manifold models of [32]. The other schemes are, to the best of our knowledge novel. The latter imposes coherence of the gradients, in addition to that of the gray levels, which implies a smoother continuation of the information across the boundary and inside the inpainting domain, thus acting as a basic local regularization.

Our work is related to recent variational formulations of non-local denoising ([2,9]) by Gilboa and Osher [20]. The image redundancy and self-similarity (measured as patch similarity) is encoded by a non-local weight function w : $O \times O^c \rightarrow \mathbb{R}$. This function serves as a fuzzy correspondence, and differs from the works [1,16], although a (eventually multivalued) correspondence map can be approximated as a limit of our model.

Notation. Images are denoted as functions $u : \Omega \rightarrow \mathbb{R}$, where Ω denotes the image domain, usually a rectangle in $\mathbb{R}^2$. Pixel positions are denoted by x, x', z, z' or y, the latter for positions inside the patch. A patch of u centered at x is denoted by $p_u(x) = p_u(x, \cdot) : \Omega_p \rightarrow \mathbb{R}$, where Ω_p is a rectangle centered at $(0,0)$. The patch is defined by $p_u(x, y) = u(x + y)$, with $y \in \Omega_p$. $O \subset \Omega$ is the hole or inpainting domain, and $O^c = \Omega \setminus O$. We still denote by u the part of the image u inside the hole, while $\hat{u}$ is the part of u in O^c: $\hat{u} = u|_{O^c}$. Additional notation will be introduced in the text.

2   Variational Framework

Our variational framework is inspired by the following non-local functional

$$F_w(u) = \int_O \int_{O^c} w(x, x')(u(x) - \hat{u}(x'))^2 \mathrm{d}x' \mathrm{d}x. \qquad (1)$$

The weight function $w : O \times O^c \rightarrow \mathbb{R}^+$ measures the similarity between patches centered in the inpainting domain and in its complement. *Gaussian* weights are commonly used, given by $w(x, x') = \exp\left(-\frac{1}{h}\|p_u(x) - p_{\hat{u}}(x')\|^2\right)$, where $\|\cdot\|$ is a weighted L_2 norm in the space of patches and h is the scale. A similar functional was proposed in [20] as a non-local regularization energy in the context of image denoising which models the non-local means filter [2,9] (see [8,35] for a different model of non-local means). An extension to super-resolution is presented in [34].

In [20] the weights are considered known and remain fixed through all the iterations. While this might be appropriate in applications where the weights can be estimated from the noisy image, in the image inpainting scenario here addressed, weights are not available and have to be inferred together with the image ([33,34]). One of the novelties of the proposed framework is the inclusion of adaptive weights in a variational setting.

For this reason, we will consider the weight function w as an additional un-kown. Instead of prescribing explicitly the Gaussian functional dependence of w w.r.t. u, we will do it implicitly, as a component of the optimization process. In doing so, we obtain a simpler functional, avoiding to deal with the complex, non-linear dependence between w and u. In our formulation, $w(x, \cdot)$ is a proba-bility density function, $\int_{O^c} w(x, x') \mathrm{d}x' = 1$, and can be seen as a relaxation of the one-to-one correspondence map of [1,16], providing a fuzzy correspondence between each $x \in O$ and the complement of the inpainting domain.

In this setting, we propose an energy which contains two terms, one of them is inspired by (1) and measures the coherence between the pixels in O and those in O^c, for a given similarity weight function w. This permits the estimation of the image u from the weights w. The second term allows us to compute the weights given the image. The complete proposed functional is

$$E(u, w) = \frac{1}{h} \widetilde{F}_w(u) - \int_{\widetilde{O}} H_w(x) \mathrm{d}x, \qquad (2)$$

where

$$\widetilde{F}_w(u) = \int_{\widetilde{O}} \int_{\widetilde{O}^c} w(x, x') \|p_u(x) - p_{\hat{u}}(x')\|_{a,\varphi} \mathrm{d}x' \mathrm{d}x, \qquad (3)$$

for a given norm-like function $\| \cdot \|_{a,\varphi}$ between patches, and

$$H_w(x) = - \int_{\widetilde{O}^c} w(x, x') \log w(x, x') \mathrm{d}x',$$

is the entropy of the probability $w(x, \cdot)$.

We take $\widetilde{O}$, the *extended inpainting domain*, as the set of centers of patches that intersect the hole, $i.e. \widetilde{O} = O + \Omega_p = \{x \in \Omega : (x + \Omega_p) \cap O \neq \emptyset\}$. Thus, patches $p_{\hat{u}}(x')$ centered in $x' \in \widetilde{O}^c$ are entirely outside O (Figure 1), simplifying the Euler-Lagrange equation for the minimizer. Accordingly, we consider that the weight function w is defined over $\widetilde{O} \times \widetilde{O}^c$ and $\int_{\widetilde{O}^c} w(x, x') \mathrm{d}x' = 1$.

For a simplified presentation, we assume that $\widetilde{O} + \Omega_p \subseteq \Omega$, $i.e.$ every pixel in $\widetilde{O}$ supports a patch centered on it and contained in Ω. This is not true if the inpainting domain reaches the boundary of the image, and details on the treatment of this situation are given in Section 5. Analogously, we also shrink $\widetilde{O}^c$ to have $\widetilde{O}^c + \Omega_p \subseteq \Omega$.

Let us now make some additional comments on the functional. The term $(u(x) - \hat{u}(x'))^2$ in F_w, penalizing differences between pixels, is substituted by $\|p_u(x) - p_{\hat{u}}(x')\|_{a,\varphi}$. This has to be understood together with the inclusion of the second term, which integrates the entropy of each probability $w(x, \cdot)$ over $\widetilde{O}$. For a given completion u, and for each $x \in \widetilde{O}$, the optimum weights minimize the mean patch error for $p_u(x)$, given by $\int_{\widetilde{O}^c} w(x, x') \|p_u(x) - p_{\hat{u}}(x')\|_{a,\varphi} \mathrm{d}x'$, while maximizing the entropy. The resulting weights are Gaussian, as can be confirmed easily by derivating the energy. This can be related to the *principle of maximum entropy* [22], widely used for inference of probability distributions. The parameter h controls the trade-off between both terms and is also the scale parameter of the Gaussian weights. Since $w(x, \cdot)$ is a probability, we discard trivial minima of E with $w(x, x') = 0$ everywhere.

The patch norm-like function. Patches are functions defined on Ω_p, and are compared using $\| \cdot \|_{a,\varphi}$. We consider a non-decreasing and continuously differentiable function $\varphi : \mathbb{R}^+ \rightarrow \mathbb{R}^+$ with $\varphi(0) = 0$ and define $\| \cdot \|_{a,\varphi}$ by

$$\|p\|_{a,\varphi} = \int_{\Omega_p} g_a(y)\varphi(|p(y)|)\mathrm{d}y,$$

where g_a is an *intra-patch* weight function, a Gaussian centered at the origin with standard deviation a. The L_1 and the squared L_2 norms are particular cases of $\| \cdot \|_{a,\varphi}$ when $\varphi(t) = t$ and $\varphi(t) = t^2$, respectively. In Section 3 we consider another norm involving derivatives of the patch. As will be described below, the patch norm determines not only the similarity criterion but also the image synthesis, and thus is a key element in the framework.

2.1 Probabilistic-Geometric Model Interpretation

The proposed model can be written in terms of the generalized Kullback-Leibler divergence [15]. Given two positive and integrable functions p, q defined over a certain measure space $\mathcal{X}$, the generalized Kullback-Leibler divergence is given by: $\mathrm{KL}(p, q) = \int_{\mathcal{X}} p(s) \log\left(\frac{p(s)}{q(s)}\right) \mathrm{d}s - \int_{\mathcal{X}} p(s)\mathrm{d}s + \int_{\mathcal{X}} q(s)\mathrm{d}s$, assuming that the integrals exist. With this notation (and taking into account that $w(x, \cdot)$ is a probability) the functional E can be written as

$$E(u, w) = \int_{\widetilde{O}} \mathrm{KL}\left(w(x, \cdot), r(x, \cdot)\right) \mathrm{d}x - \int_{\widetilde{O}} \int_{\widetilde{O}^c} r(x, x')\mathrm{d}x'\mathrm{d}x,$$

where r is the Gaussian weight function $r(x, x') = \exp\left(-\frac{1}{h}\|p_u(x) - p_{\hat{u}}(x')\|_{a,\varphi}\right)$. The first term integrates the divergence between the functions $w(x, \cdot)$ and $r(x, \cdot)$, for each $x \in \widetilde{O}$. The second term can be interpreted by noticing that

$$\tilde{q}(x) = \int_{\widetilde{O}^c} r(x, x')\mathrm{d}x' \tag{4}$$

is a density estimate (in the patch space) of the set of patches in O^c: The higher the amount of patches in $\widetilde{O}^c$ close to $p_u(x)$ (according to the scale parameter h), the higher the value of $\tilde{q}$.

The minimizers (u^*, w^*) are obtained when $w^*(x, x') = r^*(x, x')/\tilde{q}^*(x)$, (Gaussian weights normalized by (4)), and the patches of the inpainted image are in regions of high density in the patch space. This provides a geometric intuitive interpretation of our variational formulation. The image is considered as an ensemble of overlapping patches. Known patches in $\widetilde{O}^c$ are fixed, forming a patch density model used to estimate the patches in $\widetilde{O}$. The richness of the framework is given in part by the fact that different norms in the patch space induce inpainting schemes of different nature, as we are going to see next.

2.2 Minimization of E

We have formulated the inpainting problem as the constrained optimization

$$(u^*, w^*) = \arg\min_{u,w} E(u, w) \quad \text{subject to} \int_{\widetilde{O}^c} w(x, x')\mathrm{d}x' = 1 \quad \forall x \in \widetilde{O}. \tag{5}$$

To minimize the energy E, we use an alternate coordinate descent algorithm. At each iteration, two optimization steps are solved: The constrained minimization of E with respect to w while keeping u fixed; and the minimization of E with respect to u with w fixed. This procedure yields the following iterative scheme

1. [*Initial Condition*]   Given $u_0(x)$ with $x \in O$.
2. [*Weights Update*] $w_k = \arg\min_w E(u_k, w)$, subject to $\int_{\tilde{O}} w(x, x') dx' = 1$.
3. [*Image Update*] $u_{k+1} = \arg\min_u E(u, w_k)$.
4. [*Stopping Criterion*] If $\|u_{k+1} - u_k\| > \tau$, go back to step 2.

In the weights update step, the minimization of E w.r.t. w yields:

$$w_k(x, x') = \frac{1}{\tilde{q}(x)} \exp\left(-\frac{1}{h}\|p_{u_k}(x) - p_{\hat{u}}(x')\|_{a,\varphi}\right).$$

The normalizing factor $\tilde{q}(x)$ is the density estimate given by (4), for patch $p_{u_k}(x)$.

The parameter h determines the selectivity of the similarity. If h is large, maximizing the entropy becomes more relevant, yielding weights which are less selective. In the limit, when $h \rightarrow \infty$, $w(x, \cdot)$ becomes a uniform distribution over $\tilde{O}^c$. On the other hand, a small h yields weights which concentrate on the patches close to $p_u(x)$. In fact, as we will mention later on, in the limit as $h \rightarrow 0$, $w(x.\cdot)$ can be considered as an approximation to an (eventually multivalued) correspondence.

The image update step deserves more attention and is described next.

Image update step. We now detail the derivation of the image update step for the cases $\varphi(t) = t^2$ and $\varphi(t) = t$. We refer to the resulting algorithms as *patch-wise non-local means* (patch NL-means), and *medians* (patch NL-medians), respectively.

Patch-wise non-local means. If $\varphi(t) = t^2$ the image energy term is quadratic on u, and its minimum for fixed weights w can be computed explicitly leading to a non-local average:

$$u(x) = \frac{1}{C(x)} \int_{\Omega_p} g_a(y) \int_{\tilde{O}^c} w(x - y, z')\hat{u}(z' + y) dz' dy, \tag{6}$$

for each $x \in O$, where the normalization constant $C(x)$ is given by $C(x) = \int_{\Omega_p} g_a(y) dy = \mathcal{A}(\Omega_p)$, the area of the patch (measured according to g_a).

Figure 1 explains this equation. The value at x considers all patches containing x. For instance the patch $p_u(x - y)$ covers x, $p_u(x - y, y) = u(x)$. This patch is compared to all patches in the complement, $p_{\hat{u}}(z')$, yielding the weights $w(x - y, z')$. Each of these patches contributes the term $w(x - y, z')\hat{u}(z' + y)$ to the average, *i.e.* its value at position y weighted by $w(x - y, z')$.

Patch-wise non-local medians. We now consider the L_1 norm in the energy E, corresponding to $\varphi(t) = t$. The Euler equation for u, given the weights w, is

$$\delta_u E(u, w)(z) = \int_{\Omega_p} g_a(y) \int_{\tilde{O}^c} w(z - y, x')\text{sign}[u(z) - \hat{u}(x' + y)] dx' dy = 0. \tag{7}$$

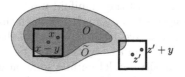

Fig. 1. Patch-wise non-local means inpainting. The value at $x \in O$ is computed using all the patches that overlap x. The patch centered at $x - y$ contributes with the term $w(x - y, z')\hat{u}(z' + y)$ to the average for each $z' \in \widetilde{O}^c$.

The solution of this equation is given by a weighted median of the values outside the hole. We can see this easily by defining $z' = x' + y$ and rewriting Eq. (7) as

$$\delta_u E(u, w)(z) = \int_{O^c} \text{sign}[u(z) - \hat{u}(z')]\rho_z(z')\mathrm{d}z',$$

where

$$\rho_z(z') := \int_{\Omega_p} \chi_{\widetilde{O}^c}(z' - y)g_a(y)w(z - y, z' - y)\mathrm{d}y. \tag{8}$$

For a given $z \in O$, the function $\rho_z : O^c \to \mathbb{R}^+$ weights the contribution of each location z' to the median. The quantity $\rho_z(z')$ is computed by integrating the similarity $w(z - y, z' - y)$ between all patches that overlap z' and those that overlap z in the *same relative position*. It tells us how much evidence there is supporting $u(z')$ as the intensity value for z. The function $\chi_{\widetilde{O}^c}$ takes the value 1 on $\widetilde{O}^c$ and 0 on $\widetilde{O}$.

2.3 Revisiting Related Work

We conclude this section by further connecting our work with previous art. The method in [38] is closely related to the patch NL-means scheme of Eq. (6). The key difference lies in the underlying theoretical model. The problem is addressed as a MRF, where pixels outside the hole are observable variables, missing pixels in the hole are the parameters, and the hidden variables are given by the correspondence $\Gamma : O \to O^c$, which assigns a patch outside the hole to each x in O. The method can be seen as an approximate EM algorithm for maximizing the log-likelihood w.r.t. the pixels in O, and some approximations have to be taken to make the optimization tractable. Based on heuristics, the authors also propose to use more robust estimators than the mean for the synthesis of pixels. With the framework here proposed, robust estimators (as the median) naturally result from particular choices of the patch norm $\| \cdot \|_{a,\varphi}$.

The patch NL-means algorithm is also related to the interesting manifold image models of [32]. Eq. (6) can be split into two steps which are analog to Peyré's manifold and image projection steps. First, for each patch centered in $\widetilde{O}$ we compute a new patch as a weighted average of all patches in the complement, according to the patch similarity weights $p_u^{MS}(z) := \int_{\widetilde{O}^c} w(z, z')p_{\hat{u}}(z')\mathrm{d}z'$ with $z \in \widetilde{O}$. Doing this for each hole position yields an incoherent ensemble of patches. The image is obtained by averaging these patches: $u(z) = \frac{1}{A(\Omega_p)} \int_{\Omega_p} p_u^{MS}(z - y, y)\mathrm{d}y$.

We use a density model, instead of the manifold model of [32]. Indeed, $p_u^{MS}(x)$ is the *mean shift* operator applied to $p_u(x)$. It is known that the iteration of this operator corresponds to an adaptive gradient ascent of the Parzen estimate of a PDF [13], which in this case is generated by the set of patches in the complement of the hole. The use of a density model entails some advantages, mainly from the computational point of view, learning a manifold model is computationally costly. Furthermore, the assumption that patches lie on a manifold is questionable (one could think for instance in a *stratification* as a more realistic model), and its dimension is hard to determine for real images.

3   Higher Order Variational Models

The proposed variational framework allows the introduction of derivatives of the image, by considering them in the patch norm used in (3). In this section we study a functional using the L_2 norm of the gradients of the patches,

$$\|p(y)\|_{a,\nabla}^2 = \int_{\Omega_p} g_a(y)\|\nabla p(y)\|_2^2 dy,$$

where $\|\cdot\|_2$ is the Euclidean norm in $\mathbb{R}^2$. Firstly, the similarity weights are now based on patch gradients, and secondly, the image update step is given by a *non-local* Poisson equation, *i.e.* a Poisson equation with non-local coefficients. The functional is obtained by substituting in (2) the image energy term $\widetilde{F}_w(u) = \int_{\tilde{O}} \int_{\tilde{O}^c} w(x,x')\|p_u(x) - p_{\hat{u}}(x')\|_{a,\nabla}^2 dx' dx$ (we assume that $u|_{O^c} = \hat{u}$).

The Euler equation w.r.t. u of the resulting functional is

$$\nabla \cdot [C(z)\nabla u(z)] = \nabla \cdot \boldsymbol{v}(z), \qquad (9)$$

for all $z \in O$, where $u|_{O^c} = \hat{u}$ and the field $\boldsymbol{v} : O \to \mathbb{R}^2$ is given by

$$\boldsymbol{v}(z) = \int_{\Omega_p} g_a(y) \int_{\tilde{O}^c} w(z - y, x')\nabla \hat{u}(x' + y)dx' dy.$$

The solutions are minimizers of $\int_{\tilde{O}} C(z)\|\nabla u(z) - \boldsymbol{v}(z)\|_2^2 dz$ (as before, $C(z) = \mathcal{A}(\Omega_p)$). Therefore, u is computed as the image with the closest gradient (in the L_2 sense) to the *guiding* vector field $\boldsymbol{v}$, which corresponds to a non-local weighted average of the gradients in the complement. The coefficients in the average have exactly the same form as in (6). The only difference is that the patch similarity weights used here are Gaussian weights of the L_2 norm of the gradients. See [30] for further uses of the Poisson equation in image editing.

This energy can be combined with the patch NL-means energy by considering a linear combination of the corresponding image energy terms. The resulting scheme computes the weights based on the image together with its gradient, and updates the image by solving a linear combination of Eqs. (6) and (9).

4   Confidence Mask

For large inpainting domains, it is useful to introduce a mask $\kappa : \Omega \to (0,1]$ which assigns a confidence value to each pixel, depending on the certainty of its information (see also [14,24]). This will help in guiding the flow of information from the boundary towards the interior of the hole, eliminating some local minima and reducing the effect of the initial condition. The resulting image energy term takes the form

$$\widetilde{F}_w(u) = \int_{\widetilde{O}} \int_{\widetilde{O}^c} \kappa(x) w(x, x') \| p_u(x) - p_{\hat{u}}(x') \|_{a,\varphi} \mathrm{d}x' \mathrm{d}x,$$

where κ modulates the penalization of the incoherences between w and the φ-norm between patches.

The effect of κ on the image update step is easier to visualize on the evidence function ρ_z, Eq. (8). Recall that this function gathers all evidence supporting $u(z')$ as a value for $u(z)$, for each $z' \in O^c$. As in Eq. (8), now taking κ into account, we obtain $\rho_{\kappa,z}(z') = \int_{\Omega_p} \chi_{\widetilde{O}^c}(z' - y) g_a(y) \kappa(z - y) w(z - y, z' - y) \mathrm{d}y$. Thus, the contribution of the patch $p_u(z - y)$ to the evidence function is now weighted by its confidence. Patches with higher confidence will support stronger evidence. In this case the weights are given by $w(x, x') = \frac{1}{q(x)} \exp\left(-\frac{\kappa(x)}{h} \| p_u(x) - p_{\hat{u}}(x') \|_{a,\varphi}\right)$.

The inclusion of the confidence mask modifies the patch space scale parameter h. If the confidence is high, the effective scale $h/\kappa(x)$ will be lower, thus increasing the selectivity of the similarity measure. If the information at x is uncertain, more patches are considered similar. The same reasoning applies to the patch NL-Poisson energy, with similar modifications to Eq. (9).

5 Experimental Results

We tested the proposed methods with gray scale and color images. The energy for the latter can be obtained by considering a norm for color patches that adds the norms of the three scalar components: $\| p_{\vec{u}}(x) \|_{a,\varphi} = \sum_{i=1}^3 \| p_{u_i}(x) \|_{a,\varphi}$, where $\vec{u} : \Omega \to \mathbb{R}^3$ is the color image, and u_i, with $i = 1, 2, 3$, its components (analogously for $\| \cdot \|_{a,\nabla}$). Thus, the weights will take into account the three channels. Given the weights, each channel is updated using the corresponding scheme for scalar images. All channels are updated using the same weights. This scheme can be applied to any Euclidean color space. We show results with RGB and CIE La*b* color spaces.

In our implementation we use a square patch domain Ω_p of side $s \in \mathbb{N}$, with the Gaussian intra-patch weights g_a centered on it. For all experiments we set $s = 3a$ (s should be chosen such that most of the effective support of the Gaussian fits in the patch, we used a smaller s to lower the computational cost). This leaves only two independent parameters, namely, the intra-patch Gaussian width a, and the patch similarity scale h. The former determines the size of the patch, a parameter inherent to all patch-based techniques. It should be large enough so as to allow the identification of the image patterns.

In the limit when $h \to 0$, we compute the weights as $\lim_{h\to 0} w_h(x, x') = \frac{1}{\#n(x)} \delta(x' - n(x))$, where $n(x) \subseteq O^c$ is the set of nearest neighbors of x, i.e. $n(x) = \{x' \in O^c : \| p_u(x) - p_{\hat{u}}(x') \|_{a,\varphi} = \delta_n\}$, where δ_n represents the nearest neighbor distance. In practice, we assume that $\#n(x) = 1$, i.e. the nearest neighbor is unique. The choice of this parameter will be addressed later.

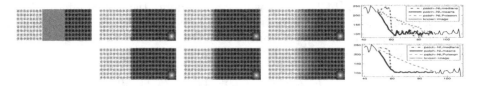

Fig. 2. Results with $s = 15$ and $a = 5$. The first four columns correspond to the initial condition, result of path NL-medians, -means and -Poisson. Top row, $h = 0$, bottom row $h = 0.01$, $h = 0.05$ and $h = 0.04$, respectively. The used intra-patch weight kernel g_a is shown in each figure on the bottom right. The fifth column shows the value of the images for a horizontal line going between the circles.

The confidence mask, when used, adds another parameter. We found good results using the following function:

$$\kappa(x) = \begin{cases} (1 - \kappa_0)e^{-\frac{d(x,\partial O)}{\tau_\kappa}} + \kappa_0 & \text{if } x \in O, \\ 1 & \text{if } x \in O^c, \end{cases}$$

which shows an exponential decay w.r.t. the distance to the boundary inside the hole $d(\cdot, \partial O)$, where $\tau_\kappa > 0$ is the decay time and $\kappa_0 > 0$ determines the asymptotic value reached far away from the boundary.

If a patch centered in the inpainting domain does not fit in the image, we mirror the image w.r.t. the boundary to complete the patch. Whenever the hole reaches $\partial\Omega$, the Poisson equation requires a different boundary condition. We have considered Neuman boundary conditions, $i.e. \nabla u \cdot \mathbf{n}(x) = 0$, for $x \in O \cap \partial\Omega$, where $\mathbf{n}(x)$ is the normal direction at the boundary. This amounts again to a reflection of u w.r.t. $\partial\Omega$.

The computational cost of each iteration is $\mathcal{O}(\mathcal{A}(O) \times \mathcal{A}(O^c) \times s^2)$. This is typical of non-local methods, and several strategies can be used for speed-up [8,27].

Figure 2 compares the results of the three methods on a texture with two different mean intensities, darker on the right half of the image. The inpainting domain hides all patches on the boundary between the dark and bright textures. With this we can test the ability of each method to *create* an interface between both regions. Situations like these are common in real inpainting problems due for instance to shadows. Moreover, when inpainting non-regular textures, a good completion may not be possible just by copying, and creating new patterns becomes necessary (see Figure 3).

We have also added Gaussian noise with standard deviation $\sigma = 10$ to show the influence of the patch space scale parameter h. Figure 2 shows two results for each scheme, one with $h = 0$, and the other with a higher h, chosen empirically for each method.

The rightmost column in Figure 2 plots the image values for a horizontal line between the circles. The interpolation done by the patch NL-Poisson method is linear, since this is a solution of the homogeneous Poisson equation. The profile shown by patch NL-means shows a smooth transition when both regions meet, whereas the use of the L_1 norm yields a sharp edge. The results using a higher h show some denoising, since for larger h, more patches are regarded as similar

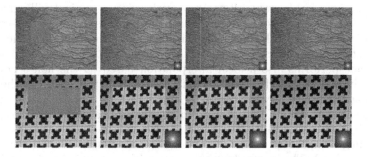

Fig. 3. From left to right: Initial condition, result of path NL-medians, -means and -Poisson. Top: Results with $s = 15$, $a = 5$ and $h = 0$, using the CIE La*b* color space. Bottom: Results with $s = 25$, $a = 8$ and $h = 0$. Gray scale image.

Fig. 4. From left to right, inpainting domain with confidence mask, result of path NL-medians, -means and -Poisson (the latter only for the first row). Top: *cylinders*- Results with $s = 27$, $a = 9$ and $h = 0$ for patch NL-medians and -means and $s = 33$, $a = 1$ and $h = 0$ patch NL-Poisson. Bottom: *elepahnt*- Results with $s = 19$, $a = 6$ and $h = 0$. The parameters of the confidence mask are $\tau_0 = 5$ and $\kappa_0 = 0.4$ in all cases except for the patch NL-medians with the bottom image, in which we set $\kappa_0 = 0.1$. Results using RGB. Please refer to [17,24] for other results on the same images.

to each patch in $\widetilde{O}$ and each pixel value is synthesized from more complement pixels. For inpainting of noiseless images, we use $h = 0$.

The top row of Figure 3 shows results with the three schemes for a non-regular texture. The result with patch NL-medians shows image regions copied without any modification. The boundaries between these regions are determined so that each patch on the boundary is close to some patch outside the hole. This does not always yields a smooth transition. Copied patterns can also be seen in the result of the patch NL-means, but the copies are less sharp, and the discontinuities less noticeable. The patch NL-Poisson shows a better continuation of the color at the boundary of the hole. However the inpainted structure looks too blurry (zoom on the pdf file for details).

The bottom row of Figure 3 depicts results on a regular texture. The regularity of the texture hides the blurring effects of the L_2 metrics (both on the image and gradients). At a stable state, all patches overlapping on a pixel will agree on its value. Notice that in this case, the patch NL-Poisson is able to reconstruct the illumination gradient of the image. This is imposed to the solution of the Poisson equation by the boundary conditions. In addition to the non-local inpainting, this scheme performs also a local interpolation based on the hole's boundary. Instead, the other methods copied the information from the bottom of the image, generating a discontinuity at the top.

The results shown in Figure 4, were computed using a confidence mask shown at the leftmost column. In both cases the patch NL-medians scheme yields the best results, comparable to state of the art (see results in [17,24] for results on the same images). The images look as a composition of copied regions (although some parts in the *elephant* image seem new). Again the patch NL-means shows blurred results, most noticeable for *elephant* due to the non-regularity of the textures. The patch NL-Poisson fails with this image. In this case the gradient is not a good feature for computing the weights. However, the results for Figure 4 are still reasonable, it did recover the structure of the image. Due to the averaging of gradients, when overlapping patches do not agree on the value of a pixel, lower gradients may appear. These generate phantom edges surrounding the cylinders. Presumably a more robust estimation of the gradient would not have this problem. We are currently developing a scheme using the L_1 norm between the gradients of patches.

The initial condition for *elephant* is the original image, whereas for *cylinders* the hole was filled with 128 as constant gray level. A confidence mask κ with low confidence inside the hole helps in diminishing the influence of the initialization. Further results are available at: http://gpi.upf.edu/static/vnli.

6   Conclusions and Future Work

In this work we present a variational framework for non-local image inpainting. The proposed energy lends itself to intuitive interpretations, and contrary to previous variational models, allows a straightforward minimization using a coordinate descent scheme. Beyond the specific application of inpainting, this framework provides also a sound variational modelling of non-local regularizers with *adaptive* weights, extending previous work in which the weights are considered known and fixed.

Starting from this model, we derived three different inpainting schemes, each one corresponding to a different norm measuring the distance between patches. We showed results on synthetic and natural images comparing their properties.

The derived *patch NL-means* provides a variational interpretation of the methods proposed by [32,38]. The *patch NL-medians* is the one that showed the best overall performance, comparable with the state of the art. The results obtained suggest a possible relation with the piece-wise traslation models of [1,16]. The *patch NL-Poisson* presents two interesting features. First, the similarity weights are computed based on the gradients, allowing the transference of information from areas with different intensity level. Second, the image completion is the result of a Poisson equation, thus incorporating some basic local regularization, meaning the completion must be differentiable and its gradient squared integrable. This traduces in a local interpolation based on the image values at the boundary of the inpainting domain. This method performs well for structured textures, but fails for non-regular textures.

We are currently exploring several additional aspects of this framework, including the use of robust norms in the general φ setting and the L_1 norm between patch gradients.

Acknoledgements. PA is supported by the FPI grant BES-2007-14451 from the Spanish MCI. VC acknowledges partial support by PNPGC project, reference MTM2006-14836 and by "ICREA Acadèmia" prize for excellence in research funded by the Generalitat de Catalunya. GS is partially supported by NSF, ONR, DARPA, NGA, and ARO.

References

1. Aujol, J.-F., Ladjal, S., Masnou, S.: Exemplar-based inpainting from a variational point of view (submitted) (2008)
2. Awate, S.P., Whitaker, R.T.: Unsupervised, information-theoretic, adaptive image filtering for image restoration. IEEE Trans. on PAMI 28(3), 364–376 (2006)
3. Ballester, C., Bertalmío, M., Caselles, V., Sapiro, G., Verdera, J.: Filling-in by joint interpolation of vector fields and gray levels. IEEE Trans. on IP 10(8), 1200–1211 (2001)
4. Bertalmío, M., Sapiro, G., Caselles, V., Ballester, C.: Image inpainting. In: Proc. of SIGGRAPH (2000)
5. Bertalmío, M., Vese, L., Sapiro, G., Osher, S.: Simultaneous structure and texture inpainting. IEEE Trans. on Image Processing 12(8), 882–889 (2003)
6. Bornard, R., Lecan, E., Laborelli, L., Chenot, J.-H.: Missing data correction in still images and image sequences. In: Proc. ACM Int. Conf. on Multimedia (2002)
7. Bornemann, F., März, T.: Fast image inpainting based on coherence transport. J. of Math. Imag. and Vis. 28(3), 259–278 (2007)
8. Brox, T., Kleinschmidt, O., Cremers, D.: Efficient nonlocal means for denoising of textural patterns. IEEE Trans. on IP 17(7), 1057–1092 (2008)
9. Buades, A., Coll, B., Morel, J.M.: A non local algorithm for image denoising. In: Proc. of the IEEE Conf. on CVPR, vol. 2, pp. 60–65 (2005)
10. Cao, F., Gousseau, Y., Masnou, S., Pérez, P.: Geometrically-guided exemplar-based inpainting (submitted) (2008)
11. Chan, T., Kang, S.H., Shen, J.H.: Euler's elastica and curvature based inpaintings. SIAM J. App. Math. 63(2), 564–592 (2002)
12. Chan, T., Shen, J.H.: Mathematical models for local nontexture inpaintings. SIAM J. App. Math. 62(3), 1019–1043 (2001)
13. Cheng, Y.: Mean shift, mode seeking and clustering. IEEE Trans. on PAMI 17(8), 790–799 (1995)
14. Criminisi, A., Pérez, P., Toyama, K.: Region filling and object removal by exemplar-based inpainting. IEEE Trans. on IP 13(9), 1200–1212 (2004)
15. Csiszár, I.: Axiomatic characterizations of information measures. Entropy 10(3), 261–273 (2008)
16. Demanet, L., Song, B., Chan, T.: Image inpainting by correspondence maps: a deterministic approach. Technical report, UCLA (2003)
17. Drori, I., Cohen-Or, D., Yeshurun, H.: Fragment-based image completion. ACM Trans. on Graphics. Special issue: Proc. of ACM SIGGRAPH 22(3), 303–312 (2003)
18. Efros, A.A., Leung, T.K.: Texture synthesis by non-parametric sampling. In: Proc. of the IEEE Conf. on CVPR, September 1999, pp. 1033–1038 (1999)

19. Esedoglu, S., Shen, J.H.: Digital image inpainting by the Mumford-Shah-Euler image model. European J. App. Math. 13, 353–370 (2002)
20. Gilboa, G., Osher, S.J.: Nonlocal linear image regularization and supervised segmentation. SIAM Mult. Mod. and Sim. 6(2), 595–630 (2007)
21. Igehy, H., Pereira, L.: Image replacement through texture synthesis. In: Proc. of the IEEE Conf. on CVPR (October 1997)
22. Jaynes, E.T.: Information theory and statistical mechanics. Physical Review 106(4), 620–630 (1957)
23. Jia, J., Tang, C.-K.: Inference of segmented color and texture description by tensor voting. IEEE Trans. on PAMI 26(6), 771–786 (2004)
24. Komodakis, N., Tziritas, G.: Image completion using efficient belief propagation via priority scheduling and dynamic pruning. IEEE Trans. on IP 16(11), 2649–2661 (2007)
25. Levin, A., Zomet, A., Weiss, Y.: Learning how to inpaint from global image statistics. In: Proc. of ICCV (2003)
26. Levina, E., Bickel, P.: Texture synthesis and non-parametric resampling of random fields. Annals of Statistics 34(4) (2006)
27. Mahmoudi, M., Sapiro, G.: Fast image and video denoising via nonlocal means of similar neighborhoods. Signal Processing Letters 12(12), 839–842 (2005)
28. Masnou, S.: Disocclusion: a variational approach using level lines. IEEE Trans. on IP 11(2), 68–76 (2002)
29. Masnou, S., Morel, J.-M.: Level lines based disocclusion. In: Proc. of IEEE ICIP (1998)
30. Pérez, P., Gangnet, M., Blake, A.: Poisson image editing. In: Proc. of SIGGRAPH (2003)
31. Pérez, P., Gangnet, M., Blake, A.: PatchWorks: Example-based region tiling for image editing. Technical report, Microsoft Research (2004)
32. Peyré, G.: Manifold models for signals and images. Comp. Vis. and Im. Unders. 113(2), 249–260 (2009)
33. Peyré, G., Bougleux, S., Cohen, L.: Non-local regularization of inverse problems. In: Forsyth, D., Torr, P., Zisserman, A. (eds.) ECCV 2008, Part III. LNCS, vol. 5304, pp. 57–68. Springer, Heidelberg (2008)
34. Protter, M., Elad, M., Takeda, H., Milanfar, P.: Generalizing the non-local-means to super-resolution reconstruction. IEEE Trans. on IP 18(1), 36–51 (2009)
35. Singer, A., Shkolnisky, Y., Nadler, B.: Diffusion interpretation of non-local neighborhood filters for signal denoising. SIAM J. on Imag. Sci. 2(1), 118–139 (2009)
36. Tschumperlé, D., Deriche, R.: Vector-valued image regularization with PDE's: a common framework for different applications. IEEE Trans. on PAMI 27(4) (2005)
37. Wei, L.-Y., Levoy, M.: Fast texture synthesis using tree-structured vector quantization. In: Proc. of the SIGGRAPH (2000)
38. Wexler, Y., Shechtman, E., Irani, M.: Space-time completion of video. IEEE Trans. on PAMI 29(3), 463–476 (2007)

Image Filtering Driven by Level Curves

Ajit Rajwade, Arunava Banerjee, and Anand Rangarajan

Department of CISE, University of Florida, Gainesville, USA
{avr,arunava,anand}@cise.ufl.edu

Abstract. This paper presents an approach to image filtering that is driven by the properties of the iso-valued level curves of the image and their relationship with one another. We explore the relationship of our algorithm to existing probabilistically driven filtering methods such as those based on kernel density estimation, local-mode finding and mean-shift. Extensive experimental results on filtering gray-scale images, color images, gray-scale video and chromaticity fields are presented. In contrast to existing probabilistic methods, in our approach, the selection of the parameter that prevents diffusion across the edge is robustly decoupled from the smoothing of the density itself. Furthermore, our method is observed to produce better filtering results for the same settings of parameters for the filter window size and the edge definition.

1   Introduction

Filtering of images has been one of the most fundamental problems studied in low-level vision and signal processing. Over the past decades, several techniques for data filtering have been proposed with impressive results on practical applications in image processing. As straightforward image smoothing is known to blur across significant image structures, several anisotropic approaches to image smoothing have been developed using partial differential equations (PDEs) with stopping terms to control image diffusion in different directions [1]. The PDE-based approaches have been extended to filtering of color images [2] and chromaticity vector fields [3]. Other popular approaches to image filtering include adaptive smoothing [4] and kernel density estimation based algorithms [5]. All these methods produce some sort of weighted average over an image neighborhood for the purpose of data smoothing, where the weights are obtained from the difference between the intensity values of the central pixel and the pixels in the neighborhood, or from the pixel gradient magnitudes. Beyond this, techniques such as bilateral filtering [6] produce a weighted combination that is also influenced by the relative location of the central pixel and the neighborhood pixels. The highly popular mean-shift procedure [7], [8] is grounded in similar ideas as bilateral filtering, with the addition that the neighborhood around a pixel is allowed to change dynamically until a convergence criterion is met. The authors prove that this convergence criterion is equivalent to finding the mode of a local density built jointly on the spatial parameters (image *domain*) and the intensity parameters (image *range*).

D. Cremers et al. (Eds.): EMMCVPR 2009, LNCS 5681, pp. 359–372, 2009.
© Springer-Verlag Berlin Heidelberg 2009

In this paper, we present a new approach to data filtering that is rooted in simple yet elegant geometric intuitions. At the core of our theory is the representation of an image as a function that is at least C_0 continuous everywhere. A key property of the image level sets is used to drive the diffusion process, which we then incorporate in a framework of dynamic neighborhoods *à la* mean-shift. We demonstrate the relationship of our method to many of the existing filtering techniques such as those driven by kernel density estimation. The efficacy of our approach is supported with extensive experimental results. To the best of our knowledge, *ours is the first attempt to explicitly utilize image geometry (in terms of its level curves)* for this particular application.

This paper is organized as follows. Section 2 presents the key theoretical framework. Section 3 presents extensions to our theory. In section 4, we present the relationship between our method and mean-shift. Extensive experimental results are presented in section 5, and we present further discussions and conclusions in section 6.

2   Theory

Consider an image over a discrete domain $\Omega = \{1, ..., H\} \times \{1, ..., W\}$ where the intensity of each discrete location (x, y) is given by $I(x, y)$. Moreover consider a neighborhood $\mathcal{N}(x_i, y_i)$ around the pixel (x_i, y_i). It is well-known that a simple averaging of all intensity values in $\mathcal{N}(x_i, y_i)$ will blur edges, so a weighted combination is calculated, where the weight of the jth pixel is given by $w^{(1)}(x_j, y_j) = g(|I(x_i, y_i) - I(x_j, y_j)|)$ for a non-increasing function $g(.)$ to facilitate anisotropic diffusion, with common examples being $g(z) = e^{-\frac{z^2}{\sigma^2}}$ or $g(z) = \frac{\sigma^2}{\sigma^2 + z^2}$, or their truncated versions. This approach is akin to the kernel density estimation (KDE) approach proposed in [5], where the filtered value of the central pixel is calculated as:

$$\hat{I}(x_i, y_i) = \frac{\displaystyle\sum_{(x_j, y_j) \in \mathcal{N}(x_i, y_i)} I(x_j, y_j) K(I(x_j, y_j) - I(x_i, y_i); W_r)}{\displaystyle\sum_{(x_j, y_j) \in \mathcal{N}(x_i, y_i)} K(I(x_j, y_j) - I(x_i, y_i); W_r)}. \tag{1}$$

Here the kernel K centered at $I(x_i, y_i)$ (and parameterized by W_r) is related to the function g and determines the weights. The major limitations of the kernel based approach to anisotropic diffusion are that the entire procedure is sensitive to the parameter W_r and the size of the neighborhood, and might suffer from a small-sample size problem. Furthermore, in a discrete implementation, for any neighborhood size larger than 3×3, the procedure depends only on the actual pixel values and does not account for any gradient information, whereas in a filtering application, it is desirable to place greater importance on those regions of the neighborhood where the gradient values are lower.

Now consider that the image is treated as a continuous function $I(x, y)$ of the spatial variables, by interpolating in between the pixel values. The earlier discrete average is replaced by the following continuous average to update the value at (x_i, y_i):

$$\hat{I}(x_i, y_i) = \frac{\displaystyle\int\int_{\mathcal{N}(x_i, y_i)} I(x, y)g(|I(x, y) - I(x_i, y_i)|)dxdy}{\displaystyle\int\int_{\mathcal{N}(x_i, y_i)} g(|I(x, y) - I(x_i, y_i)|)dxdy}. \tag{2}$$

The above formula is usually not available in closed form. We now show a principled approximation to this formula, by resorting to geometric intuition. Imagine a contour map of this image, with multiple iso-intensity level curves $C_m = \{(x, y)|I(x, y) = \alpha_m\}$ (referred to henceforth as 'level curves') separated by an intensity spacing of Δ. Consider a portion of this contour map in a small neighborhood centered around the point (x_i, y_i) (see Figure 1(a)). Those regions where the level curves (separated by a fixed intensity spacing) are closely packed together correspond to the higher-gradient regions of the neighborhood, whereas in lower-gradient regions of the image, the level curves lie far away from one another. Now as seen in Figure 1(a), this contour map induces a tessellation of the neighborhood into some K facets, where each facet corresponds to a region in between two level curves of intensity α_m and $\alpha_m + \Delta$, bounded by the rim of the neighborhood. Let the area a_k of the kth facet of this tessellation be denoted as a_k. Now, if we make Δ sufficiently small, we can regard even the facets from high-gradient regions as having constant intensity value $I_k = \alpha_m$. This now leads to the following weighted average in which the weighting function has a very clean geometric interpretation, unlike the arbitrary choice for $w^{(1)}$ in the previous technique:

$$\hat{I}(x_i, y_i) = \frac{\displaystyle\sum_{k=1}^{K} a_k I_k g(|I_k - I(x_i, y_i)|)}{\displaystyle\sum_{k=1}^{K} a_k g(|I_k - I(x_i, y_i)|)}. \tag{3}$$

As the number of facets is typically much larger than the number of pixels, and given the fact that the facets have arisen from a locally smooth interpolation method to obtain a continuous function from the original digital pixel values, we now have a more robust average than that provided by Equation 1. To introduce anisotropy, we still require the stopping term $g(|I_k - I(x_i, y_i)|)$ to prevent smearing across the edge, just as in Equation 1.

Equation 2 essentially performs an integration of the intensity function over the domain $\mathcal{N}(x_i, y_i)$. If we now perform a change of variables transforming the integral on (x, y) to an integral over the range of the image, we obtain the expression

$$
\hat{I}(x_i, y_i) = \frac{\int\int_{\mathcal{N}(x_i,y_i)} I(x,y)w^{(1)}(x,y)dxdy}{\int\int_{\mathcal{N}(x_i,y_i)} w^{(1)}(x,y)dxdy} = \frac{\int_{q=q_1}^{q=q_2} \int_{\mathcal{C}(q)} \frac{qg(|q-I(x_i,y_i)|)}{|\nabla I|}dldq}{\int_{q=q_1}^{q=q_2} \int_{\mathcal{C}(q)} \frac{g(|q-I(x_i,y_i)|)}{|\nabla I|}dldq}
$$

$$
= \frac{\lim\limits_{\Delta\to 0} \sum\limits_{\alpha=q_1}^{q_2} \int_{q=\alpha}^{\alpha+\Delta} \int_{\mathcal{C}(q)} \frac{qg(|q-I(x_i,y_i)|)}{|\nabla I|}dldq}{\lim\limits_{\Delta\to 0} \sum\limits_{\alpha=q_1}^{q_2} \int_{q=\alpha}^{q=\alpha+\Delta} \int_{\mathcal{C}(q)} \frac{g(|q-I(x_i,y_i)|)}{|\nabla I|}dldq} \tag{4}
$$

.

where $\mathcal{C}(q) = \mathcal{N}(x_i, y_i) \cap f^{-1}(q)$, $q_1 = \inf\{I(x,y)|(x,y) \in \mathcal{N}(x_i,y_i)\}$, $q_2 = \sup\{I(x,y)|(x,y) \in \mathcal{N}(x_i,y_i)\}$ and l stands for a tangent along the curve $f^{-1}(q)$. This approach is inspired by the smooth co-area formula for regular functions [9] which is given as

$$
\int_\Omega \phi(u)|\nabla u|dxdy = \int_{-\infty}^{+\infty} \text{Length}(\gamma_q)\phi(q)dq \tag{5}
$$

where γ_q is the level set of u at the intensity q and $\phi(u)$ represents a function of u. Note that the term $\int_{q=\alpha}^{q=\alpha+\Delta} \int_{\mathcal{C}(q)} \frac{dldq}{|\nabla I|}$ in Equation 4 actually represents the area in $\mathcal{N}(x_i, y_i)$ that is trapped between two contours whose intensity value differs by Δ. Previous work from [10] and [11] considers this quantity when normalized by $|\Omega|$ to be actually equal to the probability that the intensity value lies in the range $[\alpha, \alpha + \Delta]$. Bearing this in mind, Equation 3 now acquires the following *probabilistic* interpretation:

$$
\hat{I}(x_i, y_i) = \frac{\sum\limits_{\alpha=q_1}^{q_2} \Pr(\alpha < I < \alpha + \Delta|\mathcal{N})\alpha g(|\alpha - I(x_i,y_i)|)}{\sum\limits_{\alpha=q_1}^{q_2} \Pr(\alpha < I < \alpha + \Delta|\mathcal{N})g(|\alpha - I(x_i,y_i)|)}. \tag{6}
$$

As $\Delta \to 0$, this produces an increasingly better approximation to Equation 2.

It should be pointed out that there exist methods such as adaptive filtering [4], [12] in which the weights in Equation 1 are obtained as $w^{(2)}(x_j, y_j) = g(|\nabla I(x_j, y_j)|)$. These methods place more importance on the lower-gradient *pixels* of the neighborhood, but do not exploit level curve relationships in the way we do, and the choice of the weighting function does not have the geometric interpretation that exists in our technique. There also exists an extension to the standard neighborhood filter in Equation 1 reported in [13], which performs a weighted least squares polynomial fit to the intensity values (of the pixels) in the neighborhood of a location (x, y). The value of this polynomial at (x, y) is

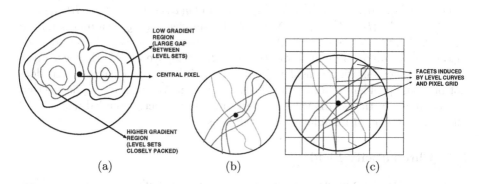

Fig. 1. (a) An image contour map with high and low gradient regions in a neighborhood around a pixel (dark dot). (b) A contour map of an RGB image in a neighborhood. The red, green and blue contours correspond to contours of the R,G,B channels respectively. The tessellation induced by the above level-curve pairs contains 19 facets. (c) A tessellation induced by RGB level curve pairs and the square pixel grid.

then considered to be the smoothed intensity value. This technique differs from the one we present here in two fundamental ways. Unlike our method, it does not use areas between level sets as weights to explicitly perform a weighted averaging. Secondly as proved in [13], its limiting behavior when $W_r \to 0$ and $|\mathcal{N}(x,y)| \to 0$ resembles the geometric heat equation with a linear polynomial, and resembles higher order PDEs when the degree of the polynomial is increased. Our method is the true continuous form of the KDE-based filter from Equation 1. This KDE-based filter limits to the Perona-Malik equation, as proved in [13].

3   Extensions of Our Theory

3.1   Color Images

We now extend our technique to color (RGB) images. Consider a color image defined as $I(x,y) = (R(x,y), G(x,y), B(x,y)) : \Omega \to \mathcal{R}^3$ where $\Omega \subset \mathcal{R}^2$. In color images, there is no concept of a single iso-contour with constant values of all three channels. Hence it is more sensible to consider an overlay of the individual iso-contours of the R, G and B channels. The facets are now induced by a tessellation involving the intersection of three iso-contour sets within a neighborhood, as shown in Figure 1(b). Each facet represents those portions of the neighborhood for which $\alpha_R < R(x,y) < \alpha_R + \Delta_R, \alpha_G < G(x,y) < \alpha_G + \Delta_G, \alpha_B < B(x,y) < \alpha_B + \Delta_B$. The probabilistic interpretation for the update on the R,G,B values is as follows

$$\hat{R}(x_i, y_i), \hat{G}(x_i, y_i), \hat{B}(x_i, y_i) = \frac{\displaystyle\sum_{\beta} \Pr[\beta < (R,G,B) < \beta + \Delta | \mathcal{N}] \beta g(R,G,B)}{\displaystyle\sum_{\beta} \Pr[\beta < (R,G,B) < \beta + \Delta | \mathcal{N}] g(R,G,B)}$$

where $\beta = (\alpha_R, \alpha_G, \alpha_B)$, $\boldsymbol{\Delta} = (\Delta_R, \Delta_G, \Delta_B)$ and $g(R, G, B) = g(|R - R(x_i, y_i)| + |G - G(x_i, y_i)| + |B - B(x_i, y_i)|)$. Note that in this case, $I(x, y)$ is a function from a subset of $\mathcal{R}^2$ to $\mathcal{R}^3$, and hence the three-dimensional joint *density* is ill-defined in the sense that it is defined strictly on a 2D subspace of $\mathcal{R}^3$. However given that the implementation considers joint cumulative interval measures, this does not pose any problem in a practical implementation. We wish to emphasize that the averaging of the R,G,B values is performed in a strictly coupled manner, all affected by the *joint* cumulative interval measure.

3.2   Chromaticity Fields

Previous research on filtering chromaticity noise (which affects only the direction and not the magnitude of the RGB values at image pixels) includes the work in [3] using PDEs specially tuned for unit-vector data, and the work in [5] (page 142) using kernel density estimation for directional data. The more recent work on chromaticity filtering in [14] actually treats chromaticity vectors as points on a Grassmann manifold $\mathcal{G}_{1,3}$ as opposed to treating them as points on $\mathcal{S}^2$, which is the approach presented here and in [5] and [3].

   We extend our theory from the previous section to unit vector data and incorporate it in a mean-shift framework for smoothing. Let $I(x, y) : \Omega \rightarrow \mathcal{R}^3$ be the original RGB image, and let $J(x, y) : \Omega \rightarrow \mathcal{S}^2$ be the corresponding field of chromaticity vectors. A possible approach would involve interpolating the chromaticity vectors by means of commonly used spherical interpolants to create a continuous function, followed by tracing the level curves of the individual unit-vector components $v(x, y) = (v_1(x, y), v_2(x, y), v_3(x, y))$ and computing their intersection. However for ease of implementation for this particular application, we resorted to a different strategy. If the intensity intervals $\boldsymbol{\Delta} = (\Delta_R, \Delta_G, \Delta_B)$ are chosen to be fine enough, then each facet induced by a tessellation that uses the level curves of the R, G and B channel values, can be regarded as having a constant color value, and hence the chromaticity vector values within that facet can be regarded as (almost) constant. Therefore it is possible to use just the R,G,B level curves for the task of chromaticity smoothing as well. The update equation is very similar to Equation 7 with the R,G,B vectors replaced by their unit normalized versions. However as the averaging process does not preserve the unit norm, the averaged vector needs to be renormalized to produce the spherical weighted mean.

3.3   Gray-Scale Video

For the purpose of this application, the video is treated as a single 3D signal (volume). The extension in this case is quite straightforward, with the areas between level curves being replaced by volumes between the level surfaces at nearby intensities. However we take into account the causality factor in defining the temporal component of the neighborhood around a pixel, by performing the averaging at each pixel over frames only from the past.

4   Level Curve Based Filtering in a Mean-shift Framework

All the above techniques are based on an averaging operation over only the image intensities (i.e. in the range domain). On the other hand, techniques such as bilateral filtering [6] or local mode-finding [15] combine both range and spatial domain, thus using weights of the form $w_j = g^{(s)}((x_i - x_j)^2 + (y_i - y_j)^2)g^{(r)}(||I(x_i, y_i) - I(x_j, y_j)|)$ in Equation 1, where $g^{(s)}$ and $g^{(r)}$ affect the spatial and range kernels respectively. The mean-shift framework [8] is based on similar principles, but changes the filter window dynamically for several iterations until it finds a local mode of the joint density of the spatial and range parameters, estimated using kernels based on the functions $g^{(r)}$ and $g^{(s)}$. Our level curve based approach fits easily into this framework with the addition of a spatial kernel. One way to do this would be to consider the image as a surface embedded in 3D (a Monge patch), as done in [16], and compute areas of patches in 3D for the probability values. However such an approach may not necessarily favor the lower gradient areas of the image. Instead we adopt another method wherein we assume two additional functions of x and y, namely $X(x, y) = x$ and $Y(x, y) = y$. We compute the joint probabilities for a range of values of the joint variable (X, Y, I) by drawing local level sets and computing areas in 2D. Assuming a uniform spatial kernel for $g^{(s)}$ within a radius W_s and a rectangular kernel on the intensity for $g^{(r)}$ with threshold value W_r (though our core theory is unaffected by other choices), we now perform the averaging update on the vector $(X(x, y), Y(x, y), I(x, y))$, as opposed to merely on $I(x, y)$ as was done in Equation 6. This is given as:

$$(X(x_i, y_i), Y(x_i, y_i), \hat{I}(x_i, y_i)) = \frac{\sum_{k=1}^{K}(x_k, y_k, I_k)a_k g^{(r)}(|I_k - I(x_i, y_i)|)}{\sum_{k=1}^{K} a_k g^{(r)}(|I_k - I(x_i, y_i)|)}. \tag{7}$$

In the above equation (x_k, y_k) stands for a representative point (say, the centroid) of the kth facet of the induced tessellation[1], and K is the total number of facets within the specified spatial radius. Note that the area of the kth facet, i.e. a_k, can also be interpreted as the joint probability for the event $\tilde{x} < X(x, y) < \tilde{x} + \Delta_x, \tilde{y} < Y(x, y) < \tilde{y} + \Delta_y, \alpha < I(x, y) < \alpha + \Delta$, if we assume a uniform distribution over the spatial variables x and y. Here Δ is the usual intensity binwidth, (Δ_x, Δ_y) are the pixel dimensions, and $(\tilde{x}, \tilde{y})$ is a pixel grid-point. The main difference between our approach and all the aforementioned range-spatial domain approaches is the fact that we naturally incorporate a weight in favor of the lower-gradient areas of the filter neighborhood. Hence the mean-shift vector in our case will have a stronger tendency to move towards the region of the neighborhood where the local intensity change is as low as possible (even if a uniform spatial kernel is used). Moreover just like conventional mean shift,

---

[1] The notion of the centroid will become clearer in Section 5.

our iterative procedure is guaranteed to converge to a mode of the local density in a finite number of steps, by exploiting the fact that the weights at each point (i.e. the areas of the facets) are positive. Hence Theorem 5 of [7] can be readily invoked. This is because in Equation 7, the threshold function $g^{(r)}$ for the intensity is the rectangular kernel, and hence the corresponding update formula is equivalent to one with a weighted rectangular kernel, with the weights being determined by the areas of the facets.

A major advantage of our technique is that the parameter Δ can be set to as small a value as desired (as it just means that more and more level curves are being used), and the interpolation gives rise to a robust average. This is especially useful in the case of small neighborhood sizes, as the intensity quantization is now no more limited by the number of available pixels. In conventional mean-shift, the proper choice of bandwidth is a highly critical issue, as very few samples are available for the local density estimate. Though variable bandwidth procedures for mean-shift algorithms have been developed extensively, they themselves require either the tuning of other parameters using rules of thumb, or else some expensive exhaustive searches for the automatic determination of the bandwidth [17], [18]. Although our method does require the selection of W_s and W_r, the filtering results are less sensitive to the choice of these parameters in our method than in standard mean shift.

5   Experimental Results

In this section we present experimental results to compare the performance of our algorithm in a mean shift framework w.r.t. conventional kernel-based mean shift, as well as to two recent algorithms that are closely related to mean-shift: UINTA [19] and NL-Means [20]. For our algorithm, we obtain a continuous function approximation to the digital image, by means of piecewise linear interpolants fit to a triple of intensity values in half-pixels of the image (in principle, we could have used any other smooth interpolant). The corresponding level sets for such a function are also very easy to trace, as they are just segments within each half-pixel. The level sets induce a polygonal tessellation. We choose to split the polygons by the square pixel boundaries as well as the pixel diagonals that delineate the half-pixel boundaries, thereby convexifying all the polygons that were initially non-convex (see Figure 1(c)). Each polygon in the tessellation can now be characterized by the x, y coordinates of its centroid, the intensity value of the image at the centroid, and the area of the polygon. Thus, if the intensity value at grid location x_i, y_i is to be smoothed, we choose a window of spatial radius W_s and intensity radius W_r around $(x_i, y_i, I(x_i, y_i))$, over which the averaging is performed. In other words, the averaging is performed only over those locations x, y for which $(x - x_i)^2 + (y - y_i)^2 < W_s^2$ and $|I(x, y) - I(x_i, y_i)| < W_r$. We would like to point out that *though the interpolant used for creating the continuous image representation is indeed isotropic in nature, this still does not make our filtering algorithm isotropic*. This is because polygonal regions, whose intensity value does not satisfy the constraint $|I(x, y) - I(x_i, y_i)| < W_r$, do not contribute

Fig. 2. Leftmost column: original images, Second from left: degraded images with zero mean Gaussian noise of std. dev. 0.003, Second from right: results obtained by our algorithm, and rightmost column: mean shift with Gaussian kernel (right column). Both both methods, $W_s = W_r = 3$. *VIEWED BEST when ZOOMED in the pdf file.*

to the averaging process (see the stopping term in Equation 3), and hence the contribution from pixels with very different intensity values will be nullified.

5.1   Gray-Scale Images

We ran our filtering algorithm over four arbitrarily chosen images from the popular Berkeley image dataset[2], and the Lena image. To all these images, zero mean Gaussian noise of variance 0.003 (per unit gray-scale range) was added. The filtering was performed using $W_s = W_r = 3$ for our algorithm and compared to mean-shift using Gaussian and Epanechnikov kernels with the same parameter. Our method produced superior filtering results to conventional mean shift with both Gaussian and Epanechnikov kernels. The results for our method, for Gaussian kernel mean shift and for UINTA are displayed in Figure 2. The visually superior appearance was confirmed objectively with mean squared error (MSE) values in Table 1. *It should be noted that the aim was to compare our method to standard mean shift for the exact same setting of the parameters W_r and W_s, as*

---

[2] http://www.eecs.berkeley.edu/Research/Projects/CS/vision/grouping/segbench/

Table 1. MSE for filtered images using (M1) = Our method with $W_s = W_r = 3$, using (M2) = Mean shift with Gaussian kernels with $W_s = W_r = 3$ and (M3) = Mean shift with Gaussian kernels with $W_s = W_r = 5$, M_{UINTA} = MSE with UINTA method with neighborhood radius 9, smoothing parameter $h = 10$ (similar to W_r), 1000 samples for density estimate and 30 iterations per pixel, and M_{NL} = MSE with NL-means with search window size 18×18, neighborhood size 5×5, and smoothing parameter $h = 5$ (similar to W_r). MSE = mean-squared error in the corrupted image. Intensity scale is from 0 to 255.

Image	M1	M2	M3	M_{UINTA}	M_{NL}	MSE
1	110.95	176.57	151.27	280.7	130.7	181.27
2	53.85	170.18	106.32	95.43	127.48	193.5
3	106.64	185.15	148.379	121.3	147.41	191.76
4	113.8	184.77	153.577	127.4	147.98	190
Lena	78.42	184.16	128.04	101.5	125.38	194.82

they have the same meaning in all these algorithms. Although increasing the value of W_r will provide more samples for averaging, this will allow more and more intensity values to leak across edges. Moreover, in Table 1, we also compare our method to NL-means [20] and UINTA [19], again for similar parameter settings. Further empirical results with our algorithm (using $W_S = W_r = 5$) were obtained on Lansel's benchmark dataset [21]. The dataset contains noisy versions of 13 different images. Each noisy image is obtained from one of three noise models: additive Gaussian, Poisson, and multiplicative noise model, for one of five different values of the noise standard deviation $\sigma \in \{\frac{5}{255}, \frac{10}{255}, \frac{15}{255}, \frac{20}{255}, \frac{25}{255}\}$, leading to a total of 195 images. Despite the fact that we did not tweak any parameters depending on the noise model (we chose $W_r = W_s = 5$), we produced excellent denoising results. The average MSE and MSSIM (an image quality metric defined in [21]) are shown in the plots in Figure 3. We have also displayed the denoised versions of a fingerprint image from this dataset under three different values of σ for additive noise in Figure 3.

5.2 Color Images

Similar experiments were run on colored versions of the same four images from the Berkeley dataset. The original images were degraded by zero mean Gaussian noise of variance 0.003 (per unit intensity range), added independently to the R,G,B channels. For our method, independent interpolation was performed on each channel and the joint densities were computed as described in the previous sections. Level sets at intensity gaps of $\Delta_R = \Delta_G = \Delta_B = 1$ were traced in every half pixel. Experimental results were compared with conventional mean shift using a Gaussian kernel. The parameters chosen for both algorithms were $W_s = W_r = 6$. Despite the documented advantages of color spaces such as Lab [5], all experiments were performed in the R,G,B space for the sake of simplicity, and also because many well-known color de-noising techniques operate in this

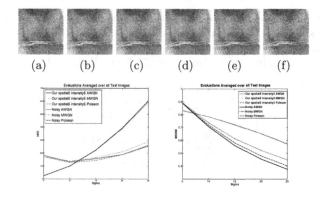

Fig. 3. First row: (a), (c) and (e): Fingerprint image subjected to additive Gaussian noise of std. dev. $\sigma = \frac{5}{255}$, $\frac{10}{255}$ and $\frac{15}{255}$ respectively. (b), (d) and (f): Denoised versions of (a), (c) and (e) respectively. Second row: A plot of the performance of our algorithm on the Lansel dataset, averaged over all images from each noise model (Additive Gaussian (AWGN), multiplicative Gaussian (MWGN) and Poisson) and over all five σ values, using MSE (left) and MSSIM (right) as the metric. *VIEWED BEST when ZOOMED in the pdf file (in color).*

Fig. 4. Left column: original images, second from left: Degraded images with zero mean Gaussian noise of std. dev. 0.003, second from right: results by our algorithm (left column) and Rightmost: mean shift with Gaussian kernel (right column). Both both methods, $W_s = W_r = 6$. *Viewed best when ZOOMED in the pdf file (in color).*

space [2]. As seen in Figure 4 and Table 2, our method produced better results than Gaussian kernel mean shift for the chosen parameter values.

Table 2. MSE for filtered images using (M1) = Our method with $W_s = W_r = 6$, using (M2) = Mean shift with Gaussian kernels with $W_s = W_r = 6$ and (M3) = Mean shift with Epanechnikov kernels with $W_s = W_r = 6$. MSE = mean-squared error in the corrupted image. Intensity scale is from 0 to 255 for each channel.

Image	M1	M2	M3	MSE
1	319.88	496.7	547.9	572.54
2	354.76	488.7	543.4	568.69
3	129.12	422.79	525.48	584.24
4	306.14	477.25	526.8	547.9

Fig. 5. Two images and their corrupted versions obtained by adding chromaticity noise (first and second columns respectively). Results obtained by filtering with our method (third column), and with Gaussian mean shift (fourth column). *Viewed best when ZOOMED in the pdf file (in color).*

Fig. 6. First two images: frames from the corrupted sequence. Third and fourth: images filtered by our algorithm. Fifth and sixth images: a slice through the tenth row of the corrupted and filtered video sequences. The images are numbered left to right, top to bottom.

5.3   Experiments with Chromaticity Vectors and Video

Two color images were synthetically corrupted with chromaticity noise altering just the direction of the color-triple vector. These images are shown in Figure 5. These images were filtered using our method and Gaussian kernel mean shift with a spatial window of size $W_s = 4$ and a chromaticity threshold of $W_r = 0.1$ radians. Note that in this case, the distance between two chromaticity vectors v_1 and v_2 is defined to be the length of the arc between the two vectors along the great circle joining them, which turns out to be $\theta = \cos^{-1} v_1{}^T v_2$. The specific expression for the joint spatial-chromaticity density using the Gaussian kernel was $e^{-\frac{(x-x_i)^2+(y-y_i)^2}{2W_s^2}} e^{-\frac{\theta^2}{2W_i^2}}$. The filtered images using both methods are shown in Figure 5. Despite the visual similarity of the output, our method produced a mean-squared error of 378 and 980.8, as opposed to 534.9 and 1030.7 for Gaussian kernel mean shift.

We also performed an experiment on video de-noising using the David sequence[3]. The first 100 frames from the sequence were extracted and artificially degraded with zero mean Gaussian noise of variance 0.006. Two frames of the corrupted and de-noised (using our method) sequence are shown in Figure 6, as also a temporal slice through the entire video sequence (for the tenth row of each frame). For this experiment, the value of Δ was set to 8 in our method.

6   Discussion

We have presented a new method for image denoising, whose principle is rooted in the notion that the lower-gradient portions of an image inside a neighborhood around a pixel should contribute more to the smoothing process. The geometry of the image level sets (and the fact that the spatial distance between level sets is inversely proportional to the gradient magnitudes) is the driving force behind our algorithm. We have linked our approach to existing probability-density based approaches, and our method has the advantage of robust decoupling of the edge definition parameter from the density estimate. In some sense, our method can be viewed as a continuous version of mean-shift. It should be noted that a modification to standard mean-shift based on simple image up-sampling using interpolation will be an approximation to our area-based method (given the same interpolant). We have performed extensive experiments on gray-scale and color images, chromaticity fields and video sequences. To the best of our knowledge, ours is the *first piece of work on denoising which explicitly incorporates the relationship between image level curves and uses local interpolation between pixel values* in order to perform filtering. Future work will involve a more detailed investigation into the relationship between our work and that in [16], by computing the areas of the contributing regions with explicit treatment of the image $I(x, y)$ as a surface embedded in 3D. Secondly, we also plan to develop topologically inspired criteria to automate the choice of the spatial neighborhood and the parameter W_r for controlling the anisotropic smoothing.

---

[3] Obtained from http://www.cs.utoronto.ca/~dross/ivt/

References

1. Perona, P., Malik, J.: Scale-space and edge detection using anisotropic diffusion. IEEE Trans. on PAMI 12(7), 629–639 (1990)
2. Tschumperle, D., Deriche, R.: Vector-valued image regularization with pdes: A common framework for different applications. IEEE Trans. on PAMI 27(4), 506–517 (2005)
3. Tang, B., Sapiro, G.: Color image enhancement via chromaticity diffusion. IEEE Trans. on Image Processing 10, 701–707 (1999)
4. Saint-Marc, P., Chen, J., Medioni, G.: Adaptive smoothing: a general tool for early vision. IEEE Trans. on PAMI 13(6), 514–520 (1991)
5. Plataniotis, K., Venetsanopoulos, A.: Color image processing and applications. Springer, New York (2000)
6. Tomasi, C., Manduchi, R.: Bilateral filtering for gray and color images. In: ICCV, pp. 839–846 (1998)
7. Cheng, Y.: Mean shift, mode seeking and clustering. IEEE Trans. on PAMI 17(8), 790–799 (1995)
8. Comaniciu, D., Meer, P.: Mean shift: a robust approach toward feature space analysis. IEEE Trans. on PAMI 24(5), 603–619 (2002)
9. Chan, T., Shen, J.: Image Processing and Analysis: Variational, PDE, wavelets, and stochastic methods. SIAM, Philadelphia (2005)
10. Hadjidemetriou, E., Grossberg, M., Nayar, S.: Histogram preserving image transformations. IJCV 45(1), 5–23 (2001)
11. Rajwade, A., Banerjee, A., Rangarajan, A.: Probability density estimation using isocontours and isosurfaces: applications to information-theoretic image registration. IEEE Trans. on PAMI 31(3), 475–491 (2009)
12. Barash, D., Comaniciu, D.: A common framework for nonlinear diffusion, adaptive smoothing, bilateral filtering and mean shift. IVC 22, 73–81 (2004)
13. Buades, A., Coll, B., Morel, J.M.: Neighborhood filters and PDEs. Numerische Mathematik 105(1), 1–34 (2006)
14. Subbarao, R., Meer, P.: Discontinuity preserving filtering over analytic manifolds. In: CVPR (2007)
15. van de Weijer, J., van den Bloomgard, R.: Local mode filtering. In: CVPR, vol. 2, pp. 428–436 (2001)
16. Sochen, N., Kimmel, R., Malladi, R.: A general framework for low level vision. IEEE Trans. on Image Processing 7, 310–318 (1998)
17. Comaniciu, D.: An algorithm for data-driven bandwidth selection. IEEE Trans. on PAMI 25, 281–288 (2003)
18. Comaniciu, D., Ramesh, V., Meer, P.: The variable bandwidth mean shift and data-driven scale selection. In: ICCV, pp. 438–445 (2001)
19. Awate, S., Whitaker, R.: Unsupervised, information-theoretic, adaptive image filtering for image restoration. IEEE Trans. on PAMI 28(3), 364–376 (2006)
20. Buades, A., Coll, B., Morel, J.M.: Nonlocal image and movie denoising. IJCV 76(2), 123–139 (2008)
21. Lansel, S.: About DenoiseLab,
http://www.stanford.edu/~slansel/DenoiseLab/documentation.htm

Color Image Restoration Using Nonlocal Mumford-Shah Regularizers

Miyoun Jung, Xavier Bresson, Tony F. Chan, and Luminita A. Vese

Department of Mathematics, University of California, Los Angeles, CA 90095, U.S.A.
{gomtaeng,xbresson,chan,lvese}@math.ucla.edu

Abstract. We introduce several color image restoration algorithms based on the Mumford-Shah model and nonlocal image information. The standard Ambrosio-Tortorelli and Shah models are defined to work in a small local neighborhood, which are sufficient to denoise smooth regions with sharp boundaries. However, textures are not local in nature and require semi-local/non-local information to be denoised efficiently. Inspired from recent work (NL-means of Buades, Coll, Morel and NL-TV of Gilboa, Osher), we extend the standard models of Ambrosio-Tortorelli and Shah approximations to Mumford-Shah functionals to work with nonlocal information, for better restoration of fine structures and textures. We present several applications of the proposed nonlocal MS regularizers in image processing such as color image denoising, color image deblurring in the presence of Gaussian or impulse noise, color image inpainting, and color image super-resolution. In the formulation of nonlocal variational models for the image deblurring with impulse noise, we propose an efficient preprocessing step for the computation of the weight function w. In all the applications, the proposed nonlocal regularizers produce superior results over the local ones, especially in image inpainting with large missing regions. Experimental results and comparisons between the proposed nonlocal methods and the local ones are shown.

1 Introduction

We consider the restoration problem of a color image formalized as

$$f = Hu + n, \quad (f^i = H^i u^i + n^i, \, i = r, g, b) \tag{1}$$

where H is a linear operator accounting for some blurring, sub-sampling or missing pixels so that the observed data $f : \Omega \to \mathbb{R}$ loses some portion of the original image u we wish to recover, and n is an additive noise. We approach the restoration problem within the variational framework: $\inf_u \{ \Phi(f - Hu) + \Psi(|\nabla u|) \}$, where Φ defines a data-fidelity term, and Ψ defines the regularization that enforces a smoothness constraint on u, depending on its gradient ∇u.

Problem (1) is ill-posed, but the regularization term Ψ alleviates this difficulty by reflecting some a priori properties. Several edge-preserving regularization terms were suggested in the literature, including [9], [20], [2], [21], [5], [4]. These traditional regularization terms are based on local image operators, which

D. Cremers et al. (Eds.): EMMCVPR 2009, LNCS 5681, pp. 373–387, 2009.
© Springer-Verlag Berlin Heidelberg 2009

denoise and preserve edges and smooth regions very well, but may not deal well with fine structures like texture during the restoration process because textures are not local in nature.

Recently, new image denoising models have been developed, based on non-local image operators, to better deal with textures. Buades et al [8] introduced the nonlocal means filter, which produces excellent denoising results. Kindermann et al [13], and Gilboa-Osher [10,11] formulated the variational framework of NL-means by proposing nonlocal regularizing functionals. Lou et al [14] used the nonlocal total variation of Gilboa-Osher (NL/TV) in grey-scale image deblurring in the presence of Gaussian noise. Moreover, Peyré et al [18] used NL/TV for grey-scale image inpainting, super-resolution of a single image, and compressive sensing. Protter et al. [19] generalized the NL-means filter to super-resolution.

The previous works on nonlocal methods have been done on the Gaussian noise model, but no study has been developed on the impulse noise model using non-local information. However, the impulsive noise model was studied in the local case. Bar et al [4] used the Ambrosio-Tortorelli and Shah approximations to Mumford-Shah regularizing functional for color image deblurring in the presence of impulse noise, producing better restorations than total variation (TV) regularizer, and moreover providing the edge set detected concurrently with the restoration process. We propose in this paper the nonlocal versions of Ambrosio-Tortorelli [2] and Shah [21] approximations to the Mumford-Shah regularizer for the multichannel case. We also propose (a) several applications of the NL/MS for color image denoising, deblurring in the presence of Gaussian or impulse noise, inpainting with large missing portion, super-resolution of a single image, and (b) an efficient preprocessing step to compute the weights w in the deblurring-denoising model in the presence of impulse noise.

2   Background

Local regularizer. We recall two approximations of the Mumford-Shah-like regularizing functionals [16] that have been used in several algorithms. The MS regularizer, depending on the image u and on its edge set $K \subset \Omega$, gives preference to piecewise smooth images: $\Psi^{MS}(u, K) = \beta \int_{\Omega \setminus K} |\nabla u|^2 dx + \alpha \int_K d\mathcal{H}^1$, where $\mathcal{H}^1$ is the 1D Hausdorff measure. The first term enforces smoothness of u everywhere except on the edge set K, and the second one minimizes the total length of edges. It is difficult to minimize in practice the non-convex MS functional. One approach is using phase field by Γ-convergence with applications to image deblurring and denoising [3], and inpainting [22]. More specifically, Ambrosio-Tortorelli [2] approximated the MS regularizer by a sequence of regular functionals Ψ_ϵ using the Γ-convergence (we call MSAT, the Ambrosio and Tortorelli approximation of MS regularizer). The edge set K is represented by a smooth auxiliary function v. Shah [21] suggested a modified version of the AT approximation to the MS functional by replacing $|\nabla u|^2$ by $|\nabla u|$ (we call it MSTV). Furthermore, Bar et al [4] used the color versions of these functionals for color image deblurring-denoising by replacing the magnitude of the gradient

$|\nabla u|$ by the Frobenius norm of the matrix ∇u, $\|\nabla u\| = \sqrt{\sum_i [(u_x^i)^2 + (u_y^i)^2]}$ with $i \in \{r, g, b\}$ in the RGB color space, suggested by Brook et al [7]:

$$\Psi_\epsilon^{MSAT}(u, v) = \beta \int_\Omega v^2 \|\nabla u\|^2 dx + \alpha \int_\Omega \left(\epsilon |\nabla v|^2 + \frac{(v-1)^2}{4\epsilon} \right) dx,$$

$$\Psi_\epsilon^{MSTV}(u, v) = \beta \int_\Omega v^2 \|\nabla u\| dx + \alpha \int_\Omega \left(\epsilon |\nabla v|^2 + \frac{(v-1)^2}{4\epsilon} \right) dx$$

where $0 \leq v(x) \leq 1$ represents the edges: $v(x) \approx 0$ if $x \in K$ and $v(x) \approx 1$ otherwise, and $\alpha, \beta > 0$, $\epsilon \to 0$ are parameters. Note that, in both regularizers, the edge map v is common for the three channels and provides the necessary coupling between colors.

Nonlocal regularizer. Nonlocal methods in image processing have been explored in many papers because they are well adapted to texture denoising while the standard denoising models working with local image information seem to consider texture as noise, which results in losing details. Nonlocal methods are generalized from neighborhood filters and patch based methods. The idea of neighborhood filter is to restore a pixel by averaging the values of neighboring pixels with a similar color value. Buades et al. [8] generalized this idea by applying the patch-based method, and proposed the famous nonlocal-means (or NL-means) filter for denoising, given by $NLu(x) = \frac{1}{C(x)} \int_\Omega e^{-\frac{d_a(u(x),u(y))}{h^2}} u(y) dy$, where $u(y)$ is the color at y, $d_a(u(x), u(y)) = \int G_a(t)\|u(x+t) - u(y+t)\|^2 dt$ is the patch distance, G_a is the Gaussian kernel with standard deviation a determining the patch size, $C(x) = \int_\Omega e^{-\frac{d_a(u(x),u(y))}{h^2}} dy$ is a normalization factor, and h is the filtering parameter corresponding to the noise level (usually the standard deviation of the noise). The NL-means not only compares the color value at a single point but the geometrical configuration in a whole neighborhood (patch).

In the variational framework, Kindermann et al [13] formulated the neighborhood filters and NL-means filters as nonlocal regularizing functionals which generally are not convex. Then, Gilboa-Osher [10], [11] formalized the convex nonlocal functional inspired from graph theory, based on the gradient and divergence definitions on graphs in the context of machine learning. Let $u : \Omega \to \mathbb{R}$ be a function, and $w : \Omega \times \Omega \to \mathbb{R}$ be a nonnegative and symmetric weight function. The nonlocal gradient vector $\nabla_w u : \Omega \times \Omega \to \mathbb{R}$ is $(\nabla_w u)(x, y) := (u(y) - u(x))\sqrt{w(x, y)}$. Hence, the nonlocal divergence $\text{div}_w \vec{v} : \Omega \to \mathbb{R}$ of the vector $\vec{v} : \Omega \times \Omega \to \mathbb{R}$ is defined as the adjoint of the nonlocal gradient, $(\text{div}_w \vec{v})(x) := \int_\Omega (v(x, y) - v(y, x))\sqrt{w(x, y)} dy$, and the norm of the nonlocal gradient of u at $x \in \Omega$ is given by $|\nabla_w u|(x) = \sqrt{\int_\Omega (u(y) - u(x))^2 w(x, y) dy}$. Based on these nonlocal operators, they introduced nonlocal regularizing functionals of the general form $\Psi(u) = \int_\Omega \phi(|\nabla_w u|^2) dx$, where $s \mapsto \phi(s)$ is a positive function, convex in $\sqrt{s}$, and $\phi(0) = 0$. By taking $\phi(s) = \sqrt{s}$, they proposed the nonlocal TV regularizer (NL/TV) which corresponds in the local case to $\Psi^{TV}(u) = \int_\Omega |\nabla u| dx$. Inspired by these ideas, we propose in the next section nonlocal versions of Ambrosio-Tortorelli and Shah approximations to the MS regularizer for color image restoration. This is also continuation or nonlocal extension of the work by Bar et al. [4], first to propose the use of local Mumford-Shah-like approximations to color image restoration. Part of this work is a generalization of [12].

3   Description of the Proposed Models

We propose the following nonlocal Mumford-Shah regularizers (NL/MS) by applying the nonlocal operators to the multichannel approximations of the MS regularizer

$$\Psi^{NL/MS}(u,v) = \beta \int_\Omega v^2 \phi(\|\nabla_w u\|^2)dx + \alpha \int_\Omega \left(\epsilon|\nabla v|^2 + \frac{(v-1)^2}{4\epsilon}\right)dx$$

where $u : \Omega \to \mathbb{R}^3$, $v : \Omega \to [0,1]$, $\phi(s) = s$ or $\phi(s) = \sqrt{s}$ correspond to the nonlocal versions of MSAT and MSTV regularizers, so called NL/MSAT and NL/MSTV, respectively. In addition, we apply these nonlocal MS regularizers to color image denoising, color image deblurring in the presence of Gaussian or impulse noise, color image inpainting, and moreover, to color image super-resolution, by incorporating proper fidelity terms. We define the weight function w using the noisy-blurry data f as $w(x,y) = e^{-\frac{d_a(f(x),f(y))}{h^2}}$, but we will see that this definition must be modified sometimes. Also, note that the nonlocal and nonconvex continuous models proposed in the following sections have not been analyzed theoretically; however, these formulations become well-defined in the discrete, finite differences case, but we prefer to present them in the continuous setting for simplicity.

3.1   Color Image Deblurring and Denoising

The degradation model for deblurring-denoising (or denoising) is given by $f_i = k * u_i + n_i$, with a (known) space-invariant blurring kernel k. First, in the case of Gaussian noise model, the L^2-fidelity term led by the maximum likelihood estimation is commonly used: $\Phi(f - k * u) = \int_\Omega \sum_i |f^i - k * u^i|^2 dx$. However, the quadratic data fidelity term considers the impulse noise, which might be caused by bit errors in transmissions or wrong pixels, as an outlier. So, for the impulse noise model, the L^1-fidelity term is more appropriate, due to its robustness of removing outlier effects [1], [17]; moreover, we consider the case of independent channels noise [4]: $\Phi(f - k * u) = \int_\Omega \sum_i |f^i - k * u^i| dx$. Thus, we design two types of total energies for color image deblurring-denoising, depending on the type of noise as follows:

Gaussian noise: $E^G(u,v) = \frac{1}{2}\int_\Omega \sum_i |f^i - k * u^i|^2 dx + \Psi^{NL/MSTV}(u,v)$ (2)

Impulse noise: $E^{Im}(u,v) = \int_\Omega \sum_i |f^i - k * u^i| dx + \Psi^{NL/MSAT}(u,v)$.   (3)

Note that, for the impulse noise model, MSAT regularizer produces better results (especially in the presence of high density of noise), while for the Gaussian noise model, MSTV regularizer produces better results.

To extend the nonlocal methods to the impulse noise case, we need a preprocessing step for the weight function w since we cannot directly use the data f to compute w. In other words, in the presence of impulse noise, the noisy pixels tend to have larger weights than the other neighboring points, so it's likely to keep the noise value at such pixel. Thus, we propose a simple algorithm to obtain a preprocessed image g, which removes the impulse noise (outliers) as well as preserves the textures as much as we can. Basically, we use the median filter, well-known for removing impulse noise. However, for the deblurring-denoising model, if we apply one-step of the median filter, then the output may be too smoothed out. In order to preserve fine structures as well as to remove noise properly, we define a preprocessing method for the deblurring-denoising model inspired by the idea of Bregman iteration [6]. Thus, we propose the following algorithm to obtain a preprocessed image g that will be used only in the computation of the weight function w:

Initialize : $r_0^i = 0$, $g_0^i = 0$, $i = r, g, b$.
do (iterate $n = 0, 1, 2, \dots$)
$\qquad g_{n+1} = median(f + r_n, [a\ a])$
$\qquad r_{n+1} = r_n + f - k * g_{n+1}$
while $\sum_i \| f^i - k * g_n^i \|_1 > \sum_i \| f^i - k * g_{n+1}^i \|_1$
[Optional] $g_m = median(g_m, [b\ b])$

where f is the given noisy-blurry data, and $median(f, [a\ a])$ is the median filter of size $a \times a$ with input f. The optional step is needed in the case when the final g_m still has some salt-and-pepper-like noise left. The preprocessed image g_m is a deblurred and denoised version of f; it will be used only in the computation of the weights w, while keeping f in the data fidelity term, thus artifacts are not introduced by the median filter. Note that for denoising only (no blurring), we apply the adaptive median filter or the median filter to the noisy image f, to produce a preprocessed image g.

3.2   Color Image Inpainting

Inpainting corresponds to the operation H of losing pixels from an image, i.e. the observed data f is given by $f = u$ on $\Omega - D$ with the region $D = D^0$ where the input data u has been damaged. Thus, we formulate the total energy functional for color image inpainting as

$$E^{Inp}(u, v) = \frac{1}{2} \int_\Omega \lambda_D(x) \sum_i |f^i - u^i|^2 dx + \Psi^{NL/MS}(u, v), \qquad (4)$$

where $\lambda_D(x) = 0$ at $x \in D$ and $\lambda_D(x) > 0$ on $x \in \Omega - D$. In addition, we update the weights w only in the damaged region D^0 in every mth iteration for u using the patch distance: $d_a^R(u(x), u(y)) = \int_{\Omega-R} G_a(t)\|u(x+t) - u(y+t)\|^2 dt$, where

$R \subset D^0$ is an un-recovered region (still missing region). Therefore, the missing region D^0 is recovered by the following iterative algorithm, producing the un-recovered regions D^i, $i = 0, 1, 2, ...$, with $D^0 \supset D^1 \supset D^2 \supset \cdots$:

1. Compute weights w for $x \in \Omega$ s.t. $P(x) \bigcap (\Omega - D^0) \neq 0$ with $d_a^{D^0}(u^0(x), u^0(y))$, $u^0 = f$, and a patch $P(x)$ centered at x.
2. Iterate $n = 1, 2, ...$ to get a minimizer (u, v) starting with $u = u^0$:
    a. For fixed u, update v in Ω to get v^n.
    b. For fixed v, update u in Ω to get u^n with a recovered region $\Omega - D^n \supset \Omega - D^0$: at every mth iteration, update weights w only in $x \in D^0$ s.t. $P(x) \bigcap (\Omega - D^{n,m}) \neq 0$ with $d_a^{D^{n,m}}(u(x), u(y))$ where $D^{n,m}$ is an un-recovered region in D^0 until mth iteration with $D^{n,m} \supset D^{n,2m} \supset \cdots \supset D^{n,n} = D^n$.

3.3   Color Image Super-Resolution

Super-resolution of a single still image corresponds to the recovery of a high resolution image from a filtered and down-sampled image, i.e., the observed data f is given by $f^i = D_k(h * u^i)$, $i \in r, g, b$ where h is a low-pass filter, $D_k : \mathbb{R}^n \to \mathbb{R}^p$, $p = \frac{n}{k^2}$, is the down-sampling operator by a factor k along each axis. We want to recover a high resolution image $u \in (\mathbb{R}^n)^3$ by minimizing

$$E^{Sup}(u, v) = \frac{1}{2} \int_\Omega \sum_i |f^i - D_k(h * u^i)|^2 dx + \Psi^{NL/MS}(u, v). \qquad (5)$$

In addition, we use a super-resolved image $g \in (\mathbb{R}^n)^3$ obtained by a bicubic interpolation of $f \in (\mathbb{R}^p)^3$ only for the computation of the weights w. We refer to [15,19] for prior relevant work.

3.4   Optimality Condition for (2)-(5)

Finally, minimizing the proposed functionals (2)-(5): E^G, E^{Im}, E^{Inp}, E^{Sup} in u and v, we obtain the Euler-Lagrange equations

$$\frac{\partial E^{G,Im,Inp,Sup}}{\partial v} = 2\beta v\phi(\|\nabla_w u\|^2) - 2\epsilon\alpha\triangle v + \alpha\left(\frac{v-1}{2\epsilon}\right) = 0,$$

$$\frac{\partial E^G}{\partial u} = \tilde{k} * (k * u - f) + L^{NL/MSTV}u = 0,$$

$$\frac{\partial E^{Im}}{\partial u} = \tilde{k} * \text{sign}(k * u - f) + L^{NL/MSAT}u = 0,$$

$$\frac{\partial E^{Inp}}{\partial u} = \lambda_D(u - f) + L^{NL/MS}u = 0,$$

$$\frac{\partial E^{Sup}}{\partial u} = \tilde{h} * (D_k^T(D_k(h * u) - f)) + L^{NL/MS}u = 0,$$

where $\tilde{k}(x) = k(-x)$, $\tilde{h}(x) = h(-x)$, $D_k^T : (\mathbb{R}^p)^3 \to (\mathbb{R}^n)^3$ is the transpose of D_k i.e. the up-sampling operator, and

$$L^{NL/MS}u = -2\int_\Omega \Big\{(u(y) - u(x))w(x,y)$$
$$\cdot \left[(v^2(y)\phi'(\|\nabla_w u\|^2(y)) + v^2(x)\phi'(\|\nabla_w u\|^2(x))\right]\Big\}dy.$$

To solve two Euler-Lagrange equations simultaneously, the alternate minimization approach is applied. Note that since the energy functionals are not convex in the joint variable (u,v), we may compute only a local minimizer. However, this is not a drawback in practice, since the initial guess for u in our algorithm is the data f (except for the super-resolution problem). Due to its simplicity, we use Gauss-Seidel scheme for v, and an explicit scheme for u using gradient descent method.

4   Experimental Results and Comparisons

The nonlocal MS regularizers proposed here, NL/MSTV and NL/MSAT, are tested on several color images corrupted by different blur kernels or different noise types, on color images with some missing parts, as well as on a sub-sampled color image. We compare them with their local versions. For comparisons with TV and NL/TV models of Rudin-Osher and Gilboa-Osher, we refer the reader to [12] for the case of gray-scale images.

First, we use noisy Lena images only corrupted by Gaussian noise or salt-and-pepper noise in Fig. 1. As expected, NL/MSTV and NL/MSAT perform better than MSTV and MSAT respectively in the sense that not only they preserve fine scales such as textures, but also in the case of NL/MSTV, the model does not produce any staircase effect (appeared in MSTV). For the restoration of the image (c) with salt-and-pepper noise, we apply one-step adaptive median filter with the maximum size 7×7 and then 5×5 median filter to the noisy image f, to produce a preprocessed image (PSNR=26.8145) only for the computation of w, and moreover the edges v detected concurrently with the restoration process are presented for the nonlocal methods.

Next, we recover a blurred image contaminated by Gaussian noise or random-valued impulse noise in Figures 2, 3, 5-7. First, we test the local and nonlocal MSTV on the Barbara image in Fig. 2 with Gaussian blur and noise, and then we test the local and nonlocal MSAT on the Lena images in Fig. 5 and the Girl images in Fig. 6 with different blur kernels and random-valued impulse noise with different noise levels. More precisely, in Fig. 3 in the presence of blur and Gaussian noise, NL/MSTV recovers well the fine scales such as textures leading to cleaner image and higher PSNR, while with MSTV, the textures are more smoothed out during the denoising process. In Fig. 5, we restore the Lena images blurred with motion blur and then contaminated by random-valued impulse noise with different noise densities $d = 0.2, 0.4, 0.5$ using MSAT and NL/MSAT. By using a preprocessed image for the weights, NL/MSAT provides

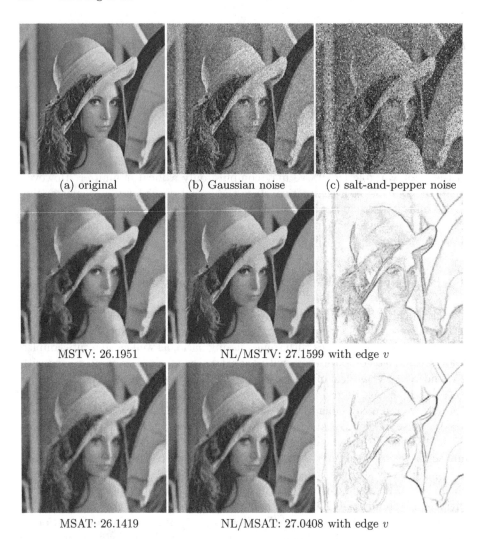

(a) original   (b) Gaussian noise   (c) salt-and-pepper noise

MSTV: 26.1951   NL/MSTV: 27.1599 with edge v

MSAT: 26.1419   NL/MSAT: 27.0408 with edge v

Fig. 1. Recovery of noisy image $f = u + n$ and PSNR values. Top row: (a) original, (b) noisy image with Gaussian noise with $\sigma_n = 0.02$, (c) noisy image with salt-and-pepper noise with noise density $d = 0.3$. Middle row: recovered images of (b) and edge set v using MSTV: $\beta = 0.17$, NL/MSTV: $\beta = 0.07$. Bottom row: recovered images of (c) and edge set v using MSAT: $\beta = 6$, NL/MSAT: $\beta = 1.8$.

better results visually and according to PSNR. Moreover, in Fig. 6, we test MSAT and NL/MSAT on the Girl images with either (b) high blur and low noise or (d) low blur and high noise. The restorations of (b) are shown in the first row in Fig. 7. Both local and nonlocal MSAT provide reasonable results, but NL/MSAT produces less ringing effects especially appeared on the cloth part. With the data (c), NL/MSAT gives cleaner result and higher PSNR while MSAT still has some noise, which are shown in the second row in Fig. 7.

Fig. 2. Left to right: original image, blurry image with Gaussian blur with $\sigma_b = 1$, noisy-blurry image corrupted by Gaussian noise with $\sigma_n = 0.004$

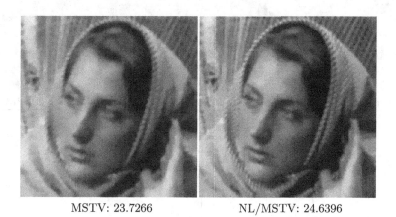

MSTV: 23.7266   NL/MSTV: 24.6396

Fig. 3. Recovery of the noisy-blurry image in Fig. 2 using (left) MSTV: $\beta = 0.04$, (right) NL/MSTV: $\beta = 0.012$

In Figures 8-9 we use the NL/MS regularizers to recover images with textures and large missing portions. In Fig. 8, we observe that both nonlocal regularizers recover the missing parts very well, and moreover NL/MSTV gives slightly better result than NL/MSAT according to PSNR even though they visually seem to produce very similar results. However, in Fig. 9 with a real image, we only present the result of NL/MSAT since it provides slightly better result than NL/MSTV (PSNR=34.2406), especially better recovering the part damaged by the circle.

In Fig. 4, we recover an image filtered with a low-pass filter and then sub-sampled. The nonlocal regularizers provide higher PSNRs and better visual qualities providing cleaner edges, while the local ones produce some artifacts especially on the edges. In addition, NL/MSTV gives the best result providing visually sharper and cleaner image, and highest PSNR.

Finally, we note that the parameters α, β and ϵ were selected manually to provide the best PSNR results. The smoothness parameter β increases with the noise level, while the other parameters α, ϵ are approximately fixed such as

382   M. Jung et al.

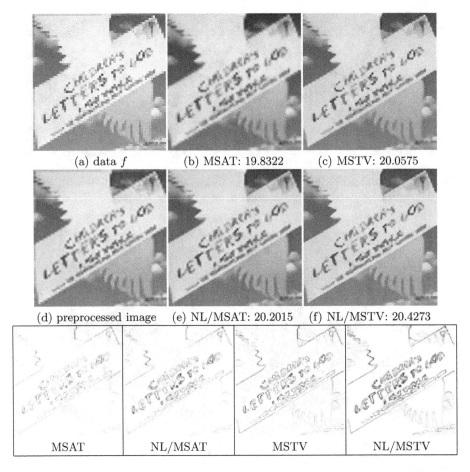

Fig. 4. Super-resolution of a low resolution image. (a) Data $f = D_k(h*u)$ of size 64×64 with uniform blur h of size 3×3 and sub-sampling factor $k = 4$, (d) a preprocessed image using bicubic interpolation (PSNR=18.0711), (b),(c),(e),(f): super-resolved images. Bottom: edge set v.

$\alpha = 0.1$, $\epsilon = 0.00000001$ for deblurring-denoising and inpainting model, and $\alpha = 0.1$, $\epsilon = 0.001$ for denoising and super-resolution (although in theory $\epsilon \to 0$, it is common in practice to work with a small fixed ϵ). For the weight function w, we use the search window $\Omega_w = \{y \in \Omega : |y - x| \leq r\}$ instead of Ω (semi-local) and the weight function w at $(x, y) \in \Omega \times \Omega$ depending on a function $g : \Omega \to \mathbb{R}^3$, $w(x, y) = exp\left(-\frac{d_a(g(x),g(y))}{h^2}\right)$. We use 11×11 search window with 5×5 patch for deblurring-denoising model (or denoising), and 21×21 search window with 9×9 patch for super-resolution, while larger search windows are needed for inpainting. For the computational time, it takes about 5 minutes for constructing the weight function of a 256×256 image with the 11×11 search window and 5×5 patch in MATLAB on a dual core laptop with 2GHz processor

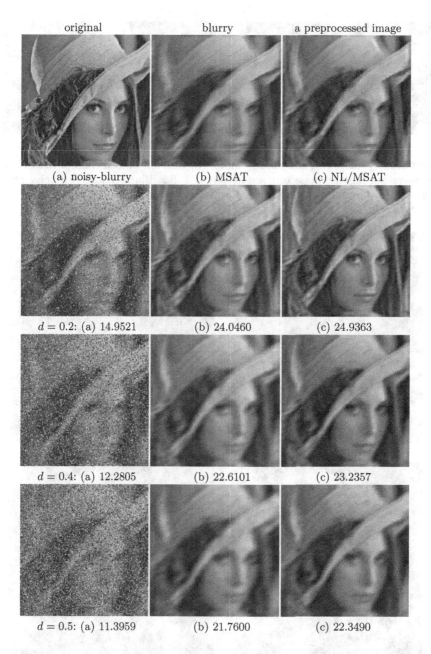

Fig. 5. Recovery of blurry Lena image with several random-valued impulse noise levels, and PSNR values. Top row: original image, blurry image with motion blur kernel of length=10, oriented at angle $\theta = 25°$ w.r.t. the horizon, a preprocessed image when $d = 0.2$ (PSNR=22.0317.) 1st column: (2nd to 4th row) noisy blurry data f with noise density $d = 0.2, 0.4, 0.5$. 2nd and 3rd columns: recovered images using (2nd) MSAT and (3rd) NL/MSAT. β (2nd to 4th row): 2, 4, 5 (MSAT), 0.3, 0.9, 1.2 (NL/MSAT).

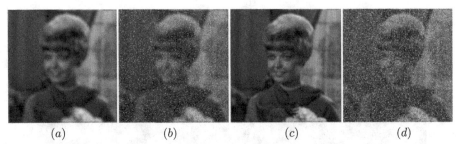

(a)   (b)   (c)   (d)

Fig. 6. Blurry image (a), noisy blurry image (b) with out-of-focus blur of radius 5 and radom-valued impulse noise of noise density $d = 0.2$. Blurry image (c), noisy blurry image (d) with out-of-focus blur of radius 3 and radom-valued impulse noise of noise density $d = 0.4$

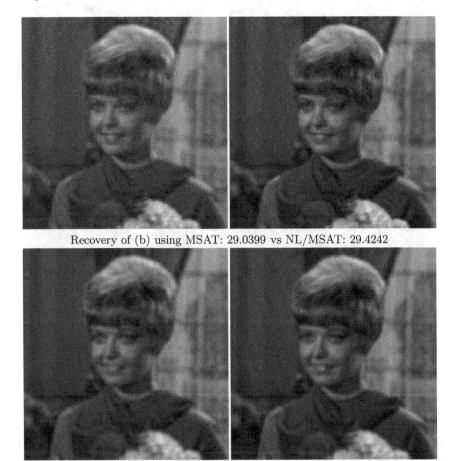

Recovery of (b) using MSAT: 29.0399 vs NL/MSAT: 29.4242

Recovery of (d) using MSAT: 28.2293 vs NL/MSAT: 28.8418

Fig. 7. Recovery of the noisy-blurry image (b) and (d) in Fig. 5 using MSAT (left), NL/MSAT (right) and PSNR values. Top: (MSAT) $\beta = 1$, (NL/MSAT) $\beta = 0.2$. Bottom: (MSAT) $\beta = 5$, (NL/MSAT) $\beta = 1.8$.

Fig. 8. Inpainting of 100×100 size image with 40×40 missing part. (left) data f, recovered using (middle) NL/MSAT: PSNR=35.6704, (right) NL/MSTV: PSNR=35.8024 (41×41 search window, 9×9 patch).

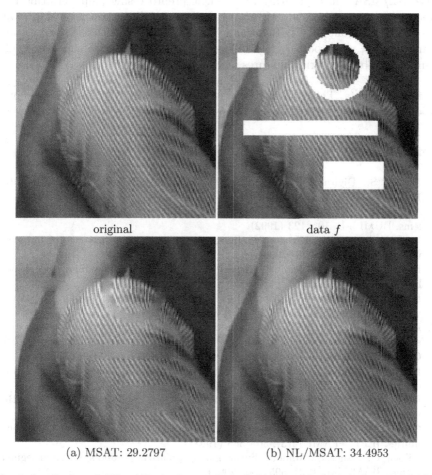

original   data f

(a) MSAT: 29.2797   (b) NL/MSAT: 34.4953

Fig. 9. Inpainting of 150×150 size image. Top: (left) original, (right) data f. (a), (b): recovered images with (a) MSAT, (b) NL/MSAT (51×51 search window, 9×9 patch).

and 2GB memory. The minimization for the (local or nonlocal) MS regularizers in the deblurring-denoising model takes about 150 seconds for the computations of both u using an explicit scheme based on the gradient descent method and v using a semi-implicit scheme with the total iterations $5 \times (2 + 100)$ (without including the computation of the weight function $w(x, y)$). For inpainting model with 150×150 size image, it takes about 20 minutes with total iteration numbers $7 \times (2+100)$ since we update the weight function at every 50th iteration for u. For super-resolution, $10 \times (2 + 200)$ iteration numbers are needed for all regularizers.

5   Summary and Conclusions

The proposed nonlocal MS regularizers, NL/MSAT and NL/MSTV, outperform the local ones; in the image denoising or deblurring models in the presence of noise, NL/MSAT incorporating an efficient preprocessing step performs best for impulse noise, while NL/MSTV works best for Gaussian noise. Moreover, for super-resolution, NL/MSTV produces the best result with the sharpest and cleanest edges, while both NL/MSAT and NL/MSTV provide superior results in inpainting by recovering textures and large missing regions.

Acknowledgments. This work has been supported by the National Science Foundation Grants DMS- 0714945 and DMS-0312222.

References

1. Alliney, S.: Digital Filters as Absolute Norm Regularizers. IEEE TSP 40(6), 1548–1562 (1992)
2. Ambrosio, L., Tortorelli, V.M.: On the approximation of free discontinuity problems. BUMI 6-B, 105–123 (1992)
3. Bar, L., Sochen, N., Kiryati, N.: Image deblurring in the presence of impulsive noise. IJCV 70, 279–298 (2006)
4. Bar, L., Brook, A., Sochen, N., Kiryati, N.: Deblurring of Color Images Corrupted by Impulsive Noise. IEEE TIP 16(4), 1101–1111 (2007)
5. Blomgren, P., Chan, T.F.: Color TV: Total variation methods for restoration of vector-valued images. IEEE TIP 7(3), 304–309 (1998)
6. Bregman, L.M.: The relaxation method for finding common points of convex sets and its application to the solution of problems in convex programming. USSR Computational Mathematics and Mathematical Physics 7, 200–217 (1967)
7. Brook, A., Kimmel, R., Sochen, N.: Variational restoration and edge detection for color images. JMIV 18, 247–268 (2003)
8. Buades, A., Coll, B., Morel, J.M.: A review of image denoising algorithms, with a new one. SIAM MMS 4(2), 490–530 (2005)
9. Geman, D., Reynolds, G.: Constrained Restoration and the Recovery of Discontinuities. IEEE TPAMI 14(3), 367–383 (1992)
10. Gilboa, G., Osher, S.: Nonlocal linear image regularization and supervised segmentation. SIAM MMS 6(2), 595–630 (2007)
11. Gilboa, G., Osher, S.: Nonlocal operators with applications to image processing. SIAM MMS 7(3), 1005–1028 (2008)

12. Jung, M., Vese, L.A.: Nonlocal variational image deblurring models in the presence of Gaussian or impulse noise. LNCS, vol. 5567, pp. 402–413. Springer, Heidelberg (2009)
13. Kindermann, S., Osher, S., Jones, P.W.: Deblurring and denoising of images by nonlocal functionals. SIAM MMS 4(4), 1091–1115 (2005)
14. Lou, Y., Zhang, X., Osher, S., Bertozzi, A.: Image recovery via nonlocal operators. UCLA C.A.M. Report 08-35 (2008)
15. Malgouyres, F., Guichard, F.: Edge direction preserving image zooming: a mathematical and numerical analysis. SIAM NA 39(1), 1–37 (2001)
16. Mumford, D., Shah, J.: Optimal approximations by piecewise smooth functions and associated variational problems. CPAM 42, 577–685 (1989)
17. Nikolova, M.: Minimizers of cost-functions involving non-smooth data-fidelity terms. Application to the processing of outliers. SIAM NA 40(3), 965–994 (2002)
18. Peyré, G., Bougleux, S., Cohen, L.: Non-local regularization of inverse problems. In: Forsyth, D., Torr, P., Zisserman, A. (eds.) ECCV 2008, Part III. LNCS, vol. 5304, pp. 57–68. Springer, Heidelberg (2008)
19. Protter, M., Elad, M., Takeda, H., Milanfar, P.: Generalizing the Non-Local-Means to super-resolution reconstruction. IEEE TIP 18(1), 36–51 (2009)
20. Rudin, L., Osher, S.: Total variation based image restoration with free local constraints. IEEE ICIP 1, 31–35 (1994)
21. Shah, J.: A common framework for curve evolution, segmentation and anisotropic diffusion. In: IEEE CVPR, pp. 136–142 (1996)
22. Shen, J., Chan, T.F.: Variational image inpainting. CPAM 58(5), 579–619 (2005)

Reconstructing Optical Flow Fields by Motion Inpainting

Benjamin Berkels[1], Claudia Kondermann[2],
Christoph Garbe[2], and Martin Rumpf[1]

[1] Institute for Numerical Simulation, Universität Bonn,
Endenicher Allee 60, 53115 Bonn, Germany
{benjamin.berkels,matrin.rumpf}@ins.uni-bonn.de
http://numod.ins.uni-bonn.de/
[2] IWR, Universität Heidelberg,
Im Neuenheimer Feld 368, 69120 Heidelberg
{Claudia.Kondermann,Christoph.Garbe}@iwr.uni-heidelberg.de
http://hci.iwr.uni-heidelberg.de/

Abstract. An edge-sensitive variational approach for the restoration of optical flow fields is presented. Real world optical flow fields are frequently corrupted by noise, reflection artifacts or missing local information. Still, applications may require dense motion fields. In this paper, we pick up image inpainting methodology to restore motion fields, which have been extracted from image sequences based on a statistical hypothesis test on neighboring flow vectors. A motion field inpainting model is presented, which takes into account additional information from the image sequence to improve the reconstruction result. The underlying functional directly combines motion and image information and allows to control the impact of image edges on the motion field reconstruction. In fact, in case of jumps of the motion field, where the jump set coincides with an edge set of the underlying image intensity, an anisotropic TV-type functional acts as a prior in the inpainting model. We compare the resulting image guided motion inpainting algorithm to diffusion and standard TV inpainting methods.

1 Introduction

Many methods have been proposed to estimate motion in image sequences. Yet, in difficult situations such as multiple motions, aperture problems or occlusion boundaries optical flow estimates are often incorrect. These incorrect flow patterns can be detected and removed from the flow field e.g. by means of confidence measures [1,2,3]. But since many applications demand a dense flow field, it would be beneficial to reconstruct a reliable dense vector field based on information from the surrounding flow field. A similar task has been addressed in the field of image reconstruction and is called inpainting, picking up a classical term from the restoration of old and damaged paintings. The digital reconstruction of corrupted images was first proposed by Masnou and Morel [4]. Over the last decade

D. Cremers et al. (Eds.): EMMCVPR 2009, LNCS 5681, pp. 388–400, 2009.

a wide range of methods has been developed for the inpainting of grayscale or color images. Edge preserving TV inpainting and curvature-driven diffusion inpainting was suggested by Chan and Shen [5,6]. Transport based methods with a fast marching type inpainting algorithm were proposed by Telea [7] and improved by Bornemann and März [8]. The relation to fluid dynamics was studied by Bertalmio et al. [9] and Chan and Shen [10] investigated texture inpainting. Already in 1993, Mumford et al. [11] proposed to study a variational approach which treats contour lines as elastic curves. In [12], Ballester et al. introduced a variational approach based on the smooth continuation of isophote lines. A variational approach based on level sets and a Perimeter and Willmore energy was presented by Ambrosio and Masnou in [13]. A combination of TV inpainting and wavelet representation was proposed in [14].

The inpainting methodology has been generalized to video sequences with occluding objects by Patwardhan [15]. The reconstruction of motion fields has lately been proposed in the field of video completion. In case of large holes with complicated texture, previously used methods are often not suitable to obtain good results. Instead of reconstructing the frame itself by means of inpainting, the reconstruction of the underlying motion field allows for the subsequent restoration of the corrupted region even in difficult cases. This type of motion field reconstruction called "motion inpainting" was first introduced for video stabilization by Matsushita et al. in [16]. The idea is to continue the central motion field to the edges of the image sequence, where the field is lost due to camera shaking. This is done by a basic interpolation scheme between four neighboring vectors and a fast marching method. Chen et al. [17] refined the approach of Matsushita et al. to obtain a robust motion inpainting approach, which can deal with sudden scene changes by means of Markov Random Field based diffusion and applied it to spatio-temporal error concealment in video coding. In [18], Kondermann et al. proposed to improve motion fields by only computing a few reliable flow vectors and filling in the missing vectors by means of a diffusion based motion inpainting approach.

In general, the variational reconstruction of optical flow fields can be accomplished by straightforward extension of inpainting functionals for images to two dimensional vector fields. However, these methods usually fail in situations where the course of motion discontinuity lines is unclear, e.g. if objects with curved boundary move or junctions occur in overlapping motion. Since image edges often correspond to motion edges the information drawn from the image sequence can be important for the reconstruction, especially in such cases where the damaged vector field does not contain enough information to determine the shape of the motion discontinuity.

In the special case of optical flow extracted from an image sequence, the underlying image sequence itself provides additional information, which can be used to guide the reconstruction process in ambiguous cases. So far, optical flow fields have already been used for the reconstruction of images in video restoration, e.g. in [15]. Here, we use the underlying image data to improve the reconstruction of the optical flow field. The resulting functional is nonlinear and can be minimized by means of the finite element method. We compare our results to diffusion based and TV inpainting methods.

To prepare the discussion of the proposed new motion field inpainting model, let us briefly review some basic image inpainting methodology. Given an image $u_0 : \Omega \to \mathbb{R}$ and an inpainting domain $D \subset \Omega$, one asks for a restored image intensity $u : \Omega \to \mathbb{R}$, such that $u|_{\Omega \backslash D} = u_0$ and $u|_D$ is a suitable and regular extension of the image intensity u_0 outside D. The simplest inpainting model is based on the construction of a harmonic function u on D with boundary data $u = u_0$ on ∂D. Based on the Dirichlet principle, this model is equivalent to the minimization of the Dirichlet functional $E_{\text{harmon}}[u] = \frac{1}{2} \int_D |\nabla u|^2 \, dx$ for given boundary data. Due to standard elliptic regularity the resulting intensity function u is smooth – even analytic – inside D but does not continue any edge type singularity of u_0 prominent at the boundary ∂D. To resolve this shortcoming the above mentioned TV-type inpainting models have been proposed. They are based on the functional $E_{\text{TV}}[u] = \frac{1}{2} \int_D |\nabla u| \, dx$. Then the minimizing image intensity is a function of bounded variation; hence characterized by jumps along rectifiable edge contours. It solves - in a weak sense - the geometric PDE $h = 0$ where $h = \text{div} \left(|\nabla u|^{-1} \nabla u \right)$ is the mean curvature on level sets or edge contours. Making use of the coarea formula (cf. [19]) one sees that minimizing E_{TV} corresponds to minimizing the lengths of the level lines of u. Thus, the resulting edges will be straight lines.

In many applications the assumption of a sharp boundary ∂D turns out to be a significant restriction. In fact, the reliability of the given image intensity gradually deteriorates from the outside to the inside of the inpainting region. This can be reflected by a relaxed formulation of the variational problem. In fact, one considers the functional

$$\mathcal{E}^\epsilon[u] = \int_\Omega |u - u_0|^2 \, H_\epsilon + \lambda(1 - H_\epsilon) \, |\nabla u|^p \, dx \, ,$$

where $\lambda > 0$, $p = 1$ or 2, and H_ϵ is a convoluted characteristic function χ_D and ϵ indicates the width of the convolution kernel [5]. In our case this blending function will depend on a confidence measure.

Contribution. In this paper, we address the restoration problem for locally corrupted optical flow fields. The underlying image information has not been exploited previously for optical flow restoration. We propose a novel anisotropic TV-type variational approach, where the anisotropy takes into account edge information of the underlying image sequence. To identify unreliable flow vectors, a confidence measure is used. This non binary measure can be taken into account as a weight in the functional. We validate our method on test data and on real world motion sequences with given ground truth.

2   The Variational Model

In this section we derive our restoration approach for optical flow fields. Given an image sequence, we denote by u_0 the image intensity and by v_0 the corresponding estimated motion field at a fixed time t. Let us suppose that a confidence measure ζ is given together with a user selected threshold θ, such that the set

$$[\zeta < \theta] := \{x \in \Omega : \zeta(x) < \theta\}$$

is the region of low confidence on the estimated optical flow field v_0. Hence, we aim at inpainting v in the region $[\zeta < \theta]$.

Design of an anisotropic prior. Let us first construct the regularizing prior that is supposed to fill in the missing parts of the vector field. We choose the function $g(s) = (1 + \frac{s^2}{\mu^2})^{-1}$ (first proposed by Perona and Malik [20]) evaluated on the slope $|\nabla u_0^\delta|$ of the image intensity as an edge-sensitive weight. To ensure robustness, the intensity gradient is regularized via convolution with a Gaussian-type kernel $G_\delta(y) = \frac{1}{2\pi\delta} \exp(-\frac{y^2}{2\delta^2})$, i. e. $\nabla u_0^\delta = G_\delta * u_0$. In the spatially discrete model, we will realize this convolution via a single time step of the discrete heat equation (cf. Section 4). Thus, the weight $g(|\nabla u_0^\delta|)$ masks out edges of u_0.

In the vicinity of edges, we use a strongly anisotropic norm $\gamma(\nabla u_0^\delta, \mathcal{D}v)$ of the Jacobian $\mathcal{D}v$ of the motion field v depending on the regularized gradient of the image intensity and defined as follows

$$\gamma(\nabla u_0^\delta, \mathcal{D}v) = \sqrt{\nu^2 \left|\mathcal{D}v\, n^\delta\right|^2 + \left|\mathcal{D}v\left(\mathbb{1} - n^\delta \otimes n^\delta\right)\right|^2}. \tag{1}$$

Here, $n^\delta = \frac{\nabla u_0^\delta}{|\nabla u_0^\delta|}$ is the regularized edge normal on the underlying image and $\mathbb{1}$ denotes the identity matrix of size 2. Furthermore, $x \otimes y := (x_i y_j)_{i,j=1,2}$ is the usual definition of a rank one matrix which renders $\mathbb{1} - n^\delta \otimes n^\delta$ as the orthogonal projection on the direction orthogonal to the normal n^δ. Hence, for a small parameter $\nu > 0$ and a point x near a motion edge the value $\gamma(\nabla u_0^\delta(x), \mathcal{D}v(x))$ will be small if the motion edge is locally aligned with the underlying image edge and vice versa. In two space dimensions, one obtains

$$\left|\mathcal{D}v\left(\mathbb{1} - n^\delta \otimes n^\delta\right)\right|^2 = \sum_{i=1}^{2} \left((n^\delta)^\perp \cdot \nabla v_i\right)^2,$$

where $(n^\delta)^\perp = (n_2^\delta, -n_1^\delta)$. This easily follows for the unit length property $(n_1^\delta)^2 + (n_2^\delta)^2 = 1$ of the normal field n^δ. Hence, the anisotropy $\gamma(\nabla u_0^\delta(x), \mathcal{D}v(x))$ simplifies to

$$\gamma(\nabla u_0^\delta, \mathcal{D}v) = \sqrt{\sum_{i=1}^{2} \left(\nu^2 \left(n^\delta \cdot \nabla v_i\right)^2 + \left((n^\delta)^\perp \cdot \nabla v_i\right)^2\right)}.$$

Finally, we obtain the following prior

$$\beta(\nabla u_0^\delta, \mathcal{D}v) = g(|\nabla u_0^\delta|)|\mathcal{D}v| + (1 - g(|\nabla u_0^\delta|))\gamma(\nabla u_0^\delta, \mathcal{D}v). \tag{2}$$

Locally minimizing this prior will favor sharp motion edges aligned with edges in the underlying image. Apart from edges, a usual TV prior is applied to the motion field. In particular, for larger destroyed regions this leads to an effective image based guidance in the reconstruction of motion edges. For ν values close to 1 there is no preference for any orientation of a motion edge and we obtain the classical TV-type inpainting model on motion fields.

Note that Nagel and Enkelmann [21] pioneered the idea of anisotropic image-driven smoothing in the context of optical flows and proposed an anisotropic prior that is closely related to the anisotropic part of β (second part of (2)), while the isotropic part of β (first part of (2)) was already proposed by Alvarez et al. [22]. In this regard, β can be seen as an interpolation between existing isotropic and anisotropic priors. However, both [21] and [22] used their corresponding priors in the context of optical flow estimation, whereas we use the combined prior to inpaint the flow field in low confidence regions of the optical flow estimation.

Dirichlet boundary conditions. Based on the prior β, we can define the energy

$$\mathcal{E}_D[v] = \int_{[\zeta < \theta]} \beta(\nabla u_0^\delta(x), \mathcal{D}v(x)) \, dx \tag{3}$$

that has to be minimized on the set of functions $\mathcal{A} := \{v | v = v_0 \text{ on } \partial[\zeta < \theta]\}$. Note that with this model, it is crucial to choose the threshold θ conservatively to ensure the validity of the values of v_0 on $\partial[\zeta < \theta]$. If the chosen threshold is too low, the values used for the Dirichlet boundary conditions are possibly corrupted and may lead to undesirable inpainting results.

Smooth overlapping blending. Surely, the criterium to identify the inpainting domain, i.e. $[\zeta < \theta]$, is not sharp. Thus, we may select a parameter $\epsilon > 0$ for the width of the transition interval between full confidence and no confidence and define the blending function $x \to H_\epsilon(\text{sdf}[\zeta - \theta](x))$, where $H_\epsilon(x) := \frac{1}{2} + \frac{1}{\pi} \arctan\left(\frac{x}{\epsilon}\right)$ (cf. the active contour approach by Chan and Vese [23]) and $\text{sdf}[f]$ denotes the signed distance function of the set $[f < 0]$. Given this diffusive weight function, we can define the total energy

$$\mathcal{E}[v] = \int_\Omega \frac{1}{2}(v(x) - v_0(x))^2 H_\epsilon(\text{sdf}[\zeta - \theta](x)) \tag{4}$$

$$+ \lambda \beta(\nabla u_0^\delta(x), \mathcal{D}v(x))(1 - H_\epsilon(\text{sdf}[\zeta - \theta](x) - \rho)) \, dx \,,$$

which consists of two terms. The first term measures the distance from the precomputed motion field v_0 and acts as a relaxed penalty to ensure that $v \approx v_0$ in the region of confidence. The second term is a spatially inhomogeneous and anisotropic prior, primarily active on the complement of the confidence set. The parameter $\rho > 0$ leads to an overlap of the regions where the first and second term are active. If omitted, there are artifacts in the inpainting, cf. Figure 1.

3   First Variation of the Energy

As a core ingredient of the minimization algorithm we have to compute descent directions of the energy functional $\mathcal{E}[\cdot]$. Thus, let us derive explicit formulas

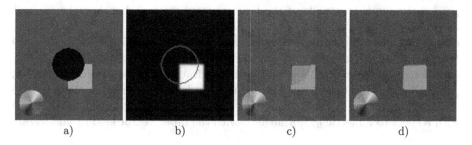

Fig. 1. Effect of the overlapping of the fidelity and the regularity energy term (4), controlled by the parameter ρ. a) Corrupted flow field, b) Underlying image and corruption indicated by the red shape, c) Reconstructed flow field with $\rho = 0$, d) Reconstructed flow field with $\rho = 9h$.

for the variation of the different terms in the integrant of $\mathcal{E}$ with respect to v. We denote by $\langle \partial_w f, \vartheta \rangle$ a variation of a function f with respect to a parameter function w in a direction ϑ. Using straightforward differentiation, for sufficiently smooth v, we obtain for $i \in \{1, 2\}$

$$\langle \partial_{v_i} \gamma(\nabla u_0^\delta, \mathcal{D}v), \vartheta \rangle = \frac{\left(\nu^2 (n^\delta \cdot \nabla v_i) n^\delta + ((n^\delta)^\perp \cdot \nabla v_i) (n^\delta)^\perp \right) \nabla \vartheta}{\gamma(\nabla u_0^\delta, \mathcal{D}v)},$$

$$\langle \partial_{v_i} \beta(\nabla u_0^\delta, \mathcal{D}v), \vartheta \rangle = g(|\nabla u_0^\delta|) \frac{\nabla v_i}{|\mathcal{D}v|} \cdot \nabla \vartheta +$$

$$\frac{1 - g(|\nabla u_0^\delta|)}{\gamma(\nabla u_0^\delta, \mathcal{D}v)} \left(\nu^2 (n^\delta \cdot \nabla v_i) n^\delta + ((n^\delta)^\perp \cdot \nabla v_i)(n^\delta)^\perp \right) \cdot \nabla \vartheta.$$

Finally, we derive the following variation $\langle \partial_{v_i} \mathcal{E}[v], \vartheta \rangle$ of the energy $\mathcal{E}[\cdot]$ with respect to the i-th component of the motion field v:

$$\langle \partial_{v_i} \mathcal{E}[v], \vartheta \rangle = \int_\Omega H_\epsilon(\mathrm{sdf}[\zeta - \theta])(v_i - v_{i,0})\vartheta$$

$$+ \lambda (1 - H_\epsilon(\mathrm{sdf}[\zeta - \theta])) \left[g(|\nabla u_0^\delta|) \frac{\nabla v_i}{|\mathcal{D}v|} \cdot \nabla \vartheta + \right. \tag{5}$$

$$\left. \frac{1 - g(|\nabla u_0^\delta|)}{\gamma(\nabla u_0^\delta, \mathcal{D}v)} \left(\nu^2 (n^\delta \cdot \nabla v_i) n^\delta + ((n^\delta)^\perp \cdot \nabla v_i)(n^\delta)^\perp \right) \cdot \nabla \vartheta \right] \mathrm{d}x.$$

The variation $\langle \partial_{v_i} \mathcal{E}_D[v], \vartheta \rangle$ is computed analogously.

4   The Algorithm

For the spatial discretization, we use the finite element (FE) method (cf. [24]): The whole domain $\Omega = [0, 1]^2$ is covered by a uniform quadrilateral mesh $\mathcal{C}$, on which a standard bilinear Lagrange finite element space is defined. We consider the image u_0 and the components of the vector fields as sets of pixels, where each pixel corresponds to a node of the finite element mesh $\mathcal{C}$. Let $\mathcal{N} = \{x_1, ..., x_n\}$ denote the nodes of $\mathcal{C}$. The FE basis function of the node x_i is defined as the continuous, piecewise bilinear function determined by $\varphi_i(x_i) = 1$ and $\varphi_i(x_j) = 0$ for $i \neq j$. To compute the integrals necessary to evaluate the energy $\mathcal{E}$ and its variations we employ a numerical Gauss quadrature scheme of order three (cf. [25]). All numerical calculations are done with double precision arithmetic.

As minimization method we use the following explicit gradient flow scheme with respect to a metric g. Initialize v^0 with the input vector field v_0 and iterate

$$v_j^{k+1} = v_j^k - \tau[\mathcal{E}, v^k, F[v^k]]G^{-1}F_j[v^k].$$

Here, G denotes the matrix representation of the metric g and the timestep width $\tau[\mathcal{E}, v^k, F[v^k]]$ is determined by the Armijo step size control [26] and depends by construction on the target functional $\mathcal{E}$, the current iterate of the solution v^k and the descent direction $F[v^k]$. Let us emphasize that the choice of g does not affect the energy landscape itself, but solely the descent path towards the set of minimizers.

The choice of the metric depends on the model used. In case of the smooth overlapping blending model (4), we chose g, inspired by the Sobolev active contour approach [27], to be a scaled version of the H^1 metric, i.e.

$$g(\vartheta_1, \vartheta_2) = \int_\Omega \vartheta_1 \cdot \vartheta_2 + \frac{\sigma^2}{2}\mathcal{D}\vartheta_1 : \mathcal{D}\vartheta_2 \, \mathrm{d}x$$

on variations ϑ_1, ϑ_2 of v and where σ represents a filter width of the corresponding time discrete and implicit heat equation filter kernel and $A : B = \mathrm{tr}(A^T B)$. The i-th component of the descent direction $F_j[v^k]$ is given by $(F_j[v^k])_i = \langle \partial_{v_j}\mathcal{E}[v], \varphi_i \rangle$.

In case of the Dirichlet boundary model (3), we choose g as the Euclidean metric, i.e. $G = \mathbb{1}$ and the i-th component of the descent direction $F_j[v^k]$ is given by

$$(F_j[v^k])_i = \begin{cases} 0 & ; x_i \text{ Dirichlet node or } x_i \notin D, \\ \langle \partial_{v_j}\mathcal{E}_D[v], \varphi_i \rangle & ; \text{else.} \end{cases}$$

Let us remark, that by construction of F in the energy descent the Dirichlet boundary values are preserved. The step size control significantly speeds up the descent and at least experimentally ensures convergence.

The absolute value function is regularized by $|z|_\eta = \sqrt{z^2 + \eta^2}$ (here $\eta = 0.1$ is used). Alternatively to the gradient descent scheme the nonlinear Euler Lagrange equation could be solved iteratively by a freezing-coefficient technique [28]. The more sophisticated and very efficient method for Total Variation Minimization based on the dual formulation of the BV norm proposed by Chambolle [29] unfortunately cannot be applied to TV inpainting directly, because the weight of the fidelity term can vanish inside the inpainting domain.

5 Numerical Experiments and Applications

As already explained in the introduction, for applications such as motion compensation, motion segmentation or the computation of divergences in fluid dynamical flows, dense motion fields are required. To demonstrate the applicability of the presented approach for the inpainting of motion fields in regions indicated by a confidence measure we apply our method to artificial and real world data.

Reconstruction of artificial motion fields. As a first test case we consider the reconstruction of a corrupted rectangular and circular motion field. Figure 2 shows the color coded ground truth flow field on the left hand side (a), the red shape indicating the region to be reconstructed in the second image column (b), the corrupted input flow field that is also used as the initialization of the image guided motion inpainting algorithm in the third column (c), and the result of the algorithm on the right hand side (d). Obviously the method successfully retrieves the motion edge along the boundary of the square (first row) and the circle (second row). We used the following set of parameters: $\mu = 50$ and $\nu = 0.1$.

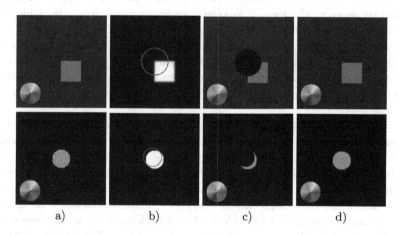

Fig. 2. a) Ground truth flow field, b) Underlying image and corruption indicated by the red shape, c) Corrupted flow field which is the initialization of the image guided inpainting algorithm, d) Reconstruction result

If the flow field to be inpainted not only contains destroyed regions, but is also corrupted by noise, enforcing Dirichlet boundary values on the boundary of the inpainting domain is not feasible. The blending model (4) on the other hand is well suited to handle such cases. In Figure 3 the motion edge is reconstructed along the boundary of the square present in the underlying image. Due to the nature of the regularization term, the reconstructed region does not contain any noise, while the noise is preserved in the complement of the inpainting domain. In between there is a smooth transition whose size is controlled by the regularization parameter of H_ϵ. Note that the regularized region is bigger than

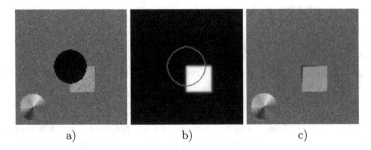

Fig. 3. Results of the blending model (4) on noisy input data. a) Corrupted flow field, b) Underlying image and corruption indicated by the red shape, c) Reconstructed flow field with $\rho = 3h$.

the inpainting domain because of the overlap induced by ρ. We used the following set of parameters: $\lambda = 0.01$, $\mu = 1$, $\nu = 0.1$.

Reconstruction of real world motion fields. Let us now consider real world examples and reconstruct the motion field of a sequence taken from the Middlebury dataset [30]. Special attention should be on the effect of the parameters μ and ν on the reconstruction result. Figure 4 shows the Rubber Whale sequence with corrupted regions indicated by a confidence measure and marked by red outlines (a), the ground truth flow field (b), the result of the image guided reconstruction algorithm (c) and the angular error (d). We used the following set of parameters: $\mu = 1$ and $\nu = 0.1$.

To investigate the effect of the parameter ν we take a closer look at two different regions in the scene: the upper left corner of the turning wheel on the left hand side and the flap of the box on the right hand side. At the upper boundary of the wheel the image contrast is low which renders the reconstruction along image edges difficult. Hence, the sensitivity of the method concerning the image gradient should be high and the method's inclination to follow image edges should be large as well, which would lead to a preference for small values μ, ν.

At the flap of the box the configuration is converse. The image contrast is large, but the motion edge does, in fact, not follow the stronger but the upper weaker edge. Hence the inclination of the method to follow image edges should be reduced, which would result in a higher value for ν.

The effect of different parameter constellations for both regions is shown in Figure 5. The results demonstrate that for low ν values the wheel can be reconstructed quite well, but the motion field also follows the sharp edge of the box flap and yields errors in that part of the sequence. In contrast, for high ν values the box flap can be reconstructed well, but the wheel is reconstructed by a straight edge which does not follow its original contour.

Comparison to diffusion and TV inpainting. We compare the image guided motion inpainting algorithm to a linear diffusion and a TV inpainting method in case of the corrupted Marble sequence. Note that we confine the comparison to these relatively simple priors, because more sophisticated image driven priors like

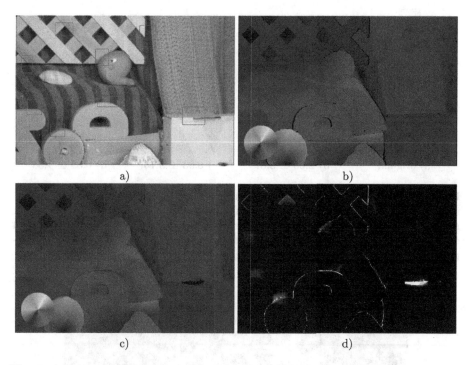

Fig. 4. a) Original Rubber Whale frame, b) Ground truth flow field, c) Reconstructed flow field, d) Resulting angular error

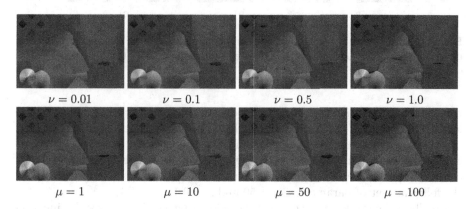

Fig. 5. Upper row: results for different values of ν for $\mu = 50$, lower row: results for different values of μ for $\nu = 0.1$

the one proposed by Nagel and Enkelmann [21] so far only have been used in the context of optical flow estimation but not for motion inpainting. Figure 6 shows the original corrupted sequence and the results of the diffusion based, the TV-based and the image guided motion inpainting methods. Not surprisingly, the diffusion based motion inpainting fails to reconstruct motion edges. In contrast,

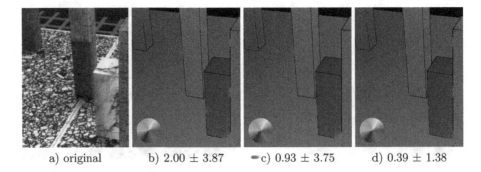

a) original   b) 2.00 ± 3.87   ●c) 0.93 ± 3.75   d) 0.39 ± 1.38

Fig. 6. Comparison of the proposed inpainting algorithm to diffusion and TV inpainting; the numbers indicate the average angular error within the corrupted regions after reconstruction; a) Original corrupted Marble sequence, b) Reconstruction result of diffusion based motion inpainting, c) Reconstruction result of TV based motion inpainting, d) Reconstruction result of image guided motion inpainting

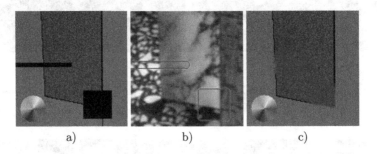

a)   b)   c)

Fig. 7. Results of the blending model (4) on noisy input data. a) Corrupted flow field, b) Underlying image and region of corrupted motion field indicated by the red shape, c) Reconstructed flow field with $\rho = 6h$.

by means of TV motion inpainting flow edges can be reconstructed. However, the lower right corner of the central marble block cannot be reconstructed properly, because the exact course of the edges near the junction is unclear. Our image guided motion inpainting uses the image gradient information to correctly reconstruct the motion boundary of the central marble block as well. Here we used the following set of parameters: $\mu = 50$ and $\nu = 0.1$.

Finally, we consider a part of the Marble sequence that shows the junction mentioned before and apply artificial noise to the corrupted input. As noted earlier, using the Dirichlet boundary model is not feasible in such a case. Hence, the blending model (4) is used for the reconstruction. In Figure 7, the motion edge junction is properly reconstructed based on the information from the underlying image. We used the following set of parameters: $\lambda = 0.01$, $\mu = 1$, $\nu = 0.1$.

6   Conclusion and Outlook

Given an image sequence and an extracted underlying motion field together with a local measure of confidence for the motion estimation, we have proposed a variational approach for the restoration of the motion field. This restoration is vital for a number of applications requiring dense motion fields. Based on a confidence measure, regions of corrupted motion can be detected. The underlying image data is still available and reliable. We make use of this information to improve the restoration of the motion field. The approach is based on an anisotropic TV-type functional, where the anisotropy takes into account edge information extracted from the underlying image data. The approach has been applied to test data and to two different real world optical flow problems. The results are compared to harmonic vector field inpainting and TV-type inpainting. We demonstrate that inpainting guided by the underlying intensity data outperforms purely flow driven approaches. We consider this as a feasibility study for the coupling of motion field and image sequence data in variational inpainting approaches. Robustness and reliability might be improved based on a fully joint approach, where the motion field and the image sequence are jointly restored. Furthermore, a restoration in space time would be promising as well.

Finally, a weakness of the proposed method is that for some motion fields the optimal performance is obtained in different locations for different parameter values (cf. Figure 5). To obtain the optimal performance in all locations, one should develop a methodology to locally adapt the parameters automatically after specifying a global set of parameters for the entire image.

References

1. Bruhn, A., Weickert, J.: In: A Confidence Measure for Variational Optic Flow Methods, pp. 283–298. Springer, Heidelberg (2006)
2. Kondermann, C., Kondermann, D., Jähne, B., Garbe, C.S.: An adaptive confidence measure for optical flows based on linear subspace projections. In: Hamprecht, F.A., Schnörr, C., Jähne, B. (eds.) DAGM 2007. LNCS, vol. 4713, pp. 132–141. Springer, Heidelberg (2007)
3. Kondermann, C., Mester, R., Garbe, C.: A statistical confidence measure for optical flows. In: Forsyth, D., Torr, P., Zisserman, A. (eds.) ECCV 2008, Part III. LNCS, vol. 5304, pp. 290–301. Springer, Heidelberg (2008)
4. Masnou, S., Morel, J.: Level lines based disocclusion. In: Proceedings of the ICIP 1998, vol. 3, pp. 259–263 (1998)
5. Chan, T.F., Shen, J.: Mathematical models for local nontexture inpaintings. SIAM J. Appl. Math. 62, 1019–1043 (2001)
6. Chan, T.F., Shen, J.: Non-texture inpainting by curvature-driven diffusions. J. Visual Comm. Image Rep. 12, 436–449 (2001)
7. Telea, A.: An image inpainting technique based on the fast marching method. Journal of graphics tools 9(1), 25–36 (2003)
8. Bornemann, F., März, T.: Fast image inpainting based on coherence transport. Journal of Mathematical Imaging and Vision 28(3), 259–278 (2007)
9. Bertalmio, M., Bertozzi, A., Sapiro, G.: Navier-stokes, fluid dynamics, and image and video inpainting. In: IEEE Proceedings of the International Conference on Computer Vision and Pattern Recognition, vol. 1, pp. 355–362 (2001)

10. Chan, T., Shen, J.: Mathematical models for local non-texture inpaintings. SIAM J. Appl. Math. 62(3), 1019–1043 (2002)
11. Nitzberg, M., Shiota, T., Mumford, D.: Filtering, Segmentation and Depth. LNCS, vol. 662. Springer, Heidelberg (1993)
12. Ballester, C., Bertalmio, M., Caselles, V., Sapiro, G., Verdera, J.: Filling-in by joint interpolation of vector fields and gray levels. IEEE Transactions on Image Processing 10(8), 1200–1211 (2001)
13. Ambrosio, L., Masnou, S.: A direct variational approach to a problem arising in image reconstruction. Interfaces and Free Boundaries 5, 63–81 (2003)
14. Chan, T.F., Shen, J., Zhou, H.M.: Total variation wavelet inpainting. Journal of Mathematical Imaging and Vision 25(1), 107–125 (2006)
15. Patwardhan, K.A., Sapiro, G., Bertalmio, M.: Video inpainting of occluding and occluded objects. IMA Preprint Series 2016 (January 2005)
16. Matsushita, Y., Ofek, E., Weina, G., Tang, X., Shum, H.: Full-frame video stabilization with motion inpainting. IEEE Transactions on Pattern Analysis and Machine Intelligence 28(7), 1150–1163 (2006)
17. Chen, L., Chan, S., Shum, H.: A joint motion-image inpainting method for error concealment in video coding. In: IEEE International Conference on Image Processing, ICIP (2006)
18. Kondermann, C., Kondermann, D., Garbe, C.: Postprocessing of optical flows via surface measures and motion inpainting. In: Rigoll, G. (ed.) DAGM 2008. LNCS, vol. 5096, pp. 355–364. Springer, Heidelberg (2008)
19. Ambrosio, L., Fusco, N., Pallara, D.: Functions of bounded variation and free discontinuity problems. Oxford Mathematical Monographs. Oxford University Press, New York (2000)
20. Perona, P., Malik, J.: Scale-space and edge detection using anisotropic diffusion. IEEE Transactions on Pattern Analysis and Machine Intelligence 12(7), 629–639 (1990)
21. Nagel, H.H., Enkelmann, W.: An investigation of smoothness constraints for the estimation of displacement vector fields from image sequences. IEEE Trans. Pattern Anal. Mach. Intell. 8(5), 565–593 (1986)
22. Alvarez, L., Monreal, L.J.E., Lefebure, M., Perez, J.S.: A pde model for computing the optical flow. In: Proc. XVI Congreso de Ecuaciones Diferenciales Aplicaciones, Universidad de Las Palmas de Gran Canaria, September 1999, pp. 1349–1356 (1999)
23. Chan, T.F., Vese, L.A.: Active contours without edges. IEEE Transactions on Image Processing 10(2), 266–277 (2001)
24. Braess, D.: Finite Elemente, 2nd edn. Springer, Heidelberg (1997); Theorie, schnelle Löser und Anwendungen in der Elastizitätstheorie
25. Schaback, R., Werner, H.: Numerische Mathematik, 4th edn. Springer, Berlin (1992)
26. Kosmol, P.: Methoden zur numerischen Behandlung nichtlinearer Gleichungen und Optimierungsaufgaben. 2nd edn. Teubner, Stuttgart (1993)
27. Sundaramoorthi, G., Yezzi, A., Mennucci, A.: Sobolev active contours. International Journal of Computer Vision 73(3), 345–366 (2007)
28. Chan, T., Shen, J.: The role of the bv image model in image restoration. AMS Contemporary Mathematics (2002)
29. Chambolle, A.: An algorithm for total variation minimization and applications. Journal of Mathematical Imaging and Vision 20(1-2), 89–97 (2004)
30. Baker, S., Roth, S., Scharstein, D., Black, M., Lewis, J., Szeliski, R.: A database and evaluation methodology for optical flow. In: Proceedings of the International Conference on Computer Vision, pp. 1–8 (2007)

Color Image Segmentation in a Quaternion Framework

Özlem N. Subakan and Baba C. Vemuri*

Department of Computer and Information Science and Engineering,
University of Florida, Gainesville, FL, 32611, USA
{ons,vemuri}@cise.ufl.edu

Abstract. In this paper, we present a feature/detail preserving color image segmentation framework using Hamiltonian quaternions. First, we introduce a novel Quaternionic Gabor Filter (QGF) which can combine the color channels and the orientations in the image plane. Using the QGFs, we extract the local orientation information in the color images. Second, in order to model this derived orientation information, we propose a continuous mixture of appropriate hypercomplex exponential basis functions. We derive a closed form solution for this continuous mixture model. This analytic solution is in the form of a spatially varying kernel which, when convolved with the signed distance function of an evolving contour (placed in the color image), yields a detail preserving segmentation.

1 Introduction

A major turning point in the field of mathematics, specifically, in algebra, was the birth of noncommutative algebra via Hamilton's discovery of quaternions. This discovery was the precursor to new kinds of algebraic structures and has had an impact in various areas of mathematics and physics, including group theory, topology, quantum mechanics etc. More recently, quaternions have found use in computer graphics [1], navigation systems [2] and coding theory [3]. In computer graphics, quaternion representation of orientations facilitated computationally efficient and mathematically robust (such as avoiding the gimbal lock in Euler angle representation) applications. In image processing, quaternions have been used to represent color images [4,5]. This representation, together with the extension of the Fourier transform to hypercomplex numbers, has led to applications in color sensitive filtering [6], edge detection in color images [7,8] and cross correlation of color images [9]. The first definition of a hypercomplex Fourier transform was reported by Delsuc [10] in nuclear magnetic resonance. Later, different definitions for the quaternionic Fourier transform (QFT) have been introduced in [11] and [12] independently. Based on their definition of QFT, Bülow and Sommer generalized the concept of analytic signal to two dimensions and introduced quaternionic Gabor filters for use with scalar images [13]. They extended the Gabor filter by using two quaternion basis i and j to replace the single complex number i in the definition of the complex Gabor filter. However, they did not apply it to color images since their definition of QFT associates the imaginary units i and j to the local orientations in the image plane, which has no relationship to the color channels in a color image. Therefore, we follow an alternative definition for QFT proposed in [14] that utilizes simple

---

* This research was in part supported by the grant NIH EB007082.

formulae for the Fourier transform of complex-valued signals that can be computed efficiently. The use of this alternative QFT allows us to introduce a novel definition for the Quaternionic Gabor Filters that can be used to extract features from color images without conflicting interpretations being assigned to the hypercomplex units. An additional key contribution of the work presented here is that – we propose to model the derived orientation information (at a pixel) using a continuous mixture of exponential basis functions. Continuous mixture models have been presented in various contexts [15,16,17,18,19]. In this paper, we propose a continuous mixture model, where the mixing density is a Bingham density on the 3-dimensional sphere $\mathbb{S}^3 \subset \mathbb{R}^4$. To solve the continuous mixture integral in a closed form, we rewrite it using the matrix Fisher distribution on the manifold of special-orthogonal group. We use this closed form solution to construct a spatially-varying kernel for feature preserving segmentation of color images.

Color image segmentation is a relatively nascent area in computer vision. The literature on color image segmentation is not as extensive as that on gray-valued image segmentation. The key issue in color image segmentation is how to couple the information contained in the given color (red, green and blue) channels. Some published methods directly apply the existing gray level segmentation methods to each channel of a color image and then combine them in some way to obtain a final segmentation result. Chan *et al.* extend the Chan-Vese algorithm for scalar valued images to the vector valued case ([20]). In their work, in addition to the Mumford-Shah functional over the length of the contour, the minimization involves the sum of the fitting error over each color component. In the color snakes model, Sapiro extends the geodesic active contour model to the color images based on the idea of evolving the contour with a coupling term based on the eigenvalues of the Riemannian metric of the underlying manifold ([21]).

In this paper, we adopt the quaternion framework for representing color images since it offers scope to process color images holistically, rather than as separate color space components, and thereby handles the coupling between the color channels. Moreover, trichromatic theory of human color vision suggests vector mathematics as a natural tool to analyze color images. For a detailed discussion and motivation on quaternion representation of color images, we refer the reader to [7,9]. The key innovation of our work here is a holistic approach to color image segmentation using a quaternion framework to extract the local orientation and to model the derived information using a continuous mixture in the unit quaternion space. The proposed segmentation kernel does not use any prior information, and yet yields high quality results. Another contribution of this paper is a quaternion Gabor filter for the use with color images. We present our experimental results on some images drawn from the Berkeley Segmentation Data Set [22] along with F-measure plots for quantitative validation. We also compare our method with the mean shift algorithm in [23].

The remainder of this paper is structured as follows: We briefly describe the quaternion algebra and quaternion Fourier transform – needed for defining the QGF – in Section 2. We also develop a novel definition for QGFs in this section. In Section 3, we introduce the continuous mixture model for quantifying the derived orientation information. Then, in Section 4, we present the experimental results along with the quantitative evaluation depicting the merits of the proposed approach. Lastly, in Section 5 we summarize our contributions.

2 Local Orientation Analysis Using QGFs

2.1 Quaternions

In this section, we present background material on quaternions and the associated algebra which will be used in developing the local orientation analysis using QGFs.

Higher dimensional complex numbers are called *hypercomplex* and defined as

$$q = q_0 + \sum_{k=1}^{N} i_k q_k , \ q_k \in \mathbb{R} , \tag{1}$$

where i_k is orthonormal to i_l for $k \neq l$ in an N+1 dimensional space. The Hamiltonian quaternions are unitary $\mathbb{R}$-algebra; the basic algebraic form for a quaternion $q \in \mathbb{H}$ is:

$$q_0 + q_1 i + q_2 j + q_3 k , \tag{2}$$

where $q_0, q_1, q_2, q_3 \in \mathbb{R}$, the field of real numbers, and i, j, k are three imaginary numbers. $\mathbb{H}$ can be regarded as a 4-dimensional vector space over $\mathbb{R}$ with the natural definition of addition and scalar multiplication. The set $\{1, i, j, k\}$ is a natural basis for this vector space. $\mathbb{H}$ is made into a ring by the usual distributive law together with the following multiplication rules:

$$i^2 = j^2 = -1 , \ ij = -ji = k . \tag{3}$$

If we denote the scalar and vector parts of a quaternion q by Sq and Vq respectively, the product of two quaternions q and p can be written as

$$qp = SqSp - Vq \cdot Vp + SqVp + SpVq + Vq \times Vp , \tag{4}$$

where the $\cdot$ and $\times$ indicate the vector dot and cross products respectively. The conjugate of a quaternion, denoted by $*$, simply negates the vector part, $q^* = q_0 - q_1 i - q_2 j - q_3 k$. The norm of a quaternion q is $\|q\| = \sqrt{qq^*} = \sqrt{q^*q} = \sqrt{q_0^2 + q_1^2 + q_2^2 + q_3^2}$. A quaternion with unit norm is called *unit quaternion*. Hamilton called a quaternion with zero scalar part a *pure quaternion*. We can give an inner product structure to $\mathbb{H}$ if we define:

$$\langle q, p \rangle = Sqp^* . \tag{5}$$

Using the inner product, the angle α between two quaternions can be defined as:

$$\cos \alpha = \frac{Sqp^*}{\|q\| \|p\|} . \tag{6}$$

Any quaternion can be written in polar form

$$q = \|q\| e^{\theta \mu} = \|q\| (\cos \theta + \mu \sin \theta) , \tag{7}$$

where μ is a unit pure quaternion.

Quaternion representation of color image pixels has been proposed independently in [4,5]. They encode the color value of each pixel in a pure quaternion. For example, a pixel value at location (n, m) in an RGB image can be given as a quaternion-valued function $f(n, m) = R(n, m)i + G(n, m)j + B(n, m)k$ where R, G and B denote the red, green and blue components of each pixel respectively. This 3-component vector representation yields a system which has well-defined and well-behaved mathematical operations to apply on color images holistically.

2.2 Quaternionic Gabor Filters

In order to develop complex Gabor filters in higher-dimensional algebras, we first need to analyze corresponding generalization of the Fourier transform. The very first definition of a hypercomplex Fourier transform was due to Delsuc [10]. Later, Ell [11] and Bülow [12] independently introduced the quaternion Fourier transform, respectively as follows:

$$H(jw, kv) = \int_{-\infty}^{\infty} \int_{-\infty}^{\infty} e^{-jwt} h(t, \tau) e^{-kv\tau} dt d\tau \,. \tag{8}$$

$$F(u, v) = \int_{\mathbb{R}^2} e^{-i2\pi ux} f(x, y) e^{-j2\pi vy} dx dy \,. \tag{9}$$

In [14], another definition for QFT was proposed with the motivation of using a simple generalization of the standard complex operational formulae for convolution in color images:

$$F[u, v] = \frac{1}{\sqrt{MN}} \sum_{m=0}^{M-1} \sum_{n=0}^{N-1} e^{-\mu 2\pi (mv/M + nu/N)} f(n, m) \,, \tag{10}$$

where μ is a unit pure quaternion. For color images in RGB space, μ is chosen as $\frac{1}{\sqrt{3}}(i + j + k)$ (note that both the luminance and the chromaticity information is still preserved; this is still a full color image processing, not a grayscale image processing.).

Following the QFT definition above, we introduce a novel Quaternionic Gabor Filter.

Definition 1 (Quaternionic Gabor Filter). *The impulse response of a quaternionic Gabor filter is a Gaussian modulated with the basis functions of the QFT:*

$$G_{\mathbb{H}}(\mathbf{x}; \mathbf{u}, \sigma, \lambda, \theta) = g(x', y') \exp(\mu 2\pi (u_0 x + v_0 y)) \,, \tag{11}$$

where $g(x, y) = N \exp\left(-\frac{x^2 + \lambda y^2}{2\sigma^2}\right)$ with N being the normalization constant, λ being the aspect ratio.

$$\begin{bmatrix} x' \\ y' \end{bmatrix} = \begin{bmatrix} \cos\theta & \sin\theta \\ -\sin\theta & \cos\theta \end{bmatrix} \begin{bmatrix} x \\ y \end{bmatrix}$$

The center frequency of the QGF is given by $\sqrt{u_0^2 + v_0^2}$ and its orientation is $\theta = \arctan(v_0/u_0)$.

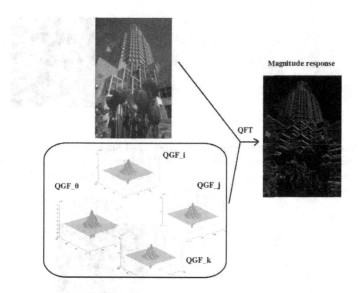

Fig. 1. Quaternion convolution of a QGF (with an orientation of π) with a color image from Berkeley Data Set ([22])

For an application of QGFs, consider the Fig. 1. If we apply a horizontally oriented QGF to an image, then we obtain high responses wherever there are horizontally oriented features. Fig. 1 illustrates the magnitude response of such a horizontally oriented QGF convolved with an image in quaternion form. Quaternion convolution is equivalently performed by using QFT. Note that all the calculations follow the rules of the quaternion algebra.

In an image, it is possible to have a color contrast without having a luminance contrast. In a black-and-white version of such an image, the two different colored objects appear blended into a single one. In Fig. 2, we demonstrate that the proposed Quaternionic Gabor Filters can extract the local orientation information from a constant luminance image as well. Fig. 2a shows a synthetic color image where all pixels have the same luminance value, but the chromaticity inside the object differs from the chromaticity outside. The luminance channel shows that all pixels have the same value (see Fig. 2b). We applied 10 QGFs to the quaternion representation of this color image. The sum of the magnitude responses of 10 QGFs is shown in Fig. 2d. Although a black-and-white version (Fig. 2c) of the input image is a uniform gray without any changes in orientation, the proposed QGFs successfully derive the orientation information in the color version, showing that they are well suited for analyzing color images and the result is not a grayscale image processing.

We have chosen the unit pure quaternion direction μ in QGF as $\frac{1}{\sqrt{3}}(i + j + k)$. However, this choice does not mean that the proposed quaternion framework is processing the sum of the RGB values. Also note that the convolution between a QGF and a quaternion representation of a color image is performed following the rules of quaternion algebra. At each pixel, the quaternion-valued filter is multiplied with the color

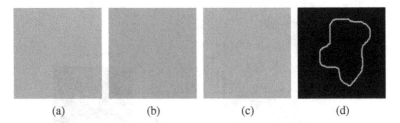

(a)   (b)   (c)   (d)

Fig. 2. Application of Quaternionic Gabor Filters across equal luminance: (a) a synthetic color image where the object and the background are of equal luminance, (b) luminance channel, (c) a grayscale version of (a), (d) the sum of the magnitude responses of QGFs applied to the color image in (a)

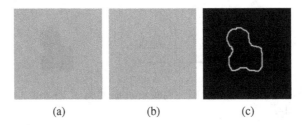

(a)   (b)   (c)

Fig. 3. (a) A synthetic color image where $(R + G + B)/3$ is the same everywhere. (b) Grayscale image which shows $(R + G + B)/3$ values. (c) The sum of the magnitude responses of QGFs applied to the color image in (a).

direction of that pixel through a quaternion product. Hence, QGF handles the coupling between the channels while, at the same time, processing all information in a color image. Fig. 3a shows a color image where $(R + G + B)/3$ is the same for all pixels. As shown in Fig. 3c, the proposed framework can accurately extract the orientation information.

In the next section, the magnitudes of the quaternion-valued filter responses are modeled using a continuous mixture of quaternionic exponential basis functions. We derive a closed form solution for this integral and use it to construct a spatially varying convolution kernel for detail preserving color image segmentation.

3   A Continuous Mixture Model on the Unit Quaternion Space

In the previous section, we introduced the QGFs to extract the local orientation information in a color image. The resulting responses over a sphere of directions are modeled in this section in a probabilistic framework. This framework is powerful and allows one to capture the complicated local geometries present in the image data and incorporate them into spatially varying segmentation kernels. We postulate that at each lattice point there is an underlying probability measure induced on the manifold of the unit quaternions.

The space of unit quaternions

$$\mathbb{S}^3 = \{q \in \mathbb{H} \mid \|q\| = 1\} \tag{12}$$

is the 3-sphere in $\mathbb{H}$, it forms a group under multiplication and preserves the hermitian inner product. An appropriate choice for the kernel functions is $\exp(-\cos(d(q,p)))$, where $d(q,p) = 2\cos^{-1}(Sq^*p)$ is the length of the shortest geodesic between quaternions q and p. It can also be called the *angle of rotation* metric for quaternions. Thus the proposed model is given by,

$$\|G_\mathbb{H}(\mathbf{x}; \cdot, \theta)\|/G_\mathbb{H}^{max} = \int_{\mathbb{S}^3} f(q)e^{-\cos(d(q,p_\theta))}\,dq, \tag{13}$$

where $dF := f(q)dq$ denotes the underlying probability measure with respect to the uniform distribution dq on $\mathbb{S}^3$. $G_\mathbb{H}^{max}$ is the maximal value among all responses at an image location. In order to avoid an ill-posed inverse problem which requires recovering a distribution defined on the manifold of unit quaternions given the measurements $G_\mathbb{H}(\mathbf{x}; \cdot, \theta)$, we impose a mixture of Bingham distributions on q as a prior. Manifold of the unit quaternions double-covers $SO(3)$. Double-coverage can be interpreted as antipodal-symmetry; thus, Bingham distribution is a natural choice for quaternion priors. For statistical purposes, Bingham distribution is characterized as the hyperspherical analogue of the n-variate normal distribution; essentially it can be obtained by the "intersection" of a zero-mean normal density with the unit sphere in $\mathbb{R}^n$. Let q be a 4-dimensional random unsigned unit direction. q is distributed as $\mathcal{B}_{L,A}$ if it has the Bingham density [24] given by,

$$_1F_1(1/2, 2, L)^{-1}\exp\{\operatorname{tr} LAqq^T A^T\}dq, \tag{14}$$

where A is a 4×4 rotation matrix, L is a diagonal matrix with concentration values (which determine the amount of clustering around the mean directions) and $_1F_1$ is a *confluent hypergeometric function of matrix argument* as defined in [25]. Using the relationship between $\mathbb{S}^3$ and $SO(3)$, Prentice [24] has shown that q has a Bingham density if and only if the corresponding rotation matrix, Q, in $SO(3)$ has a matrix Fisher distribution. A random 3×3 rotation matrix Q is said to have a matrix Fisher distribution $\mathcal{F}_F$ if it has the following pdf:

$$_0F_1(3/2; FF^T/4)^{-1}\exp\{\operatorname{tr} F^TQ\}dQ. \tag{15}$$

F is a 3×3 parameter matrix which encapsulates the concentration values and orientations, $_0F_1$ is a hypergeometric function of matrix argument and can be evaluated using zonal polynomials. By using the distance on the manifold $SO(3)$, the proposed model can be equivalently written in $SO(3)$ instead of $\mathbb{S}^3$ as:

$$\|G_\mathbb{H}(\mathbf{x}; \cdot, P)\|/G_\mathbb{H}^{max} = \int_{SO(3)} e^{-\frac{\operatorname{tr} P^TQ-1}{2}}\,dF, \tag{16}$$

where P is the rotation matrix corresponding to the orientation of the QGF, and

$$dF = \sum_{i=1}^N w_i\,_0F_1(3/2; F_iF_i^T/4)^{-1}e^{\operatorname{tr} F_i^TQ}dQ \tag{17}$$

is a discrete mixture of matrix Fisher densities over the rotation matrix Q with respect to the uniform distribution on $SO(3)$. We choose to change the prior to this mixture of matrix Fisher densities since the matrix Fisher density is unimodal and will not be able to handle orientational heterogeneity. However, note that the model in (16) is still a continuous mixture model. N here corresponds to the resolution of the $SO(3)$ discretization and not the number of dominant local orientations. We observed that the kernel of the matrix Fisher distribution can be utilized to derive a closed form solution for the right-hand side, leading to:

$$\sum_{i=1}^{N} w_i \frac{{}_0F_1\left(\frac{3}{2};\frac{1}{4}\left[\mathsf{F}_i - \frac{P}{2}\right]\left[\mathsf{F}_i - \frac{P}{2}\right]^T\right)}{{}_0F_1(3/2;\mathsf{F}_i\mathsf{F}_i^T/4)}. \tag{18}$$

We can formulate the computation of this analytic form as the solution to a linear system $\mathbf{A}\mathbf{w} = \mathbf{y}$, where $\mathbf{y} = \{\|G_{\mathbb{H}}(\mathbf{x};\cdot,\theta_j)\|\}_{j=1}^{M}/G_{\mathbb{H}}^{max}$ contains the normalized measurements obtained via an application of M QGFs to the color image, $\mathbf{A}$ is an $M \times N$ matrix with

$$A_{ji} = \frac{{}_0F_1\left(\frac{3}{2};\frac{1}{4}\left[\mathsf{F}_i - \frac{P_j}{2}\right]\left[\mathsf{F}_i - \frac{P_j}{2}\right]^T\right)}{{}_0F_1(3/2;\mathsf{F}_i\mathsf{F}_i^T/4)} \tag{19}$$

and $\mathbf{w} = (w_i)$ is the unknown weight vector. The weights in the mixture can be solved using a sparse deconvolution technique, a non-negative least squares (NNLS) minimization which yields an accurate and sparse solution. A sparse solution is what is expected at each image lattice point since local image geometry does not have a large number of edges meeting at a junction. Once $\mathbf{w}$ is estimated for the given data at each lattice point, we can construct the convolution kernel for color image segmentation. We represent an evolving curve C (in a curve evolution framework) by the zero level set of a Lipschitz continuous function $\phi\colon \Omega \to \mathbb{R}$. So, $C = \{(x,y) \in \Omega\colon \phi(x,y) = 0\}$. We choose ϕ to be negative inside C and positive outside. C is evolved using the following update equation:

$$\phi_{t+1}(\mathbf{x}) = \phi_t(\mathbf{x}) * Q(\mathbf{x}), \tag{20}$$

where $Q(\mathbf{x})$ is the convolution kernel obtained from (18) by setting the matrix P to the rotation matrix corresponding to the angle that the coordinate vector $\mathbf{x}$ makes with the x-axis. Note that this formulation yields a spatially varying convolution kernel since the $\mathbf{w}$ vector is estimated at each lattice point in an image.

4   Experiments and Comparisons

In this section, we present several experimental results of our approach (named as QGmF – **Q**uaternionic **G**abors with **m**atrix **F**isher density) and compare its performance with a state-of-the-art technique in segmentation: the mean shift algorithm presented in [23]. We compare with this algorithm since it presents a tool for feature space analysis. In the following experiments, for each algorithm the segmentations that yield the highest F-measure values are shown.

In QGmF, we can adjust the level of details/features, which reveal themselves in the output of the QGF applied to the color images. To do this, we introduce a threshold

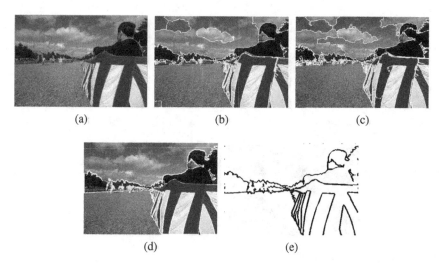

(a)   (b)   (c)

(d)   (e)

Fig. 4. (a) Segmentation performed by a human subject (from the ground truth in the Berkeley Segmentation Data Set [22]). (b) Segmentation result of the mean shift algorithm. (c) Segmentation result of the QGmF method with a low threshold value of 0.005. (d) Segmentation result of the QGmF method with a threshold value of 0.02. (e) True positives (TP) map of (d) with respect to (a).

(a)   (b)

(c)   (d)

Fig. 5. (a) Human segmentation (from the ground truth in the Berkeley Segmentation Data Set). (b) Output of the mean shift algorithm. (c) Output of the QGmF method with a threshold of 0.005. (d) Output of the QGmF method with a threshold of 0.025.

parameter on the magnitude of the filter responses. A relatively low threshold results in a segmentation capturing the low contrast details in small scales. Fig. 4c illustrates such an example where the threshold is set to 0.005. Mean shift algorithm achieves a

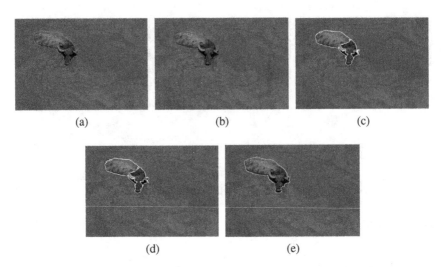

(a)   (b)   (c)

(d)   (e)

Fig. 6. (a) Human segmentation (from the ground truth data). (b) Luminance channel of the color image. (c) Output of the QGmF method for the color image (QGF threshold = 0.005). (d) Output of the mean shift segmentation (e) Output of the QGmF method (QGF threshold = 0.025).

Table 1. F_1-measure (or Dice's Coefficient) values

Image	QGmF	MeanShift
Astronauts	0.74	0.56
Starfish	0.81	0.52
Parade	0.76	0.65
Buffalo	0.86	0.67

successful result as shown in Fig. 4b. However, uniform regions are not consistently preserved, *e.g.* the sky is mis-segmented; the boundaries divide the regions which are actually composed of connected components, as can be seen between the clouds. Moreover, the barricade is mis-segmented with the pavement. Fig. 4d shows a better segmentation using our QGmF method (note that the man riding the horse and the crowd are clearly segmented, also note the accurate localization of the boundary between the barricade and the pavement). Fig. 4e shows the pixels correctly labeled as belonging to the segmentation boundary by QGmF.

Another visual comparison is provided in Fig. 5. Since the mode detection calculations in mean shift algorithm are determined by global bandwidth parameters, the algorithm tends to miss small-scale details in some places or over-segment the uniform regions (see the small areas on the starfish which are mis-segmented as being a part of the outer region in Fig. 5b). On the other hand, QGmF maintains coherence within textured regions while preserving the small scale details around the boundaries as shown in Fig. 5d. Once again, a low threshold value results in over-segmentation (see Fig. 5c).

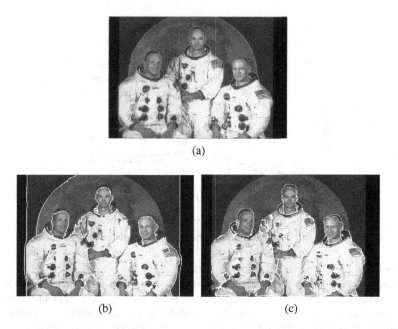

(a)

(b)   (c)

Fig. 7. (a) Segmentation performed by a human subject (from the ground truth in the Berkeley Segmentation Data Set). (b) Output of the mean shift segmentation. (c) Output of the QGmF method.

In Fig. 6b, note the regions which have almost equal luminance but different chromaticity. Both Fig. 6c and 6d are over-segmented; however, 6e shows a high quality result which is very close to the human segmentation (see Figs. 6e and 6a). In Fig. 7b, mean shift segmentation algorithm mis-segments the heads of the astronauts, and the boundaries of the astronaut on the left are missed. As visually evident, QGmF performs better than the competing method.

In order to have a quantitative evaluation of our approach, we present the highest F_1-measure (or Dice's Coefficient) scores of our method and the competing method for the above images, as shown in Table 1. Furthermore, in Fig. 8 we present a sensitivity analysis using the F_1-measures on 100 color images (including the images above) drawn from the Berkeley Segmentation Data Set [22]. F_1-measure, commonly known as the F-measure, is the evenly weighted harmonic mean of precision and recall scores. The human segmentations from the Berkeley Segmentation Data Set were used as the ground truth in the evaluation. The boundaries between two segmentations are matched by examining a neighborhood within a radius of $\epsilon = 2$. In the QGmF, we tested the effect of the threshold parameter (for values in $[0.005, 0.05)$) on the QGF responses. For the mean shift segmentation algorithm, we tested the effect of the kernel bandwidth parameters: h_s, space bandwidth; and h_r, range bandwidth. They determine the resolution of the mode selection and the clustering. We tested for 3 different h_s values in $[7, 10, 20]$. In each curve for the mean shift algorithm, x-axis shows the variations of the h_r values in $[4, 20]$ arranged in ascending order from left to right. Experimentation showed that the F-measure scores change significantly with respect to the

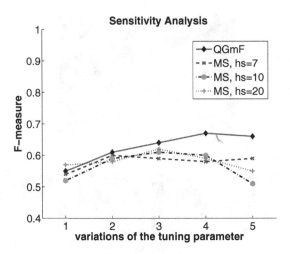

Fig. 8. F-measure plots for mean shift segmentation algorithm and QGmF convolution-based kernel method. For QGmF, x-axis shows the variations of the threshold parameter for QGF responses, arranged in order from left to right, while y-axis shows the corresponding F-measure value. The threshold for QGF varies within $[0.005, 0.05]$. For mean shift segmentation algorithm, the corresponding values for the space bandwidth parameter (hs) are shown in the plot, points along each curve correspond to the variations of the range bandwidth parameter (hr) in $[4, 20]$.

bandwidth parameters in mean shift segmentation algorithm, making it difficult to choose the range of the parameters which can provide good results. In QGmF, we observed that a low threshold value for QGF results in over-segmentation which is characterized in the curves by low F-measure, whereas any level of detail for segmentation can be achieved by tuning the threshold parameter. We notice that the scores of QGmF are higher than the competing method.

5 Conclusion

In this paper, we addressed the problem of feature/detail preserving segmentation in color images, and presented a hypercomplex representation framework for capturing the complicated local geometry contained in a color image via the use of a spatially varying convolution filter. We introduced a novel quaternionic Gabor filter to extract the local orientation information in color images. This information is then represented by a continuous mixture of hypercomplex exponential basis functions, where the mixing density is assumed to be a mixture of Bingham densities. This integral, when expressed using matrix Fisher densities on the Stiefel manifold, can be solved in a closed form leading to the QGmF kernel. Additionally, the same kernel when iteratively applied to a signed distance function representation of an active contour yields feature/detail preserving segmentations of the color images. Our method does not use any prior information to perform segmentations, and yet delivers superior performance in comparison to a state-of-the-art method. We validated the performance of the proposed method in the experimental section using color images from the Berkeley Segmentation Data Set and showed that our model yields results significantly close to the segmentations performed by human subjects.

References

1. Shoemake, K.: Animating rotation with quaternion curves. SIGGRAPH Comput. Graph. 19(3), 245–254 (1985)
2. Kuipers, J.B.: Quaternions and Rotation Sequences: A Primer with Applications to Orbits, Aerospace and Virtual Reality. Princeton University Press, Princeton (2002)
3. Sethuraman, B.A., Rajan, B.S., Member, S., Shashidhar, V.: Full-diversity, highrate space-time block codes from division algebras. IEEE Trans. Inform. Theory 49, 2596–2616 (2003)
4. Pei, S.C., Cheng, C.M.: A novel block truncation coding of color images by using quaternion-moment-preserving principle. In: IEEE International Symposium on Circuits and Systems, ISCAS 1996., 'Connecting the World', May 1996, vol. 2, pp. 684–687 (1996)
5. Sangwine, S.: Fourier transforms of colour images using quaternion or hypercomplex numbers. Electronic Lett. 32(21), 1979–1980 (1996)
6. Sangwine, S., Ell, T.: Colour image filters based on hypercomplex convolution. IEE Proceedings Vision, Image and Signal Processing 147(2), 89–93 (2000)
7. Ell, T., Sangwine, S.: Hypercomplex fourier transforms of color images. IEEE Transactions on Image Processing 16(1), 22–35 (2007)
8. Sangwine, S.: Colour image edge detector based on quaternion convolution. Electronic Lett. 34(10), 969–971 (1998)
9. Moxey, C., Sangwine, S., Ell, T.: Hypercomplex correlation techniques for vector images. IEEE Transactions on Signal Processing 51(7), 1941–(1953)
10. Delsuc, M.A.: Spectral representation of 2d nmr spectra by hypercomplex numbers. Journal of Magnetic Resonance 77(1), 119–124 (1988)
11. Ell, T.: Quaternion-fourier transforms for analysis of two-dimensional linear time-invariant partial differential systems. In: Proceedings of the 32nd IEEE Conference on Decision and Control, December 1993, vol. 2, pp. 1830–1841 (1993)
12. Bülow, T., Sommer, G.: Das konzept einer zweidimensionalen phase unter verwendung einer algebraisch erweiterten signalrepräsentation. In: Mustererkennung 1997, 19. DAGM-Symposium, London, UK, pp. 351–358. Springer, Heidelberg (1997)
13. Bülow, T., Sommer, G.: Multi-dimensional signal processing using an algebraically extended signal representation. In: Sommer, G. (ed.) AFPAC 1997. LNCS, vol. 1315, pp. 148–163. Springer, Heidelberg (1997)
14. Sangwine, S., Ell, T.A.: The discrete fourier transform of a colour image. In: Blackledge, J.M., Turner, M.J. (eds.) Image Processing II: Mathematical Methods, Algorithms and Applications, pp. 430–441 (2000)
15. Jian, B., Vemuri, B.: Multi-fiber reconstruction from diffusion mri using mixture of wisharts and sparse deconvolution. Inf. Process Med. Imaging 20 (2007)
16. Jian, B., Vemuri, B.C.: A unified computational framework for deconvolution to reconstruct multiple fibers from diffusion weighted mri. IEEE Trans. Med. Imaging 26(11), 1464–1471 (2007)
17. Jian, B., Vemuri, B.C., Özarslan, E., Carney, P., Mareci, T.: A novel tensor distribution model for the diffusion-weighted mr signal. Neuroimage 37(1), 164–176 (2007)
18. Subakan, Ö.N., Jian, B., Vemuri, B.C., Vallejos, C.E.: Feature preserving image smoothing using a continuous mixture of tensors. In: IEEE International Conference on Computer Vision, Rio de Janeiro, Brazil (October 2007)

19. Subakan, Ö.N., Vemuri, B.C.: Image segmentation via convolution of a level-set function with a Rigaut kernel. In: IEEE Conference on Computer Vision and Pattern Recognition, Anchorage, Alaska (June 2008)
20. Chan, T.F., Yezrielev, B., Vese, L.A.: Active contours without edges for vector-valued images. Journal of Visual Communication and Image Representation 11, 130–141 (2000)
21. Sapiro, G.: Color snakes. Comput. Vis. Image Underst. 68(2), 247–253 (1997)
22. Martin, D., Fowlkes, C., Tal, D., Malik, J.: A database of human segmented natural images and its application to evaluating segmentation algorithms and measuring ecological statistics. In: IEEE Intl. Conf. on Computer Vision, July 2001, vol. 2, pp. 416–423 (2001)
23. Comaniciu, D., Meer, P.: Mean shift: A robust approach toward feature space analysis. IEEE Transactions on Pattern Analysis and Machine Intelligence 24(5), 603–619 (2002)
24. Prentice, M.J.: Orientation statistics without parametric assumptions. Journal of the Royal Statistical Society. Series B (Methodological) 48(2), 214–222 (1986)
25. Herz, C.S.: Bessel functions of matrix argument. The Annals of Mathematics 61(3), 474–523 (1955)

Quaternion-Based Color Image Smoothing Using a Spatially Varying Kernel

Özlem N. Subakan and Baba C. Vemuri[*]

Department of Computer and Information Science and Engineering,
University of Florida, Gainesville, FL, 32611, USA
{ons,vemuri}@cise.ufl.edu

Abstract. Addressing the issue of feature/detail preserving color image smoothing, we propose a novel unified approach based on a quaternion framework. The main idea is to holistically extract the local orientation information at each lattice point, and then to incorporate it into the smoothing process. We introduce a new Quaternion Gabor Filter to derive the local orientation information in color images. This derived orientation information is modeled using a continuous mixture of appropriate exponential basis functions. We solve the continuous mixture integral in analytic form, and develop a spatially varying kernel which respects to the local geometry at each lattice point in a color image. Superior performance of our smoothing framework is demonstrated via comparison to competing state-of-the-art algorithms in literature.

1 Introduction

Color conveys essential information which can be employed in many vision tasks including but not limited to object recognition, tracking, segmentation, registration etc. With the advances in the computing power and memory, color image processing has attracted much interest over the past few years. In this subject area, color image denoising is still an elusive challenge. Due to the multichannel nature of the color images, the key issue is how to couple the information contained in the given color (*e.g.* red, green and blue) channels. Considering each individual channel of a color image as a separate monochrome image, the early approaches often comprise the component-wise application of the traditional gray level denoising techniques on each channel separately. However, this approach fails to notice the inherent correlation between the components and results in color artifacts or blending. To avoid this, denoising process should be performed in a common and coherent way. In order to restore color and other vector valued images, Blomgren and Chan [1] proposed to minimize a measure of *Color Total Variation* which is still similar to a channel by channel *Total Variation* diffusion, but weighted by a coupling term. To retrieve the local geometry of vector valued images, Weickert proposed to extend his coherence enhancing diffusion using a common diffusion tensor for all image channels [2]. Later, Kimmel *et al.* introduced a diffusion PDE called *Beltrami flow* which involves the minimization of the global area of the surface representing the vector valued image [3], with respect to the surface metric. In [4], Tang

---

[*] This research was in part supported by the grant NIH EB007082.

D. Cremers et al. (Eds.): EMMCVPR 2009, LNCS 5681, pp. 415–428, 2009.

et al. extended their direction diffusion framework to smoothing only the chromaticity channel of color images, and combined it with the scalar anisotropic diffusion applied to the brightness channel of the color image. More recently, Tschumperlé introduced an image regularization PDE which takes the curvature constraints into account, and applied it to multi-valued images [5]. For more on multichannel image recovery, we refer the reader to [6,7].

In this paper, we adopt the quaternion framework for smoothing color images since it offers scope to process color images holistically, rather than as separate color space components, and thereby handles the coupling between the color channels naturally. Moreover, the trichromatic theory of the human color vision suggests vector mathematics as a natural tool to analyze color images. For a detailed discussion and motivation on the quaternion representation of color images, we refer the reader to [8,9]. The key innovation of our work here is a unified approach to color image restoration using a quaternion framework to extract the local orientation and to model the derived information using a continuous mixture on the unit sphere. Continuous mixture models have been presented in various contexts [10,11,12,13,14]. Another contribution of this paper is a quaternion Gabor filter (QGF) for the use with color images.

Since their discovery by Hamilton in 1843, quaternions have had a tremendous amount of influence on various areas of mathematics and physics, including group theory, topology, quantum mechanics etc. More recently, quaternions have been employed in bioinformatics, computer graphics [15], navigation systems [16] and coding theory [17]. In computer graphics, the quaternion representation of orientations facilitated computationally efficient and mathematically robust (such as avoiding the gimbal lock in Euler angle representation) applications. In image processing, quaternions have been used to represent color images [18,19]. This representation, together with the extension of the Fourier transform to hypercomplex numbers, has led to applications in color sensitive filtering [20], edge detection [8,21] and cross correlation of color images [9]. The very first definition of a hypercomplex Fourier transform was that of Delsuc [22] in nuclear magnetic resonance. Later, different definitions for the quaternionic Fourier transform (QFT) have been introduced in [23] and [24] independently. Based on their definition of QFT, Bülow and Sommer generalized the concept of analytic signal to two dimensions and introduced quaternionic Gabor filters for use with scalar images [25]. They extended the Gabor filter by using two quaternion basis i and j to replace the single complex number i in the definition of the complex Gabor filter. However, they did not consider an application to color images since their definition of QFT associates the imaginary units i and j to the local orientations in the image plane, which has no relationship to the color channels in a color image. In [26], an alternative definition for QFT was proposed, which utilizes simple formulae for the Fourier transform of complex-valued signals that can be computed efficiently. We follow this alternative QFT to introduce a novel definition for the Quaternionic Gabor Filters which can be employed to extract features from color images without conflicting interpretations being assigned to the hypercomplex units. We further test QGFs for the optimality with respect to the two-dimensional uncertainty principle.

The rest of this paper is organized as follows: We briefly describe the quaternion algebra and quaternion Fourier transform in Section 2 and then present a novel definition for QGFs. In Section 3, we introduce a continuous mixture model for quantifying the derived orientation information. Section 4 reports on the experimental results along with a quantitative evaluation depicting the merits of the proposed approach, while a summary and an outlook for future research in Section 5 conclude the paper.

2 Local Orientation Analysis Using QGFs

2.1 Quaternions

Hypercomplex numbers are higher dimensional complex numbers defined as

$$q = q_0 + \sum_{k=1}^{N} i_k q_k \, , \ q_k \in \mathbb{R} \, , \tag{1}$$

where i_k is orthonormal to i_l for $k \neq l$ in an N+1 dimensional space. The Hamiltonian quaternions are elements with four orthogonal components; the basic algebraic form for a quaternion $q \in \mathbb{H}$ is:

$$q_0 + q_1 i + q_2 j + q_3 k \, , \tag{2}$$

where q_0, q_1, q_2, $q_3 \in \mathbb{R}$, the field of real numbers, and i, j, k are three complex operators obeying the following rules:

$$i^2 = j^2 = -1 \, , \ ij = -ji = k \, . \tag{3}$$

$\mathbb{H}$ can be regarded as a 4-dimensional vector space over $\mathbb{R}$ with the natural definition of addition and scalar multiplication. $\mathbb{H}$ is made into a ring by the usual distributive law together with the multiplication rules above.

Denoting the scalar and vector parts of a quaternion q by $Sq = q_0$ and $Vq = q_1 i + q_2 j + q_3 k$ respectively, we can give the product of two quaternions q and p as

$$qp = SqSp - Vq \cdot Vp + SqVp + SpVq + Vq \times Vp \, , \tag{4}$$

where the $\cdot$ and $\times$ indicate the usual $3D$ scalar and vector cross products respectively. For any quaternion q, there exists a conjugate quaternion, $q^* = q_0 - q_1 i - q_2 j - q_3 k$. The norm of a quaternion q is $\|q\| = \sqrt{qq^*} = \sqrt{q^*q} = \sqrt{q_0^2 + q_1^2 + q_2^2 + q_3^2}$. A quaternion with a unit norm is called *unit quaternion*, whereas a quaternion with a zero scalar part is called a *pure quaternion*.

Euler's formula for the complex exponential can be generalized to hypercomplex, yielding a polar form:

$$q = \|q\| e^{\theta \mu} = \|q\| (\cos \theta + \mu \sin \theta) \, , \tag{5}$$

where μ is a unit pure quaternion.

Quaternion representation of color image pixels was proposed independently in [18,19]. They encode the color value of each pixel in a pure quaternion. For example, a pixel value at location (n, m) in an RGB image can be given as a quaternion-valued function $f(n, m) = R(n, m)i + G(n, m)j + B(n, m)k$ where R, G and B denote the red, green and blue components of each pixel respectively. This 3-component vector representation yields a system which has well-defined and well-behaved mathematical operations to apply on color images holistically.

2.2   Quaternionic Gabor Filters

In order to develop our complex Gabor filters in higher-dimensional algebras, we follow the QFT definition in [26], which was proposed with the motivation of using a simple generalization of the standard complex operational formulae for convolution in color images:

$$F[u,v] = \frac{1}{\sqrt{MN}} \sum_{m=0}^{M-1} \sum_{n=0}^{N-1} e^{-\mu 2\pi(mv/M + nu/N)} f(n,m), \qquad (6)$$

where μ is a unit pure quaternion. For color images in RGB space, μ is chosen as $\frac{1}{\sqrt{3}}(i + j + k)$ (note that both the luminance and the chromaticity information is still preserved; this is still a full color image processing, not a grayscale image processing.). Complex Fourier transform is a special case of this transform, where $\mu = i$, and $f(n,m)$ is a complex valued function.

In the following, we introduce a novel Quaternionic Gabor Filter.

Definition 1 (Quaternionic Gabor Filter). *The impulse response of a quaternionic Gabor filter is a Gaussian modulated with the basis functions of the QFT:*

$$G_{\mathbb{H}}(\mathbf{x}; \mathbf{u}, \sigma, \lambda, \alpha) = g(x', y') \exp(\mu 2\pi(u_0 x + v_0 y)), \qquad (7)$$

where $g(x,y) = N \exp\left(-\frac{x^2 + \lambda y^2}{2\sigma^2}\right)$ *with N being the normalization constant, λ being the aspect ratio.*

$$\begin{bmatrix} x' \\ y' \end{bmatrix} = \begin{bmatrix} \cos\alpha & \sin\alpha \\ -\sin\alpha & \cos\alpha \end{bmatrix} \begin{bmatrix} x \\ y \end{bmatrix}$$

The center frequency of the QGF is given by $\sqrt{u_0^2 + v_0^2}$ and its orientation is $\alpha = \arctan(v_0/u_0)$. Let us consider the QFT of an isotropic Gaussian in $2D$. QFT of an anisotropic Gaussian can be evaluated similarly.

$$QFT\{g(x,y)\} = N \int_{\mathbb{R}^2} e^{-\frac{x^2 + y^2}{2\sigma^2}} e^{-\mu 2\pi(ux + vy)} d\mathbf{x}$$

$$= N \int_{\mathbb{R}} \left(\int_{\mathbb{R}} e^{-(x + 2\pi\mu u\sigma^2)/2\sigma^2} dx \right) e^{-\frac{y^2}{2\sigma^2}} e^{\mu^2 2\pi^2 \sigma^2 u^2} e^{-\mu 2\pi vy} dy \qquad (8)$$

After some algebraic manipulations, we obtain that $QFT\{g(x,y)\} = c e^{-2\pi^2\sigma^2(u^2 + v^2)}$, i.e. an un-normalized Gaussian in (u,v)-space, with c being a constant. Now we prove the Modulation Theorem for the continuous QFT.

Theorem 1 (Modulation Theorem for QFT). *Let f(x,y) be a quaternion-valued signal, $F_{\mathbb{H}}(u,v)$ be its quaternion Fourier transform, and $h(x,y) = f(x,y)e^{\mu 2\pi(u_0 x + v_0 y)}$. Then, $QFT\{h(x,y)\} = F_{\mathbb{H}}(u - u_0, v - v_0)$.*

Proof.

$$QFT\{f(x,y)\} = \int_{\mathbb{R}^2} f(x,y)e^{-\mu 2\pi(ux + vy)} d\mathbf{x} =: F_{\mathbb{H}}(u,v)$$

$$QFT\{h(x,y)\} = \int_{\mathbb{R}^2} f(x,y)e^{\mu 2\pi(u_0 x + v_0 y)} e^{-\mu 2\pi(ux + vy)} d\mathbf{x}$$

$$= F_{\mathbb{H}}(u - u_0, v - v_0)$$

The QFT of a Gaussian together with the Modulation Theorem can then be used to conclude that QGFs defined above are shifted Gaussian functions in the quaternionic frequency domain, i.e. if

$$f(\mathbf{x}) = e^{-\frac{x^2}{2\sigma_x{}^2} - \frac{y^2}{2\sigma_y{}^2}} e^{\mu 2\pi(u_0 x + v_0 y)} , \tag{9}$$

then the QFT of f is:

$$F_{\mathbb{H}}(\mathbf{u}) = e^{-2\pi^2 \sigma_x^2 (u - u_0)^2 - 2\pi^2 \sigma_y^2 (v - v_0)^2} . \tag{10}$$

In analogy to Gabor filters, we consider the quaternionic analytic signal which has been defined in [25] to work with QGFs. For positive frequencies u_0 and v_0, the main amount of the Gabor filter's energy in (10) is in the upper right quadrant. Hence, QGFs provide approximation to quaternionic analytic signal. In order to show that QGFs are optimally localized in both quaternionic spatial and frequency domains simultaneously, we will simply extend the definition of the uncertainties for quaternion-valued functions which has also been done in [27]. The *spatial and frequency uncertainties* Δx and Δu of a quaternion-valued signal f can be given as:

$$(\Delta x)^2 = \frac{\int_{\mathbb{R}} f(\mathbf{x}) f^*(\mathbf{x}) x^2 d\mathbf{x}}{\int_{\mathbb{R}} f(\mathbf{x}) f^*(\mathbf{x}) d\mathbf{x}} , \qquad (\Delta u)^2 = \frac{\int_{\mathbb{R}} F_{\mathbb{H}}(\mathbf{u}) F_{\mathbb{H}}^*(\mathbf{u}) u^2 d\mathbf{u}}{\int_{\mathbb{R}} F_{\mathbb{H}}(\mathbf{u}) F_{\mathbb{H}}^*(\mathbf{u}) d\mathbf{u}} . \tag{11}$$

The uncertainties of the QGF given in (9) can be evaluated using the above definitions and their analogs for Δy and Δv to be

$$\Delta x = \frac{\sigma_x}{\sqrt{2}} , \; \Delta y = \frac{\sigma_y}{\sqrt{2}} , \; \Delta u = \frac{1}{2\sqrt{2}\sigma_x \pi} , \; \Delta v = \frac{1}{2\sqrt{2}\sigma_y \pi} . \tag{12}$$

Thus, QGFs are shown to achieve the minimum product of uncertainties defined in [28]

$$\Delta x \Delta y \Delta u \Delta v \geq 1/16\pi^2 . \tag{13}$$

For an application of QGFs, consider the Fig. 1. We applied 13 oriented QGFs to the image of Barbara, by convolving the quaternion representation of the color image with the quaternion valued filter. Quaternion convolution is equivalently performed by using QFT. Calculations follow the rules of the quaternion algebra. Note that color transitions in the coupled channels GB, RB and RG show themselves in the components of the vector part of the QGF responses.

In an image, it is possible to have a color contrast without having a luminance contrast. In a grayscale version of such an image, the two different colored objects appear blended into a single one. In Fig. 2, we demonstrate that the proposed Quaternionic Gabor Filters can extract the local orientation information from a constant luminance image as well. Fig. 2a shows a synthetic color image where all pixels have the same luminance value (see Fig. 2b), but the chromaticity inside the object differs from the chromaticity outside. We applied 10 QGFs to the quaternion representation of this color image. The sum of the magnitude responses of 10 QGFs is shown in Fig. 2d. Although a black-and-white version (Fig. 2c) of the input image is a uniform gray without any changes in orientation, the proposed QGFs successfully derive the orientation information in the color version, showing that they are well suited for analyzing color images and the result is not a grayscale image processing.

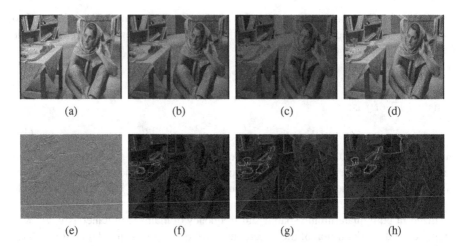

Fig. 1. Image of Barbara is quaternion-convolved with QGFs of different orientations. (a)-(d) Color image and GB, RB, RG images respectively, (e) Scalar part of the sum of the QGF responses, (f)-(h) Vector part of the sum of QGF responses.

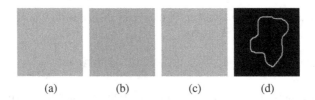

Fig. 2. Application of Quaternionic Gabor Filters across equal luminance: (a) a synthetic color image where the object and the background are of equal luminance, (b) luminance channel, (c) a grayscale version of (a), (d) the sum of the magnitude responses of QGFs applied to the color image in (a).

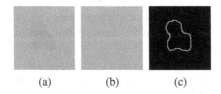

Fig. 3. (a) A synthetic color image where $(R + G + B)/3$ is the same everywhere. (b) $(R + G + B)/3$ values for each pixel. (c) The sum of the magnitude responses of QGFs applied to the color image in (a).

We have chosen the unit pure quaternion direction μ in QGF as $\frac{1}{\sqrt{3}}(i + j + k)$. However, this choice does not mean that the proposed quaternion framework is processing the sum of the RGB values. Also note that the convolution between a QGF and a quaternion representation of a color image is performed following the rules of quaternion algebra. At each pixel, the quaternion-valued filter is multiplied with the color direction of that pixel through a quaternion product. Hence, QGF handles the coupling between the channels while, at the same time, processing all information in a color image. Fig. 3a shows a color image where $(R + G + B)/3$ is the same for all pixels. As shown in Fig. 3c, the proposed framework can accurately extract the orientation information.

3   Modeling Derived Orientation Information

In the previous section, we introduced the QGFs to extract the local orientation information in a color image. The resulting responses over a circle of directions are modeled in this section in a probabilistic framework. We postulate that at each lattice point there is an underlying probability measure induced on the unit circle. An appropriate choice for the basis functions is $\exp(\cos(\theta - \alpha))$, where α is the orientation of the QGF, and θ is a random variable on the circle. The proposed continuous mixture model is given by,

$$G_{\mathbb{H}}^v(\mathbf{x}; \cdot, \alpha) = \int_{\mathbb{S}^1} e^{\cos(\theta - \alpha)} dF \,, \tag{14}$$

where $dF = f(\theta)d\theta$ denotes the underlying probability measure with respect to the uniform distribution $d\theta$ on $\mathbb{S}^1$. $G_{\mathbb{H}}^v, v = i, j, k$ denote the i, j and k components of the vector part in the quaternion-valued response, respectively. We only model the components of the vector part. Scalar part of the filter response can be regarded as a smoothed second derivative of the initial image, and can be of use for edge detection.

In order to avoid an ill-posed inverse problem which requires recovering a distribution defined on the circle given the measurements $G_{\mathbb{H}}^i(\mathbf{x}; \cdot, \alpha)$, we impose a mixture of von Mises distributions on θ as a prior. The von Mises distributions have a significant role in statistical inference on the circle, analogous to that of the normal distributions on the line. For statistical purposes, any von Mises distribution can be approximated by a normal distribution wrapped around the circumference of the circle of unit radius. θ is distributed as $f_{\mathcal{M}}(\theta; \beta, \kappa)$ if it has the von Mises density given by,

$$\frac{1}{2\pi\mathcal{I}_0(\kappa)} e^{\kappa \cos(\theta - \beta)} d\theta \,, \tag{15}$$

where β and κ are the mean direction and the concentration parameter, respectively. $\mathcal{I}_0$ is the modified Bessel function of the first kind and zeroth order ([29]).

This distribution is unimodal and symmetric about $\theta = \beta$. κ determines the degree of the clustering around the mode; i.e. the larger the value of κ, the greater the clustering around the mode. In order to handle orientational heterogeneity we need a multimodal distribution. Therefore, we choose the prior to be a discrete mixture of von Mises distributions:

$$dF = \sum_{n=1}^{N} w_n \frac{1}{2\pi\mathcal{I}_0(\kappa_n)} e^{\kappa_n \cos(\theta - \beta_n)} d\theta \,. \tag{16}$$

Plugging this measure into (14), we obtain our model given as follows:

$$G_{\mathbb{H}}^v(\mathbf{x}; \cdot, \alpha) = \int_{\mathbb{S}^1} \sum_{n=1}^{N} w_n \frac{1}{2\pi\mathcal{I}_0(\kappa_n)} e^{\kappa_n \cos(\theta - \beta_n)} e^{\cos(\theta - \alpha)} d\theta \,. \tag{17}$$

However, note that this is still a continuous mixture model. N here corresponds to the resolution of the discretization of the circle; it does not correspond to the number of modes (peaks) characterizing the local geometry or the number of dominant local orientations. We observed that the kernel of the von Mises distribution can be utilized to derive a closed form solution for the continuous mixture integral, leading to:

$$G_{\mathbb{H}}^v(\mathbf{x}; \cdot, \alpha) = \sum_{n=1}^{N} w_n \frac{\mathcal{I}_0(\sqrt{\kappa_n^2 + 1 + 2\kappa_n \cos(\beta_n - \alpha)})}{\mathcal{I}_0(\kappa_n)} . \tag{18}$$

We can formulate the computation of this analytic form as the solution to a linear system $A\mathbf{w} = \mathbf{y}$, where $\mathbf{y} = \{G_{\mathbb{H}}^v(\mathbf{x}; \cdot, \alpha_m)\}_{m=1}^{M}$ contains the measurements obtained via an application of M QGFs to the image, A is an $M \times N$ matrix with

$$A_{mn} = \frac{\mathcal{I}_0\left(\sqrt{\kappa_n^2 + 1 + 2\kappa_n \cos(\beta_n - \alpha_m)}\right)}{\mathcal{I}_0(\kappa_n)} \tag{19}$$

and $\mathbf{w} = (w_n)$ is the unknown weight vector. We solve the weights in the mixture using a sparse deconvolution technique, a non-negative least squares (NNLS) minimization which yields an accurate and sparse solution for:

$$\min \|A\mathbf{w} - \mathbf{y}\|^2 \quad \text{subject to} \quad \mathbf{w} \geq 0. \tag{20}$$

A sparse solution is what is expected at each image lattice point since local image geometry does not have a large number of edges meeting at a junction. Once $\mathbf{w}$ is estimated for the given data at each lattice point, we can construct the convolution kernel for color image smoothing. The update equation for image channel I^v, $v = R, G, B$ is given as follows:

$$I_{t+1}^v(\mathbf{x}) = I_t^v(\mathbf{x}) * Q^v(\mathbf{x}) , \tag{21}$$

$$Q^v(\mathbf{x}) = \sum_{n=1}^{N} w_n^v \frac{\mathcal{I}_0(\sqrt{\kappa_n^2 + 1 + 2\kappa_n \cos(\beta_n - \alpha)})}{\mathcal{I}_0(\kappa_n)}$$

where $Q^v(\mathbf{x})$ is the convolution kernel on the right-hand side of (18) for the corresponding $G_{\mathbb{H}}^v(\cdot)$, $\mathbf{w}^v$ is the weight vector obtained from (20) using the corresponding $G_{\mathbb{H}}^v(\cdot)$ measurements, and the orientation α is the angle that the coordinate vector $\mathbf{x}$ makes with the x-axis. This formulation yields a spatially varying convolution kernel since the $\mathbf{w}$ vector depends on location; it is estimated at each lattice point $\mathbf{x}$ in an image. Moreover, the weights $\mathbf{w}$ and hence the convolution kernel is different for each color channel I^v. Note that this framework handles the coupling between the color channels through the application of quaternionic Gabor filters to the quaternion representation of the color image.

4   Experiments and Comparisons

In this section, we evaluate the performance of the proposed framework with applications on color image denoising and inpainting. We compare our denoising results with

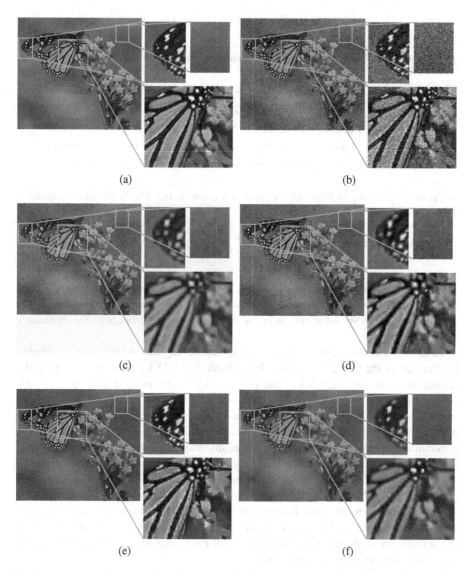

Fig. 4. (a) Original image. (b) Noisy image with a Gaussian noise of standard deviation 35. Denoised images using (c) the coherence enhancing diffusion, (d) the Beltrami flow, (e) the curvature preserving regularization, (f) our method.

three prominent techniques: Weickert's coherence enhancing diffusion (CED) for color images [2], the Beltrami flow proposed by Kimmel *et al.* [3], and the curvature preserving regularization (CPR) proposed by Tschumperlé [5]. In the denoising experiments, for each algorithm the outputs that have the highest PSNR values are shown. Parameters of each method were chosen so as to reach its best PSNR value. We compute the PSNR on the RGB channels of the color image. We also report the PSNR values on the luminance channel of the YCbCr representation of the RGB image, since the human

Table 1. PSNR Values for Denoised Images

Image	PSNR Method	CED	Beltrami	CPR	Ours	Noisy Image
Butterfly	Luminance	26.45	27.37	25.14	**28.18**	22.32
	RGB	24.48	24.84	23.11	**26.33**	17.71
Parrots	Luminance	29.01	28.95	28.91	**30.03**	22.3
	RGB	26.95	26.85	26.75	**27.70**	17.62

eye is more sensitive to luma information in a color image. PSNR for RGB domain is defined as:

$$\text{PSNR} = 10 \log_{10} \frac{255^2}{\text{MSE}}, \ \text{MSE} = \frac{1}{3|\Omega|} \sum_{x \in \Omega} \sum_{v=R,G,B} (I_0^v(x) - \hat{I}^v(x))^2 \qquad (22)$$

where Ω is the image domain of $|\Omega|$ pixels, I_0 is the noise-free ideal image, and $\hat{I}$ is its estimate obtained from the denoising method. PSNR for the luminance channel is the same except the MSE is the sum over the squared value differences of the luminance channel, divided by $|\Omega|$.

In all of our experiments, we use the same number of measurements for our model; i.e. the size of the Quaternion Gabor Filter bank, M, is 21 for all experiments. N, the resolution of the discretization of the unit circle for the mixing density, is set to 81. Hence, the size of matrix A is 21 × 81, and the unknown of this under-determined system, which is the weight vector w, is an 81-dimensional vector. Note that this size does not correspond to the expected number of different orientations at a pixel. The concentration parameter κ is the same for all distributions in the mixture of von Mises distributions. We experimented with different values of κ and in the following experiments, a value of 8, which gives a sharper mode in a von Mises distribution, yields the best PSNR values. In denoising experiments, the original images are corrupted by additive white-Gaussian noise, having a high standard deviation ($\sigma = 35$). For a quantitative evaluation of our approach, we present the highest PSNR values obtained using the abovementioned methods in Table 1. In all cases, our unsupervised and adaptive method produces the best PSNR values.

In Fig. 4, we illustrate the potential of our approach with a butterfly image corrupted by additive white-Gaussian noise, having a high standard deviation (Fig. 4b, $\sigma = 35$). Our method preserves significant geometric features and the original color contrasts without producing undesirable artifacts (see Fig. 4f). However, both in Fig. 4c and in Fig. 4d, we can notice the color artifacts in flat regions, which look like artificial texture effects. Coherence enhancing diffusion creates fiber effects on the background. Curvature preserving regularization performs better, however it creates a color bleeding around the edges of the wings (see zoomed-in view in Fig. 4e). Both visually and in terms of PSNR, our method outperforms the competing methods.

Another comparison is presented in Fig. 5 with multi-colored parrots. The noisy image has a PSNR value of 17.62 in RGB domain. In this experiment, competing methods

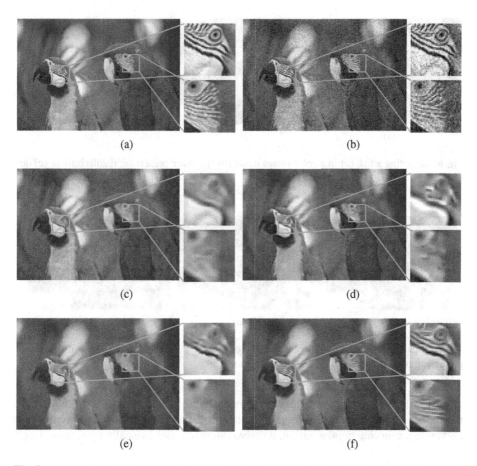

Fig. 5. (a) Original image. (b) Noisy image with a Gaussian noise of standard deviation 35. Denoised images obtained from (c) the coherence enhancing diffusion, (d) the Beltrami flow, (e) the curvature preserving regularization, and (f) our method.

generated blurred images. Although the Beltrami flow gives a slightly lower PSNR than the coherence enhancing diffusion, it smoothes the flat regions better and produces a visually more pleasing image (Fig. 5d). We can notice some color diffusing effect in Fig. 5e. Our algorithm, however, is able to remove the noise, preserve the color and the orientation details without any color blending problems (see the patch around the eye in the close-up view in Fig. 5f), as well as achieve the highest PSNR value.

In inpainting, we compare our results with the direct application of the curvature preserving PDE as proposed by Tschumperlé in [5]. To fill-in the missing/desired image regions, we apply the iterative convolution of our spatially-varying kernel on the regions to inpaint, without using any texture synthesis or reconstruction technique as a post-processing step. We illustrate how our technique can be used to remove objects from digital photographs in Fig. 6 and Fig. 7 along with the comparisons. In both

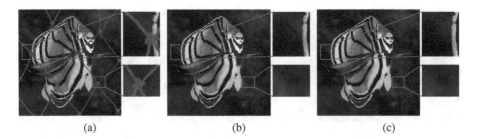

(a)   (b)   (c)

Fig. 6. Inpainting a fish net in a color image using (b) curvature preserving regularization, (c) our method

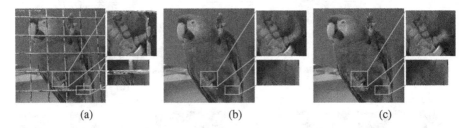

(a)   (b)   (c)

Fig. 7. Inpainting a cage in (a) a color image (courtesy of D. Tschumperlé [30]), with curvature preserving regularization (b), and with our method (c).

experiments, our method generates a better result. Note that the fish net is still noticeable in Fig. 6b, similarly the cage in Fig. 7b. In addition, parrot's toe is over-diffused by the curvature preserving regularization, whereas our result looks visually more appealing.

5   Conclusion

We described a novel feature preserving color image smoothing technique based on a quaternion framework. The main idea is to capture the complicated local geometry contained at a lattice point via a continuous mixture model, and then to incorporate this information into a spatially varying convolution filter. We first introduced a novel quaternionic Gabor filter to extract the local orientation information while appreciating the vectorial nature of a color image. Each component of this quaternion-valued data is then represented by a continuous mixture of exponential basis functions, where the mixing density is assumed to be a mixture of von Mises densities. We derived a closed form solution for this integral, which leads to a spatially varying convolution kernel. We qualitatively and quantitatively validated that our framework delivers superior performance in comparison to competing state-of-the-art methods; it produces smoother flat regions, and preserves complex geometries and texture details without any prior information.

The proposed method handles the coupling between the channels through the application of QGFs to the quaternion representation of color images; image channels do not evolve independently with different smoothing geometries, because the color and orientation spaces are linked through the QGFs. We envision that the derived orientation information, being quaternion-valued, can be modeled using quaternionic basis functions in the unit quaternion space. Moreover, depending on the formation of the convolution kernel, the update equation of the smoothing process can be modified to perform a quaternion-convolution of color image with a quaternion-valued kernel. Future research will encompass the search for such formulations to discover new valuable tools for color image processing.

References

1. Blomgren, P., Chan, T.F.: Color TV: Total variation methods for restoration of vector-valued images. IEEE Transactions on Image Processing 7(3), 304–309 (1998)
2. Weickert, J.: Coherence-enhancing diffusion of colour images. In: 7th Nat. Symp. on Patt. Rec. Img. Anal., pp. 239–244 (1997)
3. Kimmel, R., Malladi, R., Sochen, N.A.: Images as embedded maps and minimal surfaces: Movies, color, texture, and volumetric medical images. International Journal of Computer Vision 39(2), 111–129 (2000)
4. Tang, B., Sapiro, G., Caselles, V.: Color image enhancement via chromaticity diffusion. IEEE Transactions on Image Processing 10(5), 701–707 (2001)
5. Tschumperlé, D.: Fast anisotropic smoothing of multi-valued images using curvature-preserving pde's. Int. J. Comput. Vision 68(1), 65–82 (2006)
6. Galatsanos, N.P., Wernic, M.N., Katsaggelos, A.K., Molina, R.: Multichannel image recovery. In: Bovik, A. (ed.) Handbook of Image and Video Proc. Elsevier, Academic Press (2005)
7. Lukac, R., Smolka, B., Martin, K., Plataniotis, K., Venetsanopoulos, A.: Vector filtering for color imaging. IEEE Signal Processing Magazine 22(1), 74–86 (2005)
8. Ell, T., Sangwine, S.: Hypercomplex fourier transforms of color images. IEEE Transactions on Image Processing 16(1), 22–35 (2007)
9. Moxey, C., Sangwine, S., Ell, T.: Hypercomplex correlation techniques for vector images. IEEE Transactions on Signal Processing 51(7), 1941–1953 (2003)
10. Jian, B., Vemuri, B.: Multi-fiber reconstruction from diffusion mri using mixture of wisharts and sparse deconvolution. Inf. Process Med Imaging 20 (2007)
11. Jian, B., Vemuri, B.C.: A unified computational framework for deconvolution to reconstruct multiple fibers from diffusion weighted mri. IEEE Trans. Med. Imaging 26(11), 1464–1471 (2007)
12. Jian, B., Vemuri, B.C., Özarslan, E., Carney, P., Mareci, T.: A novel tensor distribution model for the diffusion-weighted mr signal. Neuroimage 37(1), 164–176 (2007)
13. Subakan, Ö.N., Jian, B., Vemuri, B.C., Vallejos, C.E.: Feature preserving image smoothing using a continuous mixture of tensors. In: IEEE International Conference on Computer Vision, Rio de Janeiro, Brazil (October 2007)
14. Subakan, Ö.N., Vemuri, B.C.: Image segmentation via convolution of a level-set function with a Rigaut kernel. In: IEEE Conference on Computer Vision and Pattern Recognition, Anchorage, Alaska (June 2008)
15. Shoemake, K.: Animating rotation with quaternion curves. SIGGRAPH Comput. Graph. 19(3), 245–254 (1985)
16. Kuipers, J.B.: Quaternions and Rotation Sequences: A Primer with Applications to Orbits, Aerospace and Virtual Reality. Princeton University Press, Princeton (2002)
17. Sethuraman, B.A., Rajan, B.S., Member, S., Shashidhar, V.: Full-diversity, highrate space-time block codes from division algebras. IEEE Trans. Inform. Theory 49, 2596–2616 (2003)

18. Pei, S.C., Cheng, C.M.: A novel block truncation coding of color images by using quaternion-moment-preserving principle. In: IEEE International Symposium on Circuits and Systems, ISCAS 1996., 'Connecting the World', May 1996, vol. 2, pp. 684–687 (1996)
19. Sangwine, S.: Fourier transforms of colour images using quaternion or hypercomplex numbers. Electronic Lett. 32(21), 1979–1980 (1996)
20. Sangwine, S.J., Ell, T.A.: Colour image filters based on hypercomplex convolution. IEE Proceedings Vision, Image and Signal Processing 147(2), 89–93 (2000)
21. Sangwine, S.: Colour image edge detector based on quaternion convolution. Electronic Lett. 34(10), 969–971 (1998)
22. Delsuc, M.A.: Spectral representation of 2d nmr spectra by hypercomplex numbers. Journal of Magnetic Resonance 77(1), 119–124 (1988)
23. Ell, T.: Quaternion-fourier transforms for analysis of two-dimensional linear time-invariant partial differential systems. In: Proceedings of the 32nd IEEE Conference on Decision and Control, December 1993, vol. 2, pp. 1830–1841 (1993)
24. Bülow, T., Sommer, G.: Das konzept einer zweidimensionalen phase unter verwendung einer algebraisch erweiterten signalrepräsentation. In: Mustererkennung 1997, 19. DAGM-Symposium, London, UK, pp. 351–358. Springer, Heidelberg (1997)
25. Bülow, T., Sommer, G.: Multi-dimensional signal processing using an algebraically extended signal representation. In: Sommer, G. (ed.) AFPAC 1997. LNCS, vol. 1315, pp. 148–163. Springer, Heidelberg (1997)
26. Sangwine, S., Ell, T.A.: The discrete fourier transform of a colour image. In: Blackledge, J.M., Turner, M.J. (eds.) Image Processing II: Mathematical Methods, Algorithms and Applications, pp. 430–441 (2000)
27. Bülow, T.: Hypercomplex Spectral Signal Representations for Image Processing and Analysis. PhD thesis, University of Kiel, Advisor-Gerald Sommer (1999)
28. Daugman, J.G.: Uncertainty relation for resolution in space, spatial frequency, and orientation optimized by two-dimensional visual cortical filters. J. Opt. Soc. Am. A 2(7), 1160–1169 (1985)
29. Mardia, K.V., Jupp, P.E.: Directional Statistics, 2nd edn. John Wiley and Sons Ltd., Chichester (2000)
30. Tschumperlè, D.: Greycstoration (2008), http://cimg.sourceforge.net/greycstoration/

Locally Parallel Textures Modeling with Adapted Hilbert Spaces

Pierre Maurel[1], Jean-François Aujol[1], and Gabriel Peyré[2]

[1] CMLA, ENS Cachan, CNRS, UniverSud, 61 avenue du Président Wilson, 94235
Cachan Cedex, France
[2] Ceremade, Université Paris-Dauphine, Place du Maréchal De Lattre De Tassigny,
75775 Paris Cedex 16, France

Abstract. This article[1] presents a new adaptive texture model. Locally
parallel oscillating patterns are modeled with a weighted Hilbert space
defined over local Fourier coefficients. The weights on the local Fourier
atoms are optimized to match the local orientation and frequency of
the texture. We propose an adaptive method to decompose an image
into a cartoon layer and a locally parallel texture layer using this model
and a total variation cartoon model. This decomposition method is then
used to denoise an image containing oscillating patterns. Finally we show
how to take advantage of such a separation framework to simultaneously
inpaint the structure and texture components of an image with missing
parts. Numerical results show that our method improves state of the art
algorithms for directional and complex textures.

1   Introduction

The analysis and modeling of textures is a central topic in computer vision and
graphics. Texture modeling is fundamental for a large number of problems, such
as image segmentation, object recognition and image restoration.

1.1   Previous Works

Image Decomposition. A variational decomposition algorithm seeks a decomposition $f = u + v$ of an image f where u should capture the sketch of the image
and v the texture content. This decomposition is often defined as the solution
of a minimization problem involving two norms, one for each component. Total
variation [1] is broadly used as a cartoon model since it allows to recover piecewise smooth functions without smoothing sharp discontinuities. On the other
hand, the norm on v, the texture component, should be small for typical texture
patterns one wants to extract.

Following [1], where Rudin, Osher and Fatemi proposed to capture the noise
of an image by using the usual L^2 norm, Yves Meyer [2] pushed forward the idea

---

[1] This work has been done with the support of the French "Agence Nationale de la
Recherche" (ANR), under grant NATIMAGES (ANR-08-EMER-009), "Adaptivity
for natural images and texture representations".

D. Cremers et al. (Eds.): EMMCVPR 2009, LNCS 5681, pp. 429–442, 2009.

of using more complex norms $\|\cdot\|_T$ to capture oscillating patterns. In particular he proposed a weak norm dual of the TV norm. This idea inspired several works [3,4,5]. An alternative to this dual norm approach has been presented in [6], the Morphological Component Analysis: it uses the ℓ^1 norm of decompositions on bases such as a local cosine dictionary for the texture component and a wavelet dictionary for the cartoon one.

Inpainting. The problem of inpainting can be stated as follows : given a region Ω to be restored, use the valid surrounding information for synthesizing the most plausible data in Ω. Several classes of methods have been considered. In the first category of approaches, the focus has been on recovering the geometry. These methods [7,8,9,10,11] use partial differential equations that propagate the information from the boundary of the missing region to its interior. The drawback of this kind of methods is their well-known incapacity to restore texture. In parallel to these geometry-oriented approaches, the exemplar-based methods [12,13] turned out to be very efficient for reconstructing isotropic and non-geometric textures. Different approaches have been proposed in combination with an exemplar-based inpainting, either based on a manual intervention by the user [14], or trying to combine texture and geometric interpolation in the most automated possible way [15]. A last class of approaches relies on sparse regularization in several transform domains (e.g. Fourier, wavelet or framelet) and also aims to deal with geometric and texture information simultaneously [16,17].

1.2   Contributions

The main contribution of this work is a new adaptive texture model. We propose methods for using this new model in some applications such as image decomposition, denoising and inpainting, and we present algorithms for solving these problems. We model locally parallel textures in order to extract oscillating patterns which present spatial and frequency variability. We start (section 2) by defining a texture norm $\|\cdot\|_T = \|\cdot\|_\xi$ depending on a parameter $\xi(x)$ which is the instantaneous frequency of the oscillating texture. For a point x in the image, $\xi(x)$ gives the local frequency $\|\xi(x)\|$ and the local orientation $\xi(x)/\|\xi(x)\|$ of the texture around x. The norm $\|\cdot\|_\xi$ is small for an oscillating pattern around x if its main frequency is close to $\xi(x)$. We then use this norm for a decomposition problem (section 3): we want to separate the image into three layers, $f = u + v + w$ where u is the geometric layer, v is the texture modeled by our norm and w is the noise. And finally the interest of such a texture norm is highlighted in section 4 by its use in an inpainting method which simultaneously inpaints the geometric and the texture layers. Numerical examples are shown for decomposition, denoising and inpainting and our results are compared with other methods.

    In the following, we suppose that $f \in \mathbb{R}^N$ is a discrete image of $N = n \times n$ pixels and the two operators gradient and divergence are discretized by forward finite difference (for example, we refer the reader to [18] for details). In this framework, we have $\|\nabla\| = \sqrt{8}$.

2   Texture Modeling Using an Adaptive Hilbert Norm

2.1   Hilbert Texture Norm

In [19], Aujol and Gilboa proposed to use a linear Hilbert norm defined by some symmetric positive kernel K: $\|v\|_T^2 = \langle Kv, v \rangle_{L^2}$. This norm can be computed using a frame $\{\psi_\ell\}_\ell$ that is a possibly redundant family of $P \geqslant N$ atoms $\psi_\ell \in \mathbb{R}^N$. The decomposition of an image in this frame reads

$$\Psi f = \{\langle f, \psi_\ell \rangle\}_{\ell=0}^{P-1} \in \mathbb{R}^P, \tag{1}$$

where $\Psi : \mathbb{R}^N \to \mathbb{R}^P$ is the frame operator.

Given a set of positive weights $\gamma_\ell \geqslant 0$, a norm can then be defined as

$$\|f\|_T^2 = \sum_\ell \gamma_\ell^2 |\langle f, \psi_\ell \rangle|^2 = \|\gamma \Psi f\|_{L^2}^2, \tag{2}$$

where $\gamma = \mathrm{diag}_\ell(\gamma_\ell)$. This corresponds to a Hilbert space associated to the kernel $K = \Psi^* \gamma^2 \Psi$.

2.2   Texture Norm over a Local Fourier Basis

Aujol and Gilboa [18] proposed to use the Fourier basis so that Ψ corresponds to the discrete Fourier transform. This defines a translation-invariant kernel K. This paper proposes to replace the global Fourier basis by a redundant local Fourier basis, to capture the spatially and frequencyly varying structures of locally parallel textures.

Local Fourier Frame. A discrete short time Fourier atom, located around a position $x_p = p\Delta_x$ and with local frequency $\xi_k = k\Delta_\xi = k/q$ is defined as

$$\psi_{p,k}[y] = q^{-1} g[y - p\Delta_x] e^{\frac{2i\pi}{q}(y_1 k_1 + y_2 k_2)}, \tag{3}$$

for $k \in \{-q/2, \dots, q/2 - 1\}^2$ and $p \in \{0, \dots, n/\Delta_x\}^2$, where g is a smooth window, centered around 0, and the size of its support is $q \times q$ pixels with $q > \Delta_x$. In this paper, we use a Haning window function: $g[x] = \sin(\pi x_1/q - \pi/2)^2 \sin(\pi x_2/q - \pi/2)^2$.

The local Fourier frame $\{\psi_{p,k}\}_{p,k}$ is a redundant family of $P = (q/\Delta_x)^2 N$ vectors of $\mathbb{R}^N$. The decomposition $\Psi f = \{\langle f, \psi_{p,k} \rangle\}_{p,k} \in \mathbb{R}^P$ of an image f in this frame can be computed with the 2D Fast Fourier Transform of the $q \times q$ image $f[y]g[\Delta_x p - y]$. The computation of Ψf thus requires $O(NQ \log_2(Q)/\Delta_x^2)$ operations.

The dual operator Ψ^* reconstructs an image $\Psi^* c \in \mathbb{R}^N$ from a set of coefficients $c[p, k] \in \mathbb{R}^{Q \times N}$

$$\Psi^* c = \sum_{p,k} c[p, k] \psi_{p,k}. \tag{4}$$

This dual operator is implemented using N/Δ_x^2 inverse Fast Fourier Transforms. The operator $\Psi^*\Psi$ is in fact diagonal, and one has

$$\Psi^*\Psi = \text{diag}_x(\sum_y g^o[\Delta_x y - x]^2). \tag{5}$$

and the norm of the operator $\Psi^*\Psi$ is $\max_x \sum_y g^o[\Delta_x y - x]^2$.

2.3   Weight Design

We define a Hilbert norm $\|\cdot\|_T$ adapted to oscillating texture as a weighted norm over the local Fourier coefficients. The general formulation (2) is instantiated using a local Fourier frame $\psi_\ell = \psi_{p,k}$ for $\ell = (p, k)$ as follow

$$\|f\|_T^2 = \sum_{p,k} \gamma_{p,k}^2 |\langle f, \psi_{p,k}\rangle|^2, \tag{6}$$

where each $\gamma_{p,k} \geqslant 0$ weights the influence of each local Fourier atom in the texture model.

Intuitively, $\gamma_{p,k}$ should be small when the texture f contains a local oscillation of frequency close to ξ_k around the point x_p. We consider a locally oscillating texture model, where typical texture patterns are locally well approximated by a single atom.

The texture norm $\|\cdot\|_T$ is therefore parametrized by a vector field $\xi : \mathbb{R}^N \mapsto \mathbb{R}^2$ which represents the local frequency of the texture component of f. For a point x of the image, the local frequency around x is given by $|\xi(x)|$ and the local orientation of the texture is given by $\xi(x)/|\xi(x)|$. The norm $\|\cdot\|_T = \|\cdot\|_\xi$ should be small for an oscillating pattern around the point x if its main frequency is close to $\xi(x)$. As a consequence the weight $\gamma_{p,k}$ should be small if ξ_k is close to $\xi(x_p)$ or to $-\xi(x_p)$. By convention, $\xi(x)$ is set to $(0,0)$ if there is no significant oriented patterns around x in the image.

The weights are therefore defined as a function of ξ:

$$\gamma_{p,k}(\xi) = \begin{cases} 1 & \text{if } \xi(x_p) = (0,0) \\ \left(1 - G_\sigma(\|\xi_k + \xi(x_p)\|)\right)\left(1 - G_\sigma(\|\xi_k - \xi(x_p)\|)\right) & \text{otherwise} \end{cases} \tag{7}$$

where $G_\sigma(x) = \exp(-(x/\sigma)^2/2))$ and σ is a scale parameter reflecting the deviation we are expecting to find in the frequency spectrum of the texture compared to $\xi(x)$ (in our numerical experiments we took $\sigma = 1$). When there is not a significant oriented texture around x_p, we choose $\gamma_{p,k} = 1$ for all k, in order not to promote an arbitrary orientation in the extraction. The texture norm is finally given by:

$$\|v\|_T^2 = \|v\|_\xi^2 = \sum_{p,k} \gamma_{p,k}(\xi)^2 |\langle v, \psi_{p,k}\rangle|^2 = \|\Gamma(\xi)\Psi v\|_{L^2}^2. \tag{8}$$

where $\Gamma(\xi) = \text{diag}_{\ell=(p,k)}(\gamma_{p,k}(\xi))$. This is actually a semi-norm since $\|v\|_T = 0$ does not imply $v = 0$ but, for the sake of simplicity, we use the term of norm in the following.

3   Image Decomposition and Denoising Using an Adaptive Hilbert Norm

Decomposing an image into meaningful components is an important problem in image processing. Using the texture norm introduced in section 2, we present an image decomposition framework which aims to separate an image f into three components: $f = u+v+w$, where u should capture the sketch of the image, v the texture content and w the noise. We define this decomposition as the solution of the following minimization problem:

$$(u, v, \xi) = \underset{\tilde{u},\ \tilde{v},\ \tilde{\xi} \in \mathcal{C}}{\operatorname{argmin}}\ \mu\|\tilde{v}\|_{\tilde{\xi}}^2 + \lambda\|\tilde{u}\|_{\mathrm{TV}} + \frac{1}{2}\|f - \tilde{u} - \tilde{v}\|_{L^2}^2, \quad w = f - u - v. \quad (9)$$

where $\|\ \|_{\mathrm{TV}}$ is the total variation norm, $\|u\|_{\mathrm{TV}} = \int |\nabla u|$ (the discrete total variation of u is then defined by $\|u\|_{\mathrm{TV}} = \sum_{1 \leqslant i,j \leqslant n} |(\nabla u)_{i,j}|$) and $\|v\|_{\xi}$ is our texture norm defined by (8).

$\mathcal{C}$ is a set of constraints on the orientation field ξ. We first force the frequency $|\xi|$ to be large enough in order not to extract low frequencies in the texture component v: $\forall p, |\xi(x_p)| > \tau$, for some real positive parameter $\tau > 0$. Furthermore, an oscillating pattern of frequency $\xi(x_p)$ is assumed to be present in the image f around the point x_p only if $|\langle f, \psi_{p,k}\rangle| > \eta_p$ where $k = \xi(x_p)/\Delta_\xi$ and $\eta_p > 0$ is a real positive parameter. In fact, one does not want to arbitrary select a frequency for an area of the image where there is no oscillating pattern. In our numerical experiments we take $\tau = 2/q$, where q is the size of the local Fourier windows, and $\eta_p = 2\overline{|\Psi f_p|}$ where $\overline{|\Psi f_p|}$ is the average value of $|\langle f, \psi_{p,k'}\rangle|$ for $k' \in \{-q/2,\ldots,q/2-1\}^2$. In short, we have:

$$\mathcal{C} = \left\{ \xi : \mathbb{R}^{N/\Delta_x} \mapsto \mathbb{R}^2 \ \begin{vmatrix} \forall p, & |\xi(x_p)| > \tau \\ \forall p, & (\forall k, |\langle f, \psi_{p,k}\rangle| \leqslant \eta_p) \Rightarrow \xi(x_p) = (0,0) \end{vmatrix} \right\} \quad (10)$$

The minimization (9) iterates between two steps: one on ξ and one on u and v. We detail these two steps in the next two sections. Although the energy is decreasing at each step, this algorithm is not guaranteed in general to converge to a minimum. However we did not encounter any optimization problems during our numerical experiments.

3.1   Minimization with Respect to the Orientation Field ξ

If u and v are fixed, we search for the frequency field ξ verifying:

$$\xi = \underset{\tilde{\xi} \in \mathcal{C}}{\operatorname{argmin}}\ \|v\|_{\tilde{\xi}}^2.$$

This requires, for each p, to compute:

$$\xi(x_p) = \underset{\tilde{\xi}(x_p) \in \mathcal{C}}{\operatorname{argmin}} \sum_k \gamma_{p,k}(\tilde{\xi}(x_p))^2 |\langle v, \psi_{p,k}\rangle|^2.$$

where $\gamma_{p,k}(\tilde{\xi}(x_p))$ is given by (7).

If σ in the weight definition (7) is small enough, this minimization boils down to compute $\max_k |\langle v, \psi_{p,k}\rangle|$, which allows us to speed up the computation by taking:

$$\xi(x_p) = \Delta_\xi \underset{k > \frac{\tau}{|\Delta_\xi|}}{\operatorname{argmax}} |\Psi v[p, k]|. \tag{11}$$

Figure 1 illustrates the underlying principle of this orientation estimation: for a given point x_p, a unique direction and frequency $\xi(x_p)$ is selected and the corresponding weights $\gamma_{p,k}(\xi)$ are constructed according to (7).

(a)   (b)   (c)   (d)

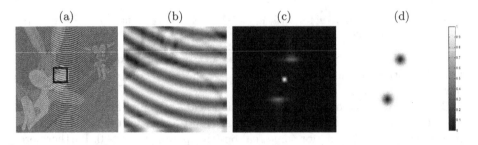

Fig. 1. Illustration of the orientation estimations. From left to right: (a) the input image f, (b) the windowed image around some point x_p, (c) the corresponding local Fourier transform and (d) the weights $\gamma_{p,k}(\xi)$ corresponding to the ξ estimated from the local Fourier transform.

3.2   Minimization with Respect to the Components u and v

If ξ is fixed, we search for u and v verifying:

$$(u, v) = \underset{\tilde{u}, \tilde{v}}{\operatorname{argmin}} \, \mu\|\Gamma(\xi)\Psi\tilde{v}\|_{L^2}^2 + \lambda\|\tilde{u}\|_{TV} + \frac{1}{2}\|f - \tilde{u} - \tilde{v}\|_{L^2}^2, \tag{12}$$

where $\Gamma(\xi)$ is defined at the end of section 2.3. This minimization is done itself iteratively on u and v. Starting from some initial $u^{(0)}$ and $v^{(0)}$, one solves:

• *v is fixed:* one minimizes

$$u^{(i+1)} = \underset{\tilde{u}}{\operatorname{argmin}} \, \lambda\|\tilde{u}\|_{TV} + \frac{1}{2}\|(f - v^{(i)}) - \tilde{u}\|_{L^2}^2. \tag{13}$$

This minimization can be solved using iterations of the original algorithm of Chambolle [20]. This algorithm is based on the observation that $u^{(i+1)} = (f - v^{(i)}) + \lambda \operatorname{div}(w)$ where w is the solution of the following constrained minimization problem

$$w = \underset{\|\tilde{w}\|_\infty \leqslant 1}{\operatorname{argmin}} \, \|(f - v^{(i)}) + \lambda \operatorname{div}(\tilde{w})\|, \tag{14}$$

where the infinite norm of a vector field $w = (w_1, w_2)$ is

$$\|w\|_\infty = \max_{i,j} \sqrt{w_1[i, j]^2 + w_2[i, j]^2}. \tag{15}$$

Chambolle proposed a fixed point algorithm to solve (14), and one can also use a projected gradient descent by initializing $w^{(0)} = 0$ and then iterating a gradient step

$$\bar{w}^{(\ell)} = w^{(\ell)} + \nu \nabla(\bar{u}^{(k)} + \lambda \operatorname{div}(w^{(\ell)})) \tag{16}$$

and a projection on the constraints

$$\forall (i,j), \quad w^{(\ell+1)}[i,j] = \frac{\bar{w}^{(\ell)}[i,j]}{\max(\|\bar{w}^{(\ell)}[i,j]\|, 1)}. \tag{17}$$

The gradient step size should satisfy $\nu < 2/\|\nabla\|^2 = 1/4$ (with the discretization used in this paper) so that $f - v^{(i)}) + \lambda \operatorname{div}(w^{(\ell)})$ converges with $\ell \to +\infty$ to $u^{(i+1)}$.

- **u is fixed:** one minimizes

$$v^{(i+1)} = \underset{\tilde{v}}{\operatorname{argmin}} \, \mu\|\Gamma(\xi)\Psi\tilde{v}\|_{L^2}^2 + \frac{1}{2}\|(f - u^{(i+1)}) - \tilde{v}\|_{L^2}^2, \tag{18}$$

Computing the gradient of (18), we obtain that $v^{(i+1)}$ satisfies:

$$(2\mu\Psi^*\Gamma^2\Psi + \operatorname{Id})v^{(i+1)} = f - u^{(i+1)} \tag{19}$$

and the solution can be obtained by conjugate gradient descent (notice that $A = \mu\Psi^*\Gamma^2\Psi + \operatorname{Id}$ is positive symmetric).

3.3 Decomposition of a Noise Free Image

If the input image f does not contain any noise, one can also decompose f into only two components, the sketch u and the texture $f - u$:

$$(u, \xi) = \underset{\tilde{u}, \, \tilde{\xi} \in \mathcal{C}}{\operatorname{argmin}} \, \lambda\|\tilde{u}\|_{\mathrm{TV}} + \frac{1}{2}\|f - \tilde{u}\|_{\tilde{\xi}}^2, \tag{20}$$

In this case, a faster algorithm can be used. The minimization step on ξ is the same as the one described in section 3.1, but the second step on v is different. One can use an extension of Chambolle's algorithm designed to deal with inverse problem, see for instance [21,22] for equivalent description of this method. From an initial texture layer $u^{(0)} \in \mathbb{R}^N$, this algorithm iterates between a gradient step of the functional $u \mapsto \|\gamma\Psi u - y\|^2$ (where $y = \gamma\Psi f$):

$$\bar{u}^{(k)} = u^{(k)} + \nu\Psi^*\gamma(y - \gamma\Psi u^{(k)}), \tag{21}$$

where $\nu > 0$ is a step size that should obey $\nu < 2/\|\gamma\Psi\|^2$, and a denoising step

$$u^{(k+1)} = \underset{\tilde{u} \in \mathbb{R}^N}{\operatorname{argmin}} \, \frac{1}{2}\|\bar{u}^{(k)} - \tilde{u}\|^2 + \lambda\nu\|\tilde{u}\|_{\mathrm{TV}}. \tag{22}$$

which is equivalent to (13) and therefore can be solved using the projected gradient descent described in section 3.2.

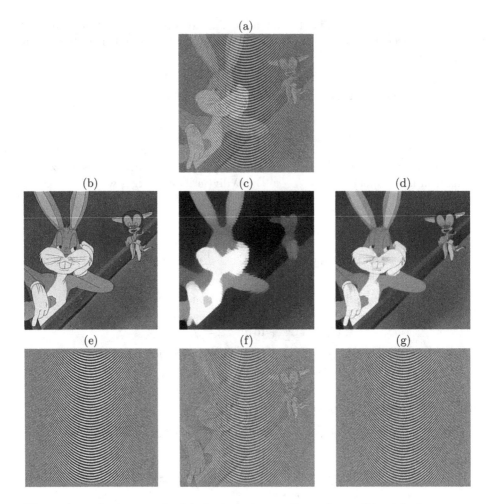

Fig. 2. A synthetic example: (a) the input image, first column: (b) original structure and (e) texture components used to produce the image, second column (c) and (f): decomposition results with $TV - L^2$ and third column (d) and (g) : decomposition results with our adapted TV-Hilbert method. The obtained result is almost perfect.

3.4 Numerical Examples

The local Fourier transform described in section 2.2 depends on two parameters q, which is the size of the local Fourier windows, and Δ_x which measure the overlapping of the windows. Let us note that an estimation of the lowest frequency ξ_{min} present in the texture component to extract is an indication for the choice of the parameter q. As a matter of fact, if q is too small, the spectrum of the local Fourier windows cannot differentiate very low frequency oscillating patterns from geometric information. In fact, we have $\xi_k = k/q$ and we can take $q = 3/\xi_{min}$ to be sure that ξ_{min} is detected. As for the parameter Δ_x, which

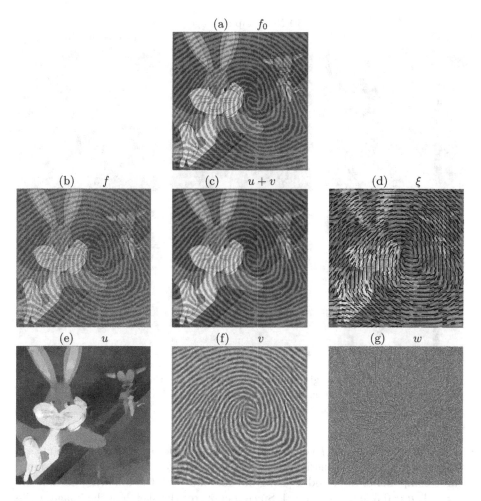

Fig. 3. First row: (a) the original noise free image. Second row: (b) the input noisy image f, (c) the restored image $u + v$ and (d) the estimated orientations of oscillating patterns ξ. Third row: the decomposition into three components, (e) the sketch u of the image, (f) the texture content v and (g) the noise w.

verifies $\Delta_x < q$, it should be taken smaller for a texture which strongly varies spatially than for a texture which is smoother. Good candidates for Δ_x are for example $q/2$ or $q/4$.

Figure 2 presents an example of the decomposition of a noise free image: the input image 256×256, shown in the first row, is generated by addition of a cartoon picture and a synthetic texture whose orientation and frequency vary spatially. These two components are shown in the first column. We applied the $TV - L^2$ method [1] and we chose the smallest parameter λ (on the total variation norm) which provides a total extraction of the texture (here $\lambda = 0.9$). For our method we chose $\lambda = 0.1$, $q = 16$, $\Delta_x = 4$.

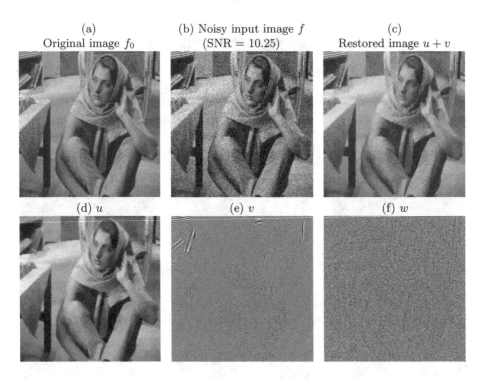

(a)
Original image f_0

(b) Noisy input image f
(SNR = 10.25)

(c)
Restored image $u + v$

(d) u (e) v (f) w

Fig. 4. first row: (a) the original noise free image f_0, (b) the noisy input image f and (c) the restored image $u + v$. second row: the decomposition (d) u the sketch of the image, (e) v the texture content and (f) w the noise.

In Figure 3 an image f composed by a cartoon picture and a fingerprint texture is degraded by a Gaussian noise. The noisy image f is then decomposed into three components u, v, and w using our method with the following parameters $\lambda = 0.1$, $\mu = 0.3$, $q = 16$, $\Delta_x = 4$. Since u captures the sketch of the image, v the locally parallel patterns and w the noise, we can reconstruct a restored version of the noisy image by addition of u and v.

With the same idea, we show in Figure 4 an example of result obtained by this decomposition and denoising process on the "barbara" image. Figure 5 compares our result with two other denoising methods. Every parameter is chosen to achieve the best SNR result. The decomposition between structure and texture provides a better reconstruction of the texture and therefore a better SNR. Let us remark that the former image decomposition frameworks (such as TV-G [2] or TV-H^{-1} [4]) are not suitable for denoising. As a matter of fact, the G and the H^{-1} norms are low for any high-frequency patterns and are then also low for a large part of the noise. On the other hand, the TV norm penalizes strongly oscillating patterns and therefore these models are not able to separate efficiently the texture from the noise. On the contrary our norm is low for patterns which presents a certain frequency and orientation and is therefore more appropriate for denoising.

(a) SNR=17.34   (b) SNR=17.98   (c) SNR=19.93

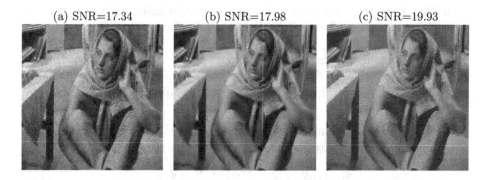

Fig. 5. Comparison with other methods. Denoising of image f from Fig. 4. (a) TV-denoising (λ is chosen to achieve the best SNR, $\lambda = 0.1$),(b) Translation Invariant Wavelet Denoising (the threshold is chosen to achieve the best SNR) and (c) our result which achieves a better SNR.

4 Inpainting with Adapted Hilbert Space

4.1 Simultaneous Cartoon and Texture Inpainting

Inpainting aims at restoring an image f from which a set $\Omega \subset \{0, \ldots, n-1\}^2$ of pixels is missing. It corresponds to the inversion of the ill posed linear problem $y = \Phi f + \varepsilon$ where Φ is defined as

$$(\Phi f)(x) = \begin{cases} 0 & \text{if } x \in \Omega, \\ f(x) & \text{if } x \notin \Omega. \end{cases} \tag{23}$$

and ε is an additive noise. We search for the image f as a decomposition $f \approx u + v$ where u has a low total variation and v has a small Hilbert texture norm. This corresponds to the solution of

$$(u, v, \xi) = \operatorname*{argmin}_{\tilde{u}, \, \tilde{v}, \, \tilde{\xi} \in \mathcal{C}} \lambda \|\tilde{u}\|_{\mathrm{TV}} + \mu \|\tilde{v}\|_{\tilde{\xi}}^2 + \frac{1}{2} \|\Phi(\tilde{u} + \tilde{v}) - y\|_{L^2}^2, \tag{24}$$

where μ and λ should be adapted to the noise level and the regularity of f.

The inpainting is done similarly to section 3 by performing the minimization iteratively on ξ, u and v. The minimization step on ξ does not change, see section 3.1. We describe here the second step, on u and v. Starting from some initial $u^{(0)}$ and $v^{(0)}$, one solves

- v *is fixed:* one minimizes

$$u^{(i+1)} = \operatorname*{argmin}_{\tilde{u}} \lambda \|\tilde{u}\|_{\mathrm{TV}} + \frac{1}{2} \|\Phi \tilde{u} - \bar{y}\|_{L^2}^2 \tag{25}$$

where $\bar{y} = y - \Phi v^{(i)}$. Similarly to (20), this minimization can again be seen as an ill-posed inverse problem from measurements $\bar{y} = \Phi(f - v^{(i)}) + \varepsilon$

(in (20), $y = \gamma \Psi f$) with a total variation regularization. We can therefore use the extension of Chambolle's algorithm described in section 3.3.

- *u is fixed:* one minimizes

$$v^{(i+1)} = \operatorname*{argmin}_{\tilde{v}} \mu \|\gamma \Psi \tilde{v}\|^2 + \frac{1}{2}\|\Phi \tilde{v} - \bar{y}\|_{L^2}^2 \qquad (26)$$

where $\bar{y} = y - \Phi u^{(i+1)}$ and where γ are the local Fourier weights. The solution is computed by conjugate gradient descent to solve the linear system

$$(2\mu\Psi^*\gamma^2\Psi + \Phi^*\Phi)v^{(i+1)} = \Phi^*\bar{y}, \qquad (27)$$

If no noise is present, then the value of $\lambda + \mu$ can be decreased during the iterations of the inpainting algorithm, in order to have a small norm for the residual term $\frac{1}{2}\|\Phi(\tilde{u} + \tilde{v}) - y\|_{L^2}^2$.

4.2   Numerical Examples

Figure 6 presents an example of inpainting reconstruction of the image from Figure 3 degraded by randomly placed holes (350 squares, 15 pixels by 15 pixels, the image is of size 512×512). We used the same parameters as in Figure 3. Let us notice that the texture is well reconstructed thanks to the estimation of

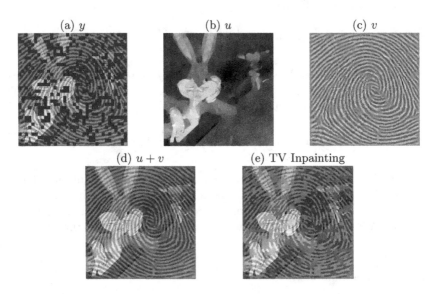

Fig. 6. First row: (a) y, the image to inpaint degraded by randomly chosen holes in black, the original image is f_0 in Fig. 3(a), (b) u the inpainted geometric component, (c) v the inpainted texture component. Second row: (d) $u + v$ the reconstruction using our method, (e) "TV Inpainting" using a simple TV diffusion.

(a)   (b)   (c)   (d)

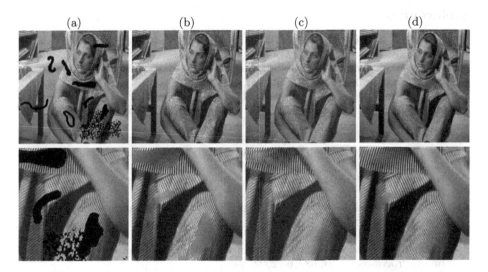

Fig. 7. Inpainting of a degraded image, the original image is f_0 in Fig. 4. First row, (a) the image to inpaint, (b) reconstruction using a TV diffusion, (c) result of MCA [17] with curvelet and local discrete cosine dictionaries, (d) our reconstruction. Our method achieves a better reconstruction of the texture directions inside the missing parts.

the orientations and to the overlapping of the local FD windows. On the other hand the reconstruction of the geometric component is only accomplished by the effect of the Total Variation norm. However since our method provides a separation into two components (geometry and texture), one can imagine to apply a post-processing on the geometric component using any method available in order to improve the final reconstruction. Figure 7 shows a second example of inpainting reconstruction for the image from Figure 4. For comparison, we also show the result of the TV diffusion process and the MCA method [17], using a curvelet dictionary for the cartoon component and a local discrete cosine transform for the texture part. For images with locally parallel patterns, our method achieves a better reconstruction of the directions of the texture inside the missing parts.

5   Conclusion

In this paper we presented a new adaptive texture model well-suited for locally parallel oscillating patterns. The use of this adaptive norm improves state of the art algorithms both in decomposition and inpainting for images which contain oriented textures. The adaptivity is in fact crucial for this kind of images where the texture is anisotropic, since it allows to take into account the texture geometry.

References

1. Rudin, L.I., Osher, S., Fatemi, E.: Nonlinear total variation based noise removal algorithms. Phys. D 60(1-4), 259–268 (1992)
2. Meyer, Y.: Oscillating Patterns in Image Processing and Nonlinear Evolution Equations. American Mathematical Society, Boston (2001)
3. Aujol, J.F., Aubert, G., Blanc-Feraud, L., Chambolle, A.: Image decomposition into a bounded variation component and an oscillating component. Journal of Mathematical Imaging and Vision 22(1), 71–88 (2005)
4. Osher, S., Solé, A., Vese, L.: Image decomposition and restoration using total variation minimization and the H^{-1} norm. Multiscale Modeling & Simulation 1(3), 349–370 (2003)
5. Nikolova, M.: A variational approach to remove outliers and impulse noise. J. Math. Imaging Vis. 20(1-2), 99–120 (2004)
6. Starck, J.L., Elad, M., Donoho, D.: Redundant multiscale transforms and their application for morphological component analysis. Advances in Imaging and Electron Physics 132 (2004)
7. Masnou, S.: Disocclusion: a variational approach using level lines. IEEE Trans. Image Processing 11(2), 68–76 (2002)
8. Bertalmio, M., Sapiro, G., Caselles, V., Ballester, C.: Image inpainting. In: Siggraph 2000, pp. 417–424 (2000)
9. Shen, J., Ha Kang, S., Chan, T.: Euler's elastica and curvature-based inpainting. SIAM Journal of Applied Mathematics 63(2), 564–592 (2003)
10. Tschumperlé, D.: Fast Anisotropic Smoothing of Multi-Valued Images using Curvature-Preserving PDE's. Int. J. of Computer Vision 68(1), 65–82 (2006)
11. Bornemann, F., März, T.: Fast image inpainting based on coherence transport. J. Math. Imaging Vis. 28(3), 259–278 (2007)
12. Efros, A.A., Leung, T.K.: Texture synthesis by non-parametric sampling. In: ICCV 1999, p. 1033 (1999)
13. Wei, L.Y., Levoy, M.: Fast texture synthesis using tree-structured vector quantization. In: SIGGRAPH 2000, pp. 479–488 (2000)
14. Sun, J., Yuan, L., Jia, J., Shum, H.Y.: Image completion with structure propagation. In: SIGGRAPH 2005, pp. 861–868 (2005)
15. Bertalmio, M., Vese, L., Sapiro, G., Osher, S.: Simultaneous structure and texture image inpainting. IEEE Transactions on Image Processing 12, 882–889 (2003)
16. Elad, M., Starck, J., Querre, P., Donoho, D.: Simultaneous cartoon and texture image inpainting using morphological component analysis (mca). Applied and Computational Harmonic Analysis 19(3), 340–358 (2005)
17. Fadili, M.J., Starck, J.L., Murtagh, F.: Inpainting and zooming using sparse representations. The Computer Journal 52, 64–79 (2007)
18. Aujol, J.F., Gilboa, G., Chan, T., Osher, S.: Structure-texture image decomposition—modeling, algorithms, and parameter selection. International Journal of Computer Vision 67(1), 111–136 (2006)
19. Aujol, J.F., Gilboa, G.: Constrained and SNR-based solutions for tv-hilbert space image denoising. Jmiv 26(1-2), 217–237 (2006)
20. Chambolle, A.: An algorithm for total variation minimization and applications. J. Math. Imaging Vis. 20, 89–97 (2004)
21. Bect, J., Blanc Féraud, L., Aubert, G., Chambolle, A.: A ℓ_1-unified variational framework for image restoration. In: Pajdla, T., Matas, J(G.) (eds.) ECCV 2004. LNCS, vol. 3024, pp. 1–13. Springer, Heidelberg (2004)
22. Aujol, J.F.: Some first-order algorithms for total variation based image restoration. J. Math. Imaging Vis. (in press)

Global Optimal Multiple Object Detection Using the Fusion of Shape and Color Information

Marek Schikora

FGAN Research Institute for Communication, Information Processing and
Ergonomics (FKIE)
D-53343 Wachtberg, Germany
schikora@fgan.de

Abstract. In this work we present a novel method for detecting multiple objects of interest in one image, when the only available information about these objects are their shape and color. To solve this task we use a global optimal variational approach based on total variation. The presented energy functional can be minimized locally due its convex formulation. To improve the runtime of our algorithm we show how this approach can be scheduled in parallel.Our algorithm works fully automatically and does not need any user interaction. In experiments we show the capabilities in non-artificial images, e.g. aerial or bureau images.

1 Introduction

To detect multiple objects of interest we use the concept of image segmentation. We will segment the image plane into two regions: foreground (objects of interest) and background. In this context we will use the minimization of an energy functional in continuous space introduced in [1] and [2]. The usage of shape information for image segmentation is normally done using the level-set representations (cf. [3] [4] [5] [6]). In this representation a shape is defined as the boundary given by the zero level set of an embedding function $\phi : \mathbb{R}^d \to \mathbb{R}$:

$$C = \left\{ \boldsymbol{x} \in \mathbb{R}^d \Big| \phi(\boldsymbol{x}) = 0 \right\}. \tag{1}$$

The shape priors in this context are then defined on a space of embedding functions using the space of signed distance functions. Although this formulation has its benefits (independency of parametrization and easy handling of topological changes) there exist two well-known drawbacks: Firstly, the space of signed distance functions is not a linear space, and secondly, the resulting cost or energy functionals are generally not convex.

Recently, an alternative to the continuous level set representation has been proposed, where the segmentation of images is formulated on the basis of convex functional minimization using the concept of *Total Variation*(TV) (c.f [7], [8]). In [9] the formulation of a globally optimal color-based image segmentation using the TV norm was shown. In this paper we extend this work by combining it with shape information.

D. Cremers et al. (Eds.): EMMCVPR 2009, LNCS 5681, pp. 443–454, 2009.

2   Shape Information

In this section we briefly describe the shape prior model, introduced in [10], which will be used in the following because of its convex and continuous formulation.

For the representation of shapes we use the shape space $\mathcal{Q}$:

Definition 1. *A* **shape** *in* $\mathbb{R}^d$ *is a function*

$$q : \mathbb{R}^d \to [0,1], \tag{2}$$

which assigns to any pixel $x \in \mathbb{R}^d$ *a probability* $q(x)$ *that* x *is part of the object. The space of all shapes will be denoted* $\mathcal{Q}$. *In our case we will only consider planar shapes, so we set* $d = 2$.

The benefit of this model lies in the independency of any parametrization. So the problem of shape alignment does not require the estimation of point correspondences. Furthermore the values of q can be easily interpreted in a probabilistic sense. Cremers et al. have shown in their paper [10] that the shape space $\mathcal{Q}$ is convex. This characteristic of $\mathcal{Q}$ leads to the conclusion that any convex combination of elements of the set

$$\chi = \{q_1, q_2, ..., q_N\} \tag{3}$$

is a valid shape. With this we can define statistic quantities such as mean, covariance matrices and eigenmodes of a training set χ.

Let $\chi = \{q_1, q_2, ..., q_N\}$ be a set of N training shapes; then the mean value $\mu : \mathbb{R}^2 \to [0,1]$ of this set is defined through

$$\mu(x) = \frac{1}{N} \sum_{i=1}^{N} q_i(x). \tag{4}$$

This is a function that assignes to each pixel $x \in \mathbb{R}^2$ the average of all probabilities. Using principal component analysis (PCA) we compute the eigenmodes of the shape set χ. We use only a subspace of χ spanned by the first $n \leq N$ eigenmodes $\{\psi_1, \psi_2, ..., \psi_n\}$. The size n follows from the cumulative energy content for each eigenmode. In experiments we used a threshold value of about 0.8. Figure 1 shows the normalized cumulative energy for our training set database. Now a subspace χ_n is given by:

$$\chi_n = \left\{ q_\alpha = \mu + \sum_{i=1}^{n} \alpha_i \psi_i \ \middle| \ q_\alpha(x) \in [0,1], \ \alpha_i \in \mathbb{R} \right\}. \tag{5}$$

In [10] it was shown that χ_n is convex. Now we can generate an shape from this space as

$$q_\alpha = \mu + \alpha^T \Psi \tag{6}$$

With this we can describe every shape only storing the vector $\alpha \in \mathbb{R}^n$. Ψ is a matrix containing the eigenmodes $\psi_1, \psi_2, ..., \psi_n$. Figure 2 shows some examples.

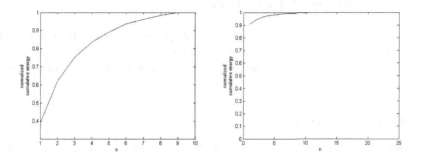

Fig. 1. Normalized cumulative energy content of eigenmodes vs. the number of eigenmodes used for the representation for a database of human hands (left figure) and for a car database segmented manually from aerial images (right).

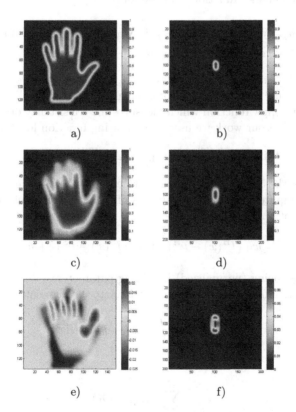

Fig. 2. Shape information: On the left side a hand database is used, on the right side we use a collection of manually-segmented cars from aerial images. a) and b) are example shapes from our database. c) and d) are the mean shapes μ from equation (4). e) and f) represent the first eigenmode ψ_1 of the database.

3   Multiple Object Detection

In this section we describe how to detect multiple object using shape and color information. First we formulate a convex energy function and show that it can be computed efficiently by parallelization. Then we describe the following steps of our algorithm.

3.1   Convex Functional

The energy function used in [9] was formulated for color-based image segmentation. We extend this approach to incorporate also shape information. The general form of a functional for a desired segmentation $u : \mathbb{R}^2 \to [0, 1]$ is

$$E(u) = E_{\text{img}}(u) + \beta \cdot E_{\text{shape}}(u). \tag{7}$$

The color based energy function is of the form

$$E_{\text{img}}(u) = \int_\Omega f(\boldsymbol{x})u(\boldsymbol{x}) \, \mathrm{d}\boldsymbol{x} + \gamma \int_\Omega |\boldsymbol{\nabla} u(\boldsymbol{x})| \, \mathrm{d}\boldsymbol{x} + \rho \int_\Omega \xi(u(\boldsymbol{x}))\mathrm{d}\boldsymbol{x}, \tag{8}$$

where $\Omega \subseteq \mathbb{R}^2$ denotes the image plane and $\beta, \gamma, \rho \in \mathbb{R}$ are weighting parameters. The function ξ penalizes values of u lying outside of the valid range of $[0, 1]$. f can be an arbitrary function which measures the consistency of a point $\boldsymbol{x}$ with the foreground. In our work we used the following function for f:

$$f(\boldsymbol{x}) = \Delta\left(I^{\text{HSV}}(\boldsymbol{x}), \boldsymbol{\nu}_{\text{obj}}\right) - \Delta\left(I^{\text{HSV}}(\boldsymbol{x}), \boldsymbol{\nu}_{\text{bgd}}\right). \tag{9}$$

Here, I^{HSV} is the input image I transformed into the HSV color space. The function Δ computes the squared distances of the single channels of I^{HSV} to the mean value $\boldsymbol{\nu}$ of a region.

$$\Delta\left(I^{\text{HSV}}(\boldsymbol{x}), \boldsymbol{\nu}\right) = w_{\text{H}}\left(I^{\text{H}}(\boldsymbol{x}) - \nu^{\text{H}}\right)^2 + w_{\text{S}}\left(I^{\text{S}}(\boldsymbol{x}) - \nu^{\text{S}}\right)^2 + w_{\text{V}}\left(I^{\text{V}}(\boldsymbol{x}) - \nu^{\text{V}}\right)^2 \tag{10}$$

$w_{\text{H}}, w_{\text{S}}$ and w_{V} being (normalized) weighting parameters.

The term introducing the shape information into the segmentation is $E_{\text{shape}}(u)$:

$$E_{\text{shape}}(u) = \int_\Omega |u(\boldsymbol{x}) - \tilde{q}_\alpha(\boldsymbol{x})| \, \mathrm{d}\boldsymbol{x} \tag{11}$$

with

$$\tilde{q}_\alpha = \sum_{k=1}^{K} \Phi_u\left(q_{\alpha k}, \Theta_k\right). \tag{12}$$

The function Φ_u projects the shape $q_{\alpha k}$ into the image plane of u using the transformation vector $\Theta_k = (t_x, t_y, \phi, \lambda)$ for every object k. K is the number of object in the image I. This number is estimated automatically. Details on this will be given later in this paper. The transformation vector Θ_k contains two parameters for the translation (t_x and t_y), one for rotation (ϕ), and one for scaling (λ). With these parameters we can perform any similarity transformation of a planar shape q. q_{α_k} is a shape generated from our database given the vector α_k:

$$q_{\alpha_k} = \mu + \sum_{i=1}^{n} \alpha_k(i) \cdot \psi_i. \tag{13}$$

Let us define the transformed version of q_{α_k} as:

$$q_{\alpha_k}^{\Theta_k} = \Phi_u(q_{\alpha_k}, \Theta_k). \tag{14}$$

Since it was shown in [9] that E_{img} is a convex functional, what remains to be shown is that $E_{\text{shape}}(u)$ is also convex.

Lemma 1. *The energy functional (11) is convex.*

Proof (of lemma 1). To show that (11) is convex with respect to u, we have to show that for all $\rho \in (0, 1)$ holds

$$\forall u_1, u_2 : E_{\text{shape}}((1 - \rho)u_1 + \rho \cdot u_2) \leq (1 - \rho)E_{\text{shape}}(u_1) + \rho \cdot E_{\text{shape}}(u_2). \tag{15}$$

So we can write

$$E_{\text{shape}}((1 - \rho)u_1 + \rho \cdot u_2) = \int_{\Omega} |(1 - \rho)u_1 + \rho \cdot u_2 - \tilde{q}_\alpha| \, d\boldsymbol{x} \tag{16}$$

$$\leq \int_{\Omega} (1 - \rho)|u_1 - \tilde{q}_\alpha| + \rho \cdot |u_2 - \tilde{q}_\alpha| \, d\boldsymbol{x} \tag{17}$$

$$= \int_{\Omega} (1 - \rho)|u_1 - \tilde{q}_\alpha| \, d\boldsymbol{x} + \int_{\Omega} \rho \cdot |u_2 - \tilde{q}_\alpha| \, d\boldsymbol{x} \tag{18}$$

$$= (1 - \rho) \cdot E_{\text{shape}}(u_1) + \rho \cdot E_{\text{shape}}(u_2) \tag{19}$$

$\square$

For the sake of completeness we write down the complete energy functional:

$$E(u) = \int_{\Omega} f(\boldsymbol{x})u(\boldsymbol{x}) \, d\boldsymbol{x} + \gamma \int_{\Omega} |\boldsymbol{\nabla} u(\boldsymbol{x})| \, d\boldsymbol{x}$$

$$+ \rho \int_{\Omega} \xi(u(\boldsymbol{x})) \, d\boldsymbol{x} + \beta \int_{\Omega} |u(\boldsymbol{x}) - \tilde{q}_\alpha(\boldsymbol{x})| \, d\boldsymbol{x}. \tag{20}$$

Since the norm function is not continuously differentiable we will replace it with a smoothed version by introducing a small offset $\epsilon \in \mathbb{R}$:

$$|u|_\epsilon = \sqrt{u^2 + \epsilon^2}. \tag{21}$$

In experiments we often used $\epsilon = 0.001$.

Now we can formulate the Euler-Lagrange equation of (20):

$$\frac{\partial E}{\partial u} = f - \gamma \text{div} \left(\frac{\nabla u}{|\nabla u|_\epsilon} \right) + \rho \xi'(u) + \beta \frac{u - \tilde{q}_\alpha}{\sqrt{(u - \tilde{q}_\alpha)^2 + \epsilon^2}} = 0 \qquad (22)$$

Without the shape term you can solve equation (22) as a system of linear equations, e.g. with successive over-relaxation (SOR). Details on this can be found in [9]. We write the new shape term in equation (22) as:

$$s(u) = \frac{u - \tilde{q}_\alpha}{\sqrt{(u - \tilde{q}_\alpha)^2 + \epsilon^2}} \qquad (23)$$

Due the fact that $s(u)$ is not linear in u we have to perform a linearization by first-order Taylor expansion:

$$s(u_t) = s\left(u^{t-1}\right) + s'\left(u^{t-1}\right) \cdot \left(u^t - u^{t-1}\right) \qquad (24)$$

$$= s\left(u^{t-1}\right) + \frac{\epsilon^2}{((u^{t-1} - \tilde{q}_\alpha) + \epsilon^2)^{3/2}} \cdot \left(u^t - u^{t-1}\right) \qquad (25)$$

Since we use a iterative solver such as SOR we know the solution of u from the last timestep $t - 1$ and denote it here as u^{t-1}. The value of $s(u^{t-1})$ can then be seen as a constant. With this we can generate a system of linear equations. For the SOR formalism we need a linear system of equations of the form $\mathbf{A}u = \mathbf{b}$. For this we write u as a vector $\mathbf{u}$, such that the columns of the image matrix are concatenated to an N-dimensional column vector with N the number of pixels. The vector $\mathbf{b}$ is given by the constant part of (22),

$$b_i = -f - \beta \cdot s\left(u^{t-1}(i)\right) - \beta \cdot s'\left(u^{t-1}(i)\right) \cdot u^{t-1}(i). \qquad (26)$$

Accordingly, $\mathbf{A}$ contains the u^t-depended part (22). It is useful to replace the function $\xi(u)$ in the actual implementation with a simple thresholding. We obtain for $\mathbf{A} = (a_{ij})$:

$$a_{ij} = \begin{cases} g_{i \sim j} & \text{if } j \in \mathcal{N}(i) \\ \beta \cdot s'(u^{t-1}(i)) - \sum_{k \in \mathcal{N}(i)} g_{i \sim k} & \text{if } i = j \\ 0 & \text{otherwise} \end{cases} \qquad (27)$$

where $g_{i \sim j}$ is the diffusivity between pixel i and its neighbor j. $\mathcal{N}(i)$ denotes the neighborhood of pixel i. The Matrix $\mathbf{A}$ is diagonally dominant. In our experiments we use a 4-connected neighborhood, so we get only five non-zero diagonals. All other entries of $\mathbf{A}$ are zero. Because the diffusivity $g = \frac{1}{|\nabla u|}$ depends on the actual solution for u, we do not really have a linear system of equations, but we make the assumption, that the diffusion is constant, and we perform a new computation of it only every L iterations.

For a speedup in the computation time we use the red-black computation scheme for SOR (see [11] for details). With this we schedule the computation parallel, so that we create a separate thread for every pixel that computes the solutions using the latest information from its neighbor. For this computation we use the NVIDIA CUDA framework, so the main computing is done in parallel on the GPU.

3.2 Estimation of the Optimal Transformation Parameters for Every Shape

Given an initial solution of u we need to determine the number of object candidates in the segmented image. Since u is almost binary this can be solved easily, e.g. through connected components. This gives us the number K of possible objects in the input image I. For each of these candidates we need to know its transformation parameters Θ_k.

Using a parallel framework we can compute the residuum

$$\mathbf{r} = \mathbf{b} - \mathbf{A}\mathbf{u}. \tag{28}$$

given the actual solution u, all transformation parameters Θ_k and all shape parameters $\boldsymbol{\alpha}_k$ for $k = 1, 2, ..., K$. The estimation of the optimal Θ_k for all k is done by computing a "branch & bound" search on the space of valid transformations parameters. For initialization we set the values of the translation parameters to the barycenter of each candidate. The norm of $\mathbf{r}$ indicates the correctness of the found parameters. In every node in the branch & bound searching tree we save the actual intervals for all parameters, the norm of $\mathbf{r}$ and an indicator holding the information which interval of a parameter has to be divided for the next level of the search. The search is stopped if a satisfying accuracy is achieved, e.g. when the residuum does not change any more. It was shown in [10] that this approach leads to a globally optimal solution. Although our derivation is more general, the extension of the proof shown there is straight-forward and will not be presented here.

3.3 Estimating the Optimal Shape Representation

Knowing the actual solution of u and the optimal transformation parameters Θ_k we have to estimate the optimal shape parameters $\boldsymbol{\alpha}_k$ for every candidate $k = 1, 2, ..., K$. This can be summarized in three steps:

1. divide $\tilde{q}_\alpha$ into $q_{\alpha_1}^{\Theta_1}, q_{\alpha_2}^{\Theta_2}, ..., q_{\alpha_K}^{\Theta_K}$, such that each $q_{\alpha_k}^{\Theta_k}$ only contains information of candidate k (cf. Figure 3),
2. transform the eigenmodes $\psi_1, ..., \psi_n$ with $\Phi_u(\psi_i, \Theta_k)$ for each eigenmode $i = 1, .., n$ and each candidate $k = 1, ..., K$, so that you get a transformed set of eigenmodes $\boldsymbol{\Psi}_k$ for each candidate,
3. solve:

$$\min_{\boldsymbol{\alpha}_k} \| \boldsymbol{\Psi}_k^T \cdot \boldsymbol{\alpha}_k - \left(q_{\alpha_k}^{\Theta_k} - \Phi_u(\mu, \Theta_k) \right) \| \tag{29}$$

for all $k = 1, ..., K$.

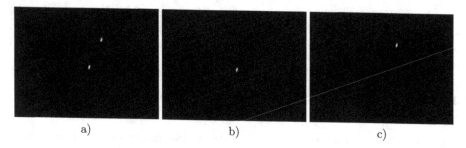

a)   b)   c)

Fig. 3. Examples of Step 1 in section 3.3: a) $\tilde{q}_\alpha$, b) $q_{\alpha_1}^{\Theta_1}$, c) $q_{\alpha_2}^{\Theta_2}$

The first two steps can be easily implemented. The third step can be solved in different ways. We use in our experiments a singular value decomposition (SVD) to obtain α_k. Due the fact that n is a very small number (in our case 3 or 5) the computation time of the SVD is short. So we do not need a more complex solving algorithm. If the training set database contains many dissimilar shapes then n will be larger and a different computation strategy for step 3 would be probably faster.

3.4   Algorithm Summary

Now we can summarize the whole algorithm:

1. solve (22) with $\beta = 0$ to get an initial solution for u only based on the color information,
2. determine the number of object candidates K,
3. estimate the optimal translation parameters Θ_k for $k = 1, 2, ..., K$ using branch & bound,
4. estimate the optimal shape parameters α_k solving (29) for $k = 1, 2, ..., K$,
5. check for each candidate $k = 1, 2, ..., K$ whether the segmented object matches the found shape representation $q_{\alpha_k}^{\Theta_k}$ and discard false responses,
6. solve (22) with $\beta \neq 0$ to get a optimal solution for u based on color and shape information
7. if the accuracy is sufficient stop, else return to step 2.

Step 5 can be realized with the following procedure. First divide the segmentation u into disjoint images $u_1, u_2, ..., u_K$, so that each u_k contains only the information of u that corresponds to candidate k. Since we have already found the optimal translation and shape parameters of the corresponding shape $q_{\alpha_k}^{\Theta_k}$, we can now simply compute the difference of u_k and $q_{\alpha_k}^{\Theta_k}$:

$$d_k\left(u_k, q_{\alpha_k}^{\Theta_k}\right) = \left| u_k - q_{\alpha_k}^{\Theta_k} \right|. \tag{30}$$

If the cumulated and normalized difference is bigger than a threshold $\tau \in \mathbb{R}$, then the candidate is discarded, and we save this information, such that the candidate will not re-appear in the segmentation. This can be realized with:

$$I(\boldsymbol{x}) = \begin{cases} u_k(\boldsymbol{x}) \cdot \boldsymbol{\nu}_{\mathrm{bgd}} + (1 - u_k(\boldsymbol{x})) \cdot I(\boldsymbol{x}) & , \ \frac{1}{\|\Omega\|} \cdot \int\limits_{\Omega} d_k(\boldsymbol{x})\mathrm{d}\boldsymbol{x} > \tau \\ I(\boldsymbol{x}) & \text{otherwise} \end{cases} \tag{31}$$

A more precise shape verification strategy, e.g. shape matching, can be applied to step 5, but was not needed in our experiments. Since shape matching generally needs high computation times we solved this problem here in a simpler way to save runtime. Some fast algorithms for shape matching are described in [12] and [13].

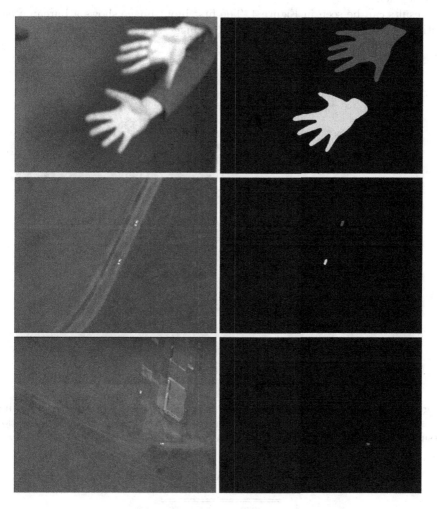

Fig. 4. Object detection results using shape and color information. Left column: input images. Right column: Detection results presented as colored version of the segmentation result u. Each color represents a label for a pixel.

4   Results

In this section we present the results obtained with the proposed algorithm. We performed the presented experiments on a Intel Core2Quad 8200 CPU with 4GB RAM and a NVIDIA GeForce GTX280 with 1GB RAM. As already described in Section 2, we use a hand database which we test on bureau images. In addition to this we created a car database from manually segmented aerial images. These images were taken from a height of about 500 meters obove ground with opening angles of 13.6 and 10.4 degrees. The resolution of both image categories is 1024 × 768 pixels.

Results can be seen in Figure 4. The first input image shows a bureau scene with hands in it. The challenge with this image is the high level of noise and

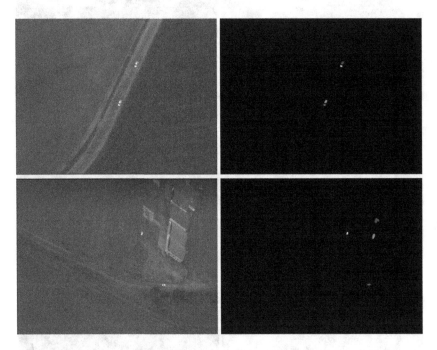

Fig. 5. Object detection results using only color information. Left column: input images. Right column: Detection results presented as colored version of the segmentation result u. Each color represents a label for a pixel.

Table 1. Running times for multiple object detection based on aerial images

size	sec
256 × 192	0.475
512 × 384	1.077
1024 × 768	3.192

strongly varying color distribution of both hands. Despite these difficulties our method yields the correct segmentation and the corresponding positions of objects of interest in this scene. The next images show aerial scenes in which cars shall be detected. Here, the challenging point is the fact that often the color of the car windows differ strongly from the rest of the car. This leads a algorithm only controlled by color information to the belief, that a car in a scene consists of two objects. Examples for this behavior can be seen in Figure 5. But the fusion of color and shape information yields the correct segmentation. Furthermore, the man-made objects in this scene (e.g. houses) have the same color distribution as the car, so they will appear as object candidates when using color information only (c.f Figure 5). In addition to this the shape representation of a car is quite unspecific (c.f Figure 2), so a shape-only algorithm will not work properly. The benefit lies here in the fusion of both approaches.

Table 1 displays the running times for our algorithm with a GPU-based solution of (22). Since these times depend on the number of objects found, we used the first aerial image from Figure 4 with different resolutions for our time measurements. We did not use a parallel version of SVD to solve (29). This would further decrease the computation time.

5   Conclusion

In this work we presented a novel method for a globally optimal multiple object detection using shape and color information. The proposed method is based on a convex energy functional for image segmentation. We showed how this functional can be parallized to improve the computation time. In experiments we demonstrated the capabilities of this approach with challenging scenes.

In future work we indend to decrease the computation time through a parallel solving of (29) and a faster solving method for the transition parameters of the objects.

References

1. Kass, M., Witkin, A., Terzopoulos, D.: Snakes: Active contour models. International Journal of Computer Vision 1(4), 321–331 (1988)
2. Mumford, D., Shah, J.: Optimal approximations by piecewise smooth functions and variational problems. CPAM XLII(5), 577–685 (1989)
3. Cremers, D.: Dynamical statistical shape priors for level set based tracking. IEEE Trans. on Patt. Anal. and Mach. Intell. 28(8) (August 2006)
4. Cremers, D., Tischhuser, F., Weicker, J., Schnörr, C.: Diffusion snakes: Introducing statistical shape knowledge into the mumford-shah functional. International Journal of Computer Vision 50(3), 295–313 (2002)
5. Rousson, M., Paragios, N.: Prior knowledge, level set representations and visual grouping. International Journal of Computer Vision 76(3), 231–243 (2008)
6. Rousson, M., Paragios, N., Deriche, R.: Implicit active shape models for 3D segmentation in mri imaging. In: MICAI. LNCS, vol. 2217, pp. 209–216. Springer, Heidelberg (2004)

7. Chambolle, A.: Total variation minimization and a class of binary MRF models. In: Rangarajan, A., Vemuri, B.C., Yuille, A.L. (eds.) EMMCVPR 2005. LNCS, vol. 3757, pp. 136–152. Springer, Heidelberg (2005)
8. Chan, T.F., Esedoglu, S., Nikolova, M.: Algorithms for finding global minimizers of image segmentation and denoising models. SIAM Journal on Applied Mathematics 66(5), 1632–1648 (2006)
9. Schikora, M., Häge, M., Ruthotto, E., Wild, K.: A convex formulation for color image segmentation in the context of passive emitter localization. In: International Conference on Information Fusion (July 2009)
10. Cremers, D., Schmidt, F.R., Barthel, F.: Shape priors in variational image segmentation: Convexity, lipschitz continuity and globally optimal solutions. In: IEEE Conference on Computer Vision and Pattern Recognition (CVPR), Anchorage, Alaska (June 2008)
11. Klodt, M., Schoenemann, T., Kolev, K., Schikora, M., Cremers, D.: An experimental comparison of discrete and continuous shape optimization methods. In: Forsyth, D., Torr, P., Zisserman, A. (eds.) ECCV 2008, Part I. LNCS, vol. 5302, pp. 332–345. Springer, Heidelberg (2008)
12. Schmidt, F.R., Töppe, E., Cremers, D.: Efficient planar graph cuts with applications in computer vision. In: IEEE Conference on Computer Vision and Pattern Recognition (CVPR), Miami, Florida (June 2009)
13. Schmidt, F.R., Farin, D., Cremers, D.: Fast matching of planar shapes in sub-cubic runtime. In: IEEE International Conference on Computer Vision (ICCV), Rio de Janeiro, Brazil (October 2007)

Human Age Estimation by Metric Learning for Regression Problems

Leting Pan

Department of Electronic Engineering, Shanghai Jiao Tong University
800 Dongchuan Road, Shanghai, China 200240
pan.leting@gmail.com

Abstract. The estimation of human age from face images has many real-world applications. However, how to discover the intrinsic aging trend is still a challenging problem. We proposed a general distance metric learning scheme for regression problems, which utilizes not only data themselves, but also their corresponding labels to strengthen the credibility of distances. This metric could be learned by solving an optimization problem. Via the learned metric, it is easy to find the intrinsic variation trend of data by a relative small amount of samples without any prior knowledge of the structure or distribution of data. Furthermore, the test data could be projected to this metric by a simple linear transformation and it is easy to be combined with manifold learning algorithms to improve the performance. Experiments are conducted on the public FG-NET database by Gaussian process regression in the learned metric to validate our framework, which shows that its performance is improved over traditional regression methods.

Keywords: Age Estimation, Metric Learning, Regression.

1 Introduction

Face-based biometric systems such as Human-Computer Interaction have great potential for many real-world applications. As an important hint for human communication, facial images comprehend lots of useful information including gender, expression, age, pose, etc. Unfortunately, compared with other cognition problems, age estimation from face images is still very challenging. This is mainly due to the fact that, aging progress is influenced by not only personal gene but also many external factors. Physical condition, living style and plenty of other things may accelerate or slower aging process. Besides, since aging process is slow and with long duration, collecting sufficient data for training is a fairly strenuous work.

[10,17] formulated human ages as a quadratic function. Yan et al. [27,28] modeled the age value as the square norm of a matrix where age labels were treated as a nonnegative interval instead of a certain fixed value. However, all of them regarded age estimation as a regression problem without special concern about the own characteristics of aging variation. As Deffenbacher [8] stated, the aging factor has its own essential sequential patterns. For example, aging is irreversible, which is expressed as a

D. Cremers et al. (Eds.): EMMCVPR 2009, LNCS 5681, pp. 455–465, 2009.
© Springer-Verlag Berlin Heidelberg 2009

trend of growing older along the time axis. Such general evolution of aging course is beneficial to age estimation, especially when training data are limited and distributed unbalanced over each age range.

Geng et al. [13,12] firstly made some pioneer research on seeking for the underlying aging patterns by projecting each face in their aging pattern subspace (AGES). Guo et al. [16] proposed a scheme based on Orthogonal Locality Preserving Projections (OLPP) [5] for aging manifold learning and get the state-of-art results. In [16], SVR (Support Vector Regression) is used to estimate ages on such a manifold and the result is locally adjusted by SVM. However, they only tested their OLPP-based method on a private large database consisting of only Japanese people, and no dimension reduction work was done to exact the so-called aging trend on the public available FG-NET database [1]. A possible reason is that, FG-NET database may not supply enough samples to recover the intrinsic structure of data. The lack of sufficient data is a prominent barrier in age estimation.

Therefore, how to dig out the underlying variation trend of data within a limited amount of samples is well worth investigation. From a generalized standpoint, manifold learning algorithm is unsupervised distance metric learning, which attempt to preserve the geometric relationships between most of the observed data. The starting point is the input data, while labels are always not taken into consideration. But labels indeed provide important cues about similarities among samples, which is crucial to construct the structure of data, especially under a small given dataset. To take full advantage of labels, a family of supervised metric learning algorithms [3,14,25,26] are developed, which adds label information as a weight to entice samples pertaining to the same class to go nearer by learning a special metric. Yet, almost all of these methods are specially designed for classification problem. For regression problems such as age estimation, there are naturally infinite classes, where the constraints in previous literatures are not practical.

We propose a new framework aiming to learn a special metric for regression problems. Age is predicted based on the learned metric rather than the traditional Euclidean distance. We accomplish this idea by formulating an optimization problem, which approximates a special designed distance that scaled by a factor determined according to the labels of data. In this way, the metric measuring the similarity of samples is strengthened. More importantly, since labels are incorporated to depict the underlying sample distribution tendency, which signifies the inclusion of more information, a smaller amount of training data is required. Unlike the nonlinear manifold learning where it is repeated to find its low dimensional embedding, a merit of our framework is that, a full metric over the input space is learned and expressed as a linear transformation, and it is easy to project a novel data into this metric. Moreover, the proposed framework may also be used as a pre-processing step to assist those unsupervised manifold learning algorithms to find a better solution.

The rest of the manuscript is arranged as follows: Section 2 gives the details of the metric learning formulation for regression problems based on labels of training data. Section 3 takes Gaussian Process Regression (GPR) as an example to explain how to make use of the learned metric. Section 4 demonstrates the experimental results of the performance of the proposed framework on FG-NET Aging Database. Section 5 comments on conclusions.

2 Metric Learning for Regression

Let $S = (X_i, y_i)$ $(1 \leq i \leq N)$ denotes a training set of N observations with inputs $X_i \in R^d$ and their corresponding non negative labels y_i. Our goal is to rearrange these data in high-dimensional space with a distinct trend as what their labels characterize. In other words, we hope to find a linear transformation $T: R^d \rightarrow R^d$, after applying which, the distances between each pair-wise observation may be measured as:

$$\hat{d}(X_i, X_j) = \| T(X_i - X_j) \|^2 \tag{1}$$

The distance $\hat{d}(X_i, X_j)$ should be reliable to measure the difference as what their labels indicate.

2.1 Problem Formulation

Metrics is a general concept, as a function giving a generalized scalar distance between two argument patterns [11]. Straightforwardly, different distances are also possible to depict the tendency of a data set. Similar to Weinberger et al. [25] and Xing et al. [26], we consider learning a distance metric of the form

$$d_A(X_i, X_j) = \sqrt{(X_i - X_j)^T A (X_i - X_j)} \tag{2}$$

But unlike their works for classification problems, in regression problems, every two observations are of different classes. Better metrics over their inputs are expected and a new metric learning strategy ought to be established.

Suppose given certain well-defined distance $\hat{d}_{ij} = \hat{d}(X_i, X_j)$ ideally delineating the data trend, our target is to approximate $\hat{d}_{ij}$ by $d_A(X_i, X_j)$ minimizing the energy function

$$\varepsilon(A) = \sum_{i,j} \left(d_A(X_i, X_j)^p - (\hat{d}_{ij})^p \right)^2 \tag{3}$$

To promise that A is a metric, A is restricted to be symmetric and positive semi-definite. For simplicity, p is assigned to be 2. This metric learning task is formulated as an optimization problem with the form below

$$\min \sum_{i,j} \left((X_i - X_j)^T A (X_i - X_j) - (\hat{d}_{ij})^2 \right)^2 \tag{4}$$

satisfying the matrix A is symmetric and positive semi- definite. And there exists a unique lower triangular L with positive diagonal entries such that $A = LL^T$ [15]. Hence learning the distance metric A is equivalent to finding a linear transform L^T projecting observation data from the original Euclidean metric to a new one by

$$\tilde{X} = L^T X \tag{5}$$

2.2 Distance with Label Information

In practical application, Euclidean distance is not always capable to guarantee the rational relationship among input data. Although manifold learning algorithms may discover the intrinsic low-dimensional parameterizations of the high dimensional data space, at the outset, it also requires Euclidean distance to apply kNN (k-Nearest Neighbors) to know the local structure of the original space. On the other hand, manifold learning demands a large amount of samples, which is not available in some circumstances. Figure 1 visualizes the age manifolds of the FG-NET Aging Database learned by Isomap [24], Locally Linear Embedding (LLE) [21] and OLPP [5] respectively. Data points of age from 0 to 69 are colored from blue to red. From the 2-D view, none of them can detect a distinctive aging trend. A possible reason is that, FG-NET database only have 1002 images, and each person only have a few images that span from 0 to 69, inadequate to approximate its underlying manifold correctly.

For many regression and classification problems, it is in fact a waste of information if only data X_i is utilized but with their associated labels y_i ignored in the training stage. Balasubramanian et al. [2] proposed a biased manifold embedding framework to estimate head poses. In their work, the distance between data is modified by a factor of the dissimilarities fetched from labels. The basic form of this modified distance is

$$d'(i, j) = \frac{\beta \times P(i, j)}{\max_{m,n} P(m,n) - P(i, j)} \times d(i, j) \qquad (6)$$

where $d(i, j)$ is the Euclidean distance between two samples X_i and X_j. $P(i, j)$ is the difference of poses between X_i and X_j.

Through incorporating the label information to adjust Euclidean distance, the modified distances are prone to give rise to the true tendency of data variation i.e. if the distance of two observations is large, then the distance of their labels is also large, vice versa. Hence it is intuitively that the biased distance is a good choice for $\hat{d}_{ij}$ in Eq.(3):

$$\hat{d}(i, j) = \left(\frac{\beta \times |L(i, j)|}{C - L(i, j)} \right)^p \times d(i, j) \qquad (7)$$

Analogously, $L(i, j)$ is the label difference between two data. C is a constant greater than any label value in a train set and p is selected to make data easier to discriminate. $d(i, j)$ is the Euclidean distance between two samples $X_{i\psi}$ and X_j.

2.3 Optimization Strategy

Since the energy function is not convex, it is a non-convex optimization and consequently it is impossible to find a closed form solution. The metric A is with the property to be symmetric and positive semi-definite, so it is natural to compute a numerical solution to Eq.(4) using the Newton's method. Similar to [26], in each iteration, a gradient descent step is employed to update A. The iteration algorithm is summarized as follows:

1. Initialize A and step length α;
2. Enforce A to be symmetric by $A \leftarrow (A+A^T)/2$;
3. The Singular Value Decomposition of $A = L^T \Delta L$, where the diagonal matrix Δ consists of the eigenvalues $\lambda_1, \ldots, \lambda_n$ of A and columns of L contains the corresponding eigenvectors;
4. Ensure A to be positive semi-definite by $A \leftarrow L^T \Delta' L$, where $\Delta' = \text{diag}(\max(\lambda_1, 0), \ldots, \max(\lambda_n, 0))$;
5. Update $A' \leftarrow A - \alpha \nabla_A \varepsilon(A)$, where $\nabla_A \varepsilon(A)$ is the gradient of the energy function in Eq.(3) w.r.t. A;
6. Compare the energy function $\varepsilon(A)$ with $\varepsilon(A')$ in Eq.(3), if $\varepsilon(A) < \varepsilon(A')$, then augment the step length α with a momentum to accelerate the optimization process; otherwise, shrink α to assure a local minimum is not overpassed.
7. If A has converged or the maximum iteration times are reached, terminate; otherwise go back to Step 2.

3 Gaussian Process Regression

GPR from the traditional Minkowski metric is extended to the learned metric. It should be clarified that, the learned metric is designed for regression problems, especially kernel based method such as GPR [20] and SVR (Support Vector Regression) [22]. GPR is preferred here because it can determine the hyper-parameters of the kernel automatically based on Bayesian model selection criterion such as Maximum A Posteriori, Markov Chain Monte Carlo method [18] etc. rather than SVR where parameters is often chosen by cross-validation. Hereby it is not necessary to partition the training set into two parts to get an extra validation set.

For target prediction such as human ages and head poses, designing an appropriate regressor is a key point to model the problem. SVR is one of the most popular and powerful tools, which adopts a Radial Basis Function (RBF) kernel computed in Euclidean space. In recent years, predictor variables in extended versions of SVR are assumed to be in the proximity of a low dimensional manifold embedded in a high dimensional input space.

Sugiyama et al. [23] modified the RBF kernels by substituting the Euclidean distance with the geodesic distance. But it has been proven that for many practical problems, geodesic distance fails to discover the intrinsic structure of data [9]. Figure 1 is just such an example. More importantly, geodesic distance is also not reliable to construct kernels even if in a proper case. Since such distance is approximated by searching the shortest path on a k nearest neighbor graph [24], and there is no guarantee that the kernel matrix is positive semi-definite. Hence it is possible that a local optimum can not be arrived and the inverse matrix could not be computed.

Many existing methods of regression on manifolds measure distance in Minkowski metric [4,19]. The evaluation of similarity by this metric is based on the assumption that, the similar point should be close to the query point in all dimensions. If the attributes of data are many enough, Euclidean distance is not credible. In [2], a biased Euclidean distance scaled by label difference is used to find the k nearest neighbors of each data and then applied to manifold learning algorithms. To construct the nonlinear relationship between the high and low dimensional space, [2] takes a Generalized Regression Neural Network to learn this nonlinear mapping. Yet, the training of a Neural Network is time consuming, and the curse of dimensionality is inevitable since the input dimension (typically raw images) is high.

Given a training set $S = (X_i, y_i)$ ($1 \leq i \leq N$) as described in Section II and a sample X^* for query, GPR predicts its output y^* by putting a Gaussian process prior on this function $f(\cdot)$, assuming that all sample points evaluated from the function have a multivariate Gaussian density [20].

Let $X=[X_1,...,X_N]$ and $Y=[y_1,...,y_N]^T$, the Gaussian predictive distribution of y^* is derived of the form

$$p(y^*|X^*,X,Y,\Theta) \sim N(\mu(X^*),V(X^*)) \tag{8}$$

The mean prediction and covariance matrix in Eq.(8) are

$$\mu(X^*)=k(X^*,X)[K+\sigma^2 I]^{-1}Y \tag{9}$$

$$V(X^*)=k(X^*,X^*)-k(X^*,X)^T[K+\sigma^2 I]^{-1}k(X^*,X) \tag{10}$$

where $k(\cdot,\cdot)$ is the covariance function, K is the covariance matrix of X and σ^2 is the variance of noise.

Another way to perceive and thus rewrite Eq.(9) is to treat the mean prediction as a linear combination of N kernel functions:

$$\mu(X^*) = \sum_{c=1}^{N} \alpha_c k(X^*,x_c) \tag{11}$$

where $\alpha=(K+\sigma^2 I)^{-1}Y$

Gaussian kernel is a good choice for the covariance function

$$k(X_i,X_j)=v^2 \exp(-\|X_i-X_j\|^2/2l^2+\sigma^2 \sigma_{X_iX_j}) \tag{12}$$

In respect that the proposed learned metric encodes label information implicitly, it is bestowed as the similarity measure and Eq.(12) becomes

$$k(X_i,X_j)=v^2 \exp(-(X_i-X_j)^T A(X_i-X_j)/2l^2+\sigma^2 \sigma_{X_iX_j}) \tag{13}$$

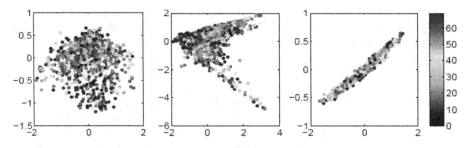

Fig. 1. The 2 dimensional embedding of FG-NET Aging Database by Isomap, LLE and OLPP methods, based on the Euclidean distance. Points of age from 0 to 69 are marked from blue to red as prescribed in the gradient ruler rightward.

4   Experimental Results

Age estimation is carried on the public FGNET Aging Database [1] by the regression strategy on the basis of the proposed metric. The database contains totally 1002 color or gray images from 82 people. Each person has around 10 face images with the ranges from 0 to 69 with labeled ground truth. These images are taken under varying lighting condition, poses and expressions. Each image is labeled by 68 points characterizing its shape features. Similar to [13,16,27,28], input features are selected to be the parameters of AAMs [6]. Figure 2 presents some typical face images and their reconstructed faces by AAM.

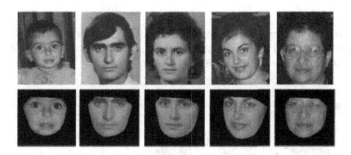

Fig. 2. Typical sample images of FG-NET Aging Database and their AAM synthetic faces

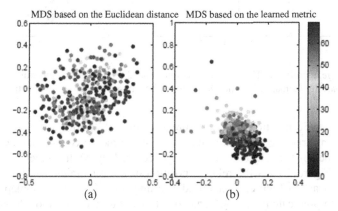

Fig. 3. 2-D view of the clustering effects of the 300 training samples by metric learning. It illustrates the 2 dimensional embedding of the training data sampled from FG-NET Aging Database by MDS. Points of age from 0 to 69 are marked from blue to red. It is seen that, the distance calculated based on our learned metric in Figure (b) preserves local proximity of samples with close labels better than that based on the traditional Euclidean distance in Figure (a).

Firstly we hope to testify that the proposed metric is able to disinter some internal patterns of human's aging progression. We randomly choose 300 images out of all the 1002 images in FG-NET Database as training samples, and the rest as test samples. The parameters in Eq.(7) are chosen as $C=100$, $\beta=1$ and $p=1$. The energy function is

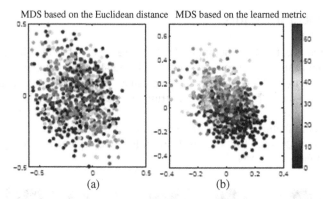

MDS based on the Euclidean distance MDS based on the learned metric

(a)   (b)

Fig. 4. 2-D view of the clustering effects of the 702 testing samples by metric learning, corresponding to Figure 3. It is obvious that, the actual aging trend is, to some extended, manifested in the hyper-space based on our learned metric.

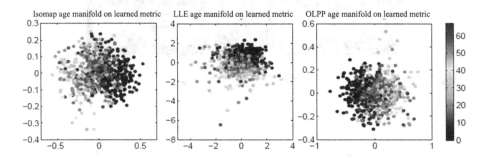

Fig. 5. 2D age manifolds. This figure illustrates the 2 dimensional embedding of FG-NET Aging Database by Isomap, LLE and OLPP algorithms based on our learned metric.

converged after 50 iterations or so. Figure 3(a) and 3(b) portrays the positional relationship among training samples in the hyper-space measured by Euclidean distance and the learned metric A. The 2D view is acquired by Multi-Dimensional Scaling (MDS) [7]. Figure 4 plots the relative position of the remaining 702 image samples for test. Contrast to Figure 1, manifold learning algorithms like Isomap, LLE and OLPP fails to predicate the aging trend sometimes. Furthermore, though only 30% of the entire data set is directed for learning the aging trend is effectually set up.

As in Eq.(5), the original parameters from AAMs can be linearly transformed into a hyper-space based on our learned metric, by multiplying L^T satisfying $A=LL^T$. Figure 5 draws the 2D aging manifold inputted with the transformed data. Compared to Figure 1, the linear transform L^T is salutary for other manifold algorithms to find an improved aging trend.

Then, age estimation of our methodology is compared with the performance of some state-of-art approaches. The Leave- One-Person-Out mode [13,16,27,28] is the mechanism for experimentation, i.e. each time we choose one person for testing and all others for training. The same as in [13,16,27, 28], two criteria are adopted for performance evaluation. One is the Mean Absolute Value (MAE), which is defined as

$$MAE = \sum_{i=1}^{N} |\widetilde{age}_i - age_i| / N \qquad (14)$$

where for each X_i, $\widetilde{age}_i$ is its labeled ground truth and age_i is the estimated age. N is the number of testing images.

Another widely acknowledged criterion is the cumulative score at error level l [13]

$$CumScore(l)=N_{error \leq l}/N \times 100\% \qquad (15)$$

In respect that, when a face image is labeled as O years old, the person is customarily thought to be $[O,O+1)$ years old [27], thus the error less than a specified number of years is by and large neglectable in practical application. Eq.(15) is an indicator of the algorithmic correct rate.

The parameters in Eq.(7) are rectified to be $C=80$, $\beta=1$ and $p=0.6$. Table 1 lists the MAE of different approaches. The MAE of the proposed method is almost the same as the best one [16]. However, unlike their LARR, we simply predict ages in a new metric by regression without any local refinement. LARR slides the estimated age up and down by checking different age values to see if it can come up with a better

Table 1. MAE comparison of different methods

Reference	Method	MAE
[13]	AGES	6.77
[12]	KAGES	6.18
[27]	RUN1	5.78
[28]	RUN2	5.33
[16]	LARR	5.07
Proposed	Metric learning+GPR	5.08

Table 2. MAEs over various age ranges on FG-NET Database for the proposed method, GPR and RUN. In the first column, the value in the parenthesis stands for the proportion (percentage) for each age group out of the whole database.

Age Range	Proposed	GPR	RUN[27]
0-9(37.0%)	2.99	3.55	2.51
10-19(33.8%)	4.19	4.34	3.76
20-29(14.4%)	5.34	5.09	6.38
30-39(7.9%)	9.28	9.04	12.51
40-49(4.6%)	13.52	14.65	20.09
50-59(1.5%)	17.79	19.77	28.07
60-69(0.8%)	22.68	31.76	42.50
Average	5.08	5.45	5.78

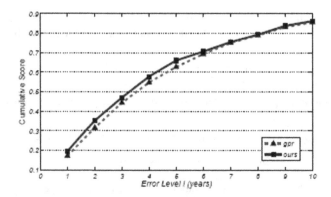

Fig. 6. Cumulative scores of our method and GPR at error levels [0,10]

prediction [16]. The parameters defining the search range is determined manually, which is at least not convenient and automatic enough, and may be laborious and not feasible in some real-world applications. Table 2 details Table 1 with separate MAEs over different age range. The MAE of our method in younger people is slightly higher than other recent methods. As compensation, an outstanding improvement is achieved in the larger age range. This trait is fairly attractive considering the fact that, people over 30 years old account for less than 15% of the whole FG-NET database. Even if there are only a few samples (for example, there are only 8 images out of 1002 over 60 years old), a relatively acceptable age prediction can be obtained.

Figure 6 displays the cumulative scores of our method and GPR. It can be found that at each level, our method is better than GP regression to some degree.

5   Conclusions

In this paper, a new metric learning framework is proposed to resolve regression problems. It is feasible to be applied to many other problems in machine learning or computer vision. No assumptions about the structure or distribution of the samples are made, and a relatively small quantity of training samples is required to learn their underlying variation trend. Experiments shows the effectiveness of the learned metric to restore the intrinsic infrastructure of input sample data and encouraging performance is acquired on a widely used public face aging database.

References

1. FG-NET Aging Database, http://www.fgnet.rsunit.com
2. Balasubramanian, V.N., Ye, J., Panchanathan, S.: Biased manifold embedding: A framework for person-independent head pose estimation. In: IEEE Conf. CVPR, pp. 1–7 (2007)
3. Bar-Hillel, A., Weinshall, D.: Learning distance function by coding similarity. In: Proc. ICML, pp. 65–72 (2007)
4. Belkin, M., Niyogi, P., Sindhwani, V.: Manifold regularization: a geometric framework for learing from labeled and unlabeled examples. Journal of Machine Learning Research 7, 2399–2434 (2006)

5. Cai, D., He, X., Han, J., Zhang, H.J.: Orthogonal laplacianfaces for face recognition. IEEE Trans. Image Processing 15, 3608–3614 (2006)
6. Cootes, T., Edwards, G., Taylar, C.: Active appearance models. IEEE Trans. Pattern Analysis & Machine Intelligence 23(6), 681–685 (2001)
7. Cox, T., Cox, M.: Multidimensional Scaling. Chapman & Hall, London (1994)
8. Deffenbacher, K.A., Vetter, T., Johanson, J., O'Toole, A.J.: Facial aging, attractiveness, and aistinctiveness. Perception 27 (1998)
9. Donoho, D.L., Grimes, C.E.: When does geodesic distance recover the true hidden parametrization of families of articulated images? In: Proc. European Symposium on Artificial Neural Networks (2002)
10. Draganova, A.L.C., Christodoulou, C.: Comparing different classifiers for automatic age estimation. IEEE Trans. Systems, Man, and Cybernetics 34(1), 621–628 (2004)
11. Duda, R.O., Hart, P.E., Stork, D.G.: Pattern Classification, 2nd edn. John Wiley & Sons, Inc., New York (2001)
12. Geng, X., Smith-Miles, K., Zhou, Z.-Z.: Facial age estimation by nonlinear aging pattern subspace. In: Proc. ACM Conf. Multimedia (2008)
13. Geng, X., Zhou, Z.H., Zhang, Y., Li, G., Dai, H.: Learning from facial aging patterns for automatic age estimation. In: Proc. ACM Conf. Multimedia, pp. 307–316 (2006)
14. Goldberger, J., Roweis, S., Hinton, G., Salakhutdinov, R.: Neighbourhood components analysis. In: NIPS (2005)
15. Golub, G.H., Loan, C.F.V.: Matrix Computations. Johns Hopkins Univ. Press (1996)
16. Guo, G., Fu, Y., Dyer, C., Huang, T.S.: Image-based human age estimation by manifold learning and locally adjusted robust regression. IEEE Trans. on Image Processing 17, 1178–1188 (2008)
17. Lanitis, A., Taylor, C.J., Cootes, T.: Toward automatic simulation of aging effects on face images. IEEE Trans. Pattern Analysis and Machine Intelligence 24(4), 442–455 (2002)
18. Neal, R.M.: Monte carlo implementation of gaussian process models for bayesian regression and classification. Technical Report CRG-TR-97-2
19. Nilsson, J., Sha, F., Jordan, M.I.: Regression on manifolds using kernel dimension reduction. In: IEEE Conf. ICML, pp. 265–272 (2007)
20. Raumussen, C.E., Williams, C.K.: Gaussian Processes for Machine Learning. MIT Press, Cambridge (2006)
21. Roweis, S.T., Saul, L.K.: Nonlinear dimensionality reduction by locally linear embedding. Science 290(5500), 2323–2326 (2000)
22. Scholkopf, B., Smola, A.J.: Learning with Kernels: Support Vector Machines, Regularization, Optimization, and Beyond. MIT Press, Cambridge (2002)
23. Sugiyama, M., Hachiya, H., Towell, C., Vijayakumar, S.: Geodesic gaussian kernels for value function approximation. Autonomous Robots 25, 287–304 (2008)
24. Tenebaum, J.B., de. Silva, V., Langford, J.C.: A global geometric framework for nonlinear dimensionally reduction. Science 290(5500), 2319–2323 (2000)
25. Weinberger, K., Blitzer, J., Saul, L.: Distance metric learning for large margin nearest neighbor classification. In: Proc. NIPS, pp. 1475–1482 (2006)
26. Xing, E., Ng, A., Jordan, M.I., Russell, S.: Distance metric learning with application to clustering with side-information. In: Proc. NIPS (2002)
27. Yan, S., ang, H.W., Huang, T.S., Tang, X.: Ranking with uncertain labels. In: IEEE Conf. Mulitimedia and Expo, pp. 96–99 (2007)
28. Yan, S., Wang, H., Tang, X., Huang, T.S.: Learning autostructured regressor from uncertain nonnegative labels. In: IEEE Conf. ICCV, pp. 1–8 (2007)

Clustering-Based Construction of Hidden Markov Models for Generative Kernels

Manuele Bicego[1,2,*], Marco Cristani[1,2], Vittorio Murino[1,2], Elżbieta Pękalska[3], and Robert P.W. Duin[4]

[1] Computer Science Department, University of Verona, Italy
[2] Istituto Italiano di Tecnologia (IIT), Italy
[3] School of Computer Science, University of Manchester, UK
[4] Delft University of Technology, The Netherlands

Abstract. Generative kernels represent theoretically grounded tools able to increase the capabilities of generative classification through a discriminative setting. Fisher Kernel is the first and mostly-used representative, which lies on a widely investigated mathematical background. The manufacture of a generative kernel flows down through a two-step serial pipeline. In the first, "generative" step, a generative model is trained, considering one model for class or a whole model for all the data; then, features or scores are extracted, which encode the contribution of each data point in the generative process. In the second, "discriminative" part, the scores are evaluated by a discriminative machine via a kernel, exploiting the data separability. In this paper we contribute to the first aspect, proposing a novel way to fit the class-data with the generative models, in specific, focusing on Hidden Markov Models (HMM). The idea is to perform model clustering on the unlabeled data in order to discover at best the structure of the entire sample set. Then, the label information is retrieved and generative scores are computed. Experimental, comparative test provides a preliminary idea on the goodness of the novel approach, pushing forward for further developments.

1 Introduction

Hidden Markov Models (HMMs) represent a powerful and ductile statistical learning framework. In the classical HMM-based classification a single HMM is built for each class and the Maximum A Posteriori (MAP) approach is used to classify an unlabeled sequence $\mathbf{O}$, thus following a pure generative classification scheme.

Even though the MAP rule represents the theoretically optimal decision rule (*i.e.* leading to the minimum probability of error [1]), in practice, generative classification may suffer from poor discriminative abilities. This is likely to occur in case of poorly estimated class models (e.g. due to insufficient learning examples), improper model topologies, (e.g. due to a bad model definition or conditional dependence of the states), or possible class overlap (as may occur

---

* Corresponding author. Strada Le Grazie, 15 - 37134 Verona, Italy, Tel.: +39 045 8027072; Fax: +39 045 8027068, `manuele.bicego@univr.it`

D. Cremers et al. (Eds.): EMMCVPR 2009, LNCS 5681, pp. 466–479, 2009.

e.g. in medical problems where patient diagnoses vary between medical doctors). Some of these issues can be addressed by improving or/and extending classical HMMs (e.g. Hierarchical HMM [2], Factorial Hidden Markov Models [3], Coupled HMM [4] and others). Alternatively, the discriminative skills can be enhanced by training HMMs with discriminative criteria. Two popular examples are based on Maximum Mutual Information (MMI) [5] and Minimum Bayes Risk (MBR) [6], but other extensions are available. One must however remember that although discriminative criteria try to reduce the recognition error directly, they require a rather large amount of training data. Furthermore, there exist generalizations of HMMs towards probabilistic discriminative models. These are Conditional Random Fields (CRFs) [7] and Hidden CRFs (HCRFs) [8], in which conditional maximum likelihood is often used to estimate the parameters. All these discriminative techniques need complex training procedures, whereas the final classification still relies on the MAP approach.

In recent years, a new direction has aroused great interest in the Pattern Recognition community: the hybrid generative-discriminative approach. The idea is to merge the description abilities of the Hidden Markov Models (and more in general of generative models) with the discriminative skills of discriminative methods, *i.e.* methods that directly model the posterior probability and, by this, focus on the class separability. Generally speaking, there is a proven complementarity of discriminative and generative estimations: asymptotically (in the number of labeled training examples), discriminative methods lead to lower classification error than the generative ones [9], when comparing logistic regression to naive Bayes classifiers. On the other hand, generative counterparts are effective with less (and possibly unlabeled) data.

Different approaches have been proposed in this context. They may roughly be divided into two classes: generative embeddings and generative kernels. In the first case, the basic idea is to employ generative models in order to embed objects to a vectorial feature space (where any discriminative classifier can be trained – [10,11,12,13,14]). In the latter case a specific kernel is designed (which may rely either on an explicit or implicit space), further used in the Support Vector Machine scenario. Such examples can be found in [15,16,17,18]. In this paper, we will focus on the second class of approaches.

The most famous and widely investigated generative kernel, defined not only for HMM but for any generative model, is the Fisher Kernel [15], first advocated in the context of protein sequence analysis. A generative model is used to build a feature space in which a kernel is defined by suitable object comparisons. In particular, the Fisher Kernel approach measures the relation between objects by comparing them in the tangent space induced by the trained generative model. In practice, each object is represented by a feature vector, whose components are called Fisher scores. These scores are defined by derivatives of the log-likelihood of the generative model with respect to all individual parameters. The resulting kernel is then defined in such a feature space; the inner product was used in [15].

In order to define a generative kernel, the first step is to employ data to build the generative model. Let us consider this problem in the Fisher Kernel case [15].

In the original scenario defined by Jaakkola and Haussler in [15], the Fisher Kernel was computed on the basis of the log-likelihood of a single generative model, representing both competing classes. Another early version, by Fine et al. [19], defined the kernel on the basis of a generative model trained on a single class (the positive class). However, if more than one generative model is established, each representing a single class, more discriminative information may be extracted. Smith and Gales [20] exploited such an idea in the binary classification case by proposing to employ in Fisher Kernel the derivatives of the log-ratio of the two likelihoods calculated from the two competing models. It has been shown in the paper that this scheme may enhance the discriminative power of Fisher Kernel. Other generalizations to multi-class models have been proposed, for example in [21], where multiple per class generative models were trained and used to derive Fisher kernel. In the paper, the authors also proposed a method to reduce the number of needed models (by randomly selecting a subset of per class models), in order to deal with the increased computational burden.

In this work a further contribution to the aforementioned scenario is made. The idea is to allow a generative framework a free discovery of natural structures or groups in the training set. This is achieved with a preliminary step of clustering, during which a large number of small hidden natural groups is extracted from the data, disregarding class label information. Subsequently, a single and simple generative model is trained for each group (as the groups tend to be small). The underlying intuition is simple: generative models are not used to discriminate between classes (this is left to the discriminative methods), but are used to finely describe the local structure of the data as an ensemble of clusters. In this way, the problem space is partitioned into small regions, each one characterized by a simple but well trained generative model.

Even if the proposed methodology may be general (and applicable to any generative kernel), here we will explore this direction focusing on the HMM-based Fisher Kernel case, showing promising and comparative results obtained from some preliminary experiments.

The remainder of the paper is organized as follows: in Sec. 2 basic theoretical notions are provided and the notation is fixed. Sec. 3 proposes our generative kernel, whose classification comparative performances are presented in Sec. 4. Some considerations about the results are discussed in Sec.5. Finally, Sec. 6 closes the paper and opens for novel research perspectives.

2   Foundations

This section describes the basics of HMM and Fisher Kernel, mainly to fix the notation.

2.1   Hidden Markov Models

A discrete-time hidden Markov model λ can be viewed as a Markov model whose states are not directly observed: instead, each state is characterized by a probability distribution function, modeling the observations corresponding to that state. More formally, an HMM is defined by the following entities [22]:

- $S = \{S_1, S_2, \cdots, S_N\}$ the finite set of possible (hidden) states;
- the transition matrix $\mathbf{A} = \{a_{ij}, 1 \leq j \leq N\}$ representing the probability of moving from state S_i to state S_j,

$$a_{ij} = P[q_{t+1} = S_j | q_t = S_i], \quad 1 \leq i, j \leq N,$$

with $a_{ij} \geq 0$, $\sum_{j=1}^{N} a_{ij} = 1$, and where q_t denotes the state occupied by the model at time t.

- the emission matrix $\mathbf{B} = \{b(o|S_j)\}$, indicating the probability of emission of symbol $o \in V$ when the system state is S_j; V can be a discrete alphabet or a continuous set (e.g. $V = I\!\!R$), in which case $b(o|S_j)$ is a probability density function.
- $\boldsymbol{\pi} = \{\pi_i\}$, the initial state probability distribution,

$$\pi_i = P[q_1 = S_i], \quad 1 \leq i \leq N$$

with $\pi_i \geq 0$ and $\sum_{i=1}^{N} \pi_i = 1$.

For convenience, we represent an HMM by a triplet $\boldsymbol{\lambda} = (\mathbf{A}, \mathbf{B}, \boldsymbol{\pi})$.

Learning the HMM parameters, given a set of observed sequences $\{\mathbf{O}_i\}$, is usually performed using the well-known Baum-Welch algorithm [22], which is able to determine the parameters by maximizing the likelihood $P(\{\mathbf{O}_i\}|\boldsymbol{\lambda})$. One of the steps of the Baum-Welch algorithm is an evaluation step, where it is required to compute $P(\mathbf{O}|\boldsymbol{\lambda})$, given a model $\boldsymbol{\lambda}$ and a sequence $\mathbf{O}$; this can be computed using the *forward-backward procedure* [22].

2.2 Fisher Kernel

Fisher Kernel [15] was first advocated in the context of protein sequence analysis and proposed as a general way of mixing generative and discriminative models for classification. The basic idea is to employ a generative model to define feature vectors and project objects to the resulting feature space. There a meaningful similarity/distance measure is defined, leading to a kernel. In particular, the Fisher kernel approach measures the relation between the objects by comparing them in the tangent space induced by the trained generative model, which is considered as a point in the Riemannian manifold defined by a family of generative models. This space has a number of desirable characteristics, such as the possibility of measuring geodesic distances between points along the manifold (leading to the concept of natural gradients [23]). In practice, each object is represented by a feature vector, whose components are called Fisher Scores, defined by derivatives of the log-likelihood of the generative model with respect to all parameters. The dimensionality of this space equals the number of parameters. A kernel can be defined in various ways in the resulting space; the inner product was used in [15].

The general formulation is as follows: given two observations $\mathbf{O}_i$ and $\mathbf{O}_j$, and a generative model $\mathcal{P}(\mathbf{O}|\theta)$ — with θ being the vector of parameters of the generative model — Fisher Kernel is defined as:

$$FK(\mathbf{O}_i, \mathbf{O}_j) = < FS(\mathbf{O}_i, \theta), FS(\mathbf{O}_j, \theta) >$$

where $< \cdot, \cdot >$ is the inner product, and $FS(\mathbf{O}, \theta)$ is called Fisher Score and is defined as

$$FS(\mathbf{O}, \theta) = \nabla_\theta \log \mathcal{P}(\mathbf{O}|\theta)$$

In the HMM case, θ is replaced with $\boldsymbol{\lambda}$, representing the trained HMM. The vector parameter is composed by the transition probabilities, the emission probabilities (the mean and the covariance in case of Gaussian models) and the initial state probabilities. The derivation of such derivatives is not complex, and is omitted here. Interested readers are referred to [21].

3 Methodology

The construction of our HMM-based generative kernel is realized in three steps: (1) discovering the data groups, (2) building single HMM for each group, and (3) calculating and exploiting the related generative scores in the kernel definition. The three phases are reviewed in detail in the following.

3.1 HMM-Based Clustering of Sequences

The first step is to discover natural groups in the data by performing sequence clustering. It is well known that data clustering is inherently a more difficult task than supervised classification, and this difficulty worsens if sequential data are considered: the structure of the underlying process is often difficult to infer, and typically different length sequences have to be dealt with.

The sequence clustering step represents the most crucial part of the proposed methodology: it seems very reasonable to employ a process able to explicitly consider the generative model employed in the Fisher Kernel. In such sense, the clustering methodology employed here is based on HMM. HMMs have not been extensively employed for clustering sequences, with only a few papers exploring this direction. More specifically, early approaches related to speech recognition were presented in [24,25,26]. A relevant contribution was made by Li and Biswas [27,28,29,30,31]). Basically, in their approach [27], the clustering problem is addressed by focusing on the model selection issue, *i.e.* the search for the HMM topology best representing data, and the clustering structure issue, *i.e.* finding the most likely number of clusters.

More advanced techniques have been proposed by Smyth [32] (see also the more general and more recent [33]), where a series of sequential steps permits to realize a block-wise HMM, modeling the whole data set. Other interesting examples can be found in [34], where HMMs are used as cluster prototypes, with the clustering obtained with the *rival penalized competitive learning* (RPCL) algorithm, and in [35], where HMMs were employed to derive a feature space where standard vector-based clustering algorithms have been applied.

In any case, the simplest and most widely used class of approaches for HMM-based clustering is the proximity-based clustering, where the main effort of the clustering process lies in devising good similarity or distance measures between sequences. With such measures, any standard distance-based method (as agglomerative clustering) can be applied. Within this context, HMMs are employed to compute similarities between sequences, using different approaches (see for example [36,37]), and standard pairwise distance-based approaches (as agglomerative hierarchical) are then used for the final data clusters.

In our study, we chose a simple yet effective clustering method, belonging to the class of proximity-based clustering approaches. Considering a given set of N sequences $\{\mathbf{O}_1...\mathbf{O}_N\}$ to be clustered, the algorithm performs the following steps:

1. Train a single HMM $\boldsymbol{\lambda}_i$ for each sequence $\mathbf{O}_i$ (the details of the training are in the next section).
2. Compute the distance matrix $D = \{D(\mathbf{O}_i, \mathbf{O}_j)\}$, representing a matrix of similarities between sequences or between models. This is typically obtained either by calculating the model-likelihood probabilities, or by devising a measure of distances between models. In the past, few authors have proposed approaches to compute these distances: early approaches were based on the Euclidean distance of the discrete observation probability, while other approaches were based on entropy, or on co-emission probability of two models, or, very recently, on the Bayes probability of error (see [37] and the references therein). Here we use the following distance, employed in [32,36]. First, given the sequences $\{\mathbf{O}_j\}$ and the models $\{\boldsymbol{\lambda}_i\}$, we compute the following matrix:

$$L_{ij} = P(\mathbf{O}_j|\boldsymbol{\lambda}_i) \tag{1}$$

The similarity matrix $S(\mathbf{O}_i, \mathbf{O}_j)$ is then obtained by symmetrizing the matrix L_{ij}. Thus we define

$$S(\mathbf{O}_i, \mathbf{O}_j) = \frac{1}{2}\left[L_{ij} + L_{ji}\right]. \tag{2}$$

Clearly the choice of this distance is crucial for the effectiveness of the clustering: readers interested in this aspect may refer to [36], where different Likelihood-based distances have been considered and tested in an EEG clustering scenario.

3. Use a hierarchical agglomerative clustering method (with the Complete Link rule [38]) on $S(\mathbf{O}_i, \mathbf{O}_j)$ to perform the clustering.

Clearly the choice of the best number of clusters represents a problem, even though different indices/strategies have been proposed (see for example [38]). In our experimental evaluation we let it vary in a proper range, and report the different results.

3.2   HMM Training

Once estimated the natural groups inside the data set, a single HMM is trained for each group. HMM training is performed by using the Baum-Welch re-estimation procedure, stopping at the likelihood convergence. We assume that we deal with fully ergodic HMMs. Initialization is random both for the transition probabilities and initial state probabilities. In case of continuous signals, the emission probability models are initialized by a Gaussian Mixture clustering. In case of discrete symbol sequences, 20 independent training runs are performed, starting from a random initialization, picking the best likelihood model as the representative. In the experimental part, the best number of states has been determined with a preliminary experimental evaluation.

3.3   Kernel Computation

The goal in this phase is to compute the Fisher Kernel given a set of trained models. Here we adopt the scheme proposed in [21], and recently adopted also in [14] – where a set of different HMM-based generative embeddings have been proposed. The idea is to concatenate the scores obtained from each model. Here we adopt the same strategy, the difference is that the models are not built on the classes but on the extracted clusters.

More formally, given two sequences $\mathbf{O}_i$ and $\mathbf{O}_j$, and the set of K trained HMMs $\{\boldsymbol{\lambda}_k\}$ (with K being the number of clusters), the kernel is then determined as the inner product in the vector space being a Cartesian product of the Fisher score spaces resulting from all individual models, in the same way the Fisher Kernel is built. In other words, given a sequence $\mathbf{O}_i$, the Fisher scores $FS(\mathbf{O}_i, \boldsymbol{\lambda}_k)$, are computed using each trained model $\boldsymbol{\lambda}_k$, concatenating them in a single vector:

$$CFS(\mathbf{O}_i, \{\boldsymbol{\lambda}_k\}) = [FS(\mathbf{O}_i, \boldsymbol{\lambda}_1), FS(\mathbf{O}_i, \boldsymbol{\lambda}_2), \cdots, FS(\mathbf{O}_i, \boldsymbol{\lambda}_K)].$$

Given two concatenated vectors $CFS(\mathbf{O}_i, \{\boldsymbol{\lambda}_k\})$ and $CFS(\mathbf{O}_j, \{\boldsymbol{\lambda}_k\})$, relative to two sequences $\mathbf{O}_i$ and $\mathbf{O}_j$, the kernel is then computed as:

$$FK(\mathbf{O}_i, \mathbf{O}_j) = <CFS(\mathbf{O}_i, \{\boldsymbol{\lambda}_k\}), CFS(\mathbf{O}_j, \{\boldsymbol{\lambda}_k\}) >$$

where $< \cdot, \cdot >$ represents the inner product.

Given the kernel, the classification task may be solved by using standard SVM.

4   Experimental Evaluation

The proposed methodology has been tested using a 2D shape recognition problem. Recognition of 2D shapes is an unconventional application of HMMs, even though promising results have been reported [39,40,41]. The idea, in this case, is to extract the contour of the shape, transforming it to a sequence that is modeled by an HMM.

In particular, we studied the Chicken database, a very nasty problem: the results published in [42] report a baseline leave-one-out accuracy of $\approx 66\%$ by using the 1-NN on the Levenshtein (non-cyclic) edit distance. In our experiments, two different sequence representations are used to model contours, chain codes and curvature angles. In the first case, a standard 8-direction chain encoding procedure is applied to each image. Discrete HMMs are used to model these classes of symbol sequences. In the second case, we derive curvature sequences as in [41,43]. First, contours are extracted by using the *Canny* edge detector; the boundary is then approximated by segments of approximately fixed length. Then, at any given point x the curvature value is derived as an angle between two consecutive segments intersecting at x. The initial point is the rightmost point lying on the horizontal line passing through the object centroid, following the boundary in a counterclockwise manner. Classes of curvature sequences are finally modeled by continuous Gaussian HMMs.

Four methodologies to build the Fisher Kernel have been tested and compared:

1. *One model for the whole data set.* This is the standard methodology, proposed by Jaakkola and Haussler in their original paper [15]. One single model is built using all the data present in the training set. At the end only one model is trained.
2. *One class-model.* This is a generalization of the method proposed in [19], where a single model was trained using the data of the positive class. Since in that paper only binary problems were addressed, here we extend it to deal with the multi class case. To do that, we just select one of the classes, build the model for that class, and use this model to compute the Fisher Kernel. At the end only one model is trained. Clearly, depending on the chosen class, results may vary. Here we tried all the possibilities (reported in the tables as "Method 2 (class k)", indicating that an HMM has been trained using only the examples of the class k).
3. *C models: one per class* (C number of classes). This is the scheme proposed in [21], where one model per class is built. As explained before, the resulting Fisher Kernel is then defined as the inner product in the space obtained as a Cartesian product of the spaces resulting from each model (namely concatenating all Fisher Scores of all models). At the end, C models are trained, where C is the number of classes.
4. *K models: one per cluster* (K number of clusters). This represents the proposed approach.

In all cases the Fisher Score space has been normalized before the training of the linear SVM: this is required in order to make the Fisher Kernel work (see for example [20]). Accuracies have been computed by using the Averaged Holdout: the data set has been split in two random partitions, one used for training and one for testing. The process is repeated 10 times and the results are averaged. The results for continuous and discrete HMMs are shown in Tables 1 and 2, respectively.

Table 1. Continuous HMMs applied to the Chicken Database with curvature sequences: averaged accuracies (and standard errors) for different methods – see above. In the Clustering-based Fisher Kernel case (method 4), only the best result among the different clusterings is shown (in the range of 2-25 clusters).

Method	# models	Accuracy (Std error of the mean)
1	1	0.759 (0.003)
2 (class 1)	1	0.755 (0.004)
2 (class 2)	1	0.758 (0.002)
2 (class 3)	1	0.755 (0.004)
2 (class 4)	1	0.734 (0.004)
2 (class 5)	1	0.752 (0.002)
3	5	0.775 (0.004)
4	19	0.798 (0.002)

Table 2. Discrete HMMs applied to the Chicken Database with chain code sequences: averaged accuracies (and standard errors) for different methods – see above. In the Clustering-based Fisher Kernel case (method 4), only the best result among the different clusterings is shown (in the range of 2-25 clusters).

Method	# models	Accuracy (Std error of the mean)
1	1	0.706 (0.006)
2 (class 1)	1	0.629 (0.004)
2 (class 2)	1	0.697 (0.009)
2 (class 3)	1	0.725 (0.006)
2 (class 4)	1	0.695 (0.005)
2 (class 5)	1	0.662 (0.004)
3	5	0.815 (0.004)
4	18	0.858 (0.002)

5   Discussion

As a general comment, it is evident from the tables that the clustering-based building of the HMM pool results in a positive increase in the accuracy of the SVM based on Fisher Kernel; this is more evident in the discrete case.

Moreover, the obtained results are remarkable, considering the difficulty of the data set. As a reference, we put in Table 3 some results on the same dataset: it is evident how the proposed approach performs very competitively with respect to the state of the art.

We also want to emphasize again that normalization of the Fisher Score spaces is essential (in whatever version, concatenated or not). Without normalization, the classification performance deteriorates significantly. This confirms the intuition provided in [20].

Table 3. Comparative Results on the Chicken dataset

Methodology	Protocol	Accuracy	Reference
1-NN + Levenshtein edit distance	Leave One Out	≈ 67%	[42]
1-NN + approximated cyclic distance	Leave One Out	≈ 78%	[42]
KNN + cyclic string edit distance	Train/Test/Valid	74.3%	[43]
SVM + Edit distance-based kernel	Train/Test/Valid	81.1%	[43]
1-NN + mBm-based features	Leave One Out	76.5%	[44]
1-NN + Hmm-based distance	Leave One Out	73.77%	[44]
SVM + Hmm-based entropic features	Leave One Out	81.21%	[45]

Concerning the Fisher Kernels defined on a single model (methods 1 and 2), it is interesting to observe that it does not make a significant difference to train the HMM either on the whole data set or on a single class. These models have discriminative powers which are, apparently, different, as combining appears to be useful. This may rise the following question: is it reasonable to train a single "general shape HMM", namely to train an HMM on a different data set of shapes (or on a large collection of many databases)? In this way, the proposed approach may be considered as a pure feature extraction process.

The fact that Fisher Kernels built with only one model perform always worse than when built with more models confirms the intuition of [21], and is even more evident when using the clustering approach. Clearly, the resulting space may be very high-dimensional when several models are used, and the question arises of how to manage such a space. Here we solve this by using an SVM based on the Fisher Kernel defined for the underlying Fisher Score spaces. As can seen from the definitions in section 3.3, the Fisher Kernel of the combined space is just the average of the Fisher Kernels of the individual spaces. The kernel matrices have, of course, the same size determined by the size of the training set and are independent of dimensionalities. Reasonably, this aspect would have become drastically crucial when studying the presented approach from a "generative embedding" point of view, namely when employing other classifiers (more sensitive to the curse of dimensionality) in the vector space derived from the generative model. Another important observation is that in all the experiments we made, the best number of states in the clustering-based approach was two, indicating very small models. What we are doing in such a case is to increase the number of models while reducing the size of each model. This may alleviate the dimensionality issue.

In Figure 1 we plot the performances of the proposed approach in the chain code experiment while varying the number of clusters. As the presented approach can be understood as based on averaging kernels that are different, but that all make sense in one way or another, the number of kernels (and thereby clusters) should be sufficiently large to cover all aspects of the class distributions. After that the performance may stabilize, or may deteriorate somewhat as cluster sizes will shrink, resulting in models that may be more specific and thereby less useful for the following discriminative step.

The interplay between model size and number of clusters needs further study. In our approach, both are unsupervised procedures and may be based on other data than those available in a training set. Together they determine the kernel. The final performance however obtained in the discriminative step by the SVM depends on the relation between this kernel and the size of the training set. So all three have to be studied together: model size, number of models (clusters) and training set size.

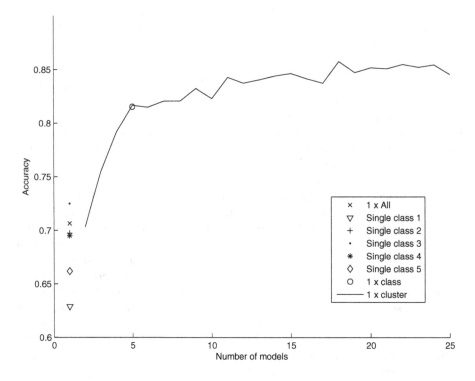

Fig. 1. Performances on chain code experiment with varying number of clusters

Looking at the whole procedure, there is a heavy data re-use: HMM training per sequence, clustering, HMM-training per group, parameter setting for SVM and classifier training. Since we over-use the data a lot, we can benefit from weak and/or simple models, which is in fact confirmed in the experimental evaluation.

In summary, we should observe that the best results for clustering are obtained with small models (two state models) and a large number of clusters. The effect of such a result is that the problem space is partitioned into small regions, each characterized by a simple but well trained generative model. Consequently, we can obtain an optimal description of the feature space (via a generative model), which is then discriminated via a discriminative method.

6   Conclusions and Future Perspectives

In this paper we furnished a novel way to build a generative kernel based on Hidden Markov Models and Fisher Kernel. In the typical generative kernel building process, a generative model is fit on the data, considering class-label information: this generates two main fit directions, namely, one model for class or one model for all the data. Then, scores are extracted from the trained models which highlight the generative correspondence among points and the single model parameters. Finally, discriminative reasoning exploits data separability. Our contribution is to suggest an alternative way to build the HMM generative framework from the data. The idea is to consider all the data together, allowing possible data structural information to better emerge. This information is captured by performing model clustering, which provide few HMM models encoding all the data as surrogates of many initial simple models fit on local data points. Class label information is then recovered in the score building process, and, subsequently, by the discriminative machinery. This paper represents a preliminary work toward a novel research direction for the manufacture of generative kernels. Our work consists on several comparative tests which show what are the potentialities to take into account and the possible issues to face. The promising results allow us to further investigate this research direction.

Acknowledgements

We acknowledge financial support from the FET programme within the EU FP7, under the SIMBAD project (Contract 213250).

References

1. Duda, R., Hart, P., Stork, D.: Pattern Classification, 2nd edn. John Wiley & Sons, Chichester (2001)
2. Fine, S., Singer, Y., Tishby, N.: The hierarchical hidden markov model: Analysis and applications. Machine Learning 32, 41–62 (1998)
3. Ghahramani, Z., Jordan, M.: Factorial hidden markov models. Machine Learning 29, 245–273 (1997)
4. Brand, M., Oliver, N., Pentland, A.: Coupled hidden markov models for complex action recognition. In: Proc. of IEEE Conf. on Computer Vision and Pattern Recognition (1997)
5. Bahl, L., Brown, P., de Souza, P., Mercer, R.: Maximum mutual information estimation of hidden Markov model parameters for speech recognition. In: IEEE International Conference on Acoustics, Speech and Signal Processing, Tokyo, Japan, vol. I, pp. 49–52 (2000)
6. Kaiser, Z., Horvat, B., Kacic, Z.: A novel loss function for the overall risk criterion based discriminative training of HMM models. In: International Conference on Spoken Language Processing, Beijing, China, vol. 2, pp. 887–890 (2000)
7. Lafferty, J., McCallum, A., Pereira, F.: Conditional random fields: probabilistic models for segmenting and labelling sequence data. In: International Conference on Machine Learning, pp. 591–598 (2001)

8. Gunawardana, A., Mahajan, M., Acero, A., Platt, J.: Hidden conditional random fields for phone classication. In: Interspeech, Lisbon, Portugal, pp. 1117–1120 (2005)
9. Ng, A., Jordan, M.: On discriminative vs generative classifiers: A comparison of logistic regression and naive Bayes. In: Advances in Neural Information Processing Systems (2002)
10. Bicego, M., Murino, V., Figueiredo, M.: Similarity-based classification of sequences using hidden markov models. Pattern Recognition 37(12), 2281–2291 (2004)
11. Bicego, M., Pękalska, E., Duin, R.P.W.: Group-induced vector spaces. In: Haindl, M., Kittler, J., Roli, F. (eds.) MCS 2007. LNCS, vol. 4472, pp. 190–199. Springer, Heidelberg (2007)
12. Layton, M., Gales, M.: Augmented statistical models: Exploiting generative models in discriminative classifiers. In: Advances in Neural Information Processing Systems (2005)
13. Smith, N.: Using Augmented Statistical Models and Score Spaces for Classification. PhD thesis, Engineering Departement, Cambridge University (2003)
14. Bicego, M., Pekalska, E., Tax, D., Duin, R.: Component-based discriminative classification for hidden markov models. Pattern Recognition (in press, 2009)
15. Jaakkola, T., Haussler, D.: Exploiting generative models in discriminative classifiers. In: Advances in Neural Information Processing Systems, pp. 487–493 (1999)
16. Tsuda, K., Kin, T., Asai, K.: Marginalised kernels for biological sequences. Bioinformatics 18, 268–275 (2002)
17. Jebara, T., Kondor, I., Howard, A.: Probability product kernels. Journal of Machine Learning Research 5, 819–844 (2004)
18. Moreno, P., Ho, P., Vasconcelos, N.: A kullback-leibler divergence based kernel for svm classification in multimedia applications. In: Proc. of Advances in Neural Information Processing., vol. 16 (2003)
19. Fine, S., Navratil, J., Gopinath, R.: A hybrid gmm/svm approach to speaker identification. In: IEEE Int. Conf. on Acoustics, Speech, and Signal Processing, pp. 417–420 (2001)
20. Smith, N., Gales, M.: Speech recognition using svms. In: Advances in Neural Information Processing Systems, pp. 1197–1204 (2002)
21. Chen, L., Man, H., Nefian, A.: Face recognition based on multi-class mapping of fisher scores. Pattern Recognition, 799–811 (2005)
22. Rabiner, L.: A tutorial on Hidden Markov Models and selected applications in speech recognition. Proc. of IEEE 77(2), 257–286 (1989)
23. Amari, S.: Natural gradient works efficiently in learning. Neural Computation 10, 251–276 (1998)
24. Rabiner, L., Lee, C., Juang, B., Wilpon, J.: HMM clustering for connected word recognition. In: Proc. Int. Conf. on Acoustics, Speech and Signal Processing (ICASSP), pp. 405–408 (1989)
25. Lee, K.: Context-dependent phonetic hidden Markov models for speaker-independent continuous speech recognition. IEEE Transactions on Acoustics, Speech and Signal Processing 38(4), 599–609 (1990)
26. Kosaka, T., Matsunaga, S., Kuraoka, M.: Speaker-independent phone modeling based on speaker-dependent hmm's composition and clustering. In: Int. Proc. on Acoustics, Speech, and Signal Processing, vol. 1, pp. 441–444 (1995)
27. Li, C.: A Bayesian Approach to Temporal Data Clustering using Hidden Markov Model Methodology. PhD thesis, Vanderbilt University (2000)

28. Li, C., Biswas, G.: Clustering sequence data using hidden Markov model representation. In: Proc. of SPIE 1999 Conf. on Data Mining and Knowledge Discovery: Theory, Tools, and Technology, pp. 14–21 (1999)

29. Li, C., Biswas, G.: A bayesian approach to temporal data clustering using hidden Markov models. In: Proc. Int. Conf. on Machine Learning, pp. 543–550 (2000)

30. Li, C., Biswas, G.: Applying the Hidden Markov Model methodology for unsupervised learning of temporal data. Int. Journal of Knowledge-based Intelligent Engineering Systems 6(3), 152–160 (2002)

31. Li, C., Biswas, G., Dale, M., Dale, P.: Matryoshka: A HMM based temporal data clustering methodology for modeling system dynamics. Intelligent Data Analysis Journal (2002)

32. Smyth, P.: Clustering sequences with hidden Markov models. In: Mozer, M., Jordan, M., Petsche, T. (eds.) Advances in Neural Information Processing Systems, vol. 9, p. 648. MIT Press, Cambridge (1997)

33. Cadez, I., Gaffney, S., Smyth, P.: A general probabilistic framework for clustering individuals. In: Proc. of ACM SIGKDD 2000 (2000)

34. Law, M., Kwok, J.: Rival penalized competitive learning for model-based sequence. In: Proc. Int. Conf. Pattern Recognition, vol. 2, pp. 195–198 (2000)

35. Bicego, M., Murino, V., Figueiredo, M.: Similarity-based clustering of sequences using hidden Markov models. In: Perner, P., Rosenfeld, A. (eds.) MLDM 2003. LNCS (LNAI), vol. 2734, pp. 86–95. Springer, Heidelberg (2003)

36. Panuccio, A., Bicego, M., Murino, V.: A hidden markov model-based approach to sequential data clustering. In: Caelli, T.M., Amin, A., Duin, R.P.W., Kamel, M.S., de Ridder, D. (eds.) SPR 2002 and SSPR 2002. LNCS, vol. 2396, pp. 734–742. Springer, Heidelberg (2002)

37. Bahlmann, C., Burkhardt, H.: Measuring hmm similarity with the bayes probability of error and its application to online handwriting recognition. In: Proc. Int. Conf. Document Analysis and Recognition, pp. 406–411 (2001)

38. Jain, A., Dubes, R.: Algorithms for clustering data. Prentice-Hall, Englewood Cliffs (1988)

39. He, Y., Kundu, A.: 2-D shape classification using Hidden Markov Model. IEEE Trans. Pattern Analysis Machine Intelligence 13(11), 1172–1184 (1991)

40. Arica, N., Yarman-Vural, F.: A shape descriptor based on circular Hidden Markov Model. In: IEEE Proc. Int Conf. Pattern Recognition, vol. 1, pp. 924–927 (2000)

41. Bicego, M., Murino, V.: Investigating Hidden Markov Models' capabilities in 2D shape classification. IEEE Trans. on Pattern Analysis and Machine Intelligence - PAMI 26(2), 281–286 (2004)

42. Mollineda, R., Vidal, E., Casacuberta, F.: Cyclic sequence alignments: Approximate versus optimal techniques. Int. Journal of Pattern Recognition and Artificial Intelligence 16(3), 291–299 (2002)

43. Neuhaus, M., Bunke, H.: Edit distance-based kernel functions for structural pattern classification. Pattern Recognition 39, 1852–1863 (2006)

44. Bicego, M., Trudda, A.: 2D shape classification using multifractional brownian motion. In: da Vitoria Lobo, N., Kasparis, T., Roli, F., Kwok, J.T., Georgiopoulos, M., Anagnostopoulos, G.C., Loog, M. (eds.) S+SSPR 2008. LNCS, vol. 5342, pp. 906–916. Springer, Heidelberg (2008)

45. Perina, A., Cristani, M., Castellani, U., Murino, V.: A new generative feature set based on entropy distance for discriminative classification. In: Proc. of Int. Conf. on Image Analysis and Processing, ICIAP 2009 (2009)

Boundaries as Contours of Optimal Appearance and Area of Support

Christina Pavlopoulou and Stella X. Yu

Computer Science Department
Boston College, Chestnut Hill, MA 02467
{pavlo,syu}@cs.bc.edu

Abstract. Bayesian boundary models often assume that the evidence for each contour is derived from the entire image. Consequently, the normalization term in the Bayes rule is the same for every contour and becomes irrelevant when seeking the optimal. However, in practice these models only use the vicinity of a contour, making the normalization term contour-specific. We propose a formulation that acknowledges the normalization term and includes it in the optimization. We show that it can be interpreted as a confidence measure promoting contours which are far better than other nearby candidate contours. We validate our approach in an interactive boundary delineation setting and demonstrate that complex boundaries can be extracted with significantly smaller amount of user input than when traditional Bayesian models are employed.

1   Introduction

The Bayesian formulation for finding boundaries in the image seeks the contour C with the maximal posterior probability $P(C|O)$ given observations O:

$$P(C|O) = \frac{P(O|C)\, P(C)}{P(O)} \tag{1}$$

Our work concerns the normalization term $P(O)$ in the formulation (Fig 1).

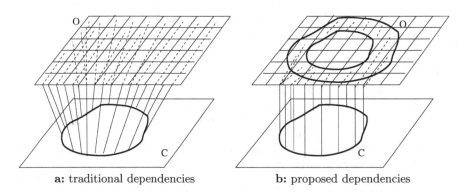

a: traditional dependencies   b: proposed dependencies

Fig. 1. The observations O for a contour C depend on the entire image in traditional Bayesian models (**a**), and only on the vicinity of C in our model (**b**)

D. Cremers et al. (Eds.): EMMCVPR 2009, LNCS 5681, pp. 480–491, 2009.

Previous work has assumed that O always consists of features extracted from the entire image and it is the same for every C (Fig. 1a). Hence the normalization term $P(O)$ becomes irrelevant and can be ignored during the optimization of $P(C|O)$. However, this assumption does not typically hold in implementations: each contour is often characterized with locally extracted features (Fig. 1b). In other words, O and thus $P(O)$ are different for different contour C's, and $P(O)$ cannot be omitted from the optimization; it affects the maximum of $P(C|O)$ and the optimal contour C.

Our formulation adopts the same criterion in Eqn. 1, but acknowledges the normalization term $P(O)$, so that boundaries are optimal contours in terms of both appearance and area of support. Optimal contours not only best explain the image evidence $P(O|C)$ (e.g. delineating the intensity discontinuities in the image), and have the desired properties $P(C)$ (e.g. smooth), but they are also the best candidates in their vicinity with respect to their local evidence $P(O)$ (e.g. weak but distinctive boundaries).

The normalization term not only helps promote distinctively better contours, but also addresses the length bias problem which plagues energy models for boundaries: short contours automatically have a lower energy than longer ones.

The bias problem has been long recognized and tackled in a number of ways. One approach is to provide a good initialization of boundaries [1,2], or guide the user during the delineation process [3,4,5]. A significant amount of human intervention is often required in these cases. Other approaches have incorporated heuristics in the optimization method [6,7,8]. These methods essentially extract the boundary in a piecewise fashion and it is unclear whether the collection of these boundary segments is optimal. Additional image features [9,10,11] and stronger contour priors [12,13,14] have also been explored. Such methods impose additional constraints but do not fundamentally tackle the bias problem. The most direct attempt to solving the bias problem has been to normalize the contour goodness score by the length of the contour [15,16]. These approaches however are applicable to closed contours and do not admit user interaction.

Our formulation by design does not favor degenerate solutions such as short contours. The normalization term serves as a confidence measure and only favors contours which are significantly better than other candidates.

Including the normalization term results in a more complex criterion. However, the criterion can be globally optimized using dynamic programming in polynomial time, with little loss in computational efficiency.

We develop and analyze our Bayesian formulation in Section 2, address computational issues related to the global optimization in Section 3, evaluate our method in Section 4, and conclude in Section 5.

2   Bayesian Formulation

We first develop our formulation and explain why the contour-specific normalization term favors contours distinctive in their vicinity. In this sense, our method is similar to non-maximum suppression used in edge detection [17]. While both methods enhance boundaries, non-maximum suppression does so in a heuristic and local fashion, whereas our method is principled and operates globally.

We then relate our criterion to entropy and show that the normalization term can be seen as a confidence measure of the quality of a boundary.

2.1   Criterion: Contour-Dependent Observations

Let O_C denote the observations associated with candidate contour C (Fig. 1b). We assume that contour points C_i are conditionally independent and that the observations are conditionally independent given the contour points. We have:

$$\log P(C|O_C) = \sum_i \{\log P(O_{C_i}|C_i) + \log P(C_i|C_{i-1})\} - \log P(O_C) \qquad (2)$$

$$P(O_C) = \sum_C P(O_C|C)\, P(C) \qquad (3)$$

The difference of Eqn. 2 with all previous Bayesian formulations is the term $P(O_C)$. In traditional formulations, it is the same for all contours and does not play any role in the optimization. When this term is not present, the probability of a contour decreases monotonically with its length; short contours are significantly more likely than long ones.

In contrast, our criterion contains two competing terms: While the term $\log P(O_C|C) + \log P(C)$ expresses the quality of a boundary in terms of features and smoothness and the term $P(O_C)$ sums the probabilities of all possible contours in the vicinity of the contour C. The favored contours are not just the ones with high $\log P(O_C|C) + \log P(C)$, but also the ones with low $P(O_C)$. The latter occurs when all the contours in the vicinity of C have very low probabilities, or in other words, when C is the best contour in a given image area.

We show how the normalization term promotes certain contours in the simple case where all the candidate contours except the optimal one have the same cost (Fig. 2). Assume the desired contour connecting points A and B is the straight

a: single best C b: multiple good C's

Fig. 2. The thickness of the line indicates the quality of the boundary. a) There is a single best candidate connecting points A and B (straight line). b) There exist several good candidate contours connecting points A and B.

line C^*. In Fig. 2a, C^* is a much better candidate than other possible contours connecting A and B, unlike Fig. 2b where other good candidates exist as well. In both cases the quantity $P(O_C|C^*)\,P(C^*)$ is α. The corresponding quantities $P(O_C|C)P(C)$ for all other candidate contours C are β and γ for Fig. 2a and 2b respectively, with $\gamma > \beta$. If there are $n+1$ possible ways of connecting points A and B, then:

$$\log(n\,\beta + \alpha) < \log(n\,\gamma + \alpha)$$
$$\log \alpha - \log(n\,\beta + \alpha) > \log \alpha - \log(n\,\gamma + \alpha)$$

Our criterion favors contours for which there are no other competitors in the same neighborhood. Degenerate solutions of very short contours in image areas with no characteristic features will receive very low probability.

2.2   Analysis: Entropy Interpretation

We show that our criterion (Eqn. 2) can be understood from an entropy point of view. Let $P(O_C|C)\,P(C) = \beta_j$, where each β_j corresponds to a different candidate contour C. β_j is the probabilistic cost of a contour according to Eqn. 1 without the normalization term and we will refer to it as "cost". We also have $P(O_C) = \sum_j \beta_j$. The log probability of Eqn. 2 is a function of β_j:

$$E(\beta_j) = \log \beta_j - \log \sum_i \beta_i \tag{4}$$

The maximum of $E(\beta_j)$ is obtained at 0, since $E(\cdot)$ is the log of a probability distribution:

$$E(\beta_j) = 0 \Rightarrow \beta_j = \beta_j + \sum_{i \neq j} \beta_i \Rightarrow \sum_{i \neq j} \beta_i \simeq 0 \tag{5}$$

The last condition holds when all the contours, except the j-th one, have costs close to 0. The minimum of $E(\beta_j)$ is achieved at $-\infty$ and this occurs when

$$\sum_i \beta_i \simeq \infty \tag{6}$$

i.e., when there are many strong candidate paths in the given image area.

The behavior $E(\beta_j)$ is reminiscent of the inverse behavior of the entropy of a distribution. Most informative or high-entropy distributions are the ones who do not favor any particular data points. For example, the most informative one-dimensional distribution is the uniform distribution. On the other hand, least informative distributions are the ones favoring a single value. $E(\beta_j)$ is maximized when a single candidate contour is assigned high cost and is minimized when all the candidate contours have very high costs.

In fact, we can find an entropy lower bound for $E(\beta_j)$, which offers an interesting interpretation of the normalization term $P(O_C)$. We have:

$$E(\beta_j) \geq \log \beta_j - \sum_i \log \beta_i \geq \log \beta_j - \sum_i \beta_i \log \beta_i \tag{7}$$

where the above holds for $\beta_i < 1$, for all i. The term $-\sum_i \beta_i \log \beta_i$ is a pseudo-entropy term since the costs β_i do not sum up to 1. The entropy of a distribution can be seen as a measure of the uncertainty of the distribution. Thus, the normalization factor $P(O_C)$ can be seen as a confidence measure regarding the image location a contour belongs to. Contours that belong to low-uncertainty image regions, that is, they are the sole candidates, are assigned high cost. On the other hand, contours from high-uncertainty image regions are assigned low costs.

3   Optimization Using Dynamic Programming

Our criterion can be globally optimized with dynamic programming. The algorithm proposed merges the optimization scheme employed by the computer vision community for $P(O_C|C)\ P(C)$ ([18,2,19,3,4,5]) with the algorithm used to calculate $P(O_C)$ [20]. We first show how to calculate $P(O_C)$ using scaling according to [20]. We then describe some additional approximations needed. Finally, we show how to optimize our criterion in a graph-based framework using Dijkstra's algorithm in low-order polynomial time. Dijkstra's algorithm can globally optimize our criterion when points on the boundary to be extracted are known (either automatically or via user input).

3.1   Calculation of $P(O_C)$

We calculate $P(O_C)$ using the well-known forward-backward algorithm employed in HMM inference problems [20]. Let $C = (c_1, \ldots c_n)$ be the hidden random variables corresponding to a contour of n points. The set of the values each of these random variables can take is $D = \{0, \ldots, 7\}$ corresponding to the possible directions employed by the chain code curve representation. Let also $O_C = \{O_{c_1}, \ldots, O_{c_n}\}$ be the observations associated with the individual contour points.

For the forward-backward algorithm, we define

$$\alpha_i(d_k) = P(O_{c_1}, \ldots, O_{c_i}, q_i = d_k)$$

where q_i is the label assigned to the i-th contour point c_i and d_j takes values from $D = \{0, \ldots, 7\}$. Recursively we can compute:

$$\alpha_1(d_k) = P(O_{c_1}|c_1 = d_k) \tag{8}$$

$$\alpha_{i+1}(d_k) = \left(\sum_{d_j} \alpha_i(d_j) P(c_{i+1} = d_j | c_i = d_k) \right) P(O_{c_{i+1}} | c_{i+1} = d_k) \tag{9}$$

The probability of the observations is now given as:

$$P(O_C) = \sum_{d_j \in D} \alpha_n(d_j) \tag{10}$$

3.2 Scaling

The calculation of $P(O_C)$ involves multiplications of very small quantities and very quickly the results are outside the range of machine precision. To this need we need to apply the scaling procedure in [20], so that each time an $\alpha_i(d_j)$ value is computed, it is scaled by s_i:

$$s_i = \frac{1}{\sum_{d_j} \alpha_i(d_j)} \tag{11}$$

$$\hat{\alpha}_i(d_j) = \frac{\alpha_i(d_j)}{\sum_{d_j} \alpha_i(d_j)} \tag{12}$$

With this scaling method, $\log P(O_C)$ is given by::

$$\log P(O_C) = \sum_{i=1}^{n} \log \frac{1}{s_i} \tag{13}$$

3.3 Approximations

$\log P(O_C)$ is usually computed as $\log \sum_{i=1}^{n} e^{-x_i}$. The summation of exponentials often approaches 0 very fast. To calculate it reliably, we have:

$$\sum_{i=1}^{n} e^{-x_i} = e^{-x_m} \left(1 + \sum_{x_i \neq x_m} e^{-(x_i - x_m)} \right) = e^{-x_m} (1 + S) \tag{14}$$

$$\text{where} \quad x_m = \min_i x_i \tag{15}$$

$$\text{Therefore,} \quad \log \sum_{i=1}^{n} e^{-x_i} = -x_m + \log(1 + S) \tag{16}$$

$$\text{where} \quad S = \sum_{x_i \neq x_m} e^{-(x_i - x_m)} \tag{17}$$

When $|S| < 0.1$, we can use the approximation $\log(1 + S) \simeq S$.

3.4 Graph-Based Optimization

We use Dijkstra's algorithm [21] to simultaneously compute $P(O_C)$ and find the optimal boundary. We assume that points $\{A_1, A_2, \cdots, A_p\}$ on the desired boundary are given. The optimal boundary passing from $\{A_1, A_2, \cdots, A_p\}$ is the concatenation of the optimal contours connecting A_1 to A_2, A_2 to A_3 and so on.

To find the optimal contour connecting two given points, we represent the image with a graph, where each pixel corresponds to a graph node and each node is connected with its 8 neighbors. The weight between adjacent nodes u, v consists of two terms. The first term is a constituent of $P(O_C|C) P(C)$ and the second of $P(O_C)$. Assuming u, v are points on the desired boundary, we have:

$$w(u, v) = \log P(O_u, O_v | u, v) + \log P(v|u) - \log \sum_{d_j} \hat{\alpha}_v(d_j) \tag{18}$$

where the summation takes place over all possible directions d_j. The calculation of $\hat{a}_v(d_j)$ depends on values calculated at neighboring nodes. Thus, the complexity of Dijkstra's algorithm increases by a small multiplicative factor, equal to the number of different directions d_j (in our case 8).

4   Experimental Validation

We first explain our choices regarding the calculations of $P(O_{c_i}|c_i)$ and $P(c_{i-1}|c_i)$, and then present boundary delineation results on a variety of images.

4.1   Feature Calculations

For reliable computations, we will assume second-order dependencies among the contour points and we will compute $P(O_{\{c_i\}}|c_{i+1}, c_i, c_{i-1})$ and $P(c_{i+1}|c_i, c_{i-1})$, where $O_{\{c_i\}}$ denotes the observations associated with contour points c_{i+1}, c_i, c_{i-1}.

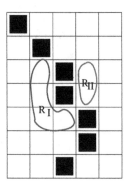

Fig. 3. The data term is calculated based on how likely the pixels in the vicinity of the contour points belong to the two sides of the desired boundary.

The term $P(O_{\{c_i\}}|c_{i+1}, c_i, c_{i-1})$ is computed by estimating how well the pixels in the vicinity of c_{i+1}, c_i, c_{i-1} belong to the two sides of the boundary. The statistical model required for this task is computed based on pixel labeling provided by the user in small image areas. Contour points c_{i+1}, c_i, c_{i-1} divide the pixels in their vicinity into two regions R_I and R_{II}, as shown in Fig. 3. If $M_I(p)$ and $M_{II}(p)$ are two functions estimating how well a pixel p is classified as belonging to side I or side II of the desired boundary, then

$$P(O_{\{c_i\}}|c_{i+1}, c_i, c_{i-1}) = \sum_{p \in R_I} M_I(p) + \sum_{p \in R_{II}} M_{II}(p) \qquad (19)$$

The prior $P(c_{i+1}|c_i, c_{i-1})$ is defined so that it takes higher values for contour points forming a straight line than for contour points that form an angle.

4.2   Boundary Delineation Results

We evaluate our formulation in an interactive boundary finding application where the user places seed points sequentially in a manner similar to [3]. Figures 4, 5, and 6 show in three columns the part of the image used to statistically characterize the interior of the object and the background, the delineation results obtained using Eqn. 1 without and with the normalization term. All the results were obtained using the same parameters $\lambda_c = 0.2$, $\lambda_s = 0.1$, and the same training data acquired at the beginning of the delineation process.

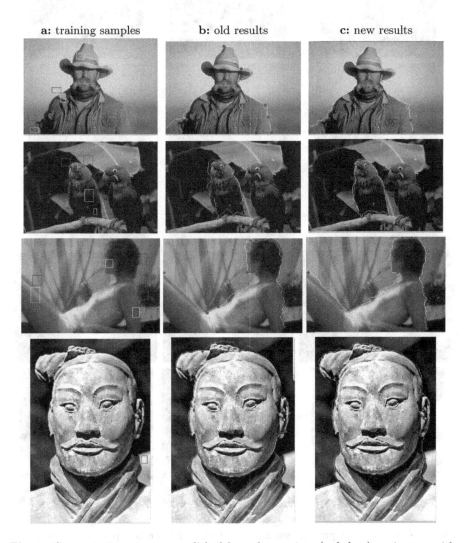

Fig. 4. Segmentation given user clicked boundary points (red dots) on images with complex boundaries. **a)** Windows mark training samples for foreground (yellow) and background (red). **b)** Results from traditional models. **c)** Our results.

a: training samples   **b:** old results   **c:** new results

Fig. 5. Color image segmentation given user clicked boundary points (red dots). **a)** Windows mark training samples for foreground (yellow) and background (red). **b)** Results from traditional models that ignore the normalization term. **c)** Our results.

a: training samples   **b:** old results   **c:** new results

Fig. 6. Texture image segmentation results. Same convention as Fig. 5.

Table 1. Number of mouse clicks required to delineate the various boundaries using the traditional formulation and our probabilistic criterion

image name	# clicks in old method	# clicks in our method
cheetah	16	**5**
zebra	12	**7**
cowboy	7	**5**
parrots	10	**4**
iris	8	**2**
pink flower	2	2
fuchsia flower	24	**2**
white flower	3	**2**
statue	12	**9**
woman	9	**5**
peppers1	2	2
peppers2	7	**4**
cover1	2	2
cover2	7	**3**
liver1	4	4
liver2	4	5
liver3	6	**3**

Table 1 summarizes the number of mouse clicks required to delineate the object boundaries, as a measure of the method effectiveness. Since the user is always part of the interactive segmetnation system, the correctness of the output of the segmentation algorithm is not to be contested.

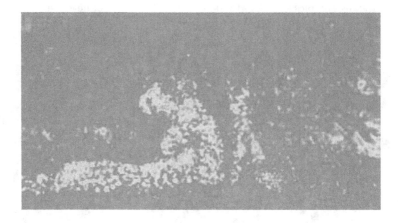

Fig. 7. Classification results using the classifier built from the training data in Fig. 6

Our criterion significantly outperforms traditional Bayesian formulations. The most drastic difference between the amount of mouse clicks required to delineate a boundary is for the image "cheetah". The reason for this can be seen in Fig. 7, where the classification of the image pixels is shown. These results were obtained using the classifier trained on the data shown in Fig. 6. The results are very noisy and the original criterion has trouble localizing the object boundary. On the other hand, the observation-dependent criterion is quite effective in eliminating the noise and tracking the desired discontinuities. In general, the probabilistic criterion consistently produces contours that adhere to the object boundary more faithfully, with fewer mouse clicks.

5   Conclusions

Traditional Bayesian criteria for boundaries assume that the evidence for a candidate contour is derived from the entire image, and the normalization term can thus be omitted during optimization. In practice however, evidence in the vicinity of the contour is employed, and the normalization term is contour-specific and cannot be ignored. Our formulation explicitly acknowledges this term and extacts boudaries optimal in both appearance and area of support.

The normalization term helps promote contours that are better than alternatives in their vicinity. Consequently, it alleviates the length bias problem present in traditional Bayesian formulations. Degenerate solutions such as short contours in featureless image areas are no longer favored by design.

Our formulation has the same asymptotic complexity as previous Bayesian formulations, as it can be globally optimized with dynamic programming.

We validate our method with an interactive boundary delineation application, where significantly fewer mouse clicks are needed to extract complex boundaries.

References

1. Kass, M., Witkin, A., Terzopoulos, D.: Snakes: Active Contour Models. In: Int'l Conference on Computer Vision. IEEE, Los Alamitos (1987)
2. Geiger, D., Gupta, A., Costa, L., Vlotzos, J.: Dynamic Programming for Detecting, Tracking, and Matching Deformable Contours. IEEE Transactions on Pattern Analysis and Machine Intelligence 17(3), 294–302 (1995)
3. Mortensen, E., Barrett, W.: Intelligent Scissors for Image Composition. In: SIGGRAPH (1995)
4. Mortensen, E., Barrett, W.: Interactive Segmentation with Intelligent Scissors. Graphical Models and Image Processing 60(5) (1998)
5. Falcao, A., Udupa, J., Samarasekera, S., Sharma, S.: User-Steered Image Segmentation Paradigms: Live Wire and Live Lane. Graphical Models and Image Processing 60, 233–260 (1998)
6. Neuenschwander, W., Fua, P., Iverson, L., Szekely, G., Kubler, O.: Ziplock Snakes. Int'l Journal of Computer Vision 25(3), 191–201 (1997)
7. Mortensen, E., Barrett, W.: A Confidence Measure for Boundary Detection and Object Selection. In: Proc. Computer Vision and Pattern Recognition. IEEE, Los Alamitos (2001)
8. Mortensen, E., Jia, J.: A Bayesian Network Framework for RealTime Object Selection. In: Proc. Workshop on Perceptual Organization on Computer Vision. IEEE, Los Alamitos (2004)
9. Cohen, L.: On Active Contour Models and Balloons. Computer Vision Graphics and Image Processing: Image Understanding 52(2), 211–218 (1991)
10. Paragios, N., Deriche, R.: Geodesic Active Contours and Level Set Methods for Supervised Texture Segmentation. International Journal of Computer Vision 46(3), 223–247 (2002)
11. Gérard, O., Deschamps, T., Greff, M., Cohen, L.D.: Real-time Interactive Path Extraction with on-the-fly Adaptation of the External Forces. In: European Conference in Computer Vision (2002)
12. Sullivan, J., Blake, A., Isard, M., MacCormick, J.: Bayesian Object Localisation in Images. Int. J. Comput. Vision 44(2), 111–135 (2001)
13. Allili, M., Ziou, D.: Active contours for video object tracking using region, boundary and shape information. Signal, Image and Video Processing 1(2), 101–117 (2007)
14. Joshi, S.H., Srivastava, A.: Intrinsic bayesian active contours for extraction of object boundaries in images. Int. J. Comput. Vision 81(3), 331–355 (2009)
15. Jermyn, I., Ishikawa, H.: Globally optimal regions and boundaries as minimum ratio weight cycles. IEEE Transactions on Pattern Analysis and Machine Intelligence 23(10), 1075–1088 (2001)
16. Schoenemann, T., Cremers, D.: Globally optimal image segmentation with an elastic shape prior. In: IEEE International Conference on Computer Vision (ICCV) (October 2007)
17. Rosenfeld, A., Kak, A.C.: Digital Picture Processing. Academic Press, London (1982)
18. Amini, A., Weymouth, T., Jain, R.: Using Dynamic Programming for Solving Variational Problems in Vision. IEEE Transactions on Pattern Analysis and Machine Intelligence 12(9), 855–867 (1990)
19. Cohen, L., Kimmel, R.: Global Minimum for Active Contour Models: A Minimal Path Approach. International Journal of Computer Vision 24(1), 57–78 (1997)
20. Rabiner, L.: A Tutorial on Hidden Markov Models and Selected Applications in Speech Recognition. Proceedings of the IEEE 77(2) (1989)
21. Cormen, T., Leiserson, C., Rivest, R.: Introduction to Algorithms. McGraw-Hill, New York (1990)

Author Index

494   Author Index